Lecture Notes in Computer Science   6091

Commenced Publication in 1973
Founding and Former Series Editors:
Gerhard Goos, Juris Hartmanis, and Jan van Leeuwen

Mark Crovella   Laura Marie Feeney
Dan Rubenstein   S.V. Raghavan (Eds.)

NETWORKING 2010

9th International IFIP TC 6 Networking Conference
Chennai, India, May 11-15, 2010
Proceedings

 Springer

Volume Editors

Mark Crovella
Boston University Department of Computer Science
111 Cummington St., Boston, MA 02215, USA
E-mail: crovella@cs.bu.edu

Laura Marie Feeney
Swedish Institute of Computer Science
Communication Networks and Systems Laboratory
Box 1263, 16429 Kista, Sweden
E-mail: lmfeeney@sics.se

Dan Rubenstein
Columbia University, 450 Computer Science Bldg. MC 0401
1214 Amsterdam Ave, New York, NY 10027, USA
E-mail: danr@cs.columbia.edu

S.V. Raghavan
Indian Institute of Technology Madras
Department of Computer Science and Engineering
Chennai 600036, India
E-mail: svr@cs.iitm.ernet.in

Library of Congress Control Number: 2010925850

CR Subject Classification (1998): C.2, H.4, K.6.5, D.4.6, E.3, D.2

LNCS Sublibrary: SL 5 – Computer Communication Networks
and Telecommunications

ISSN   0302-9743
ISBN-10   3-642-12962-5 Springer Berlin Heidelberg New York
ISBN-13   978-3-642-12962-9 Springer Berlin Heidelberg New York

springer.com

© IFIP International Federation for Information Processing 2010
Printed in Germany

Typesetting: Camera-ready by author, data conversion by Scientific Publishing Services, Chennai, India
Printed on acid-free paper   06/3180

Preface

This book constitutes the refereed proceedings of the 9th IFIP-TC6 Networking Conference, Networking 2010.

Papers were solicited in three broad topic areas: applications and services, network technologies, and internet design. All papers were considered on their merits by a unified Technical Program Committee (TPC); there was no attempt to enforce a quota among topic areas. We believe the resulting program is an excellent representation of the breadth of recent advances in networking research.

This year, the conference received 101 full paper submissions from 23 countries on five continents, reflecting a strong diversity in the networking community. Similarly, the 92 members of the TPC are from 21 countries and include a mix of academic, industry, and governmental affiliations.

The TPC members, aided by some 50 external reviewers, provided a total of 470 reviews and follow-up discussions totaling more than 200 messages. The final selections were made at a TPC meeting hosted by Columbia University in New York City, with both in-person and remote participation. In total, authors of accepted papers have academic and industry affiliations in 15 countries.

We finally selected 24 papers for presentation during the conference technical sessions. A small number of papers were assigned a shepherd from the TPC to assist in paper revision. These statistics represent an acceptance rate of just under 24%, comparable to that of previous years.

The TPC also identified several papers that reflect particularly promising early results; these papers were selected for presentation as work-in-progress papers and are identified as such in the proceedings.

The Networking 2010 program would not have been possible without the contribution of many people. We would like to thank the TPC members and all the reviewers for their careful and conscientious work. A special thanks to TPC members who gave additional time to shepherding, as well as to the (anonymous) subset of the TPC who took on the challenge of selecting the winner of the Best Paper Award. Not least, we would like to thank all of the authors who submitted their work to Networking 2010 for their interest and effort. It has been our privilege to collaborate with you in creating this program.

May 2010

Dan Rubenstein
Mark Crovella
Laura Feeney
S.V. Raghavan

Message from the General Chairman

The 9th International IFIP-TC 6 Networking Conference was organized and hosted by The Indian Institute of Technology Madras in collaboration with the Computer Society of India, Chennai Chapter. Networking 2010 once again brought together members of the networking community from both academia and industry to revel in the recent advances, to explore the unexplored in the broad and fast-evolving field of telecommunications and to highlight key issues, identify trends and turn visions to reality.

IFIP Networking is known for its high-quality papers from various networking research communities. A total of 101 papers were submitted to the conference from six out of the seven continents and after many rigorous reviews and much scanning only 24 regular and 9 work-in-progress papers were accepted. The program covered a variety of research topics in the area of P2P and overlay networks, performance measurement, quality of service, ad hoc and sensor networks, wireless networks, addressing and routing and applications and services.

The Indian Institute of Technology, Madras, incidentally a reserve forest for Black Bucks (a rare species of deer) is one of the premier institutes of India. Its state-of-the-art facilities, education system and the co-existence of two different species is a sheer joy to experience. Chennai (Earlier Madras) is a city of culture and heritage. The capital city of Tamilnadu has it roots deeply set in culture, and nurtures growth with technology. It is one of the largest and oldest metropolitan cities in India. This beach side city is famous for its old and beautifully architected temples. The once mystic land of snake charmers and astrology has found its true identity as a nest of technology and science and no better way to reflect it but Networking 2010.

In closing we would like to thank the members of the Program Committee and the army of reviewers that helped us in selecting the best papers for publication.

May 2010   S.V. Raghavan

Organization

IFIP Networking-2010 was organized by the Indian Instutute of Technology Madras and Computer Society of India, Chennai Chapter in cooperation with IFIP TC6.

Executive Committee

General Chair:	S.V. Raghavan, IIT Madras, Chennai, India
Technical Program Co-chairs	Mark Crovella, Boston University, USA
	Laura Feeney, SICS, Sweden
	Dan Rubenstein, Columbia University, USA
Steering Committee	George Carle, TU Munich, Germany
	Marco Conti, IIT-CNR, Pisa, Italy
	Pedro Cuenca, University of Castilla-la-Mancha, Spain
	Guy Leduc, University of Liège, Belgium
	Henning Schulzrinne, Columbia University, USA
Local Arrangements	S. Ramasamy Chairman, CSI Chennai Chapter
	S. Ramanathan, CSI Chennai Chapter
	Wg Cdr M. Murugesan (Retd.), CSI Chennai Chapter
	H.R. Mohan, CSI Chennai Chapter
	Sanand Sasidharan, IIT Madras
	V. Umadevi, IIT Madras
	Raktim Bhattacharjee, IIT Madras

Technical Program Committee

Rui Aguiar	University of Aveiro
Aditya Akella	University of Wisconsin-Madison
Ehab Al-Shaer	University of North Carolina Charlotte
Kevin Almeroth	University of California
Fan Bai	General Motors
Ernst Biersack	EURECOM
Olivier Bonaventure	Université catholique de Louvain
Raouf Boutaba	University of Waterloo
Torsten Braun	University of Bern
Fabian Bustamante	Northwestern University
Georg Carle	Technische Universität München

Claudio Casetti   Politecnico di Torino
Augustin Chaintreau   Thomson
Marco Conti   IIT-CNR
Jun-Hong Cui   University of Connecticut
Jaudelice de Oliveira   Drexel University
Edmundo de Souza e Silva   Federal University of Rio de Janeiro
Jordi Domingo-Pascual   Technical University of Catalunya (UPC)
Constantine Dovrolis   Georgia Institute of Technology
Lars Eggert   Nokia Research Center
Anja Feldmann   TU-Berlin
Wu-chi Feng   Portland State University
Daniel Figueiredo   UFRJ
Luigi Fratta   Politecnico di Milano
Timur Friedman   UPMC Paris Universitas and CNRS
Zihui Ge   AT&T Labs - Research
Erol Gelenbe   Imperial College London
Brighten Godfrey   University of Illinois at Urbana-Champaign
Timothy Griffin   University of Cambridge
Minaxi Gupta   Indiana University
Guenter Haring   Universität Wien
Markus Hofmann   Bell Labs/Alcatel-Lucent
David Hutchison   Lancaster University
Gianluca Iannaccone   Intel Corporation
Ping Ji   John Jay College of Criminal Justice
Holger Karl   University of Paderborn
Peter Key   Microsoft Research
Kimon Kontovasilis   NCSR Demokritos
Aleksandar Kuzmanovic   Northwestern University
Guy Leduc   University of Liege
Patrick Pak-Ching Lee   The Chinese University of Hong Kong
Kenji Leibnitz   Osaka University
Douglas Leith   Hamilton
Jorg Liebeherr   University of Toronto
Benyuan Liu   University of Massachusetts Lowell
Yong Liu   Polytechnic University
John Chi Shing Lui   Chinese University of Hong Kong
Gaia Maselli   University of Rome La Sapienza
Laurent Mathy   Lancaster University
Martin May   Thomson
Ketan Mayer-Patel   University of North Carolina
Michael Menth   University of Wuerzburg
Jelena Misic   Ryerson University
Vishnu Navda   Microsoft Research
Erik Nordstrm   Uppsala University
Ilkka Norros   VTT Technical Research Centre of Finland

Philippe Owezarski	LAAS
Maria Papadopouli	University of Crete
Christos Papadopoulos	Colorado State University
Kaustubh Phanse	Lule University of Technology
Marcelo Pias	Cambridge University
Ana Pont	Polytechnic University of Valencia
Konstantinos Psounis	University of Southern California
Ramon Puigjaner	UIB
Guy Pujolle	University of Paris 6
Raj Rajendran	Columbia University
Sanjay Rao	Purdue University
James Roberts	France Telecom
George Rouskas	North Carolina State University
Sambit Sahu	IBM Research
Theodoros Salonidis	Thomson Technology Paris Laboratory
Henning Schulzrinne	Columbia University
Aruna Seneviratne	NICTA
Krishna Sivalingam	Indian Institute of Technology Madras
Robin Sommer	ICSI and LBNL
Otto Spaniol	RWTH Aachen University
Cormac Sreenan	University College Cork
David Starobinski	Boston University
Ioannis Stavrakakis	National and Kapodistrian University of Athens
Yutaka Takahashi	Kyoto University
Phuoc Tran-Gia	University of Wuerzburg
Piet Van Mieghem	Delft University of Technology
Wenye Wang	NC State University
Bing Wang	University of Connecticut
Carey Williamson	University of Calgary
Tilman Wolf	University of Massachusetts
Adam Wolisz	Technical University of Berlin
Guoliang Xue	Arizona State University
Daniel Zappala	Brigham Young University
Rong Zheng	University of Houston
Gil Zussman	Columbia University

External Reviewers

Gahng-Seop Ahn	Radovan Bruncak
Issam Aib	Raffaele Bruno
Muhammad Qasim Ali	Lukasz Budzisz
Anteneh Beshir	Lin Cai
Berk Birand	Niklas Carlsson
David Black	Costas Courcoubetis
Mathias Bohge	Franca Delmastro

Christian Doerr
Christopher Dunn
Carol Fung
Matthias Hartmann
David Hock
Gavin Holland
Tobias
Fida Hussain
Luigi Iannone
Aravind Iyer
Emmanouil Kafetzakis
Andrew Kalafut
Syed Ali Khayam
Alexander Klein
Dominik Klein
Frank Lehrieder
Tianji Li
Noura Limam
Michael Mackay
Francesco Malandrino
David Malone

Constantine Murenin
Huy Nguygen
Christoforos Ntantogian
Simon Oechsner
Jasmina Omic
Oluwasoji Omiwade
Katsunori Ori
Ramjit Pillay
Gerasimos Pollatos
Rastin Pries
Muntasir Raihan Rahman
Rajiv Ramdhany
Massimo Reineri
Ehssan Sakhaee
Damien Saucez
Simone Silvestri
Thomas Staub
Vijay Subramanian
Dan-Cristian Tomozei
Yongge Wang
Johan Wikman

Table of Contents

Wireless Networks

Addressing and Routing

Applications and Services

Ad Hoc and Sensor Networks

Work in Progress

Using Torrent Inflation to Efficiently Serve the Long Tail in Peer-Assisted Content Delivery Systems

Niklas Carlsson[1], Derek L. Eager[2], and Anirban Mahanti[3]

[1] University of Calgary, Calgary, Canada
niklas.carlsson@cpsc.ucalgary.ca
[2] University of Saskatchewan, Saskatoon, Canada
eager@cs.usask.ca
[3] NICTA, Alexandria, Australia
anirban.mahanti@nicta.com.au

Abstract. A peer-assisted content delivery system uses the upload bandwidth of its clients to assist in delivery of popular content. In peer-assisted systems using a BitTorrent-like protocol, a content delivery server seeds the offered files, and active torrents form when multiple clients make closely-spaced requests for the same content. Scalability is achieved in the sense of being able to accommodate arbitrarily high request rates for individual files. Scalability with respect to the number of files, however, may be much more difficult to achieve, owing to a "long tail" of lukewarm or cold files for which the server may need to assume most or all of the delivery cost. This paper first addresses the question of how best to allocate server resources among multiple active torrents. We then propose new content delivery policies that use some of the available upload bandwidth from currently downloading clients to "inflate" torrents for files that would otherwise require substantial server bandwidth. Our performance results show that use of torrent inflation can substantially reduce download times, by more than 50% in some cases.

Keywords: Peer-assisted, multi-torrent, torrent inflation, long tail.

1 Introduction

Peer-assisted content delivery has the key advantage that the bandwidth available for serving client requests scales with the client population, potentially allowing a system to accommodate increasing file request rates with fixed server resources. This has led to considerable research interest in such systems, as well as increasing numbers of deployments by content providers.[1]

We consider here peer-assisted content delivery systems in which BitTorrent-like protocols are used, with files seeded only by a content provider server, and in which peers contribute upload bandwidth to assist in file delivery only when

---

[1] The clients of such a system are generically termed "peers" in the following, regardless of whether the context is one concerning peer-to-peer, or peer-server interaction.

M. Crovella et al. (Eds.): NETWORKING 2010, LNCS 6091, pp. 1–14, 2010.

downloading a file from the service. In addition to making a conservative assumption about peer participation, such an architecture may provide the content provider with greater content control, and require less functionality at peers, than alternative approaches relying on longer-term peer storage or caching of files.

A potential problem with this type of system, however, arises from the fact that file popularity distributions typically have a power law form, with a "long tail" of lukewarm/cold files that has a substantial aggregate request rate (e.g., [1, 2, 3]). These files may not be sufficiently popular to enable torrents consisting of multiple concurrent requesters to form, resulting in a server load that grows unsustainably with the number of offered files.

This paper devises new protocols for such systems, with the goal of enhancing their scalability. First, we propose new server scheduling policies for selecting which peers to upload to next, among multiple peers requesting different files. We find that, in comparison to a baseline "random peer" policy, the new policies improve both the overall average download time and the download times seen by the requesters of the lukewarm/cold files. However, these improvements are quite modest, less than 10% in the scenarios considered.

Second, we propose *torrent inflation*, whereby some of the peer upload bandwidth available from actively downloading peers is applied in torrents for files other than those requested. These other files are ones for which the number of concurrently active requesters may be insufficient to enable self-sustaining torrents. (We refer to a torrent as *fully self-sustained* if all peer requests can be served by the peers themselves, with negligible server assistance.) Note that since we consider systems in which there is no seeding of files by peers, use of torrent inflation entails the overhead of delivering to actively downloading peers some minimal number of blocks from files other than those requested, for those peers to then upload during their respective download sessions to other peers. Torrent inflation policies are designed that seek to harvest only peer upload bandwidth that could otherwise go unused within the peer's own torrent.[2] We find that torrent inflation policies provide significant reduction in both the overall average download time and the download times for the lukewarm/cold files (achieving reductions of more than 50% in some cases). In addition, we find that inflation policies can be designed that are remarkably insensitive to the server capacity and therefore can be used to replace server bandwidth capacity in some systems.

Torrent inflation entails utilizing upload bandwidth of actively downloading peers for improving a system-wide performance objective. The content provider can build the desired policies into software required for accessing the service. It may also be possible to devise incentives to directly reward peers for the desired behaviour. Investigation of these options is outside the scope of this work.

The remainder of the paper is organized as follows. Related work is discussed in Section 2. Section 3 describes our system assumptions, simulation model, and baseline policies, and presents simulation results for the baseline "random

---

[2] As in the peer-to-peer context [4, 5, 3], while peers in popular torrents, with high piece diversity, typically are able to fully utilize their upload bandwidth, this is not necessarily the case in lukewarm/cold torrents.

peer" and "random file" server scheduling policies. Alternative server scheduling policies are defined and their performance is compared with baseline policies in Section 4. Section 5 proposes torrent inflation policies that seek to harvest only peer upload bandwidth that could otherwise go unused, and presents performance results for these policies. Conclusions are given in Section 6.

2   Related Work

To the best of our knowledge, prior work has not directly addressed the problem of efficient delivery of potentially large numbers (or the so called "long tail") of lukewarm/cold files, in the context of peer-assisted download systems that do not rely on longer-term peer storage or caching of files. There has, however, been related work on multi-torrent systems in which peers seed files and some peers concurrently participate in distributing more than one file (e.g., [6, 7, 8, 9, 10]). Of course, any policy presented here would further benefit from additional seed capacity available to the system.

2.1   Multi-torrent Systems

Guo *et al.* [6] use measurements and analytical models to illustrate that torrent lifetimes can be extended if a node that acts as a downloader in one torrent also acts as a seed in another torrent. Yang *et al.* [7] propose a cross-torrent tit-for-tat scheme in which unchoking is done based on the aggregate download rate from each candidate peer across all torrents (rather than just within a single torrent). Other prior work has proposed that inter-swarm exchanges be incentivized through propagation of peer reputation [11], incentive-based token schemes [12], and/or history-based priority rules [13]. These works are complimentary to ours, and could perhaps be used to incentivize peer participation in torrent inflation.

To increase torrent lifetimes and file availabilities, Menasche *et al.* [8] propose that the original seeder (or content provider) "bundle" a number of its files for distribution, rather than offering the files individually. While static bundling is shown to be effective for the purpose of improved file availability, it can easily result in increased download times, especially of otherwise popular content. In contrast, we consider a peer-assisted context (in which file availability is not an issue since files are not seeded by peers, but by a content delivery server) and focus on reducing download times. Other related work has investigated the advantages offered by dynamically inflating swarm sizes by merging small swarms of the same torrent (rather than different torrents) [3].

As in our work, Wu *et al.* [9,10] propose techniques wherein peers contribute to the distribution of content they may not be interested in. The assumed context is live streaming, however, rather than our context of content download, and therefore the proposed techniques are quite different than those considered here.

2.2   Server Scheduling

The importance of server scheduling policies has been discussed in the context of both peer-assisted download [14] and live streaming systems [15]. However, to the best of our knowledge, the only previous work on server scheduling in multi-torrent file download systems is that by Sun *et al.* [16,17]. They use an analytic steady-state fluid-style model to compare the performance of equal server bandwidth allocation among all peers, and equal server bandwidth allocation among all files, with an analytic lower bound [16]. A version of the second of these policies has been implemented in FS2You [17]. We note that these two policies are very similar to the baseline server scheduling policies discussed in Section 3.

In addition, there are some similarities between the problems of allocating server bandwidth among the requesters of multiple files in a peer-assisted system, and server scheduling in a batched service content delivery system [18, 19]. In both types of systems, multiple clients can be served using a single operation by the server. In a peer-assisted system, peers are able to replicate among themselves any piece injected into a torrent by the server. In a batched service system, multiple clients can be served at once using broadcast or multicast. On the other hand, there are also obvious significant differences; for example, at high request rates, torrents can become largely self-sustaining and require minimal allocation of server bandwidth.

3   Baseline Performance

3.1   Baseline Server Scheduling Policies

Our baseline *random peer* and *random file* policies closely correspond to two policies modeled by Sun *et al.* [16], and are motivated by the goals of allocating server bandwidth fairly among active peers and active torrents, respectively.

With random peer, a peer is randomly selected from the set of all peers that the server is not currently uploading to. With random file, the server chooses a file randomly from the group of files that: (1) the server is not currently uploading data from, and (2) has at least one peer with an outstanding piece request. If there are no such files, a file is selected randomly among the files for which there is a piece request from a peer that the server is not currently uploading to, if any. Finally, within the set of peers that are downloading the selected file but that are not currently being uploaded to by the server, a peer is selected at random.[3]

3.2   Multi-torrent Simulation Model

For policy evaluation, an existing event-based simulator of BitTorrent-like systems [20] was modified to support simulation of systems with multiple files. A single server is assumed that stores all N files, and varying numbers of active clients (peers). For the results presented here, it was assumed that peers

---

[3] A version of random file is implemented in FS2You [17]. In their implementation, the server maintains a file popularity index for each file (based on the number of references for the file within some time window), and serves peers with a probability which is inversely proportional to the requested file's popularity index.

have connections to all other peers in the same torrent, as well as to the server.[4]

As with BitTorrent, a *rarest-first* piece selection policy is used with *tit-for-tat* peer selection and optimistic unchoke. It is assumed that each peer concurrently uploads to at most six peers, with one of the peer's six upload connections, on average, used for optimistic unchoke. The maximum number of server upload connections is chosen as the total server bandwidth divided by the peer upload rate (an integer value owing to the parameters chosen in our experiments).

3.3   Performance Comparisons of Baseline Policies

The simulation results that we present in this paper are for a relatively simple workload scenario. Peers arrive according to a constant-rate Poisson process, and request files according to a Zipf popularity distribution (i.e., the request rate λ_i of the ith most popular file is proportional to $\frac{1}{i^\alpha}$). Zipf and Zipf-like popularity distributions have been observed in many contexts such as Web servers, media-on-demand servers, and video sharing services (often, however, with deviations at the head and/or tail) [1,2,3]. Each peer requests a single file upon its arrival to the system, and leaves the system once its download is complete. Each simulation run is for 40,000 requests, with the initial 10,000 and the last 2,000 requests removed from the measurements.

For simplicity, peers are assumed to be homogenous, and motivated by the observed asymmetry in upload-download bandwidth (e.g., [21]), it is assumed that peers have a download-upload bandwidth capacity ratio (D/U) equal to 3. Files are assumed to be of identical size with 256 pieces each. Without loss of generality, the file size L and upload bandwidth capacity U are fixed at one. With these normalized units, the volume of data transferred is measured in units of the file size and the download time is measured in units of the minimum time it takes for a peer to fully upload a file.

Other parameters, together with their default values, are as follows: server bandwidth $B = 10$, hottest file request rate $\lambda_1 = 40$, and Zipf parameter $\alpha = 1$. For a system with $N = 200$ files, these choices imply a total request rate of roughly 235 (i.e., on average about 235 requests arrive during the time it takes a peer to upload a file at its upload capacity rate), and that the server can upload at ten times the rate of a peer, but at most can satisfy less than 5% of the total demand without peer assistance. For $N = 1$ and $N = 25$, the server can satisfy at most 25% and 7% of the total demand, respectively.

---

[4] Note that the default parameters in recent versions of the mainline BitTorrent client allow peers to be connected to up to 80 other peers, which is often achieved in practice [5], and that typically many fewer peers than this are participating in cold/lukewarm torrents. Furthermore, peers not satisfied with their performance are able to request additional peers from the tracker, or directly from other peers using the peer exchange protocol (PEX). For simulating the transmission rates of piece transfers, it is assumed that connection bottlenecks are located at the end points (i.e., the transmission rate is limited either by the upload bandwidth at the sender or by the download rate at the receiver) and the network operates using max-min fair bandwidth sharing (using TCP, for example).

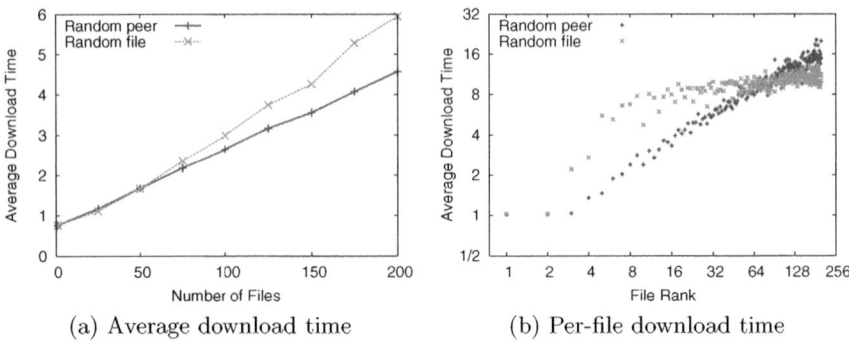

(a) Average download time   (b) Per-file download time

Fig. 1. Baseline policies ($B = 10$, $\lambda_i = 40/i$, $L = 1$, $U = 1$, $D/U = 3$, $K = 256$, $N = 200$)

Figure 1 shows performance with the baseline server scheduling policies. Figure 1(a) shows the average download time for systems with varying numbers of files, while Figures 1(b) shows the average download time for each file individually (ordered accorded to the file popularity rank), for $N = 200$ files.

Note that for both policies, the overall average download time increases approximately linearly with increasing numbers of offered files. From Figure 1(b), it can be seen that the high overall average download time in systems with large numbers of files is caused by high download times for relatively unpopular files. The most popular files, in contrast, have largely self-sustaining torrents for these parameter values, and thus the download times for these files are only minimally affected when the number of offered files increases.

Although the random peer policy achieves a better overall average download time than the random file policy, for these parameter values, it is less fair in the sense that it yields greater differences in download times between files. For example, the average download times for the least popular files are about 1.5 to 2 times as big with the random peer policy than with random file.

4   Prioritized Server Scheduling

This section develops prioritized server scheduling policies that use additional state information to make their scheduling decisions. Our primary goal in this section is to find policies that achieve an overall average download delay similar to, or better than, that achieved with random peer, and fairness across files similar to or better than that achieved with random file.

Towards this goal, we investigate policies similar to ones previously developed for scheduling in batching systems [18,19]. Similar to the random file policy, we use a probabilistic two-step approach in which a file is first selected at random (but now with weighted probabilities), and then a peer is selected at random (with uniform probability) from within the set of peers requesting pieces for this file. For file selection we consider three different weighting functions.

While we only present results for probabilistic policies, we have also evaluated several deterministic policies including deterministic versions of the policies defined in this section. However, these deterministic policies were significantly outperformed by their probabilistic counterparts (in most cases).

The first policy, called *weight by number of excess-wait peers* (NEWP), weights a file according to the number of requesting peers whose time-in-system exceeds that for a self-sustaining torrent. This time-in-system threshold is equal to the expected download time if each peer is downloading at its own upload rate (i.e., L/U). In contrast, the second policy, called *weight by maximum excess wait* (EW), weights files not by the *number* of "excess-wait" requesting peers, but according to the *longest* excess wait time (i.e., time-in-system beyond the expected download time for a self-sustaining torrent), among such peers, with weight zero given if there is no such peer.[5]

Finally, we define a policy called *weight by product of the maximum excess wait and number of excess-wait peers* (EW x NEWP) that uses the product of the two preceding metrics to determine its weights. Similar to RxW [19], this policy attempts to strike a balance between considerations of time in system, and number of requesters, in determining its weights. In contrast to RxW, however, this policy is probabilistic rather than deterministic, and considers only requesting peers whose time-in-system exceeds that for a self-sustaining torrent.[6]

Figure 2 shows results for the above prioritized server scheduling policies, using the same workload scenarios as in Figure 1. Compared to the other prioritized policies as well as to the baseline policies, EW x NEWP achieves the best overall average download time. Also, EW x NEWP achieves similar fairness with respect to the per-file download times as that achieved by random file. (This is exemplified by the similarity in average download times of the least popular files, for the two policies.) Therefore, EW x NEWP achieves our objective with respect to improving on the baseline server scheduling policies.

Figure 3 shows the average server bandwidth usage for each file individually, for $N = 200$ files. Note from Figure 3(a) that the random peer policy wastes significant server resources on the two most popular files, relative to random file; random file achieves the same performance for these files (as shown in Figure 1(b)) with considerably lower server bandwidth usage. Comparing Figure 3(a) and Figure 3(b), it can be seen that the prioritized server scheduling policies further substantially reduce the server bandwidth used for the most popular files. This bandwidth can be allocated to torrents in greater need of server assistance. Of course, dynamic server bandwidth allocation is able to make substantially better use of the server resources than policies in which server bandwidth is statically allocated among files. In this paper we focus on the relative

---

[5] Denoting the set of peers currently downloading file i by P_i, and the time-in-system of peer p by w_p, the weight given to file i is $|\{p \in P_i | w_p > L/U\}|$ using NEWP and $\max[\max_{p \in P_i}[w_p] - L/U, 0]$ using EW.

[6] Other policies attempting to achieve a compromise of similar form were considered, such as a policy that used the total number of requesting peers in place of the NEWP component, but did not perform as well as this particular version.

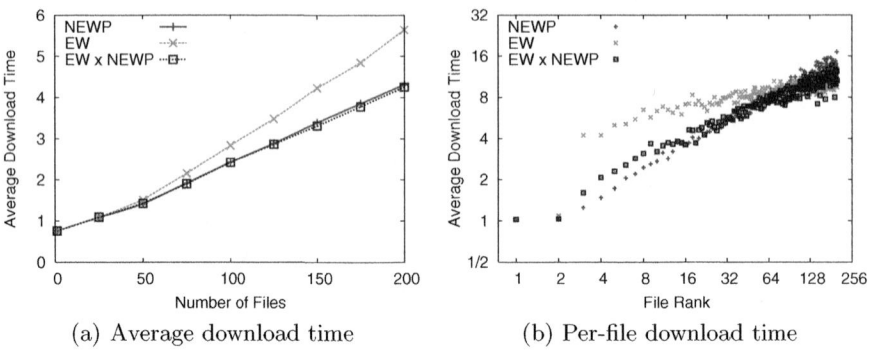

(a) Average download time (b) Per-file download time

Fig. 2. Prioritized server scheduling policies ($B = 10$, $\lambda_i = 40/i$, $L = 1$, $U = 1$, $D/U = 3$, $K = 256$, $N = 200$)

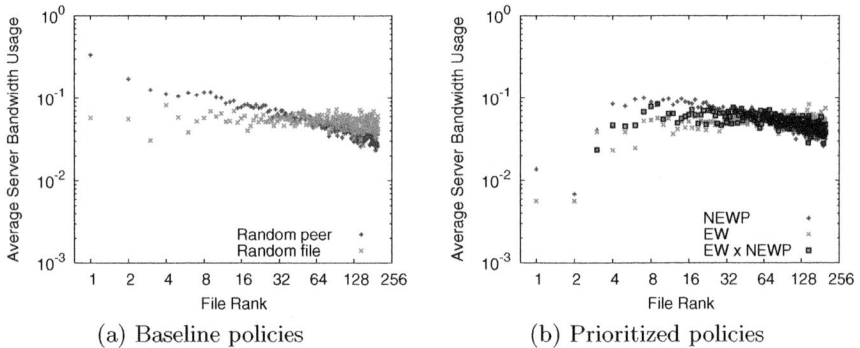

(a) Baseline policies (b) Prioritized policies

Fig. 3. Server bandwidth usage ($B = 10$, $\lambda_i = 40/i$, $L = 1$, $U = 1$, $D/U = 3$, $K = 256$, $N = 200$)

performance of dynamic allocation policies, and do not quantify the performance differences relative to static bandwidth allocation.

5 Torrent Inflation

In the results shown in Figures 1 and 2, requests for the lukewarm/cold files suffer relatively high download times, even though requests for such files receive a greater average per-request share of the server bandwidth than the hot files. This is because self-sustaining torrents are not formed for these files, and the upload bandwidth of the requesting peers often can not be utilized. In this section, new *torrent inflation* policies are described that harvest peer upload bandwidth to *inflate* the number of active peers in a torrent, thus making the torrent more self-sustaining. Torrent inflation is achieved by assigning each active peer an "inflation file". Some number of pieces from the inflation file are downloaded in parallel with the requested file. These pieces may then be uploaded to other peers, providing torrent inflation. Torrent inflation requires a policy for assigning inflation files, as well as a policy for peers to choose among piece uploads from their requested file, versus from their inflation file, and the peer to upload to.

5.1 Inflation File Selection Policies

This section defines three alternative inflation file selection policies. As with server scheduling policies, probabilistic policies were found to outperform deterministic policies, and therefore, only probabilistic policies are considered in this section. With each of these policies, the decision of which inflation file is assigned to each peer is made by the server at the time of the peer's original file request. The information needed to make these decisions would be available to any BitTorrent-like system with tracker functionality.

The simplest policy considered here, *random active torrent* (AT), randomly (with uniform probability) selects one of the files with an "active" torrent (i.e., such that there is at least one currently active peer for which the file is either the peer's requested file, or inflation file, and the peer has acquired at least one piece). Motivated by the success of the server scheduling policy EW x NEWP, the same policy is also considered for inflation file selection. Finally, we define a policy called *conditional weight by number of peers* (CNP) that makes a weighted random selection among those files with at least one requesting peer whose time-in-system exceeds that for a self-sustaining torrent. The weights are assigned according to the current number of active requesters. In our simulations of the EW x NEWP and CNP selection policies, we default to the AT policy if there are no peers whose time-in-system exceeds that for a self-sustaining torrent.

Finally, with torrent inflation, we increase the time-in-system threshold value (previously chosen as L/U) that is used when identifying a torrent as non-self-sustaining, as needed in server scheduling and the EW x NEWP and CNP inflation file selection policies, by a factor f (chosen as 1.2 in all our experiments). This change is motivated by the potential benefit of being somewhat more cautious when identifying a torrent as non-self-sustaining, when using torrent inflation.[7]

5.2 Upload Prioritization

We now describe the policy used to choose among the various options that may exist when a peer wishes to initiate a new upload operation. In the upload policy proposed here, highest priority is given to uploads in which the piece is from the common requested file of both uploader and downloader. Uploads of pieces from the requested file of the downloader, when this file is the inflation file of the uploading peer, are given the next highest priority. These two rules ensure that the peer upload bandwidth is used as much as possible for uploads of pieces wanted by the downloader.

---

[7] The particular choice of 1.2 has been found to provide a reasonable tradeoff between responsiveness to high download delays, and minimizing provision of service to torrents that are largely self-sustaining. We note that this choice roughly corresponds to peers being able to use one out of six upload connections to help other torrents.

Table 1. Upload priority levels with inflation

Uploader file	File type at downloader		
	Requested	Inflation (case 1*)	Inflation (case 2)
Requested	1st	3rd	5th
Inflation	2nd	4th	6th

* Case 1 holds if at least one of the downloader's six upload connections is not currently in use, even though the peer has at least one downloaded piece; otherwise, case 2 holds. Note that in case 1 the downloader has upload capacity that may otherwise go unused.

Two additional rules govern selection of upload targets in the event the data uploaded is for the inflation file of the recipient. Such uploads are not directly productive, and are useful only to the extent that the downloader can later pass the data to a peer that needs it. Our first rule is to favor uploads of this type only to peers that are not fully utilizing their upload connections, as this suggests that such peers are unable to fully utilize their upload bandwidth in their respective requested file torrent. (As noted earlier and discussed elsewhere in the peer-to-peer context [4,3], the upload bandwidth of peers in lukewarm/cold torrents may not be utilized as effectively as that of peers in larger torrents.) Our second rule is used as a tiebreaker for the first rule, and is to favor uploads of pieces of a file that at least one of the two peers has requested (in this case the uploader).

The above prioritizations result in the upload policy with the priority levels enumerated as in Table 1. Comparisons and preliminary experiments using alternative priority rules suggest that these priority levels work well together.

5.3   Principal Performance Comparisons

Figure 4 shows results for the inflation policies used in combination with the EW x NEWP server scheduling policy. We find that both AT and CNP, when used in conjunction with the peer upload policy described in Section 5.2, are able to provide major improvements in both the overall average download time and the download times for cold files (achieving reductions of more than 50% in some cases). While both the average delays and fairness across files with the EW x NEWP inflation file selection policy are slightly better than without inflation (Figures 1 and 2), they are not as good as with AT and CNP. This is likely due to the adverse effects of strongly correlating the choice of inflation file with the choice of which file the server will deliver data from next. For example, such correlations may result in undesirable scenarios where the peers that have requested a particular file complete their downloads fairly soon after that file becomes highly ranked as an inflation file choice, causing ineffective torrent inflation. This is seen in the results of Figure 5(b), which shows the average rate at which data from each file is uploaded by the peers for which that file was their inflation file (i.e., the "inflation bandwidth usage" for that file). Note that with EW x NEWP, these bandwidth usage values are considerably lower than with the other policies.

Figure 6(a) shows, for each file, the average amount of data uploaded by each peer selecting that file as its inflation file. Not surprisingly, the average amount

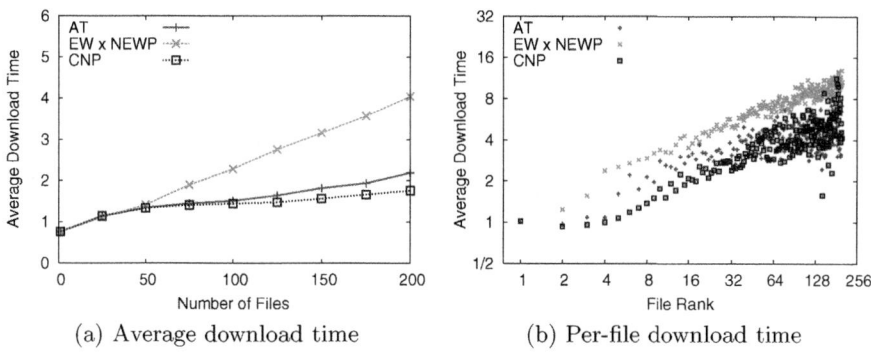

Fig. 4. Inflation policies ($B = 10$, $\lambda_i = 40/i$, $L = 1$, $U = 1$, $D/U = 3$, $K = 256$, $N = 200$)

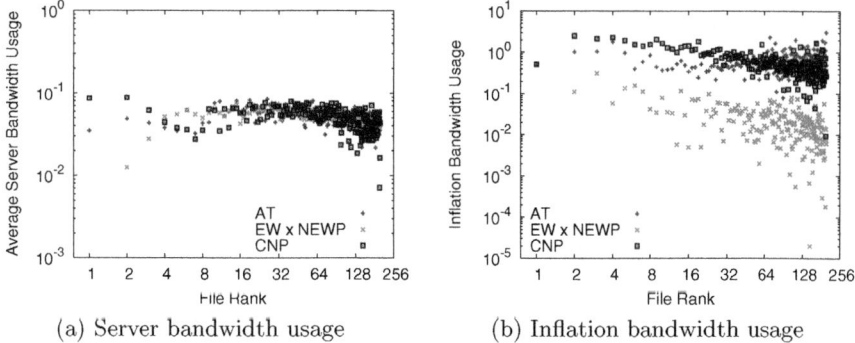

Fig. 5. Bandwidth usage for inflation ($N = 200$, $B = 10$, $\lambda_i = 40/i$, $L = 1$, $U = 1$, $D/U = 3$, $K = 256$)

uploaded is generally higher for more popular files. This is since more popular files are likely to have a greater number of concurrently active peers that have requested the file, and that can be uploaded to. Note that the case of the most popular file is an exception. This file is likely sufficiently hot that pieces are only rarely uploaded to peers with that file as their inflation file, according to our upload policy, and therefore such peers only rarely obtain any inflation file data to upload to others. Overall, the amount of peer upload bandwidth used for inflation file uploads is relatively modest, and yet, as seen, when judiciously applied can yield substantial benefits with respect to download times and fairness.

Figure 6(b) shows, for each file, the ratio of the average amount of data from the file that is uploaded by each peer selecting that file as its inflation file, to the average amount that is downloaded by such a peer. Higher ratios indicate greater efficiency of torrent inflation. In particular, ratios greater than one indicate that each downloaded piece tends to be uploaded to multiple other peers. Note that the cases in Figure 6(b) where the ratio is substantially less than 1, correspond to cases where not much data is uploaded (as seen from Figure 6(a)).

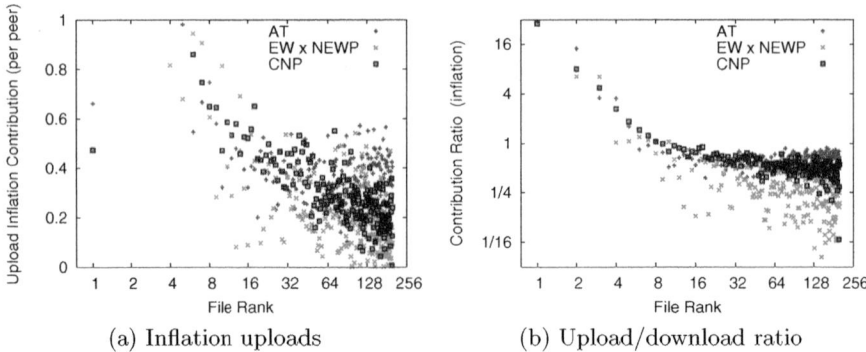

(a) Inflation uploads   (b) Upload/download ratio

Fig. 6. Contribution statistics for inflation ($N = 200$, $B = 10$, $\lambda_i = 40/i$, $L = 1$, $U = 1$, $D/U = 3$, $K = 256$)

5.4   Impact of Workload and Server Parameters

Figure 7(a) shows results for different values of the total request arrival rate λ. As in the other figures, the parameters not being varied are set to their default values. Figure 7(b) shows per-file results for a lower total request arrival rate ($\lambda = 50$) than used in previous experiments.

As seen in Figure 7(a), the average download time is largest for intermediate request rates. With lower request rates, the server is more able to serve requests itself, without peer assistance. For example, with an arrival rate $\lambda = 12.5$ the server has the capacity to satisfy as much as 80% of the demand itself. With higher request rates, for which the server can only satisfy a small fraction of the demand without peer assistance (e.g., at most 2.5% when $\lambda = 400$), self-sustaining torrents develop for a larger fraction of the files. As observed by comparing Figures 7(b) and 4(b), with higher request rates, a larger fraction of the files are able to develop self-sustaining torrents.

Figures 7(c) and 7(d) show results for different values of the Zipf parameter α, and server bandwidth B, respectively. Figure 7(c) shows that the average download time quickly decreases towards that with self-sustaining torrents, as α increases beyond 1. This is since greater skewness in the popularity distribution results in a larger fraction of the requests being directed towards a few highly popular files. Figure 7(d) shows that a system using torrent inflation with the CNP policy is remarkably insensitive to server capacity. Evidently, torrent inflation in such a system can effectively replace server bandwidth capacity.

In summary, Figures 4 through 7 show that the CNP inflation file selection policy typically outperforms the AT and EW-NEWP policies with respect to the average download time. However, the AT policy yields quite competitive performance, and as seen in Figure 4(b) may yield somewhat lower variability among the download times for popular versus unpopular files.

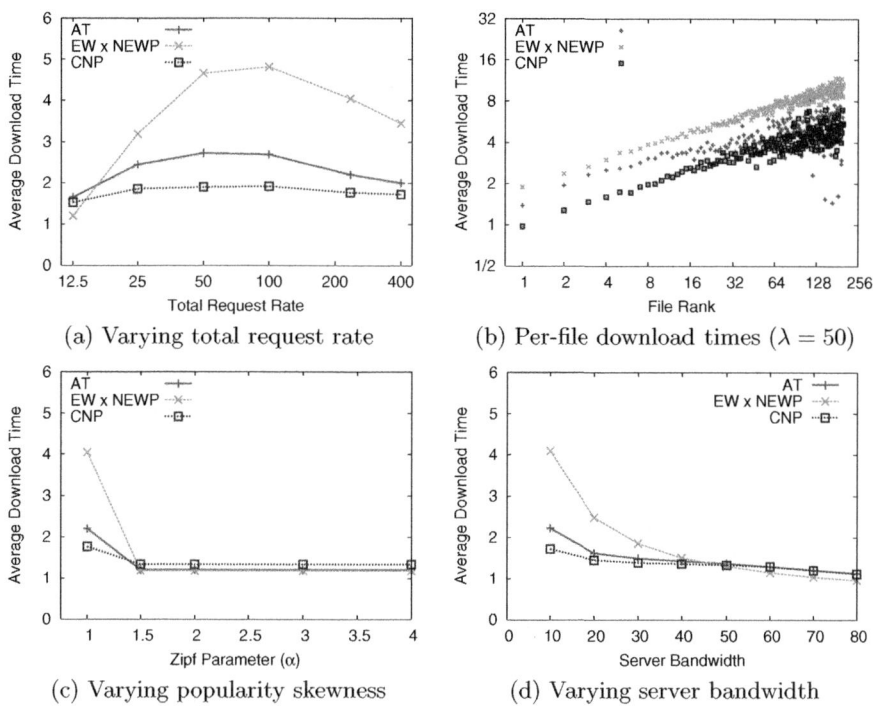

(a) Varying total request rate   (b) Per-file download times ($\lambda = 50$)

(c) Varying popularity skewness   (d) Varying server bandwidth

Fig. 7. Impact of Workload and Server Parameters (Default scenario: $B = 10$, $N = 200$, $\lambda_i = 40/i$ ($\lambda \approx 235$), $L = 1$, $U = 1$, $D/U = 3$, $K = 256$)

6   Conclusions

This paper has considered peer-assisted delivery using BitTorrent-like protocols, with files seeded only by a content provider server. A fundamental problem in such systems is that of how to efficiently support delivery of the "long tail" of lukewarm/cold files for which there are insufficient concurrent requesters for active torrents to form. To address this problem, "torrent inflation" was proposed, in which some of the available upload bandwidth from currently downloading peers is used to "inflate" torrents for other files, making them more self-sustaining. Current work is investigating analytic modelling approaches that would allow us to evaluate larger systems than we can evaluate with simulation.

Acknowledgments

This work was supported by the Natural Sciences and Engineering Research Council (NSERC) of Canada, the Informatics Circle of Research Excellence (iCORE) in the Province of Alberta, and the National Information and Communications Technology Australia (NICTA).

References

1. Breslau, L., Cao, P., Fan, L., Phillips, G., Shenker, S.: Web caching and Zipf-like distributions: Evidence and implications. In: Proc. IEEE INFOCOM, New York (March 1999)
2. Cha, M., Kwak, H., Rodriguez, P., Ahn, Y., Moon, S.: I tube, you tube, everybody tubes: Analyzing the world's largest user generated content video system. In: Proc. ACM IMC, San Deigo, CA (October 2007)
3. Dan, G., Carlsson, N.: Dynamic swarm management for improved BitTorrent performance. In: Proc. IPTPS, Boston, MA (April 2009)
4. Yang, X., de Veciana, G.: Service capacity of peer-to-peer networks. In: Proc. IEEE INFOCOM, Hong Kong, China (March 2004)
5. Legout, A., Urvoy-Keller, G., Michiardi, P.: Rarest first and choke algorithms are enough. In: Proc. ACM IMC, Rio de Janeiro, Brazil (October 2006)
6. Guo, L., Chen, S., Xiao, Z., Tan, E., Ding, X., Zhang, X.: Measurement, analysis, and modeling of BitTorrent-like systems. In: Proc. ACM IMC, Berkley, CA (2005)
7. Yang, Y., Chow, A.L.H., Golubchik, L.: Multi-torrent: A performance study. In: Proc. MASCOTS, Baltimore, MD (September 2008)
8. Menasche, D.S., Rocha, A.A.A., Li, B., Towsley, D., Venkataramani, A.: Content availability and bundling in swarming systems. In: ACM CoNEXT, Rome, Italy (December 2009)
9. Wu, D., Liu, Y., Ross, K.W.: Queuing network models for multi-channel p2p live streaming systems. In: Proc. IEEE INFOCOM, Rio de Janeiro, Brazil (April 2009)
10. Wu, D., Liang, C., Liu, Y., Ross, K.W.: View-upload decoupling: A redesign of multi-channel p2p video systems. In: Proc. IEEE INFOCOM Mini-conference, Rio de Janeiro, Brazil (April 2009)
11. Piatek, M., Isdal, T., Krishnamurthy, A., Anderson, T.: One hop reputations for peer to peer file sharing workloads. In: Proc. NSDI, San Francisco, CA (2008)
12. Ramachandran, A., das Sarma, A., Feamster, N.: Bitstore: An incentive compatible solution for blocked downloads in BitTorrent. In: Proc. NetEcon, San Diego, CA (2007)
13. Carlsson, N., Eager, D.L.: Modeling priority-based incentive policies for peer-assisted content delivery systems. In: Proc. IFIP Networking, Singapore (2008)
14. Das, S., Tewari, S., Kleinrock, L.: The case for servers in a peer-to-peer world. In: Proc. IEEE ICC, Istanbul, Turkey (June 2006)
15. Wu, C., Li, B., Zhao, S.: Multi-channel live p2p streaming: Refocusing on servers. In: Proc. IEEE INFOCOM, Phoenix, AZ (April 2008)
16. Sun, Y., Liu, F., Li, B., Li, B.: Peer-assisted online storage and distribution: Modeling and server strategies. In: Proc. NOSSDAV, Williamsburg, VA (2009)
17. Sun, Y., Liu, F., Li, B., Li, B., Zhang, X.: Fs2you: Peer-assisted semi-persistent online storage at a large scale. In: Proc. IEEE INFOCOM, Rio de Janeiro, Brazil (April 2009)
18. Aggarwal, C.C., Wolf, J.L., Yu, P.S.: On optimal batching policies for video-on-demand storage servers. In: Proc. IEEE ICMCS, Hiroshima, Japan (June 1996)
19. Aksoy, D., Franklin, M.: RxW: A scheduling approach for large-scale on-demand data broadcast. IEEE/ACM Trans. on Networking 6(7), 846–860 (1999)
20. Carlsson, N., Eager, D.L.: Peer-assisted on-demand streaming of stored media using BitTorrent-like protocols. In: Proc. IFIP Networking, Atlanta, GA (2007)
21. Saroiu, S., Gummadi, K.P., Gribble, S.D.: A measurement study of peer-to-peer file sharing systems. In: Proc. MMCN, San Jose, CA (January 2002)

Network Distance Prediction Based on Decentralized Matrix Factorization

Yongjun Liao[1], Pierre Geurts[2,3], and Guy Leduc[1]

[1] Research Unit in Networking (RUN), University of Liège, Belgium
{liao,leduc}@run.montefiore.ulg.ac.be
[2] Systems and Modeling, University of Liège, Belgium
[3] Research associate, FRS-F.N.R.S. (Belgium)
p.geurts@ulg.ac.be

Abstract. Network Coordinate Systems (NCS) are promising techniques to predict unknown network distances from a limited number of measurements. Most NCS algorithms are based on metric space embedding and suffer from the inability to represent distance asymmetries and Triangle Inequality Violations (TIVs). To overcome these drawbacks, we formulate the problem of network distance prediction as guessing the missing elements of a distance matrix and solve it by matrix factorization. A distinct feature of our approach, called Decentralized Matrix Factorization (DMF), is that it is fully decentralized. The factorization of the incomplete distance matrix is collaboratively and iteratively done at all nodes with each node retrieving only a small number of distance measurements. There are no special nodes such as landmarks nor a central node where the distance measurements are collected and stored. We compare DMF with two popular NCS algorithms: Vivaldi and IDES. The former is based on metric space embedding, while the latter is also based on matrix factorization but uses landmarks. Experimental results show that DMF achieves competitive accuracy with the double advantage of having no landmarks and of being able to represent distance asymmetries and TIVs.

Keywords: Network Coordinate System, Matrix Factorization, Decentralized Matrix Factorization, Regularization.

1 Introduction

Predicting network distances (e.g. delay) between Internet nodes is beneficial to many Internet applications, such as overlay routing [1], peer-to-peer file sharing [2], etc. One promising approach is Network Coordinate Systems (NCS), which construct models to predict the unmeasured network distances from a limited number of observed measurements [3].

Most NCS algorithms embed network nodes into a metric space such as Euclidean coordinate systems in which distances between nodes can be directly computed from their coordinates. For example, GNP [4] is the first system that models the Internet as a geometric space. It first embeds a number of landmarks into the space and a non-landmark host determines its coordinates with respect

M. Crovella et al. (Eds.): NETWORKING 2010, LNCS 6091, pp. 15–26, 2010.

to the landmarks. Vivaldi [5] is a decentralized NCS system that extends GNP by eliminating the landmarks. It simulates a system of springs where each edge is modeled by a spring and the force of the spring reflects the approximation error.

However, network distances do not derive from the measurements of a metric space. For example, Triangle Inequality Violations (TIVs) have been frequently observed and the distances between two nodes are not necessarily symmetric due to the network structure and routing policy [6,7,8,5,9]. No algorithm based on metric space embedding can model such distance space. IDES [10] is one of the few algorithms using non-metric space embedding techniques. It is based on matrix factorization which approximates a large matrix by the product of two small matrices. A drawback of IDES is that, similar to GNP, IDES also relies on landmarks. It factorizes a small distance matrix built from the landmarks at a so-called information server and other non-landmark nodes compute their coordinates with respect to the landmarks. Phoenix extends IDES by adopting a weight model and a non-negativity constraint in the factorization to enforce the predicted distances to be positive [11]. In practice, NCS systems with landmarks are less appealing. They suffer from landmark failures and overloading. Furthermore, the number of landmarks and their placement affect the performance of NCS.

In this paper we propose a novel approach, called Decentralized Matrix Factorization (DMF), to predicting network distance. Unlike IDES, we seek to factorize a distance matrix built from all nodes in a fully decentralized manner. Each node retrieves distance measurements from and to a small number of randomly selected nodes[1] and updates its coordinates simultaneously and iteratively. There are no special nodes such as landmarks or a central node where distance measurements are collected and stored. In doing so, our DMF algorithm overcomes the drawbacks of metric space embedding and is able to represent asymmetric distances and TIVs, while it is fully decentralized and requires no landmarks. Experimental results show that DMF is stable and achieves competitive performance compared to IDES and metric space embedding based algorithms such as Vivaldi.

The rest of the paper is structured as follows. Section 2 formulates the problem of network distance prediction by matrix factorization. Section 3 describes the DMF algorithm. Section 4 presents the evaluation and the comparison of DMF with other competing methods. Section 5 gives the conclusions and discusses future work.

2   Matrix Factorization for Network Distance Prediction

Matrix Factorization seeks an approximate factorization of a large matrix, i.e.,

$$D \approx \hat{D} = XY^T,$$

where the number of columns in X and Y is typically small and is called the dimension of the embedding space. Generally, the factorization is done by minimizing $||D - \hat{D}||^2$, which can be solved analytically by using singular value

---

[1] We will refer to the selected nodes as neighbors in the rest of the paper.

decomposition (SVD) [12]. In many cases, constraints can be imposed in the minimization. A popular and useful constraint is that elements of X and Y are non-negative. This so-called non-negative matrix factorization (NMF) can only be solved by iterative optimization methods such as gradient descent [13]. Note that matrix factorization has no unique solution as

$$D \approx \hat{D} = XY^T = XGG^{-1}Y^T,$$

where G is any arbitrary non-singular matrix. Therefore, replacing X by XG and Y by $G^{-1}Y$ will not increase the approximation error.

In using matrix factorization for network distance prediction, assuming n nodes in the network, a $n \times n$ distance matrix D is constructed with some distances between nodes measured and the others unmeasured. To guess the missing elements in D, we factorize D into the form $D \approx XY^T$ by solving

$$\mathbf{min} \ ||W. * (D - XY^T)||^2, \tag{1}$$

where $.*$ is element-wise product and W is the weight matrix with $w_{ij} = 1$ if d_{ij}, the distance from i to j, is measured and 0 otherwise. X and Y are of the same size $n \times l$ with $l \ll n$. l is referred to as the dimension of the embedding space and is a parameter of the factorization algorithm.

With missing elements, the minimization of eq. 1 can only be solved by iterative optimization methods. After the factorization, each node is then associated with two coordinates x_i and y_i, where x_i is the ith row of X, called *outgoing vector*, and y_i is the ith row of Y, called *incoming vector*. The estimated distance from i to j is

$$\hat{d}_{ij} = x_i \cdot y_j, \tag{2}$$

where $\cdot$ is the dot product. If done properly, the estimated distance, $\hat{d}_{ij}$, approximates the measured distance, d_{ij}, within a limited error range. Note that $\hat{d}_{ij}$ is not necessarily equal to $\hat{d}_{ji}$.

The above process is centralized and requires a large number of distance measurements to be collected and stored at a central node. To solve this problem, IDES [10] proposed to select a small number of landmarks and compute, at a so-called information server, the factorization (by using SVD or NMF) of a small distance matrix built only from measured distances between the landmarks. Once the landmark coordinates have been fixed, a non-landmark host can determine its coordinates by measuring its distance to and from each of the landmarks and finding coordinates that most closely match those measurements. As mentioned earlier, the use of landmarks is a weakness of IDES. In the next section, we will propose our approach based on a decentralized matrix factorization that eliminates the need for landmarks.

3   Decentralized Matrix Factorization for Network Distance Prediction

The problem is the same as in eq. 1, but we seek to solve it in a decentralized manner. Similar to IDES, each node records its outgoing vector x_i and incoming

vector y_i and computes distances from and to other nodes by using eq. 2. The
difference is that x_i and y_i are initialized randomly and updated continuously
with respect to some randomly-selected neighbors.

In particular, to update x_i and y_i, node i randomly selects k neighbors, measures its distances from and to them, and retrieves the outgoing and incoming
vectors. Denote $X_i = [x_{i_1}; \ldots; x_{i_k}]$ and $Y_i = [y_{i_1}; \ldots; y_{i_k}]$ the outgoing and incoming matrices built from the neighbors of i, i.e., x_{i_j} and y_{i_j} are the outgoing
and incoming vectors of the jth neighbors of i. Let $d_{to}^i = [d_{i,i_1}, \ldots, d_{i,i_k}]$ and
$d_{from}^i = [d_{i_1,i}, \ldots, d_{i_k,i}]$ the distance vectors to and from the neighbors of i.
Then, x_i and y_i are updated by

$$x_i = \arg\min_x \; ||xY_i^T - d_{to}^i||^2, \tag{3}$$

$$y_i = \arg\min_y \; ||X_i y^T - d_{from}^i{}^T||^2. \tag{4}$$

Eqs. 3 and 4 are standard least square problems of the form $min||Ax - b||^2$,
which has an analytic solution of the form:

$$x = (A^T A)^{-1} A^T b. \tag{5}$$

To increase the numerical stability of the solution, instead of solving eqs. 3 and 4
with eq 5, we penalize x_i and y_i and solve regularized least square problem of
the form $min||Ax - b||^2 + \lambda||x||^2$, which also has an analytic solution

$$x = (A^T A + \lambda I)^{-1} A^T b, \tag{6}$$

where λ is the coefficient of the regularization term. In the experimental section,
we will show the influence of the regularization terms on the performance of
DMF.

Input: D, l, k, λ
D: distance matrix with missing elements
l: dimension of the embedding space
k: number of neighbors of each node
λ: regularization coefficient
Output: X,Y
foreach *node i* **do**
    Randomly select k neighbors from the network.
    Randomly initialize x_i and y_i.
    while *forever* **do**
        retrieve $d_{to}^i, d_{from}^i, X_i, Y_i$;
        update x_i by eq. 7
        update y_i by eq. 8
        sleep some time
    end
end

Algorithm 1. DMF: Decentralized Matrix Factorization with Regularization

To summarize, the update equations of x_i and y_i are

$$x_i = d_{to}^i Y_i (Y_i^T Y_i + \lambda I)^{-1} \qquad (7)$$
$$y_i = d_{from}^i X_i (X_i^T X_i + \lambda I)^{-1} \qquad (8)$$

The DMF algorithm is given in Algorithm 1[2]. We initialize the coordinates with random numbers uniformly distributed between 0 and 1. Empirically, we found that DMF is insensitive to the random initialization of the coordinates.

4   Experiments and Evaluations

In this section, we evaluate DMF[3] and compare it with two popular NCS algorithms: Vivaldi and IDES. The former is based on metric space embedding, while the latter is also based on matrix factorization but uses landmarks. All the experiments are performed on two typical data sets collecting real Internet measurements: the P2psim [14] data set which contains the measured distances between 1740 Internet DNS servers, and the Meridian [15] data set which contains the measured distances between 2500 nodes. While DMF can in principle handle asymmetric distance matrices, in our experiment, we took $d_{i,j} = d_{j,i}$ and defined these distances as the half of the round-trip-time between nodes i and j. The same assumption is adopted in Vivaldi and has the advantage of greatly simplifying the implementation of the algorithm, as measuring one-way delay is difficult in practice.

In the simulations, we randomly selected a node and updated its coordinates at each step. An iteration of a simulation is defined by a fixed round of node updates. Since Vivaldi updates its coordinates with respect to only one neighbor in contrast to DMF that does it with respect to all neighbors, an iteration in Vivaldi is defined by $n \times k$ node updates whereas in DMF an iteration is n node updates, where n is the number of nodes and k is the number of neighbors. In doing so, we ensure that, on average, all nodes have a chance to update their coordinates with respect to all neighbors. Note that IDES is not an iterative method. The coordinates of the nodes are unchanged.

We examine the following classical evaluation criteria.

– *Cumulative Distribution of Relative Estimation Error* Relative Estimation Error (REE) is defined as

$$REE = \frac{|\hat{d}_{i,j} - d_{i,j}|}{d_{i,j}}.$$

---

[2] Note that we can also adopt the weight model and the non-negativity constraint as in [11]. However, the constrained minimization in eqs 3 and 4 have no more closed form solutions and has to be solved by iterative optimization methods. As claimed in [10], which is confirmed by our experiments, the non-negativity constraint does not improve the accuracy a lot, but significantly increases the computing time.

[3] A matlab implementation of DMF used to generate the results in the paper can be downloaded from http://www.run.montefiore.ulg.ac.be/~liao/DMF.

- *Stress* measuring the overall fitness of the embedding is defined as

$$stress = \sqrt{\frac{\sum_{i,j} (d_{i,j} - \hat{d}_{i,j})^2}{\sum_{i,j} d_{i,j}^2}}.$$

- *Median Absolute Estimation Error* (MAEE) is defined as

$$MAEE = median_{i,j}(|d_{i,j} - \hat{d}_{i,j}|).$$

Note that our DMF algorithm utilizes only a small percentage of the distance measurements in the datasets to estimate the coordinates of the nodes, but the evaluation of the above criteria is done using all distance measurements.

4.1   Parameter Tuning

DMF has three parameters to be defined: l, the dimension of the embedding space, k, the number of neighbors of each node, and λ, the regularization coefficient. We study the influence of these parameters on the performance of DMF. To this end, we tune one parameter at a time while fixing the other two. Results are shown in Figures 1, 2 and 3.

It can be seen that l does not seem to affect the performance of DMF as long as $l \geq 3$ which coincides with the conclusion drawn in [5] about Vivaldi. We nevertheless recommend $l = 10$ as it does not pose any problem and as the same number is used in IDES. On the other hand, k has a clear impact, as a larger k gives better accuracy, which is obvious because a larger k means

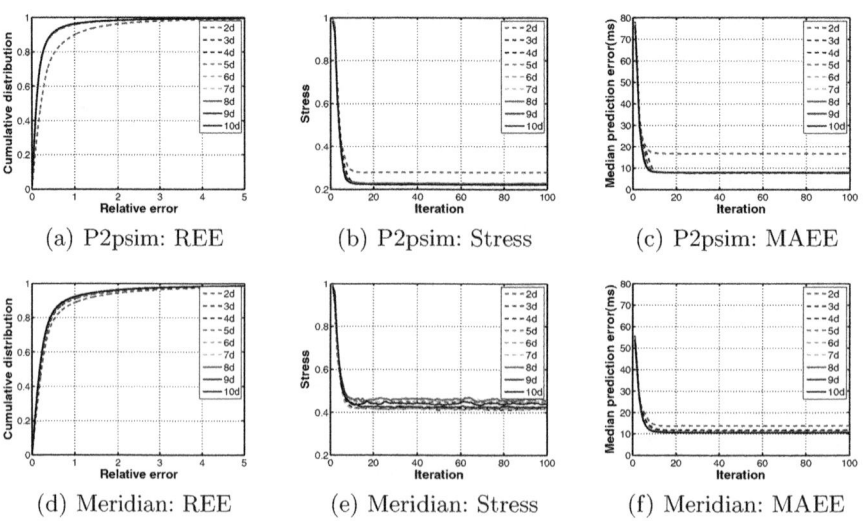

(a) P2psim: REE   (b) P2psim: Stress   (c) P2psim: MAEE

(d) Meridian: REE   (e) Meridian: Stress   (f) Meridian: MAEE

Fig. 1. The effect of the dimension (l) on the performance of DMF. ($k = 32$, $\lambda = 50$).

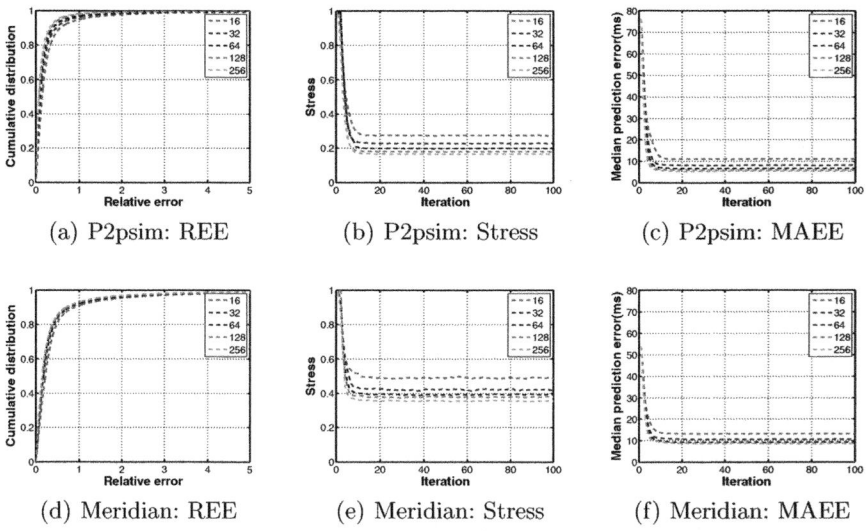

Fig. 2. The effect of the number of neighbors (k) on the performance of DMF. ($l = 10$, $\lambda = 50$).

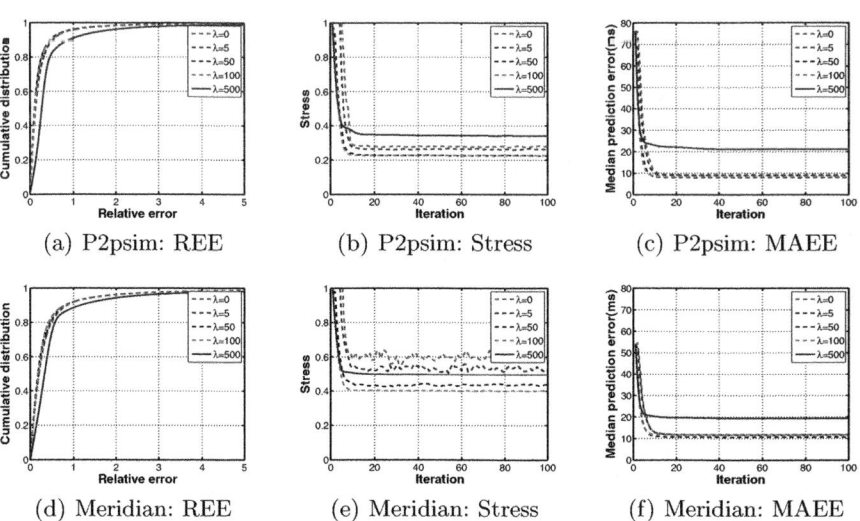

Fig. 3. The effect of regularization coefficient (λ) on the performance of DMF. ($l = 10$, $k = 32$).

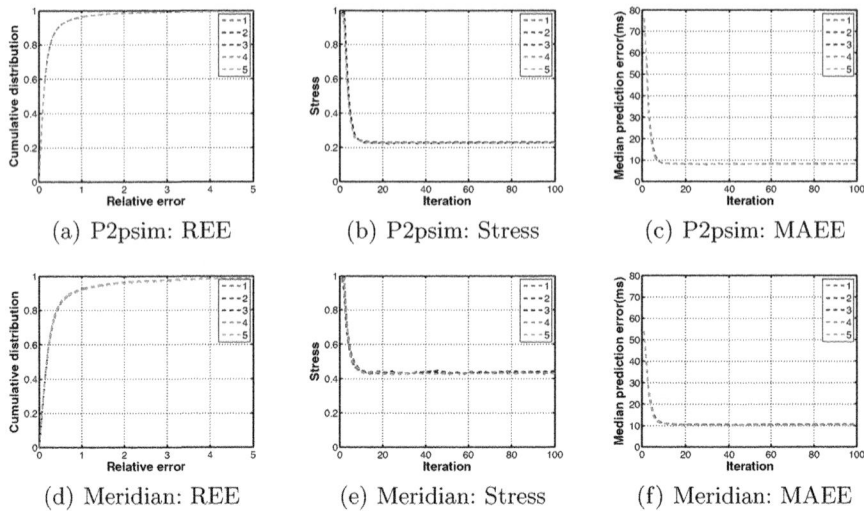

(a) P2psim: REE   (b) P2psim: Stress   (c) P2psim: MAEE

(d) Meridian: REE   (e) Meridian: Stress   (f) Meridian: MAEE

Fig. 4. Results of different simulations. The simulations differ in the initializations of the coordinates, in the selections of the neighbors by each node and in the orders in which the nodes are updated. It can be seen that the results are insensitive to these differences.

fewer missing elements thus better estimation of the coordinates. However, a larger k also means more probe traffic and a higher overhead. Following Vivaldi, we suggest $k = 32$ as a good tradeoff between accuracy and measurement overhead. For λ, too little or too much regularization only decreases the accuracy of DMF, and 50 seems to be a good choice for both P2psim and Meridian datasets.

Note that the results are very stable from one simulation to another, as highlighted in Figure 4. The algorithm does not seem very sensitive to the random initialization of the coordinates and to the particular selection of neighbours. In the following, unless otherwise stated, $l = 10$, $k = 32$ and $\lambda = 50$ are used by default and the results of all the experiments are derived from one simulation.

4.2   Analysis of Convergence and Stability

We further evaluate the convergence and the stability of DMF. From Figures 1, 2, 3 and 4, it is clear that DMF converges fast, empirically in less than 20 iterations for both P2psim and Meridian datasets.

To further verify the stability of DMF, we performed a 2D factorization ($l = 2$) and plotted the X and Y coordinates at different times of the simulation, shown in Figure 5. It can be seen that the coordinates are very stable with little drift after the embedding errors become stable. Figure 6 shows the histogram of the differences between the predicted distance matrix at the 20th and the 100th iterations.

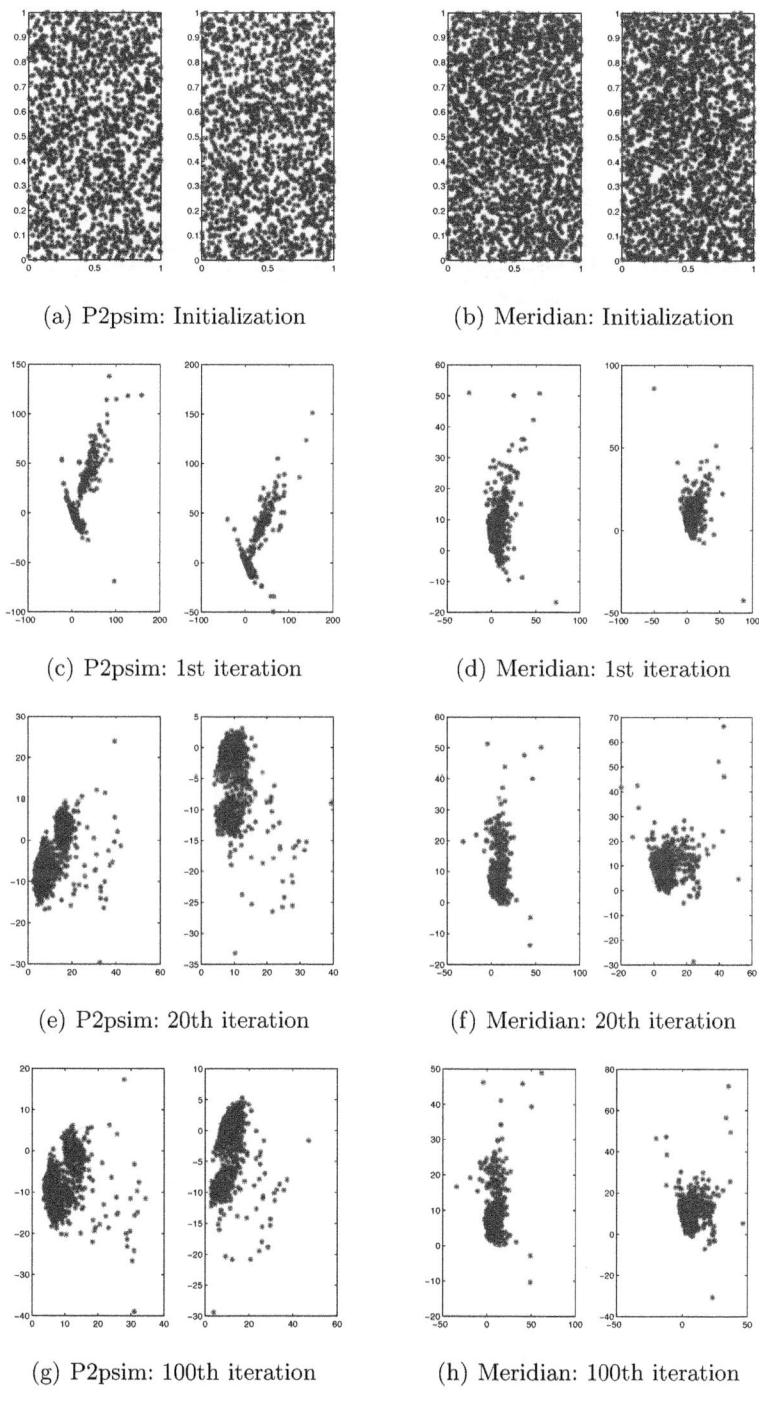

(a) P2psim: Initialization

(b) Meridian: Initialization

(c) P2psim: 1st iteration

(d) Meridian: 1st iteration

(e) P2psim: 20th iteration

(f) Meridian: 20th iteration

(g) P2psim: 100th iteration

(h) Meridian: 100th iteration

Fig. 5. The evolution of the coordinates, X (left subplot) and Y (right subplot)

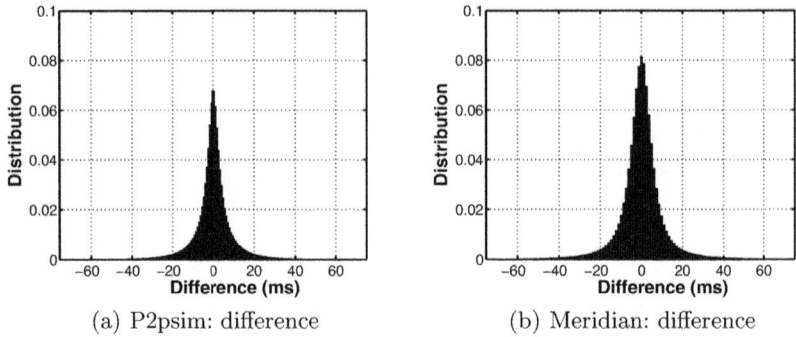

(a) P2psim: difference   (b) Meridian: difference

Fig. 6. The differences between the predicted distance matrix at the 20th and the 100th iterations

4.3 Comparisons with Vivaldi and IDES

Lastly, we compare DMF with Vivaldi and IDES, as shown in Figure 7. Vivaldi is a decentralized NCS algorithm based on Euclidian embedding. Similar to DMF, each node updates its coordinates with respect to k randomly selected neighbors. Here, we took $k = 32$ following the recommendation in Vivaldi. For IDES, a number of landmarks are needed. Although [16,10] claimed that 20 randomly selected landmarks are sufficient to achieve desirable accuracy, we nevertheless deployed 32 landmarks in our experiments for the purpose of comparison. The dimensions of the embedding space are 10 for all algorithms.

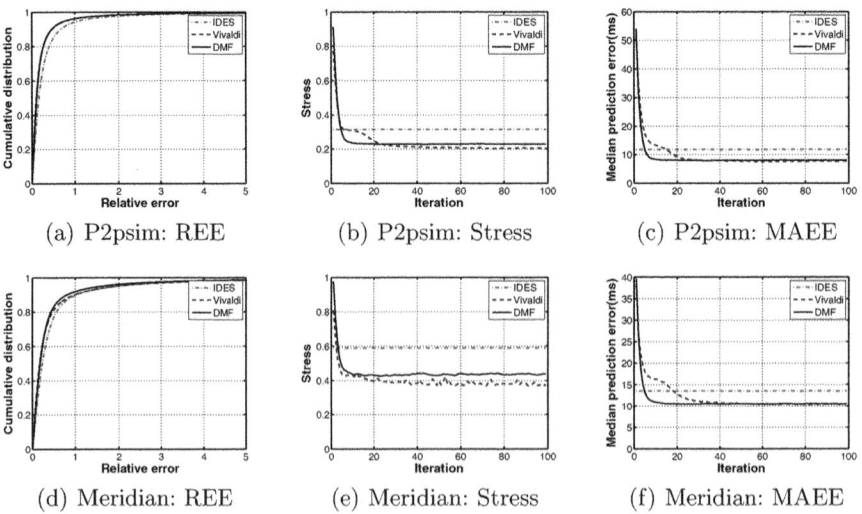

(a) P2psim: REE   (b) P2psim: Stress   (c) P2psim: MAEE

(d) Meridian: REE   (e) Meridian: Stress   (f) Meridian: MAEE

Fig. 7. Comparison with IDES and Vivaldi. ($l = 10$, $k = 32$, $\lambda = 50$ for DMF).

From Figure 7, it can be seen that both DMF and Vivaldi achieve similar accuracy and slightly outperform IDES. The worse performance by IDES is likely due to the use of the landmarks. Since in IDES, a non-landmark node only communicates with landmarks, no links between non-landmark nodes are used by the NCS. In contrast, DMF and Vivaldi are completely decentralized with links between nodes randomly selected and evenly distributed in the whole network.

5   Conclusions and Future Works

In this paper, we proposed a novel approach, called DMF, to predicting unknown network distances. Essentially, we consider it as a learning problem where coordinates of network nodes are learned from partially observed measurements and the unknown distances are approximated from the learned coordinates. Different from all previous works, the learning of the coordinates is done by DMF which requires no landmarks. Since DMF is not based on metric space embedding, it has the potential to overcome common limitations such as the inability to represent TIVs and asymmetric distances. Experimental results show that the performance of our approach is comparable with two popular NCS algorithms: Vivaldi and IDES.

The reason why DMF does not outperform Vivaldi given the above-mentioned advantages remains an open question that needs further investigation. It has to be noted that both P2psim and Meridian datasets are symmetric with $d_{ij} = d_{ji}$. We would like to test DMF on more datasets, especially those with heavily asymmetric distances.

Acknowledgements

This work has been partially supported by the EU under projects FP7-Fire ECODE, by the European Network of Excellence PASCAL2 and by the Belgian network DYSCO (Dynamical Systems, Control, and Optimization), funded by the Interuniversity Attraction Poles Programme, initiated by the Belgian State, Science Policy Office. The scientific responsibility rests with its authors.

References

1. Hari, D.A., Andersen, D., Balakrishnan, H., Kaashoek, F., Morris, R.: Resilient overlay networks, 131–145 (2001)
2. Azureus Bittorrent, http://azureus.sourceforge.net
3. Donnet, B., Gueye, B., Kaafar, M.A.: A survey on network coordinates systems, design, and security. To appear in IEEE Communication Surveys and Tutorial (December 2010)
4. Ng, T.S.E., Zhang, H.: Predicting Internet network distance with coordinates-based approaches. In: Proc. IEEE INFOCOM, New York, NY, USA (June 2002)

5. Dabek, F., Cox, R., Kaashoek, F., Morris, R.: Vivaldi: A decentralized network coordinate system. In: Proc. ACM SIGCOMM, Portland, OR, USA (August 2004)
6. Zheng, H., Lua, E.K., Pias, M., Griffin, T.: Internet Routing Policies and Round-Trip-Times. In: Proc. the PAM Conference, Boston, MA, USA (April 2005)
7. Lee, S., Zhang, Z., Sahu, S., Saha, D.: On suitability of euclidean embedding of internet hosts. SIGMETRICS 34(1), 157–168 (2006)
8. Wang, G., Zhang, B., Ng, T.S.E.: Towards network triangle inequality violation aware distributed systems. In: Proc. the ACM/IMC Conference, San Diego, CA, USA, October 2007, pp. 175–188 (2007)
9. Banerjee, S., Griffin, T.G., Pias, M.: The interdomain connectivity of PlanetLab nodes. In: Barakat, C., Pratt, I. (eds.) PAM 2004. LNCS, vol. 3015, pp. 73–82. Springer, Heidelberg (2004)
10. Mao, Y., Saul, L., Smith, J.M.: Ides: An internet distance estimation service for large networks. IEEE Journal on Selected Areas in Communications (JSAC), Special Issue on Sampling the Internet, Techniques and Applications 24(12), 2273–2284 (2006)
11. Chen, Y., Wang, X., Song, X., Lua, E.K., Shi, C., Zhao, X., Deng, B., Li., X.: Phoenix: Towards an accurate, practical and decentralized network coordinate system. In: Proc. IFIP Networking Conference, Aachen, Germany (May 2009)
12. Golub, G.H., Van Loan, C.F.: Matrix computations, 3rd edn. Johns Hopkins University Press, Baltimore (1996)
13. Lee, D.D., Seung, H.S.: Algorithms for non-negative matrix factorization. In: NIPS, pp. 556–562. MIT Press, Cambridge (2001)
14. A simulator for peer-to-peer protocols, http://www.pdos.lcs.mit.edu/p2psim/index.html
15. Wong, B., Slivkins, A., Sirer, E.: Meridian: A lightweight network location service without virtual coordinates. In: Proc. the ACM SIGCOMM (August 2005)
16. Tang, L., Crovella, M.: Geometric exploration of the landmark selection problem. In: Barakat, C., Pratt, I. (eds.) PAM 2004. LNCS, vol. 3015, pp. 63–72. Springer, Heidelberg (2004)

Topology-Awareness and Reoptimization Mechanism for Virtual Network Embedding

Nabeel Farooq Butt[1], Mosharaf Chowdhury[2], and Raouf Boutaba[1,3]

[1] David R. Cheriton School of Computer Science,
University of Waterloo,
Waterloo, ON N2L 3G1, Canada
nfbutt@uwaterloo.ca, rboutaba@uwaterloo.ca
[2] Computer Science Division,
University of California, Berkeley,
Berkeley, CA 94720, USA
mosharaf@cs.berkeley.edu
[3] Division of IT Convergence Engineering, POSTECH,
Pohang, KB 790-784, Korea

Abstract. Embedding of virtual network (VN) requests on top of a shared physical network poses an intriguing combination of theoretical and practical challenges. Two major problems with the state-of-the-art VN embedding algorithms are their indifference to the underlying substrate topology and their lack of reoptimization mechanisms for already embedded VN requests. We argue that topology-aware embedding together with reoptimization mechanisms can ameliorate the performance of the previous VN embedding algorithms in terms of acceptance ratio and load balancing. The major contributions of this paper are twofold: (1) we present a mechanism to differentiate among resources based on their importance in the substrate topology, and (2) we propose a set of algorithms for reoptimizing and re-embedding initially-rejected VN requests after fixing their bottleneck requirements. Through extensive simulations, we show that not only our techniques improve the acceptance ratio, but they also provide the added benefit of balancing load better than the previous proposals.

Keywords: Topology-awareness, Re-embedding, Virtual Network Embedding, Network Virtualization.

1 Introduction

Network virtualization is widely considered to be a potential candidate for providing the essential basis for future Internet architectures [1, 2, 3, 4]. One major challenge in network virtualization environment (NVE) is the allocation of substrate network resources, managed by infrastructure providers (InPs), to online VN requests by different service providers (SPs) [3, 4]. Unfortunately, the VN embedding problem reduces to the multi-way separator problem, which has been shown to be NP-hard [5]. From the point of view of an InP, a preferable allocation scheme should increase long-term revenue while reducing the cost of hosting

M. Crovella et al. (Eds.): NETWORKING 2010, LNCS 6091, pp. 27–39, 2010.

individual VNs. Consequently, mechanisms that can increase acceptance ratio are of great interest to InPs, because acceptance ratio is directly proportional to their revenue.

While embedding a VN request, existing proposals [6, 7, 8, 9, 10] do not distinguish between different substrate nodes and links. However, in practice, some substrate nodes and links are more *critical* than others. For example, resource depletion in bridges and articulation points in the substrate topology are expected to have deeper impact on future embeddings than a random resource near the edge of the network. Choosing amongst feasible VN embeddings, the one that uses fewer critical resources can be a key to improving acceptance ratio in the long run.

In this paper, we investigate how differentiating between substrate resources can increase the acceptance ratios of the existing VN embedding algorithms [6, 7, 8, 9, 10]. For differentiating resources, we design a mechanism for assigning weights to substrate nodes and links based on their residual capacities and their importance in the substrate topology. These weights help us in prioritizing certain VN embeddings over others. Other than links' or nodes' resource depletion being able to partition the network, attributes of the created partitions are also an important factor in weight assignment. In case of a substrate network partitioning, any virtual link from a VN request has to be turned down if its two ends can only be mapped in different partitions. We argue that distinguishing substrate resources is a key for better load balancing and for ameliorating the acceptance ratio.

Over time, as new VNs are embedded and old ones expire, resources in the substrate network become fragmented. The obvious consequence is the rejection of many VN requests lowering the acceptance ratio and revenue. This could be amended had the fragmented resources been consolidated using reoptimization techniques. We propose algorithms for identification and re-embedding of VNs that cause VN request rejection. Our reoptimization mechanism consists of two stages: detection of bottleneck links and nodes, followed by relocation of virtual links and nodes to free up resources on bottlenecks.

The remainder of this paper is organized as follows. Section 2 summarizes background and related work. Our proposals for distinguishing and reoptimizing critical resources are discussed in Section 3. Section 4 presents performance metrics, our experimental setup, and their results. We conclude in Section 5.

2   Background and Related Work

The intra-domain VN embedding problem refers to the mapping of VN requests from different SPs on top of substrate network resources managed by an InP. A substrate network in this problem formulation is defined as an undirected graph with substrate nodes and links with CPU and bandwidth capacities, respectively. Similarly, a VN request is defined as an undirected graph with virtual nodes and links with CPU and bandwidth requirements and additional constraints.

The long-term objective of a VN embedding algorithm is to maximize the revenue of an InP while minimizing its cost of embedding online VN requests. Any virtual node can be embedded onto at most substrate node, whereas a virtual link can refer to a physical path in the substrate network. Further details on the basic VN embedding problem formulation can be found in [7, 8].

Allocating resources for Virtual Private Networks [11, 12] is the earliest, but simpler, incarnation of the VN embedding problem. It is different in its consideration of only the bandwidth constraints on virtual links. Since constraints on virtual nodes are ignored, this problem often simplifies to selecting paths that satisfy the given constraints.

VN embedding proposals in the existing literature address the problem by simplifying constraints in different dimensions. Szeto et al. [13] have presented algorithms for embedding only the virtual links, with an assumption that the virtual nodes have already been mapped. Off-line version of the VN embedding problem had been the focus of some of the existing research [6, 9]. Precluding admission control complexities by assuming infinite capacities of substrate nodes and links have been addressed in [6, 9, 14]. Lu et al. [9] have developed algorithms for specific topologies. Yu et al. [7] have provided a two stage algorithm for mapping the VNs. They start off by mapping the virtual nodes first and then proceed to map the virtual links using shortest paths and multi-commodity flow (MCF) algorithms. Chowdhury et al. [8] have recently presented a solution to the VN embedding problem by coordinating the node and the link mapping phases. by using mixed integer programming formulation. Backtracking algorithms in conjunction with graph isomorphism detection to map VN requests are used in a dynamic fashion in [15]. Distributed algorithms for simultaneously embedding virtual links and virtual nodes [10] have shown issues with scalability and poorer performance than their centralized counterparts.

Although all these approaches strive for improving VN embedding by various degrees, none of them foresee the possibility of narrowing down the chances for accepting future VN requests. They treat all the substrate nodes and links equally, which is not practical. Their indifference to the underlying topology, allows uninformed decisions while mapping resources on 'critical' substrate links and nodes that can affect the acceptance ratio in the future. Another potential candidate for improving the accepting ratio is the effective use of reoptimization mechanisms. By detecting and relocating bottleneck link and node embeddings to better alternates, we can achieve improvements. Ideally, we want to keep all the VNs optimized at all times (e.g., through periodic updating [6]). But this will result in relocation of a lot of virtual links and virtual nodes and incur significant overhead. A better way to deal with this problem is to take action only when it is inevitable [7]. However, unlike [7], we propose algorithms for relocation of virtual nodes as well as links.

Before proceeding to the next section, we want to mention that we use the same formulation for revenue as in the previous literature [6, 7, 8], but in upcoming sections we will provide a slightly better cost function.

3   Improving Acceptance Ratio and Balancing Load

Indifference to the attributes of the substrate network resources and lack of re-optimization mechanisms are the two major candidates for improvement in the existing VN embedding algorithms [6, 7, 8, 9, 15]. In the following sections, we present techniques to incorporate topology-awareness and reoptimization mechanisms for bottleneck embeddings in these algorithms to improve their overall performance.

3.1   Topology-Aware Embedding

We differentiate between substrate network resources by introducing scaling factors to their costs. The *scaling factor (SF)* of a resource refers to its likelihood of becoming a bottleneck. It is calculated as a combination of two weight factors, critical index and popularity index, explained later in this section.

We aim to incorporate topology-awareness in existing algorithms with minimal changes. In our solution, proposals using MCF only need to update their objective functions (constraints require no change) and proposals using shortest path algorithms should select paths with minimum total scaled costs to upgrade themselves.

Critical Index. The first weight function we define is the *critical index (CI)* of a substrate resource. The CI of a resource measures the likelihood of a residual substrate network partition[1] due to its unavailability. To understand the impact of such partitioning, consider the following scenario: suppose that a substrate network is partitioned into two - almost equal-sized - components. The probability of both the end nodes of any new virtual link to be embedded in different partitions is 0.5. In this case, we might have to reject almost 50% of the VN requests.

We denote CI by ζ ($\zeta : x \to [0,1)$), where x is either a substrate link or a node. Higher value of $\zeta(x)$ means its highly likely that unavailability of x will partition the substrate graph and vice versa. The mathematical definition of CI is given in equations 1 and 2 for links and nodes respectively.

$$\zeta(e_s) = \begin{cases} \left(\dfrac{1 - |c_1 - c_2|}{2} \right) + \dfrac{1}{2} & e_s \in \text{cut-edges} \\[3ex] \dfrac{\phi + \psi}{4} & e_s \notin \text{cut-edges} \end{cases} \tag{1}$$

$$\zeta(n_s) = \begin{cases} \dfrac{1}{2} \left(\prod_{c \in C} P\{c\} \right)^{\frac{1}{|C|}} + \dfrac{1}{2} & n_s \in \text{cut-vertices} \\[3ex] \dfrac{\phi + \psi}{4} & n_s \notin \text{cut-vertices} \end{cases} \tag{2}$$

---

[1] We refer to residual substrate network while discussing partitioning in this paper. A residual substrate network is composed of residual capacities of substrate resources.

In equation 1, if e_s is a bridge (cut-edge) then removing it partitions the graph. c_1 and c_2 are the fractions of substrate nodes in each of these partitions. When e_s is not a bridge, then after removing it we get ϕ and ψ, where ϕ and ψ are the fractions of new cut-nodes or cut-edges to the total cut-nodes or cut-edges respectively. In equation 2, C is the set of components we get when we remove n_s. $P\{c\}$ here is the fraction of nodes present in c. In both equations 1 and 2, we make sure that a cut-resource has its CI value ≥ 0.5. Computing ζ is required only once in the beginning and it is updated whenever a substrate network is extended. Also, it is not computationally intensive.

Popularity Index. Resource saturation is the other major factor that can cause substrate network partition. *Popularity Index (PI)* measures "how many different VNs are affected when a link or a node is unavailable?" To be more specific, PI ($\rho : x \rightarrow [0, 1)$) is the time weighted average of occupied resources on a particular link or a node. It also takes into account the number of VNs that are mapped on that resource. In equation 3, $R_{E_{i-1}}$ is the percent of reserved bandwidth on link x, where the index $i - 1$ means the previous value of R_E; similarly, $R_{N_{i-1}}$ is the percent of reserved CPU capacity at the node x. The variables a and b ($a + b = 1$) are used to assign different weights to current and previous values. In equation 3, ν is the number of VNs mapped on top of x. The higher the value of ρ the higher the probability that mapping onto this link or node will saturate it and create a bottleneck.

$$\rho(x) = \begin{cases} \left(aR_{E_{i-1}} + bR_{E_i}\right)^{\frac{1}{\nu}} & x \in E_s \\ \\ \left(aR_{N_{i-1}} + bR_{N_i}\right)^{\frac{1}{\nu}} & x \in N_s \end{cases} \tag{3}$$

Scaling Factor. Both CI and PI values are very crucial when mapping a particular link or node. Together they are referred to as the scaling factor and denoted by $\aleph$. In equation 4, we have multiplied both CI and PI by α and β respectively. We also use ω, which is a parameter and can be used to further tune the cost. The objective functions of the LP formulations of previous proposals [8] can be updated by scaling the cost of resources by $\aleph$.

$$\aleph(x) = 1 + \omega\big(\alpha\zeta(x) + \beta\rho(x)\big) \tag{4}$$

We use a slightly different cost function, $C(G_V)$, than that of the previous literature. In the modified formula, we are scaling up the cost by the scaling factor to avoid embedding onto critical resources whenever possible. In the formula below, n_s is the substrate node on which a virtual node n_v is mapped, $f_{e_s}^{e_v}$ is the flow on substrate link e_s for the virtual link e_v, and $c(n_v)$ is the cost of the virtual node n_v.

$$C(G_V) = \sum_{e_V \in E_V} \sum_{e_S \in E_S} \aleph(e_s) f_{e_s}^{e_V} + \sum_{n_V \in N_V} \aleph(n_s)c(n_v) \tag{5}$$

3.2 Reoptimizing Bottleneck Embeddings

Although efficient algorithms exist for embedding of VNs onto a substrate network, arbitrary lifetimes of VNs can cause some embeddings to drift toward inefficiency over time. Such embeddings can be reoptimized by relocating virtual nodes and reassigning virtual links to better alternate embeddings. From a practical point of view, it has been shown how to migrate virtual nodes (virtual routers) from one physical node to another [16]. Link migration can also be achieved by dynamically setting up and tearing down physical links; it is already achieved using programmable transport networks [17].

Our proposed solution can be divided into two stages. First, we detect the virtual links and nodes that caused a VN request to be rejected. Second, we relocate/reassign these nodes/links to less critical regions of the substrate network. We illustrate the basic idea behind our relocation and reassignment mechanism using the example in Fig. 1 and Fig. 2. In Fig. 1 (a), VN request VN-6 is being rejected because one of its virtual nodes cannot be mapped, i.e., node U which

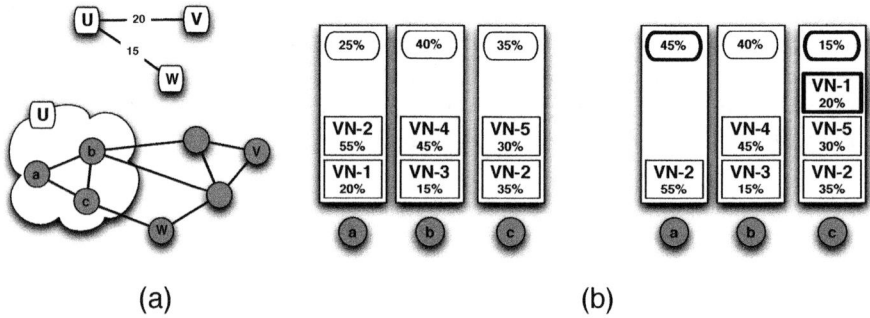

(a) (b)

Fig. 1. Relocating a virtual node of VN-1 to make room for virtual node U

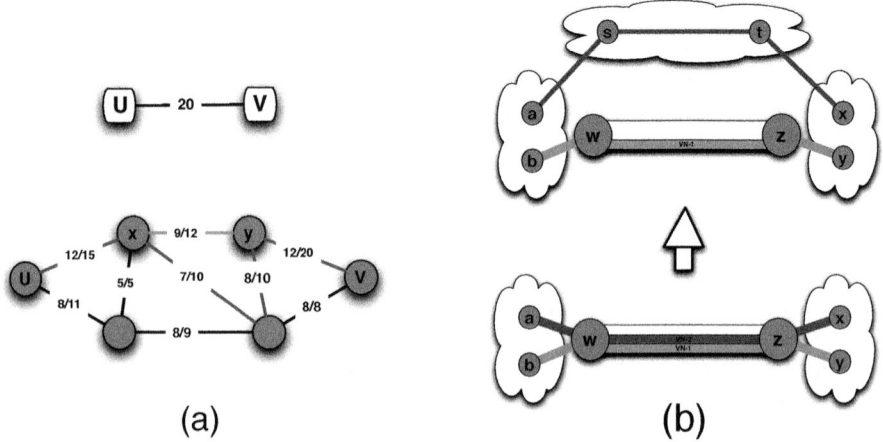

(a) (b)

Fig. 2. Reassigning links to make room for virtual link

Algorithm 1. Detecting V-Nodes that cannot be mapped

1: $N_{bad} \leftarrow$ virtual nodes that cannot be mapped
2: **for all** virtual nodes $v_n \in N_{bad}$ **do**
3:     $C_{N_S}(v_n) \leftarrow$ candidate nodes
4:     **for all** substrate node $n \in C_{N_S}(v_n)$ **do**
5:       $L_{low} \leftarrow$ All virtual nodes mapped on n that have lower priority than VN_{rej}
6:       **if** capacity$(L_{low}) >$ required capacity **then**
7:         Save (n, L_{low}), for relocating later
8:       **end if**
9:     **end for**
10:    fails if v_n cannot be fixed
11: **end for**

Algorithm 2. Detecting V-Links that cannot be mapped

1: $L_{bad} \leftarrow$ All virtual links v such that MaxFlow(v) $<$ ReqCap(v)
2: **for all** virtual links $v \in L_{bad}$ **do**
3:     **repeat**
4:       **for all** substrate link $e \in MinCut(v)$ **do**
5:         $L_{low} \leftarrow$ All virtual links mapped on e that have lower priority than VN_{rej}
6:         Save (e, L_{low})
7:       **end for**
8:     **until** $Capacity(MinCut(v)) < ReqCapacity(v)$
9:     fails if v cannot be fixed
10: **end for**

---

requires 43% of CPU capacity. There are three possible embeddings for this node (i.e., nodes a, b, and c), but none with the required capacity for U. In Fig. 1 (b), relocating a VN-1 node from node a to node c makes enough room to accommodate U. All the associated virtual links of VN-1 are also reassigned to the new location. Now suppose that we want to embed a virtual link U-V. We can see that it cannot be mapped because of the bottleneck substrate link w-z (Fig. 2 (a)). As shown in Fig. 2 (b), VN-2 is reassigned to free up some bandwidth on the substrate link w-z so that the virtual link U-V can be accommodated.

Detecting Bottleneck Nodes and Links. For bottleneck detection, we deal with VN requests that cannot be embedded due to lack of residual capacities in substrate resources. Contrary to the periodic reconfiguration scheme presented in [6], we have taken a reactive approach in detecting bottlenecks – we reoptimize whenever a VN request gets rejected.

Algorithm 1, finds a list of virtual nodes that cannot be mapped, and for each one of them it stores corresponding bottleneck substrate nodes. After detecting the candidate substrate nodes for an unmapped virtual node, it figures out whether enough capacity can be salvaged on any of the candidate nodes or actually on L_{low}. L_{low} is the set of virtual nodes of other VN requests with less priority than the rejected VN request. If a single virtual node cannot be mapped

---

Algorithm 3. Relocating a Virtual Node

1: Q_{v_n} = v-nodes that are adjacent to v_n
2: C_{v_n} = candidate s-nodes for v_n
3: status = **false**
4: **for all** substrate node $n \in C_{v_n}$ **do**
5:     status = **true**
6:     **for all** virtual link $v \in Q_{v_n}$ **do**
7:         Map-Link($Substrate(v)$, v_n)
8:         **if** map fails **then**
9:            status = **false**
10:            break
11:         **end if**
12:     **end for**
13:     **if** status **then**
14:         Relocate v_n to n
15:     **end if**
16: **end for**

---

we do not proceed any further. Finally, L_{low} and associated substrate node n are stored for later use in the relocation phase. The worst case running time of algorithm 1 is $O(n^2)$, but on the average it would be much faster than that.

Next, we use algorithm 2 to determine the unmapped virtual links, L_{bad}, and corresponding bottleneck links. A virtual link is rejected if there is not enough capacity between its two end nodes. The maximum flow from the source to the sink[2] and the minimum s-t cut can be computed in $O(V^3)$ using the Edmond-Karp maximum flow algorithm [18]. Algorithm 2 iterates through L_{bad} and for every virtual link it finds the minimum cut[3]. Re-assigning some virtual links on the minimum cut can increase the maximum flow between source and sink. Note that even if we made enough room in the first minimum cut it might still not guarantee the required flow between the source and the sink. Since the minimum cut can change over time, we have to iteratively compute the minimum cut. To keep the running time of algorithm 2 small, we repeat this only a constant number of times. Next we re-assign links in L_{low} to free up some capacity on these links. It is useless to proceed if we fail to map only a single virtual link; in this case, we can just reject this request immediately.

Nodes and Links Selection and Placement. Algorithm 3, relocates a virtual node v_n which is currently mapped on a substrate node s_n to some other substrate node, provided all the virtual links incident on v_n can also be reassigned. Q_{v_n} is a set of virtual nodes adjacent to v_n and C_{v_n} is a list of substrate nodes which are potential hosts for v_n. Note that C_{v_n} is sorted according to the overhead cost. The algorithm iterates on C_{v_n} and for every substrate node n in C_{v_n}, it checks whether it can map all the virtual links after v_n is relocated to n.

---

[2] Source and sink are the end nodes of the rejected virtual link.
[3] For information on Max-Flow Min-Cut theorem, readers are referred to [18].

Algorithm 4. Reassigning a Virtual Link

1: $R_E(e_S) = 0$
2: (u, v) be the virtual nodes of e_V
3: $c = MaxFlow(u, v)$
4: **if** $c = 0$ **then**
5:     **return false**
6: **end if**
7: **if** $ReqCap(e_V) <= c$ **then**
8:     Re-embed virtual link e_V
9: **else**
10:     Free c amount of flow from e_S
11:     Augment c amount of flow elsewhere
12: **end if**
13: Restore and Adjust $R_E(e_S)$
14: **return true**

If it can map all these links then v_n can be relocated to the substrate node used to map these links; finally, it relocates v_n.

For reassigning a virtual link, algorithm 4 takes the virtual link e_V which is mapped to substrate link e_S. First it sets the available bandwidth on e_S to zero, so that the freed capacity on e_S is not allocated to e_V, when e_V is reassigned. It then finds the maximum flow between the end nodes of e_V and if its zero than nothing can be done. If the maximum flow is greater than the required bandwidth capacity of the link e_V then it just free the old mapping of e_V and remaps it again. If the maximum flow is less than the required capacity then we can free c amounts of capacity from e_S and map c units of capacity elsewhere by using the computed flow. Finally, we restore and adjust the freed bandwidth capacity on e_S (which was set to zero at the beginning). These link mappings should be done using topology-aware LP formulation, to keep the overhead cost of reassigning e_V to minimum.

4   Evaluation

Our evaluation focuses primarily on quantifying the effectiveness of our techniques when applied to the VN embedding algorithms in the existing literature. We consider the following four algorithms in our evaluation: D-ViNE and R-ViNE [8] map nodes deterministically and randomly, respectively; in both the algorithms, links are embedded using a modified MCF. G-SP [6] and G-MCF [7] both use greedy node mapping; for link embedding, the former uses shortest path algorithm, whereas the latter uses MCF.

4.1   Experimental Setup and Performance Metrics

We have extended the simulator of [8] to include our proposed techniques and algorithms. We have used GT-ITM [19] for generating topologies for the underlying substrate networks. Substrate networks in our experiments have 100 nodes

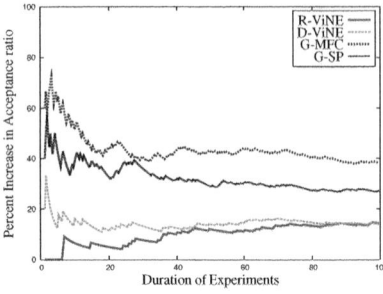 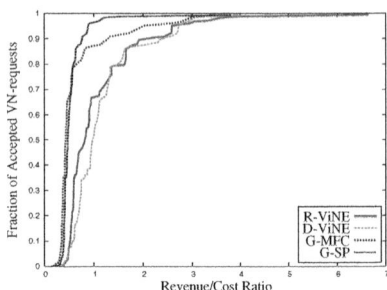

Fig. 3. Improvement in Acceptance Ratios for Different Algorithms   **Fig. 4.** Revenue-Cost Ratio for Different Algorithms

and around 400 links, on the average. Node CPU capacities and link bandwidth capacities are randomly chosen between 50 to 100. For VN requests, we used a similar setup as previously used in [6, 7, 8]. VN requests arrive in a Poisson process with an inter-arrival time of 20 time units. Lifetimes of the VN requests follow an exponential distribution with mean 1000 time units. Virtual nodes for VN requests are chosen uniformly between 5 to 10. CPU and bandwidth requirements are distributed uniformly between 0 to 20 and 0 to 50 respectively. The metrics we use to measure performance are increase in acceptance ratio, revenue-cost ratio, incurred cost, and distribution of utilization of resources.

4.2 Evaluation Results

Improvement in Acceptance Ratio. In our first set of experiments, we compare the increase in acceptance ratios of all the algorithms. As shown in Fig. 3, in steady state, our mechanisms improve acceptance ratios by almost 40% for G-MCF and just below 35% for G-SP. For D-ViNE and R-ViNE, the improvements are smaller but still a sizeable 17%. We believe that the lower increase for D-ViNE and R-ViNE is due to the fact that they already have much higher acceptance ratio than G-SP and G-MCF without the proposed improvements [8].

Revenue/Cost Ratio. We present the per request revenue/cost ratio for the compared algorithms in Fig. 4. We see that quite a few readings are close to zero which indicates that those VN requests are being mapped to one or more very critical links or nodes. On the average, the revenue/cost ratio of D-ViNE and R-ViNE is slightly above 2 for 90% of the accepted VN requests, which is significantly higher than that of G-SP and G-MCF. This shows that our techniques not only improve the acceptance ratios (Fig. 3), but they also keep the revenue/cost ratio within acceptable range (specially for D-ViNE and R-ViNE).

Acceptance Ratio vs Incurred Cost. An important factor in evaluating the effectiveness of our techniques is the cost incurred by increasing the acceptance

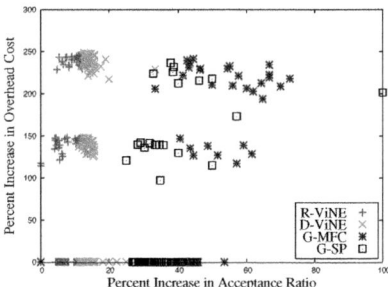
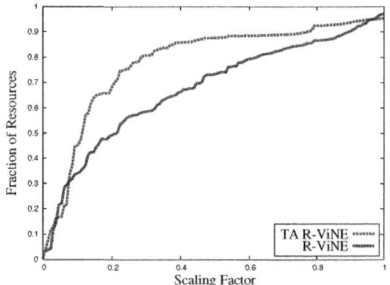

Fig. 5. Acceptance Ratio vs Incurred Cost   **Fig. 6.** CDF of SFs of all links and nodes

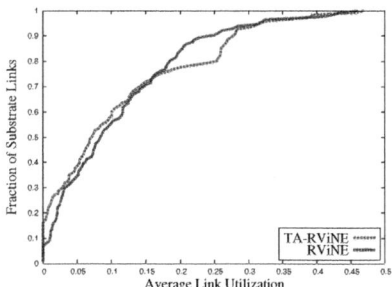
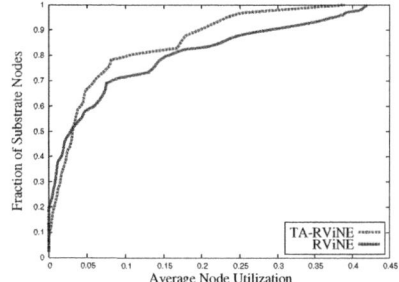

Fig. 7. Average link utilization   **Fig. 8.** Average node utilization

ratio. In Fig. 5, we have plotted the percentage increase in acceptance ratio against the percentage increase in cost for making room for these VN requests. Here the incurred cost is the sum of the cost of moving already embedded VN requests and accepting VN requests rejected earlier. Fig. 5 shows that the maximum increase in cost is close to 250%, but the range of acceptance ratio improvement varies from $7 - 77\%$ for various algorithms. For all the algorithms there are several readings with very low or even zero incurred cost. This is because the VNs required to move to make room for rejected VN requests have a lot of alternative embeddings of equal cost and incurred no additional cost.

Differentiating Resources. In Fig. 6, we have plotted the CDF of the scaling factors of all links and nodes for R-ViNE [8] and R-ViNE incorporated with topology-aware embedding (hereinafter referred to as "TA-R-ViNE"). The other algorithms, compared with their extended versions, also show similar characteristics. SF is a good measure because it gives an indication of potential bottlenecks; an SF value close to 1 means that a resource is more likely to become a bottleneck. We can see here that TA-R-ViNE has significantly fewer resources being indicated as bottlenecks than R-ViNE. With TA-R-ViNE, only 15% resources have SF values above 0.4; whereas with R-ViNE, it is almost 35%. This graph shows that topology-awareness plays a significant role in identifying bottlenecks.

Link and Node Utilization. A comparison between link utilization of TA-R-ViNE and R-ViNE is shown in Fig. 7. It presents an average of multiple runs of carefully designed experiments that made sure that both techniques could map the same VN requests. We can see that for R-ViNE, almost 90% of the links are utilizing less than 25% of bandwidth, whereas for TA-R-ViNE it is almost 80%. An explanation would be that these links are alternative links to bottlenecks; hence they are used more often. We can also see that TA-R-ViNE used almost 8% fewer substrate links than R-ViNE. In Fig. 8, almost 50% of the nodes have less utilization than 50% of the nodes mapped by TA-R-ViNE, but the other 50% have much higher utilization. This figure shows that to avoid bottlenecks TA-R-ViNE has distributed load among nodes.

5 Conclusion

In this paper, we have presented mechanisms for differentiating various resources (links and nodes) according to their impact on the connectivity of the substrate network, and their utilization coupled with the number of VNs utilizing them. We have also proposed algorithms for reoptimizing the already embedded VN requests in order to improve the overall acceptance ratio. Through simulation we have shown that our techniques significantly improve the performance of the existing algorithms in terms of increase in acceptance ratio, revenue-cost ratio, cost incurred by these techniques and load balancing based on criticalness of resources. Experimental results endorse the fact that our techniques have enough potential to be seriously considered as an integral part of any future VN embedding algorithm design.

Acknowledgment

This work was jointly supported by the Natural Science and Engineering Council of Canada (NSERC) under its Discovery program, Cisco Systems, and WCU (World Class University) program through the National Research Foundation of Korea funded by the Ministry of Education, Science and Technology (Project No. R31-2008-000-10100-0).

References

1. Anderson, T., Peterson, L., Shenker, S., Turner, J.: Overcoming the internet impasse through virtualization. Computer 38(4), 34–41 (2005)
2. Turner, J., Taylor, D.: Diversifying the internet. In: Global Telecommunications Conference, 2005. GLOBECOM 2005, November-2 December, vol. 2, p. 6. IEEE, Los Alamitos (2005)
3. Feamster, N., Gao, L., Rexford, J.: How to lease the Internet in your spare time. ACM SIGCOMM Computer Communication Review 37(1), 61–64 (2007)
4. Chowdhury, N.M.M.K., Boutaba, R.: A survey of network virtualization. Computer Networks (2010) (in press), http://dx.doi.org/10.1016/j.comnet.2009.10.017

5. Andersen, D.: Theoretical approaches to node assignment.,
   http://www.cs.cmu.edu/~dga/papers/andersen-assign.ps
6. Zhu, Y., Ammar, M.: Algorithms for assigning substrate network resources to virtual network components. In: IEEE INFOCOM, April 2006, pp. 1–12 (2006)
7. Yu, M., et al.: Rethinking virtual network embedding: substrate support for path splitting and migration. SIGCOMM CCR 38(2), 17–29 (2008)
8. Chowdhury, N.M.M.K., Rahman, M.R., Boutaba, R.: Virtual network embedding with coordinated node and link mapping. In: IEEE INFOCOM (2009)
9. Lu, J., Turner, J.: Efficient mapping of virtual networks onto a shared substrate. Washington University, Tech. Rep. WUCSE-2006-35 (2006)
10. Houidi, I., Louati, W., Zeghlache, D.: A distributed virtual network mapping algorithm. In: IEEE ICC, pp. 5634–5640 (2008)
11. Gupta, A., et al.: Provisioning a virtual private network: A network design problem for multicommodity flow. In: ACM STOC, pp. 389–398 (2001)
12. Ricci, R., Alfeld, C., Lepreau, J.: A solver for the network testbed mapping problem. ACM SIGCOMM CCR 33(2), 65–81 (2003)
13. Szeto, W., Iraqi, Y., Boutaba, R.: A multi-commodity flow based approach to virtual network resource allocation. In: IEEE GLOBECOM, pp. 3004–3008 (2003)
14. Fan, J., Ammar, M.: Dynamic topology configuration in service overlay networks: A study of reconfiguration policies. In: IEEE INFOCOM (2006)
15. Lischka, J., Karl, H.: A virtual network mapping algorithm based on subgraph isomorphism detection. In: ACM SIGCOMM VISA Workshop, pp. 81–88 (2009)
16. Wang, Y., et al.: Virtual routers on the move: Live router migration as a network-management primitive. In: ACM SIGCOMM, pp. 231–242 (2008)
17. Agrawal, M., et al.: Routerfarm: towards a dynamic, manageable network edge. In: ACM SIGCOMM INM Workshop, pp. 5–10 (2006)
18. Diestel, R.: Graph theory. Springer, New York (1997)
19. Zegura, E.: How to model an Internet. In: IEEE INFOCOM, pp. 594–602 (1996)

Survivable Virtual Network Embedding

Muntasir Raihan Rahman[1], Issam Aib[1], and Raouf Boutaba[1,2]

[1] David R. Cheriton School of Computer Science,
University of Waterloo,
Waterloo, Ontario, Canada N2L 3G1
[2] Division of IT Convergence Engineering, POSTECH,
Pohang, Korea KB 790-784
{mr2rahman,iaib,rboutaba}@cs.uwaterloo.ca

Abstract. Network virtualization can offer more flexibility and better manageability for the future Internet by allowing multiple heterogeneous virtual networks (VN) to coexist on a shared infrastructure provider (InP) network. A major challenge in this respect is the VN embedding problem that deals with the efficient mapping of virtual resources on InP network resources. Previous research focused on heuristic algorithms for the VN embedding problem assuming that the InP network remains operational at all times. In this paper, we remove that assumption by formulating the survivable virtual network embedding (SVNE) problem and developing a hybrid policy heuristic to solve it. The policy is based on a fast re-routing strategy and utilizes a pre-reserved quota for backup on each physical link. Evaluation results show that our proposed heuristic for SVNE outperforms baseline heuristics in terms of long term business profit for the InP, acceptance ratio, bandwidth efficiency, and response time.

Keywords: Survivability, Virtual Network Embedding, Network Virtualization.

1 Introduction

Network virtualization has been proposed as a diversifying attribute of the future inter-networking paradigm that can enable seamless integration of new features to the current Internet resulting in rapid evolution of the Internet architecture [4, 8, 5]. By allowing multiple heterogeneous network architectures to cohabit on a shared physical infrastructure, network virtualization promises better flexibility, security, manageability and decreased power consumption for the Internet. In a network virtualization environment (NVE), the traditional role of the Internet Service Provider (ISP) has been divided into two separate entities: (1) the infrastructure providers (InP) who are responsible for deploying and maintaining physical network resources (routers, links etc.) and the (2) service providers (SP) who implement various network protocols and heterogeneous network architectures on virtual networks (VNs) composed from physical network resources leased from one or more infrastructure providers.

M. Crovella et al. (Eds.): NETWORKING 2010, LNCS 6091, pp. 40–52, 2010.
© IFIP International Federation for Information Processing

Virtual Network Embedding (VNE) is the central resource allocation problem in network virtualization. It deals with the efficient mapping of virtual networks onto physical network resources. More specifically, for each virtual network creation request, the VNE is responsible for mapping virtual nodes onto physical nodes and virtual edges onto one or more physical paths. The VNE problem, with constraints on virtual nodes and virtual links, can be reduced to the $\mathcal{NP}$-hard multi-way separator problem, even if the schedule of VN requests is known beforehand [3]. Even when all the virtual nodes are already mapped, the virtual link embedding problem remains $\mathcal{NP}$-hard. In order to reduce the hardness of the VN embedding problem and enable efficient heuristics, existing research has been restricting the problem space in different dimensions, e.g., considering the off-line version of the problem [13, 22], ignoring either node or link requirements [7, 13], assuming infinite capacity of the substrate nodes and links to obviate admission control [7, 13, 22], and focusing on specific virtual topologies [13]. Recently the authors in [12, 6] have proposed VNE heuristics that combine the node and link embedding phases. The authors in [9] have proposed a distributed algorithm that simultaneously maps virtual nodes and virtual links without any centralized controller. However, a limitation of all these heuristics is that they assume the substrate network to be operational at all times, which is not realistic. The existing heuristics are not capable of handling substrate node and link failures, which may lead to poor performance and increased frustration for the SP.

In this paper, we formulate the survivable virtual network embedding (SVNE) problem to incorporate single substrate link failures in VNE and propose an efficient heuristic for solving it. To the best of our knowledge, this is the first work to consider survivability strategies in the network virtualization environment.

The rest of the paper is organized as follows. Section 2 provides an overview of the related work. Section 3 formalizes the network model and formulates the virtual network embedding and survivable virtual network embedding problems. In Section 4, we present our proposed hybrid policy heuristic as a solution to the survivable virtual network embedding problem. Section 5 presents simulation results that evaluate the proposed hybrid policy heuristic compared to base-line heuristics. Section 6 concludes by identifying future research directions.

2   Related Work

Node and link failure survivability problems have been investigated extensively for optical and multi-protocol label switched (MPLS) networks [16], and real time systems [21]. Our work differs in a number of aspects, due to unique challenges introduced by the network virtualization environment. First, the VNE problem is on-line in nature, whereas the survivable logical topology design problem in optical and mpls networks [17, 11, 10] is off-line. Secondly, in NVEs, we need to ensure that all virtual links are intact in the presence of failures. This restriction is not present, for example, in optical networks where the goal is to only ensure that all virtual nodes remain connected in the presence of failures, even if they are not connected via a direct virtual link. Our contribution also differs from existing work in terms of the objective formulation. Our aim is to develop a survivable virtual network embedding solution that simultaneously maximizes the long term revenue for the InP and, at the same time minimizes the long term penalty incurred by the InP due to service violations caused by failures. This dual nature of the objective function in the presence of failures is absent both in the existing research on optical and mpls networking domains and the existing VNE heuristics. Another novel aspect of our work is that we utilize path-flow based optimization formulations for solving the SVNE problem. The path formulation allows control over the characteristics of the paths selected for embedding and survivability against failures. For instance, we can directly control the total number of paths, number of hops per path, and impose delay constraints on virtual links for QoS purposes. This is not possible with a link-flow based formulation which has been used for the previous VNE heuristics [6, 20, 12, 22].

3   Problem Formulation

3.1   Substrate Network

We model the substrate network as a weighted graph $G^S(N^S, E^S)$, where N^S and E^S represent the set of substrate nodes and links respectively. Each substrate node $x \in N^S$ has an associated cpu capacity $cpu(x)$ and a geographical location value $loc(x)$. A substrate link $s = (s_x, s_y) \in E^S$ between substrate nodes $s_x, s_y \in N^S$ has a bandwidth capacity $b(s)$. From now on, we denote the endpoints of any substrate link s as s_x and s_y.

3.2   Virtual Network Request

A Virtual Network (VN) request $G^V(N^V, E^V)$ is also modeled as a weighted graph. VN requests are associated with constraints and QoS requirements embodied into service level agreements (SLA) [2]. A virtual nodes $y \in N^V$ has a cpu capacity requirement $cpu(y)$ and geographical location requirement $loc(y)$. A virtual link $v \in E^V$ is characterized by a bandwidth capacity requirement $b(v)$ and a delay constraint $d(v)$. $d(v)$ is used to preselect the set of admissible simple substrate paths that can be used to embed v. An example of a typical substrate network and two virtual network topologies are shown in figure 1. The numerical values beside the substrate nodes and links represent cpu and bandwidth constraints of those nodes and links respectively.

3.3   Resource Usage Metrics

We assume that substrate network resources are finite. As a result, the amount of residual substrate network resources diminishes as new VN requests are processed. We keep track of the residual substrate node and link capacities in order

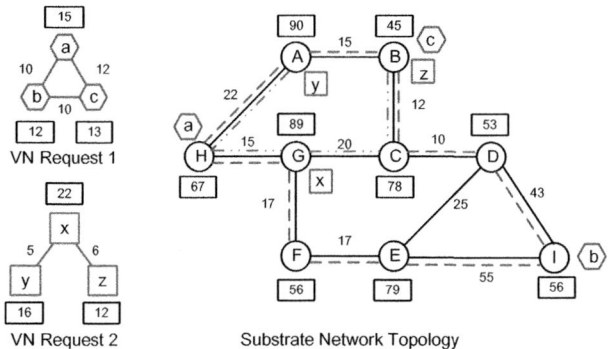

Fig. 1. Mapping of VN requests onto a shared substrate network

to make sure we don't accept a request unless there are adequate resources to serve it. The residual capacity of a substrate node $x \in N^S$ is defined as:

$$R_N(x) = cpu(x) - \sum_{y \in V(x)} cpu(y), \qquad (1)$$

where $V(x)$ denotes the set of virtual nodes mapped onto x. Similarly the residual capacity of a substrate link $s \in E^S$ is defined as:

$$R_E(s) = b(s) - \sum_{\{v : \exists p \in \Gamma_E(v),\, s \in p\}} b(v), \qquad (2)$$

where, $\Gamma_E(v)$ defines the set of paths in the InP that are used to embed the virtual link v (Section 3.4). The residual capacity values are updated after each new VN request has been successfully mapped on top of the substrate network as long as there remains adequate residual resources. The values are also updated after a VN departs and link failure arrivals and departures.

In order to protect against single substrate link failures, we dedicate a certain percentage of bandwidth resources on each substrate link for backup purposes. For a substrate link s with total bandwidth $b(s)$, $\alpha(s)b(s)$ bandwidth is reserved for primary flows, whereas $\beta(s)b(s)$ is reserved for backup flows, where $\alpha(s) + \beta(s) = 1$. The residual bandwidth measure is accordingly decomposed into primary and backup residual bandwidth measures $\mathcal{R}_\alpha(s)$ and $\mathcal{R}_\beta(s)$ respectively. As a result, we need to keep track of these two residual bandwidth measures separately.

3.4   VN Embedding

The VN Embedding process refers to the mapping of the virtual network topology (logical topology) on top of the substrate network topology (physical topology) subject to certain constraints. The constraints are normally manifested in

terms of the residual resource availability of the substrate network and the QoS
parameters specified by the VN request. An example of a VN embedding can be
seen in figure 1. From graph theoretic standpoint, the VN embedding process
can be separated into two separate stages:

1-Node Embedding Phase: Each virtual node from a VN request is mapped to
a single distinct substrate node by a one-to-one mapping:

$$\Gamma_N : N^V \leftarrow N^S, \tag{3}$$

such that $\Gamma_N(x) = \Gamma_N(y)$, iff $x = y$ $\forall x, y \in N^V$, subject to the cpu capacity
constraints: $cpu(x) \leq R_N(\Gamma_N(x))$ $\forall x \in N^V$.

2-Link Embedding Phase: Each virtual link is mapped to either an unsplittable
substrate path or a splittable multi-commodity flow based set of multiple paths
between the substrate nodes corresponding to the endpoints of the virtual link.
Mathematically, we have a mapping:

$$\Gamma_E : E^V \leftarrow \mathcal{P}^S, \tag{4}$$

such that $\forall v = (v_x, v_y) \in E^V$, and $\mathcal{P}^S$ is the set of simple paths of G^S. We have
$\Gamma_E(v) \subseteq \mathcal{P}(\Gamma_N(v_x), \Gamma_N(v_y))$, subject to the bandwidth capacity constraints:
$b(v) \leq R_E(p)$, $\forall p \in \Gamma_E(e^V)$, where $\mathcal{P}(z, w)$ denotes the set of simple substrate
paths between substrate nodes z and w, and $R_E(p) = \min_{s \in p} R_E(s)$. For any
virtual link $v \in E^V$, we specify the set of QoS constrained substrate paths for v
as $\mathcal{P}(v) = \{p \in \mathcal{P}^S | delay(p) \leq d(v)\}$.

3.5   Formulation of SVNE

We represent the input to SVNE as a tuple $< G^S, G^V, j, l, \{\alpha(s)\}_{s \in E^S} >$, where
G^S and G^V represent the substrate and virtual networks respectively, j represents the service class of the SP owning G^V, $l \in E^S$ is the failed substrate link,
and $\beta(s) = 1 - \alpha(s)$, such that $\beta(s)$ represents the percentage of bandwidth
on each substrate link s reserved for backups. Let $\Pi(G^V)$ denote the revenue
generated from G^V, where

$$\Pi(G^V) = T(G^V)[C_1 \sum_{v \in E^V} b(v) + C_2 \sum_{x \in N^V} cpu(x)] \tag{5}$$

C_1 and C_2 are weight factors which represent the relative importance of bandwidth and cpu to the generated revenue respectively. $T(G^V)$ represents the lifetime of the VN characterized by G^V. Each service class j is associated with a
penalty function $\mathcal{S}_j(.)$, where $\mathcal{S}_j(v)$ represents the monetary penalty incurred if
the bandwidth contract of virtual link v is violated.

Let $\mathcal{V}$ denote the set of all virtual links affected by the failure of l. Then the
expected total penalty incurred by the InP to the corresponding SP is:

$$\mathcal{X}(G^V; l) = MTTR(l) \sum_{v \in \mathcal{V} \cap E^V} \mathcal{S}_j(v) \frac{db(v)}{b(v)} \tag{6}$$

$MTTR(l)$ is the mean time to repair for l. The difference between the bandwidth requested for v, and the actual bandwidth supplied by the InP is represented as $db(v)$. Let $G_1^S, G_2^S, G_3^S, \ldots$ be the sequence of VN requests, and $l_1, l_2, l_3, \ldots$ be the sequence of substrate link failure events. Then the objective of SVNE is to maximize long term business profit expressed as:

$$\Pi_\infty = \sum_{p=1}^{\infty} \sum_{q=1}^{\infty} [\Pi(G_q^V) - \mathcal{X}(G_q^V; l_p)] \qquad (7)$$

4   HYBRID Policy Heuristic for SVNE

We propose a *hybrid policy* heuristic for solving SVNE. The heuristic consists of three separate phases. In the first phase, before any VN request arrives, the InP pro-actively computes a set of possible backup detours for each substrate link using a path selection algorithm. Therefore, for each substrate link l, we have a set D_l of candidate backup detours. The InP is free to utilize any path selection algorithm that suits its purposes, e. g. k-shortest path algorithm [18], column generation or primal dual methods [1]. The second phase is invoked when a VN request arrives. In this phase, the InP performs a node embedding using existing heuristics [22,6] and a multi-commodity flow based link embedding, that we denote as HYBRID_LP_LE. Finally, in the event of a substrate link failure, a reactive backup detour optimization solution HYBRID_LP_BDO is invoked which reroutes the affected bandwidth along candidate backup detours selected in the first phase. The pseudo-code for the hybrid policy is shown in the following algorithm (Figure 2).

```
 1: procedure HRP(G^S(N^S, E^S))
 2:     for all s ∈ E^S do
 3:         pre-compute candidate detour set D_s.
 4:     end for
 5:     for all event arrivals do
 6:         if event type == VN arrival then
 7:             compute node embedding for VN G^V(N^V, E^V).
 8:             solve HYBRID_LP_LE.
 9:             update R_α(s), ∀s involved in HYBRID_LP_LE.
10:         end if
11:         if event type == Failure arrival then
12:             solve HYBRID_LP_BDO.
13:             update R_β(s), ∀s involved in HYBRID_LP_BDO.
14:         end if
15:     end for
16: end procedure
```

Fig. 2. Hybrid Recovery Policy

We now show the formulations of HYBRID_LP_LE and HYBRID_LP_BDO.

4.1   Formulation of **HYBRID_LP_LE**

In this phase we use a path flow based multi-commodity flow to embed all the virtual links simultaneously. For each pair $(x, y) \in V^S \times V^S$, we have a set of preselected end-to-end paths $\mathcal{P}(x, y)$. For a virtual link $v \in E^V$, we denote $\mathcal{P}(v) = \mathcal{P}(v_x, v_y)$ as the set of pre-selected QoS constrained simple paths for embedding v, where v_x and v_y are the end-points of v. Since the node embedding phase precedes the link embedding phase, we already know which virtual node is mapped to which substrate node. For any virtual link $v = (x', y') \in E^V$, we denote this as $x' \to \Gamma_N(x') = x$ and $y' \to \Gamma_N(y') = y$. HYBRID_LP_LE can be expressed as the following linear program:

HYBRID_LP_LE
-Objective Function

$$\text{Minimize} \sum_{v \in E^V} \sum_{p \in \mathcal{P}(v)} b(p, v) \tag{8}$$

Subject to
-Primary Capacity Constraint

$$\sum_{v \in E^V} \sum_{p \in \mathcal{P}(v)} \delta_s(p) b(p, v) \leq \mathcal{R}_\alpha(s), \quad \forall s \in E^S. \tag{9}$$

-Primary Bandwidth Constraint

$$\sum_{p \in \mathcal{P}(v)} b(p, v) = b(v), \quad \forall v \in E^V \tag{10}$$

Remarks

1. $\delta_s(p)$ is the link-path indicator variable, that is, $\delta_s(p) = 1$ if $s \in p$, 0 otherwise.
2. The objective function 8 corresponds to the revenue function Π for the VN.
3. $b(p, v)$ is the amount of bandwidth allocated on path p for virtual link v. A strictly positive value for $b(p, v)$ will indicate that p is a substrate path used for v. The values of $b(p, v)$ are stored and later used in the subsequent phase of the heuristic.
4. Constraint 9 is the primary capacity constraint which states that the total primary bandwidth allocated for all virtual links must be within the primary residual capacity of each substrate link.
5. Constraint 10 is the primary bandwidth constraint which specifies that the total bandwidth requirement of each virtual link must be distributed among all the QoS constrained paths allows for that virtual link.

4.2   Formulation of HYBRID_LP_BDO

HYBRID_LP_BDO can be expressed as the following linear program.

HYBRID_LP_BDO
-Objective Function

$$\text{Minimize} \sum_{v \in E^V} \mathcal{S}_j(v) \sum_{p \in \mathcal{P}(v)} \delta_l(p) \lceil b(p,v) \rceil [1 - \sum_{d \in \mathcal{P}_l} \frac{b(d,p,v)}{b(p,v)}] \tag{11}$$

Subject to
-Backup Capacity Constraint

$$\sum_{v \in E^V, p \in \mathcal{P}(v), d \in \mathcal{D}_l} \lceil b(p,v) \rceil \; \delta_s(d) \; b(d,p,v)\delta_l(p) \leq \mathcal{R}_\beta(s) \forall s \in E^S \tag{12}$$

-Recovery Constraint

$$\sum_{d \in \mathcal{D}_l, v \in E^V, p \in \mathcal{P}(v)} \delta_l(p) \; \lceil b(p,v) \rceil \; \delta_s(d) \; b(d,p,v) \leq \sum_{d \in \mathcal{D}_l, v \in E^V, p \in \mathcal{P}(v)} \delta_l(p) \; b(p,v) \tag{13}$$

Remarks

1. j represents the service class associated with the VN. Subsequently $\mathcal{S}_j(v)$ denotes the penalty incurred for violating the bandwidth reservation for a virtual link v belonging to a VN of service type j.
2. $\lceil x \rceil$ denotes the ceiling of x, that is $\lceil x \rceil = 1$ iff $x > 0$. So $\lceil b(p,v) \rceil = 1$ indicates that p is a path used for the embedding of v. Note that the $b(p,v)$ values are calculated and stored in the HYBRID_LP_LE phase.
3. For the failed substrate link l, we have the set of candidate backup detours, $\mathcal{D}_l = \mathcal{P}(l_x, l_y) \setminus \{l\}$.
4. $b(d,p,v)$ denotes the amount of rerouted bandwidth on detour $d \in \mathcal{D}_l$ for $b(p,v)$, that is for the primary path p allocated for virtual link v.
5. The objective (equation 11) refers to the penalty function formulated in equation 6.
6. Constraint 12 is the backup capacity constraint which states that the total backup flow on all the detours passing through a substrate link must be within the backup residual capacity of that substrate link.
7. Constraint 13 is the recovery constraint and it signifies that the total disrupted primary bandwidth must be allocated along the precomputed set of detours. The objective function ensures that the virtual links that have higher penalty values will be given priority in the recovery constraint.

Both HYBRID_LP_LE and HYBRID_LP_BDO are linear programs, as a result our proposed HYBRID policy is a polynomial time heuristic for SVNE. Another important feature of HYBRID is that it decouples primary and backup bandwidth provisioning. As a result, we don't need complex disjoint constraints in our solution which would have resulted in a hard mixed integer program. The objective functions of HYBRID_LP_LE and HYBRID_LP_BDO jointly solve the long term objective of SVNE as expressed in equation 7.

5   Performance Analysis

In this section, we first describe our simulation environment, then present evaluation results. Our evaluation is aimed at quantifying the performance of the proposed HYBRID policy approach to SVNE in terms of long term business profit for the InP by maximizing revenue earned from VN's and minimizing the penalty incurred due to substrate link failures.

5.1   Simulation Environment

We implemented a discrete event simulator for SVNE adapted from our ViNE-Yard simulator [6]. Since network virtualization is still not widely deployed, the characteristics of VN's and failure are not well understood. Specifically there are no analytical or experimental results on the substrate and virtual network topology characteristics, VN arrival dynamics or link failure dynamics in network virtualization. As a result, we use synthetic network topologies, and poisson arrival processes for VN's and link failures in our simulations. However our choice of substrate and virtual topologies and VN arrival process parameters are chosen in accordance with previous work on this problem [6, 20]. We used glpk [14] to solve the linear programs.

The substrate network topologies in our experiments are randomly generated with 50 nodes using the GT-ITM tool [18] in 25 x 25 grids. Each pair of substrate nodes is randomly connected with probability 0.5. The cpu and bandwidth resources of the substrate nodes and links are real numbers uniformly distributed between 50 and 100. We assume that both VN requests and substrate link failure events follow a Poisson process with arrival rates λ_V and λ_F. The ratio $\gamma = \frac{\lambda_F}{\lambda_V}$ is a parameter that we vary in our simulations. We use realistic values for $MTTR(l)$ based on failure characteristics of real ISP networks [15] which represent InP networks in a NVE. In each VN request, the number of virtual nodes is a uniform variable between 2 and 20. The average VN connectivity is fixed at 50%. The bandwidth requirement of a virtual link is a uniform variable between 0 and 50, and the penalty value $S_j(v)$ for a virtual link v is set to a uniform random variable between 2 and 15 monetary units. In our simulations, we set $\alpha(s) = \alpha, \forall s$ belonging to the substrate network and vary α, where $0 \leq \alpha \leq 1$. For each set of experiments conducted, we plotted the average of 5 values for the performance metrics.

5.2   Comparison Method

Comparing our heuristics with previous work is difficult since the earlier heuristics do not consider substrate resource failures. As a result we evaluate our proposed *hybrid* policy against two base-line policies. The first one (we call it a *blind* policy) recomputes a new embedding for each VN affected by the substrate link failure. The second one is a *proactive* policy which pre-reserves both primary and backup bandwidth for each virtual link on link disjoint substrate paths. We omit details of these baseline policies due to space limitation. For

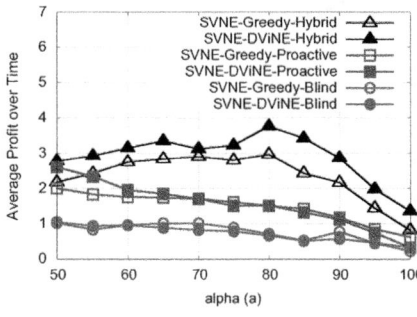
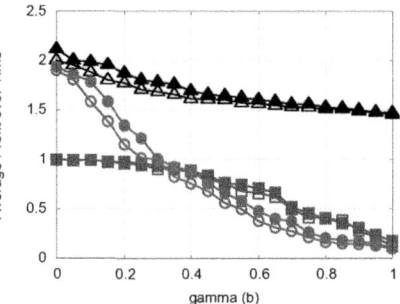

Fig. 3. Business profit against α **Fig. 4.** Business profit against γ

node embedding, we use greedy [22] and DViNE heuristics [6]. In our evaluation, we have compared six algorithms that combine different node embedding strategies [22,6] with our proposed survivable link embedding strategies, namely, SVNE-Greedy-Hybrid, SVNE-DViNE-Hybrid, SVNE-Greedy-Proactive, SVNE-DViNE-Proactive, SVNE-Greedy-Blind, and SVNE-DViNE-Blind.

5.3   Evaluation Results

We use several performance metrics for evaluation purposes in our experiments. We measure the long term average profit earned by the InP by hosting VN's. The profit function depends on both the revenue earned from VN's by leasing resources and penalties incurred due to service disruption caused by substrate link failures. The penalty depends on both the amount of bandwidth violated due to a failure and the time it takes to recover from a failure as expressed in equations 6 and 7. We also measure the long term average acceptance ratio, percentage of backup bandwidth usage and response time to failures. We present our evaluation results by summarizing the key observations.

Acceptance Ratio and Business Profit. The hybrid policy leads to higher acceptance ratio and increased business profit in the presence of failures. Figures 3 shows the long term business profit against the percentage α of substrate link bandwidth for primary flows, while Figure 4 does it against the ratio of failure and VN rate γ. We notice that over the range of values for α and γ, the hybrid policy outperforms both the blind and proactive policies. Also the hybrid policy generates the highest profit at $\alpha = 80\%$, whereas the proactive and blind policies generate lesser profit with increased values of α. As α increases, there is less bandwidth available for backups on the substrate link and this hinders the performance of these policies. This also affects the hybrid policy, but it still has better performance due to its reactive nature. The profit and acceptance ratio for the blind policy drops more rapidly than the hybrid policy against increase in γ as shown in Figures 4 and 6. Although, the profit for the proactive policy

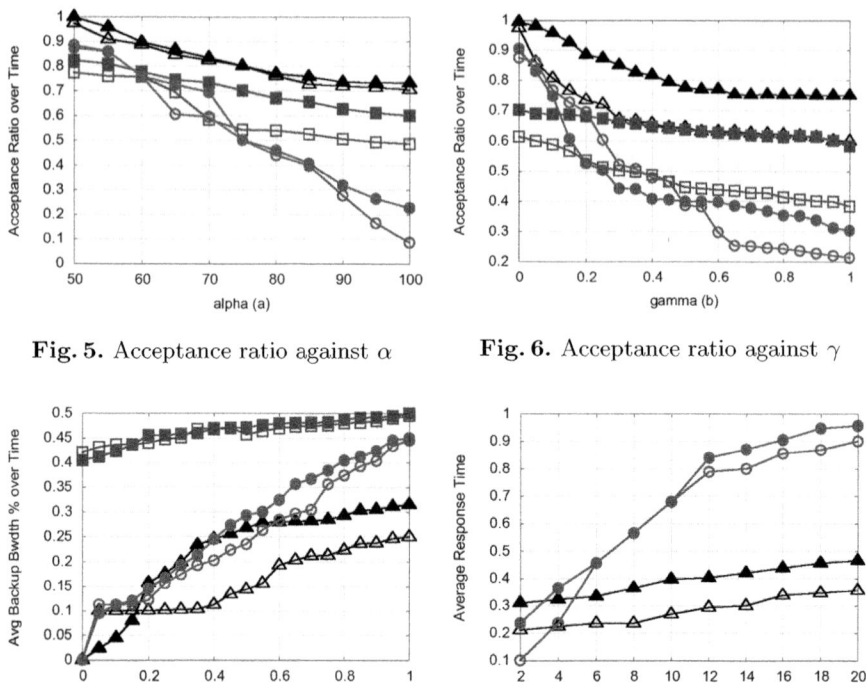

Fig. 5. Acceptance ratio against α **Fig. 6.** Acceptance ratio against γ

Fig. 7. Backup resource usage against γ **Fig. 8.** Response time against VN size

increases with γ, it is still outperformed by the hybrid policy for the range of the simulation parameters.

Responsiveness to Failures. The hybrid policy has faster reaction time to failures than its counterparts. In Figure 8, we notice that the hybrid policy reacts faster than the blind policy when a failure occurs. When a substrate link fails, the blind policy recomputes the entire embedding, which is time consuming. The hybrid policy, on the other hand, only re-routes the bandwidth of the affected virtual links which results in faster response time and ultimately lower penalty values for the InP.

Bandwidth Efficiency. The hybrid policy is bandwidth efficient. The proactive policy pre-reserves additional bandwidth for each virtual link during the instantiation phase. On the other hand, the hybrid policy does not pre-reserve any backup bandwidth during the link embedding phase. It pre-selects the candidate paths for re-routing and allocates backup bandwidth only when an actual failure occurs. As a result, the average bandwidth usage increases less rapidly with γ compared to the blind policy. This is shown in Figure 7.

6 Conclusion

In this paper, we have formulated the SVNE problem to incorporate substrate failures in the virtual network embedding problem. We have also proposed an efficient HYBRID policy heuristic to solve SVNE. To the best of our knowledge this is the first attempt to add survivability to virtual network embedding algorithms along with support for business profit oriented optimization. Moreover, our proposed heuristic can be extended to deal with multiple link failures, and subsequently combined with a node migration strategy [19] to solve the single substrate node failure problem. However, there are a number of future research directions that can be pursued. Survivability in a multi-domain NVE could raise further challenges since it involves both intra and inter domain link failures. It would also be interesting to extend survivability to recursive NVE, where the first level VNs can act as InPs to a second level of VNs. Resource allocation, protection, and restoration issues in such recursive environments could be investigated under cross layer optimization or network utility maximization (NUM) frameworks.

Acknowledgment

This work was jointly supported by the Natural Science and Engineering Council of Canada (NSERC) under its Discovery program, Cisco Systems, and WCU (World Class University) program through the Korea Science and Engineering Foundation funded by the Ministry of Education, Science and Technology (Project No. R31-2008-000-10100-0).

References

1. Ahuja, R.K., Magnanti, T.L., Orlin, J.B.: Network Flows: Theory, Algorithms, and Applications. Prentice Hall, Englewood Cliffs (1993)
2. Aib, I., Boutaba, R.: On leveraging policy-based management for maximizing business profit. IEEE Transactions on Network and Service Management (TNSM) 4(2), 14 (2007)
3. Andersen, D.: Theoretical approaches to node assignment. Unpublished Manuscript (2002), http://www.cs.cmu.edu/~dga/papers/andersen-assign.ps
4. Anderson, T., Peterson, L., Shenker, S., Turner, J.: Overcoming the internet impasse through virtualization. Computer 38(4), 34–41 (2005)
5. Chowdhury, N., Boutaba, R.: Network virtualization: state of the art and research challenges (topics in network and service management). IEEE Communications Magazine 47(7), 20–26 (2009)
6. Chowdhury, N., Rahman, M., Boutaba, R.: Virtual network embedding with coordinated node and link mapping, April 2009, pp. 783–791. IEEE INFOCOM (2009)
7. Fan, J., Ammar, M.: Dynamic topology configuration in service overlay networks - a study of reconfiguration policies. In: IEEE INFOCOM (2006)
8. Feamster, N., Gao, L., Rexford, J.: How to lease the internet in your spare time. SIGCOMM Comput. Commun. Rev. 37(1), 61–64 (2007)

9. Houidi, I., Louati, W., Zeghlache, D.: A distributed virtual network mapping algorithm. In: Proceedings of IEEE ICC, pp. 5634–5640 (2008)
10. Kurant, M., Thiran, P.: Survivable Routing of Mesh Topologies in IP-over-WDM Networks by Recursive Graph Contraction. IEEE Journal on Selected Areas in Communications 25(5), 922–933 (2007)
11. Lee, K., Modiano, E.: Cross-layer survivability in wdm-based networks. In: INFOCOM 2009, April 2009, pp. 1017–1025. IEEE, Los Alamitos (2009)
12. Lischka, J., Karl, H.: A virtual network mapping algorithm based on subgraph isomorphism detection. In: Proceedings of ACM SIGCOMM VISA (2009)
13. Lu, J., Turner, J.: Efficient mapping of virtual networks onto a shared substrate. Washington University, Tech. Rep. WUCSE-2006-35 (2006)
14. Makhorin, A.: GNU Linear Programming Kit, Moscow Aviation Institute, Russia (2008), http://www.gnu.org/software/glpk/
15. Markopoulou, A., Iannaccone, G., Bhattacharyya, S., Chuah, C.-N., Ganjali, Y., Diot, C.: Characterization of failures in an operational ip backbone network. IEEE/ACM Trans. Netw. 16(4), 749–762 (2008)
16. Stern, T.E., Bala, K.: Multiwavelength Optical Networks: A Layered Approach. Addison-Wesley Longman Publishing Co., Inc., Boston (1999)
17. Thulasiraman, K., Javed, M.S., Xue, G.L.: Circuits/cutsets duality and a unified algorithmic framework for survivable logical topology design in ip-over-wdm optical networks. In: IEEE INFOCOM (2009)
18. Topkis, D.M.: A k shortest path algorithm for adaptive routing in communications networks. IEEE Transactions on Communications 36(7) (July 1988)
19. Wang, Y., Keller, E., Biskeborn, B., van der Merwe, J., Rexford, J.: Virtual routers on the move: live router migration as a network-management primitive. In: SIGCOMM, pp. 231–242. ACM, New York (2008)
20. Yu, M., Yi, Y., Rexford, J., Chiang, M.: Rethinking virtual network embedding: Substrate support for path splitting and migration. ACM SIGCOMM CCR 38(2), 17–29 (2008)
21. Zheng, Q., Shin, K.G.: Fault-tolerant real-time communication in distributed computing systems. IEEE Trans. Parallel Distrib. Syst. 9(5), 470–480 (1998)
22. Zhu, Y., Ammar, M.: Algorithms for assigning substrate network resources to virtual network components. In: Proceedings of IEEE INFOCOM (2006)

Toward Efficient On-Demand Streaming with BitTorrent

Youmna Borghol[1,2], Sebastien Ardon[1], Niklas Carlsson[3], and Anirban Mahanti[1]

[1] NICTA, Locked Bag 9013, Alexandria, NSW 1435, Australia
[2] School of Electrical Engineering and Telecommunications, University of New South Wales, NSW 2030 Sydney, Australia
[3] University of Calgary, Calgary, AB, Canada
{youmna.borghol,sebastien.ardon,anirban.mahanti}@nicta.com.au

Abstract. This paper considers the problem of adapting the BitTorrent protocol for on-demand streaming. BitTorrent is a popular peer-to-peer file sharing protocol that efficiently accommodates a large number of requests for file downloads. Two components of the protocol, namely the rarest-first piece selection policy and the tit-for-tat algorithm for peer selection, are acknowledged to contribute toward the protocol's efficiency with respect to time to download files and its resilience to free riders. Rarest-first piece selection, however, is not suitable for on-demand streaming. In this paper, we present a new adaptive window-based piece selection policy that balances the need for piece diversity, which is provided by the rarest-first algorithm, with the necessity of in-order piece retrieval. We also show that this simple modification to the piece selection policy allows the system to be efficient with respect to utilization of available upload capacity of participating peers, and does not break the tit-for-tat incentive scheme which provides resilience to free riders.

Keywords: Peer-to-Peer, Video-on-Demand, BitTorrent.

1 Introduction

BitTorrent [5], probably the most popular peer-to-peer (P2P) file sharing system, is a loosely coupled distributed system that allows peers to opportunistically exchange file pieces. One important aspect of BitTorrent is the *rarest-first* piece selection algorithm used by each peer to select which piece to download from another peer. This algorithm provides a simple, fully decentralized, mechanism for system-wide replication of pieces such that file download times are minimized. Another important component is the *tit-for-tat* algorithm. Using rate-based reciprocation, this algorithm provides peers with upload incentives and provides resilience to free riders. Note that the symmetrical nature of tit-for-tat results in the average aggregate download rate (in systems with limited server or seed resources) being capped by the average peer upload capacity. BitTorrent can therefore be seen as a system allowing upload capacity sharing. It has been shown that these two components are fundamental to the scalability and performance of BitTorrent [9] [12].

We consider the problem of adapting BitTorrent to support on-demand streaming and more specifically, a view-as-you-download service. From a user experience point-of-view, we desire that the system provides low latency to begin playback as well as

M. Crovella et al. (Eds.): NETWORKING 2010, LNCS 6091, pp. 53–66, 2010.

jitter-free playback. Networked streaming systems attempt to attain these properties by encoding the media at a bitrate lower than the average available download bandwidth, and using a playout buffer to hide variation in available bandwidth.

Designing a system that satisfies the desired user experience, while retaining the salient features of BitTorrent, namely scalability with respect to swarm size and resilience to non-cooperative peers, is challenging. BitTorrent's file download efficiency is, in part, due to the use of a rarest-first piece selection policy; this policy increases piece diversity in the swarm, and therefore allows efficient use of available upload capacity of peers. Rarest-first piece selection cannot satisfy the user requirements mentioned above. Replacing the rarest-first piece selection policy with a strict in-order policy, however, makes the system sluggish as piece diversity is significantly reduced and tit-for-tat becomes largely irrelevant [11].

This paper presents a new adaptive window-based piece selection policy that achieves a balance between the system scalability provided by the rarest-first algorithm and the necessity of in-order pieces for seamless media playback. In particular, we propose that each peer maintains a sliding window wherein the window size is adapted based on the amount of in-order data available in the peer's buffer and its current playback position. Within the window, a variant of the rarest-first piece selection policy is used. Thus, when the window is small, near in-order piece retrieval occurs, whereas when the window is large, near rarest-first piece retrieval occurs.

Our simulations show that the simple adaptive window-based piece selection policy allows the system to be efficient with respect to utilization of available upload capacity, and also does not break the tit-for-tat incentive scheme. A swarm-based on-demand streaming protocol needs to be resilient to non-cooperative peers (free riders); in particular, the performance of the cooperating peers should not degrade because of the presence of free riders. Note that non-cooperating peers are not necessarily the result of malicious activity: a misconfigured firewall, NAT router, or other middle boxes can be the source of such problem as these may block the upload traffic from a peer.

The remainder of the paper is structured as follows. Section 2 discusses related work. The simulation tool used and our experiment methodology are discussed in Section 3. The adaptive window-based piece selection policy is presented in Section 4, along with a comparaison with prior work on fixed-sized window-based piece retrieval policies. Section 5 presents a detailed performance evaluation. A proposal that can substantially reduce the latency to begin playback is outlined in Section 6, followed by conclusions in Section 7.

2   Related Work

There has been much work on the design of peer-assisted video-on demand systems [8] using BitTorrent-like protocols. Accounting for the real-time needs of the streaming application has been the key challenge. In order to tackle the problem, modifications have been mainly proposed to BitTorrent's piece and peer selection policies.

Related studies on the piece selection algorithms have concentrated on finding a policy that can achieve a good trade-off between meeting the sequential requirement of playback and maintaining a high level of piece diversity in the system. Prior research can be broadly classified into probabilistic [3, 6], segment-based [1], and window-based [13, 14, 15, 16] approaches. With probabilistic piece policies, pieces are selected based on some probability distribution, which biases the selected pieces toward earlier pieces. With window-based approaches, a sliding window is typically used from within which pieces are given preference. Naturally, a smaller window ensures close to sequential piece retrieval, whereas a larger window, within which rarest-first can be used, for example, can ensure higher piece diversity. While all the above approaches provide a reasonable tradeoff between downloading pieces in roughly sequential order and ensuring sufficient diversity, we note that none of these approaches adaptively increases the use of rarest-first when possible.

While we are not aware of any adaptive window-based piece selection policies, we note that Carlsson *et al.* [4] have studied how a Zipf-based probabilistic piece selection policy can be enhanced to take current buffer conditions into account. They show that there are significant advantages to quickly provide newly arrived peers with rare pieces that can be shared with many users, as well as prioritize pieces that have urgent deadlines. Based on these insights they propose new policies that better utilize the server bandwidth to quickly bootstrap the tit-for-tat behavior of newly arrived peers. In contrast, we propose an adaptive-window approach which is used for both peer-to-peer or peer-to-server communication, and most importantly allows peers to dynamically help increase the piece diversity when conditions are favourable, making it more and more tit-for-tat compatible the higher download rates are available to the peers.

Another proposal has assumed in-order piece selection, and proposed performance-based peer selection policies in which peers are selected based on the urgency of the pieces they request (i.e., how soon they need the requested pieces for uninterrupted playback) and the availability of serving peers [7]. One key drawback of such design is its exposure to selfish peers as no incentive mechanism is provided for contribution. Parvez *et al.* [11] showed that upload utilization (and system throughput) can be significantly improved in systems using in-order piece selection, by giving upload preference to peers with similar playback points. This is particularly important for older peers which typically have less potential uploaders. In this paper, we show that a window-based approach in combination with rate-based tit-for-tat can achieve much of the same attractive clustering characteristics, in which peers with similar playback points (or overlapping windows) share pieces with each other.

To discourage free-riding, mechanisms have been suggested to favour peers who are verified to be good uploaders through some feedback mechanism [10]. As such techniques rely on externally gathered information about peer contributions, these systems potentially are more vulnerable to malicious peers than tit-for-tat schemes (in which the peers themselves can easily measure the rate at which pieces are exchanged). Due to the sequential nature of streaming there has, however, been some debate regarding whether or not tit-for-tat should be used. For example, it has been argued that peers will be downloading from older peers and will not have much to give in return. This is the case provided the piece selection policy is in-order or a close variant. Naturally, the effectiveness of tit-for-tat in such systems is (at best) limited. In this paper, we develop a policy that supports on-demand streaming while retaining the tit-for-tat component of BitTorrent.

3   Methodology

3.1   Simulation Environment

We use the Microsoft Research BitTorrent simulator [2]. The simulator simulates both uplink and downlink peer capacities without incurring the overhead of a packet-level simulator, and has been widely used.

All simulations have a single seed with uplink capacity of 6Mbps. The file size is 300MB with a video playback bitrate of 800Kbps (approximately 50 minute long). The file consists of 1200 pieces, each of size 256KB. Unless stated otherwise, all peers have an upload capacity of 1Mbps, and a download capacity of 2Mbps. In our simulations, peers leave the system as soon as they finish downloading the file. We use BitTorrent's default settings for the remaining parameters.

Two typical scenarios, as discussed below, are used in the experiments reported in this paper.

- Flash crowd scenario: In this scenario, we assume that all peers join the swarm nearly simultaneously (uniformly within 30s). The flash crowd scenario is often seen as a benchmark to evaluate swarm-based streaming systems, and provides a measuring point with regards to how the system handles extremely bursty request loads. With all peers typically progressing together in the download the seed may be the only peer that has pieces. In this scenario, the simulation ends when the last peer completes the download. Unless noted otherwise, results are shown for experiments with a population of 200 peers.
- Poisson scenario: In this scenario, peers arrive to the torrent according to a Poisson arrival process with rate λ. The Poisson scenario captures the performance of a system in which peers arrive to a swarm where there are peers that have been in the system for different time durations, and such peers already may have accumulated some number of pieces, depending on how long they themselves have been in the system. The arrival rate λ reflects the current file popularity. For these scenarios, the simulation time is set to five times the movie duration, and statistics from the second last duration were presented. To verify that we had reached steady state these results where compared with those reached in the second duration. For evaluation, we used peer arrival rates of $0.01, 0.1, 1$ peers per second. Assuming that peers download at roughly the play rate and leave immediately after completing their downloads, these arrival rates translate to swarm size of 30, 300, and 3000 peers, respectively.

3.2   Evaluation Metrics

For performance evaluation and policy comparisons, we use three primary metrics:

- *Upload Capacity Utilization*: The upload capacity utilization is defined as the fraction of the peers' upload capacity that is utilized during their download duration. Upload utilization is an important metric as higher utilization means higher overall system efficiency, and higher resilience to changing network available bandwidth.
- *Success Ratio*: The success ratio is defined as the ratio of pieces that were downloaded before their playback deadline, over the total number of pieces. This metric aims at measuring the playback continuity. This metric is also referred to as the continuity index [15].

- *Initial Buffering Time*: We assume that each peer must have received all pieces in its initial buffer, of size B, before playback can commence. We refer to the duration of this process as the initial buffering time.

Unless stated otherwise, results presented in Section 5 and 6 represent performance metrics using whiskers-plot diagrams which show the median, the 25 and 75 percentiles, as well as the minimum and maximum values, across all peers.

4   Adaptive Window for On-Demand Streaming

Prior work has proposed using a *fixed-size* sliding window to limit the set of pieces which can be requested by a peer [13, 14, 15]. Within the window, rarest-first or a variant has been used, with the objective of balancing the need of in-order downloads for streaming and the need of piece diversity for maintaining system efficiency. With these approaches, the window has to be kept small relative to the file size to support on-demand streaming with relatively small startup delay. A by-product of this requirement, however, is that the number of pieces available for exchange is limited, and therefore, peer's upload bandwidth utilization cannot be maximized [13].

We propose an *adaptive* window strategy where each peer dynamically computes its window size, depending on how well the peer's download is progressing. In particular, the adaptive window grows or shrinks depending upon how much additional data is available in the playback buffer, with respect to the playback position of the peer. A peer that has a large amount of in-order data available in its playback buffer will increase its window, and therefore, contribute to the system by increasing piece diversity. A peer close to stalling with a near-empty playout buffer, will shrink its window and request pieces closest to its playback position. As we demonstrate in this paper, this adaptive window mechanism is able to maximize utilization of the peers' upload capacities, and is also able to retain the tit-for-tat behavior of BitTorrent.

Fig. 1 illustrates the adaptive window mechanism. Each peer computes its window w at each piece selection event using:

$$w = \max\left[k(d - p - \theta), 0\right] + w_{min}, \tag{1}$$

where w is the effective window size, w_{min} is a pre-defined minimum window size, k is a scaling coefficient, d is the peer download position (i.e., the last contiguous piece available in the playout buffer), p is playback position (i.e., the piece that is currently being played back), and θ is a threshold that places a lower bound on the number of contiguous pieces required in the playout buffer before the window can grow. When a peer joins the swarm, its window size is initialized to w_{min}; this value needs to be sufficiently large to introduce enough diversity without playback interruption and yet ensure small buffering time. (Note that the larger the window, the more time it takes to obtain a set of contiguous pieces starting with the first piece of the file because of the use of rarest-first piece selection policy or its variants, as discussed below.) Once playback starts, and if a peer enjoys a faster download rate than its media playback rate, for every extra piece beyond threshold θ the window is incremented by a scale factor k. Following extensive experiments, we found $k = 1$, $w_{min} = 20$, and $\theta = 2.5\,w_{min}$ to give the best performance, and we use these values in the remainder of the paper.

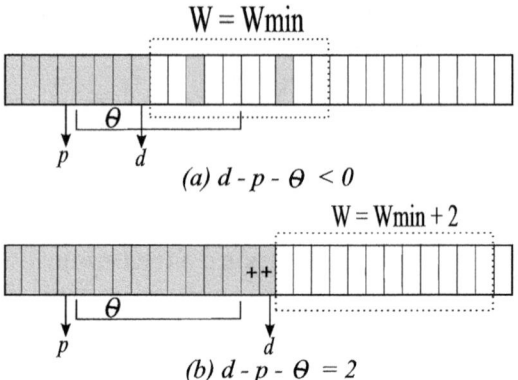

Fig. 1. Illustration of the adaptive window mechanism

For the purpose of defining the window boundaries, we use two definitions: one in the initial buffering state and another once playback starts. In the initial buffering state, to define the boundaries of the window, we count the arrived and non-arrived pieces, as in [14]. In the playback state, we count only the non-arrived pieces, as in [15].

Another design consideration is the piece selection policy inside in the window. We experimented with a number of alternative piece selection policies [3]. We found that rarest-first (within a window) performs reasonably well; it attains very high success ratios (e.g., in excess of 99%), with piece misses distributed nearly uniformly over the file. The piece misses tend to happen at the beginning of the window as the pieces toward the end of the window appear to be rarer in the system. Our experiments suggest that a variant of the Portion policy [3], where each peer independently applies, within the window, rarest-first selection with probability q and in-order selection with probability $(1 - q)$, achieves similar upload utilization as rarest-first but with fewer misses. In all applications of the adaptive window policy in this paper, Portion with $q = 0.1$ is used.

Fig. 2 presents illustrative results for the proposed *adaptive window* approach along with results for two recently proposed fixed-window approaches, namely *fixed-size window* [14] and *Bitos* [15]. The window size for the fixed-size window protocol is set to 20 pieces. We report results from two flash crowd experiments; one with the ratio of upload bandwidth (U) over playback rate (PBR) equal to 1.25, and another with U/PBR = 2. We also report results from two Poisson scenarios with U/PBR = 1.25 and arrival rates λ of 0.01 and 0.1.

From Fig. 2, we observe that the adaptive-window policy is generally more successful for streaming across both flash crowd and Poisson scenarios. In the flash crowd scenario, the fixed window policy fails to provide good upload bandwidth utilization, and as a result fails to provide a viable streaming experience when available upload bandwidth is not abundant (e.g., illustrated by the low median success ratio in the U/PBR=1.25 experiment). The main reason for the lower system throughput is the conservative definition of the window boundaries. In the Poisson scenario, the Bitos policy typically results in poor success ratios, with wide variations across peers (e.g., for $\lambda = 0.01$, 25% of peers have as low as 50% success ratio). Bitos typically achieves

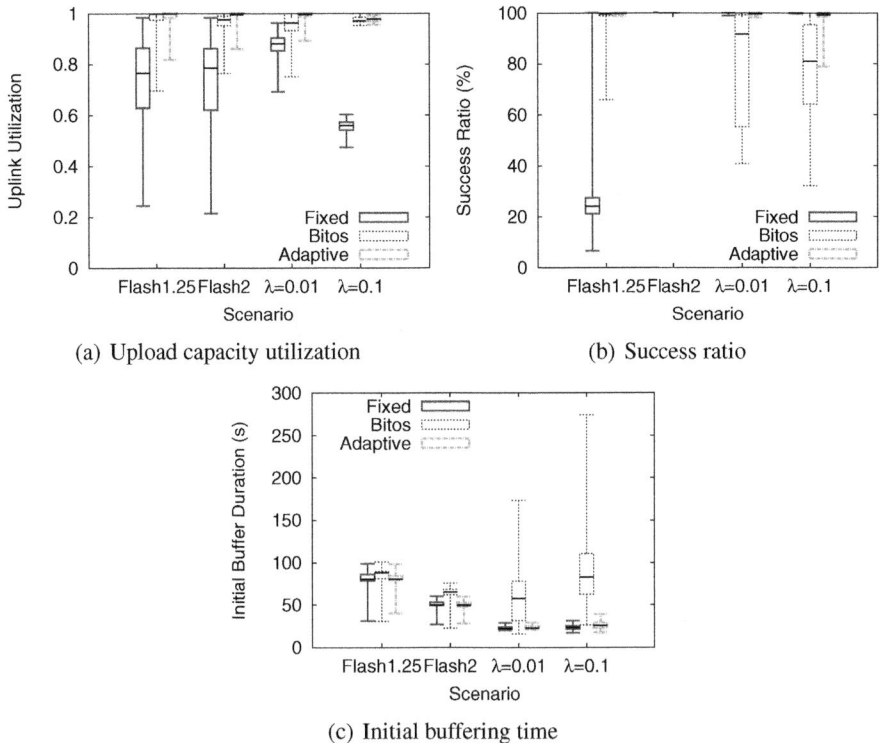

(a) Upload capacity utilization   (b) Success ratio

(c) Initial buffering time

Fig. 2. Comparison of adaptive and fixed window policies

reasonable upload utilization. As Bitos clients always download 20% of pieces outside the window, their rate of in-order pieces retrieved is affected. Meanwhile, the adaptive-window policy always takes the current buffer conditions into account when adapting the rate at which non-in-order pieces is being retrieved. Overall, the adaptive-window policy consistently achieves high upload capacity utilization, low initial buffering times, and success ratios close to 100% in nearly all cases, with very little variations across peers.

From Fig. 2(c), we can also see that the buffer time in the flash crowd scenario, were all peer are synchronized, is approximately the same for all three policies. In the Poisson scenarios, the buffer time is much lower with the fixed-size and adaptive-window policies. This is because the fixed-size window policy (also used for the initial buffering in the adaptive-window policy), does not allow peers to move the boundary of the window until the first piece is retrieved, ensuring small buffering times.

5   System Evaluation

BitTorrent's rate-based tit-for-tat algorithm selects which peer to serve next at each choking interval [5]. The four peers that provide the best download rates are selected, while a fifth peer is selected at random. This rate-based peer selection algorithm acts as

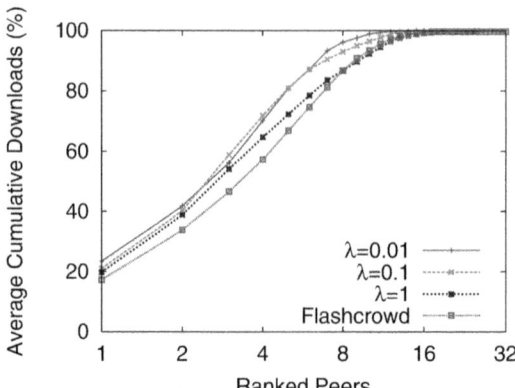

Fig. 3. Fraction of file data from peers ranked by their contribution (U/PBR = 3)

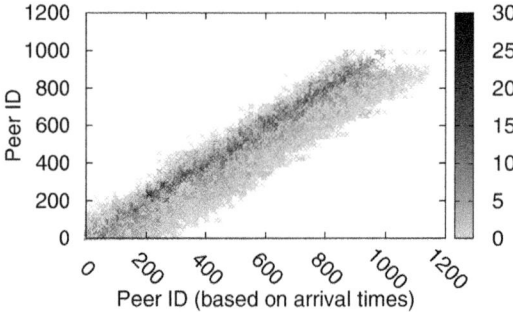

Fig. 4. Time correlation for connected pairs of peers (Poisson scenario, $\lambda = 0.1$, U/PBR = 3)

an incentive mechanism for peers to share their upload bandwidth. Legout *et al.* [9] have shown that tit-for-tat ensures reciprocation among peers, and results in relatively stable connections among peers. In this section, we evaluate these system-level properties in the context of our adaptive window-based approach for on-demand streaming.

5.1   Stability of Connections

In general, infrequent chocking of peers leads to higher upload bandwidth utilization as less time is wasted waiting for unchoke messages. Thus, if peer connections are stable for long time periods, we expect a large fraction of the file to be obtained from a small set of peers relative to the swarm size.

We present example results in Fig. 3. For each peer i, we calculate the amount downloaded from any other peer j in the swarm, d_{ij}, and maintain a rank-ordered list of d_{ij}'s where the highest download is assigned rank one. Fig. 3 shows, on average, the fraction of the total file retrieved from peers in the swarm as a function of peer rank. The figure shows that in the simulations considered, on average, 50-65% of the video file was consistently downloaded from four peers. This confirms that connections using our approach are relatively stable.

We also analysed which peers connect with each other, and more importantly, which peers exchange pieces with each other. Fig. 4 shows the correlation in-time between the data transferred between connected peers in an experiment with Poisson arrivals at rate $\lambda = 0.1$. Peer IDs are assigned in ascending order based on a peer's arrival time to the swarm. For every peer i, the scatter plot shows the percentage of the file downloaded from any other peer j. We observe a darker line in the middle of the graph, suggesting that most of the data is retrieved from a few peers that joined the swarm close together in-time (i.e peers with close playback points).

5.2   Reciprocation

We define the Reciprocation Ratio, R_i, for each peer i as $R_i = \frac{\sum_j \min[d_{ij}, u_{ij}]}{\sum_j d_{ij}}$, where u_{ij} is the amount of bytes i uploaded to j, and d_{ij} is the amount of bytes i downloaded from j. Fig. 5 shows the cumulative distribution of the Reciprocation Ratio in different scenarios. Overall, we observe good reciprocation among peers, especially for the flash crowd experiments, and the Poisson experiments with arrival rate $\lambda \geq 0.1$. Note that for lower arrival rates there typically are very few peers in the system.

Our results confirm that the reciprocation enforced by tit-for-tat in the original Bit-Torrent is maintained in the adaptive window algorithm. We show that peers tend to maintain stable connections with few peers close to their download position, as they have overlapping windows and thus pieces to exchange with each other. The swarm topology is, therefore, constrained by the slack allowed by the window. This is an interesting side-effect. Our adaptation policy naturally tends to group peers based on their playback positions, in contrast to other approaches that create similar 'daisy chains' structures using more complicated peer selection policies [11].

5.3   Heterogeneous Peer Upload Capabilities

Next, we consider the impact of peer heterogeneity on the system. In particular, we are interested in understanding the performance degradation in the presence of peers that have upload capacity less than the media playback rate. Fig. 6 shows the performance of capable peers (U/PBR = 1.25 in these simulations) as a function of the percentage of slow peers in the swarm. Results are shown for cases in which the slow peers have U/PBR = 0.25 and 0.5. We observe that even with 30% slow peers, capable peers miss only between 2-3% of the data, while still achieving reasonable buffering times.

5.4   Free-Riding Peers

BitTorrent's tit-for-tat mechanism manages free riding peers (i.e., peers that only download but do not contribute any upload bandwidth) by penalising them. Here we investigate the effectiveness of tit-for-tat in the context of our proposed adaptive window. Figs. 7(a) and (b) show the initial buffering time and success ratios, respectively, attained by both the free riders and the contributing peers. As desired, our results show that free riders are punished while the contributing peers are not substantially affected. These results demonstrate the ability of the adaptive window policy to retain the effectiveness of tit-for-tat, which is regarded as a fundamental component of BitTorrent.

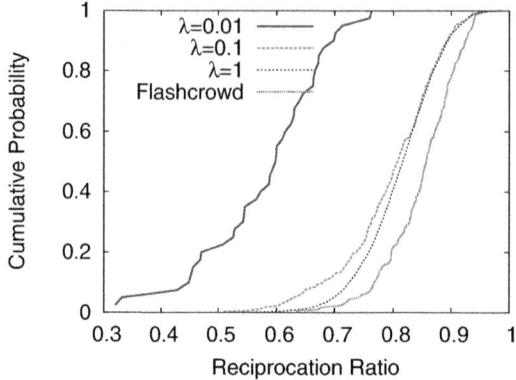

Fig. 5. Reciprocation among peers (Poisson and flash crowd scenarios, U/PBR =3)

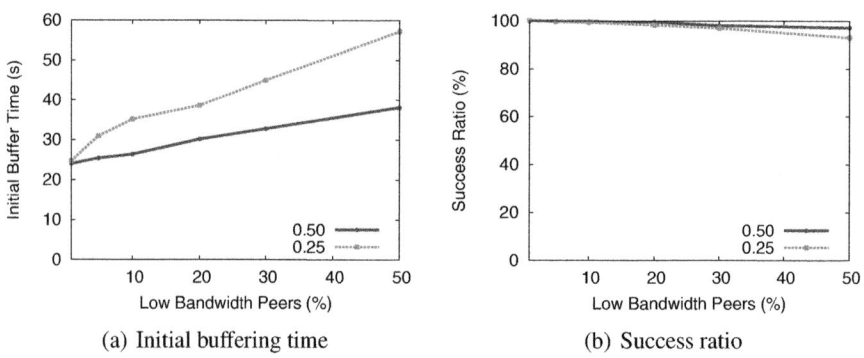

(a) Initial buffering time   (b) Success ratio

Fig. 6. Impact of network heterogeneity (Poisson scenario, $\lambda = 0.1$, U/PBR =1.25)

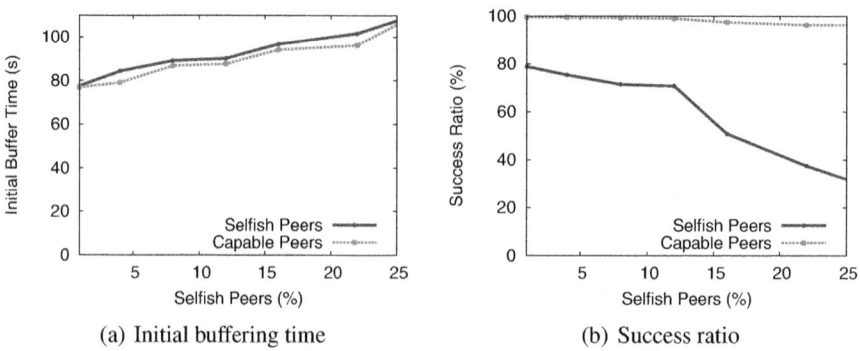

(a) Initial buffering time   (b) Success ratio

Fig. 7. Resilience to free-riding peers (Flash crowd scenario, U/PBR = 1.25)

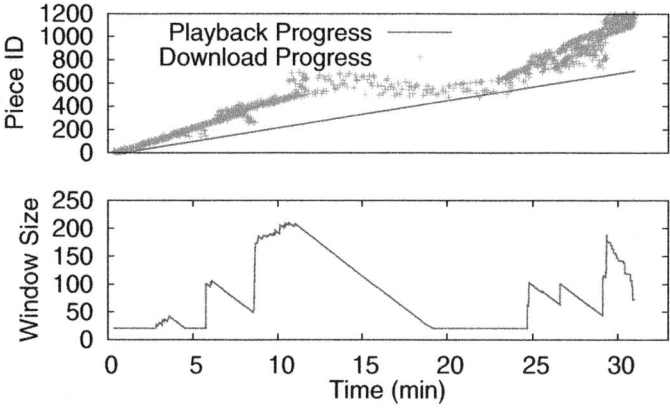

Fig. 8. Responsiveness under changing network conditions

5.5 Changing Network Conditions

Finally, we consider an example scenario in which one peer has time varying upload capacity. We use the default flash crowd scenario with U/PBR=2 and 200 peers. During the course of the simulations, one peer's upload rate is reduced to 200Kbps at 700s after its arrival to the system, and remains reduced for the next 700s (i.e., from roughly 12 to 23 minutes). Fig. 8 shows the window progress and piece retrievals for this peer. We note that the policy is responsive to the changing network conditions, and reduces the window size sufficiently quickly such that almost no pieces are retrieved late. (This particular peer had a success ratio of 99.8%.) We also note that the reduced download rate as the peer's upload bandwidth changes, provides clear evidence that tit-for-tat is working.

6   Bootstrapping and Buffering

In this section, we take a closer look at the factors contributing to the initial buffering time and propose a simple mechanism that significantly reduces the initial buffering time.

6.1 Buffering Time Analysis

As previously defined, in Section 3.2, the *initial buffering time* is the time a peer requires to receive all pieces in its initial buffer, before starting playback. One key component of this delay is the time to obtain the first piece, which the peer then can use to upload to others; we refer to this time as the *bootstrap time*. Furthermore, the amount of data buffered during the initial buffering time can directly impact the success ratio. Clearly, the larger the initial buffer size B is, the more chance of maintaining seamless playback. However, as B increases, so does the initial buffering time. Through experimentation with different swarm sizes, scenarios, network capacities, and media bitrate values, we have found that using B equal to ten pieces provides a good compromise. In particular, this choice typically allows us to achieve a success ratio above 99%. Of course, more conservative choices or more advanced startup rules (e.g., as used in [3]) could easily be augmented to further improve the success ratios.

Fig. 9(a) shows the ratio of the initial buffering time over the bootstrap time in our baseline flash crowd scenario, for different upload capacities. Although the bootstrap time only requires a single piece to be downloaded, we note that the bootstrap time is roughly equal to 50% of the buffering time (i.e., the time to receive the first ten pieces). Clearly, the bootstrap time is the primary cause to the startup delay.

We explain this high bootstrap time by observing that in the flashcrowd scenario, peers initially have no piece to exchange with each other, and the seed is the only initial source. The probability of parallel download is therefore very low, as every peer is trying to get its first piece. Because the total number of concurrent uploads per peer is limited in BitTorrent, peers during this phase effectively form an *k-ary tree*, where k is the maximum number of upload connections per peer. Peers arriving early in the flash crowd are connected closer to the seed, and later peers gradually are forced to connect to other peers further down the tree. The further down a tree a peer is, the longer it will take for it to obtain the first piece as it first has to travel down every other peer in the tree. From this observation, and the fact that the average node depth in an *k-ary tree* is approximately $\log_k N$, we can estimate the average time to obtain the first piece to be:

$$\log_k(N)\frac{kC}{U}, \tag{2}$$

where N is the swarm size, C is the piece size, k is the maximum number of concurrent uploads per peer, and U is the average upload capacity. An interesting observation is that the bootstrap time is proportional to logarithm of the swarm size. To confirm this analysis, Fig. 9(b) shows the evolution of the bootstrap duration as a function of the swarm size. We note that the above approximation (2), provides a reasonable fit, and more importantly that the bootstrap time scales logarithmically with the swarm size, as indicated by the straight-line appearance of the graph with the swarm size on a logarithmic scale. The results are from our flash crowd experiments with varying sizes of the flash crowd.

6.2 Reducing Startup Time through Pre-fetching

In order to reduce the initial buffering time, we propose that each peer should be given a randomly chosen piece from within the initial window (of size w_{min}). From a practical point, a server could quickly upload this piece to each newly arriving peer, (or the tracker, in the case such nodes would be willing to store and serve a small subset of pieces, by piggybacking the piece in the initial reply). Alternatively, the server (or voluntary peers) could push random pieces from media files of interest to each user. For some applications, e.g. video-on-demand systems, it could be practical to drive this pre-fetching from a content recommendation engine, and allow the pre-fetching operation to take place at time of low system utilization. Another option would be to implement the pre-fetching as part of the content browser; in this case, one 256KB piece could be pushed along with the content thumbnail.

While quantifying the overhead of this pre-fetching is difficult and application and/or implementation-specific, we note that this approach can reduce the average bootstrap

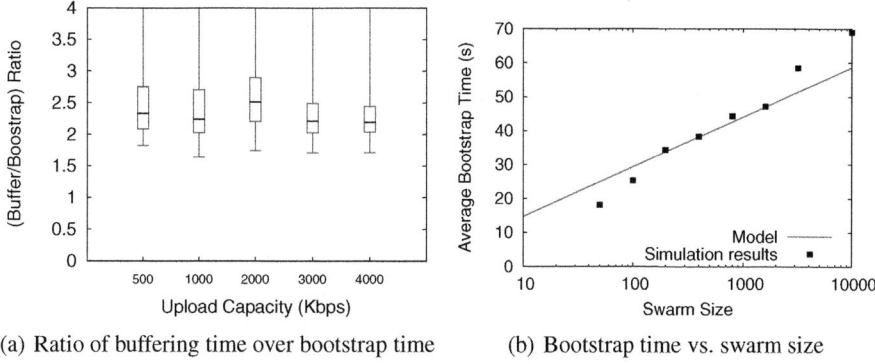

(a) Ratio of buffering time over bootstrap time   (b) Bootstrap time vs. swarm size

Fig. 9. Bootstrap time analysis in flash crowds

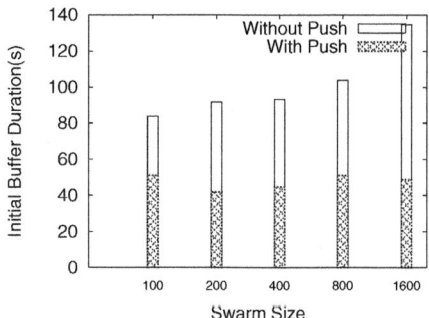

Fig. 10. Impact of pre-fething on startup delay

time to nearly zero, as peers immediately have blocks to exchange. Equation (2) suggests this reduction should be proportional to the logarithm of the swarm size. We implemented this pre-fetching scheme in the simulator. We ran a number of simulations with varying flash crowd sizes, both with and without prefetching. Our simulation parameters are as discussed earlier in Section 3. Fig. 10 shows the overall difference with and without pre-fetching. The figure shows that a large reduction in the initial buffering time is possible with prefetching a single piece, and further suggests that removing the bootstrap time contribution makes the initial buffering time roughly independent of the swarm size.

7   Conclusion

This paper presented a new adaptive window-based policy that effectively balances the need for piece diversity with the necessity of in-order piece retrieval. The policy makes effective use of the available upload capacity to improve piece diversity, and increase the overall system efficiency. Compared with previously proposed window-based approaches, the policy consistently achieves high upload capacity utilization, low initial buffering times, and high success ratio. We showed that our adaptive window-based policy ensures effective use of tit-for-tat, is robust in the presence of free riders, and is resilient to variations in network available bandwidth. Finally, we showed that the latency to begin playback can be be reduced by 50%, on average, by pre-fetching a single random piece from within the initial window.

Acknowledgements

This work was supported by the National Information and Communications Technology Australia (NICTA) and the Informatics Circle of Research Excellence (iCORE) in the Province of Alberta, Canada. NICTA is funded by the Australian Government as represented by the Department of Broadband, Communications and the Digital Economy and the Australian Research Council through the ICT Centre of Excellence program.

References

1. Annapureddy, S., Gunawardena, D.: Is high-quality vod feasible using p2p swarming. In: Proc. WWW, Banff, Canada (May 2007)
2. Bharambe, A.R., Herley, C., Padmanabhan, V.N.: Analyzing and improving a bittorrent network's performance mechanisms. In: Proc. IEEE INFOCOM, Barcelona, Spain (2006)
3. Carlsson, N., Eager, D.L.: Peer-assisted on-demand streaming of stored media using bittorrent-like protocols. In: Proc. IFIP Networking, Atlanta, GA (May 2007)
4. Carlsson, N., Eager, D.L., Mahanti, A.: Peer-assisted on-demand video streaming with selfish peers. In: Proc. IFIP Networking, Aachen, Germany (May 2009)
5. Cohen, B.: Incentives build robustness in bittorrent. In: Proc. Workshop on Economics of Peer-to-Peer Systems, Berkeley, CA (2003)
6. Garbacki, P., Epema, D.H.J., Pouwelse, J.: Offloading servers with collaborative video on demand. In: Proc. IPTPS, Tampa Bay, FL (February 2008)
7. Guo, Y., Mathur, S., Ramaswamy, K., Yuy, S., Patel, B.: Ponder: Performance aware p2p video-on-demand service. In: Proc. IEEE GLOBECOM, Washingston, DC (November 2007)
8. Huang, Y., Fu, T.Z.J., Chiu, D.M., Lui, J.C.S., Huang, C.: Challenges, design and analysis of a large-scale p2p-vod system. In: Proc. ACM SIGCOMM (August 2008)
9. Legout, A., Urvoy-Keller, G., Michiardi, P.: Rarest first and choke algorithms are enough. In: Proc. ACM IMC, Rio de Janeiro, Brazil (October 2006)
10. Mol, J., Pouwelse, J., Meulpolder, M., Epema, D., Sips, H.: Give-to-get: free-riding resilient video-on-demand in p2p systems. In: Proc. MMCN, San Jose, CA (January 2008)
11. Parvez, K., Williamson, C., Mahanti, A., Carlsson, N.: Analysis of bittorrent-like protocols for on-demand stored media streaming. In: Proc. ACM SIGMETRICS, Annapolis, MD (2008)
12. Qiu, D., Srikant, R.: Modeling and performance analysis of bittorrent-like peer-to-peer networks. In: Proc. SIGCOMM (2004)
13. Savolainen, P., Raatikainen, N., Tarkoma, S.: Windowing bittorrent for video-on-demand: Not all is lost with tit-for-tat. In: Proc. IEEE GLOBECOM, New Orleans, LA (2008)
14. Shah, P., Pris, J.-F.: Peer-to-peer multimedia streaming using bittorrent. In: Proc. IEEE IPCCC, New Orleans, LA (April 2007)
15. Vlavianos, A., Iliofotou, M., Faloutsos, M.: Bitos: enhancing bittorrent for supporting streaming applications. In: Proc. IEEE Global Internet Symposium, Barcelona, Spain (2006)
16. Zhou, Y., Chiu, D.M., Lui, J.C.: A simple model for analyzing p2p streaming protocols. In: Proc. IEEE ICNP, Beijing (2007)

Synapse: A Scalable Protocol for Interconnecting Heterogeneous Overlay Networks*

Luigi Liquori[1,**], Cédric Tedeschi[2], Laurent Vanni[1],
Francesco Bongiovanni[1], Vincenzo Ciancaglini[1], and Bojan Marinković[3]

[1] Institut National de Recherche en Informatique et Automatique, France
firstName.lastName@sophia.inria.fr
[2] Université de Rennes I/INRIA, France
Cedric.Tedeschi@inria.fr
[3] Mathematical Institute of the Serbian Academy of Sciences and Arts, Serbia
bojanm@turing.mi.sanu.ac.rs

Abstract. This paper presents Synapse, a scalable protocol for information retrieval over the inter-connection of heterogeneous overlay networks. Applications on top of Synapse see those intra-overlay networks as a unique inter-overlay network.

Scalability in Synapse is achieved via co-located nodes, *i.e.* nodes that are part of multiple overlay networks at the same time. Co-located nodes, playing the role of *neural synapses* and connected to several overlay networks, allow a larger search area and provide alternative routing. Synapse can either work with "open" overlays adapting their protocol to synapse interconnection requirements, or with "closed" overlays that will not accept any change to their protocol.

Results from simulation and experiments show that Synapse is scalable, with a communication and state overhead scaling similarly as the networks interconnected. Thanks to alternate routing paths, Synapse also gives a practical solution to network partitions.

We precisely capture the behavior of traditional metrics of overlay networks within Synapse and present results from simulations as well as some actual experiments of a client prototype on the Grid'5000 platform. The prototype developed implements the Synapse protocol in the particular case of the inter-connection of many Chord overlay networks.

Keywords: Peer-to-peer, overlay networks, information retrieval.

1 Introduction

Context. The interconnection of overlay networks has recently been identified model showing great promise when it comes to dealing with the issues of the Internet of today, such as scalability, resource discovery, failure recovery or routing efficiency and, in particular, in the context of information retrieval. Several recent research efforts have focused on the design of mechanisms for building bridges

---

* Supported by AEOLUS FP6-IST-15964-FET Proactive and DEUKS JEP-41099 TEMPUS.
** Corresponding author. Thanks to anonymous referees and Ernst Biersack for the precious discussions.

M. Crovella et al. (Eds.): NETWORKING 2010, LNCS 6091, pp. 67–82, 2010.

between heterogeneous overlay networks for the purpose of improving cooperation between networks which have different routing mechanisms, logical topologies and maintenance policies. However, more comprehensive studies of such interconnections for information retrieval and both quantitative and experimental studies of its key metrics, such as satisfaction rate or routing length, are still missing. During the last decade, different overlay networks were specifically designed to answer well-defined needs such as content distribution through unstructured overlay networks (Kazaa) or through structured networks, mainly utilizing concepts such as Distributed Hash Tables [14, 15, 17] and publish/subscribe systems [2, 12].

An overview of the current problem. Many disparate overlay networks may not only simultaneously co-exist in the Internet but also compete for the same resources on shared nodes and underlying network links. One of the problems of the overlay networking area is how heterogeneous overlay networks may *interact* and *cooperate* with each other. Overlay networks are heterogeneous and basically unable to cooperate each other in an effortless way, without merging, an operation which is very costly since it not scalable and not suitable in many cases for security reasons. However, in many situations, distinct overlay networks could take advantage of cooperating for many purposes: collective performance enhancement, larger shared information, better resistance to loss of connectivity (network partitions), improved routing performance in terms of delay, throughput and packets loss, by, for instance, cooperative forwarding of flows.

As a basic example, let us consider two distant databases. One node of the first database stores one $(key, value)$ pair, which is then requested by a node of the second database. Without network cooperation those two nodes will never communicate together. In another example, we have an overlay network in which a number of nodes got isolated by an overlay network failure, leading to a partition: if some or all of those nodes are reachable via an alternate overlay network, than the partition "could" be recovered via an alternate routing.

In the context of large scale information retrieval, several overlays may want to offer an aggregation of their information/data to their potentially common users without giving up control over it. Imagine two companies wishing to share or aggregate information contained in their distributed databases, while, obviously, keeping their proprietary routing and their exclusive right to update it. Finally, in terms of fault-tolerance, cooperation can increase the availability of the system, if one overlay becomes unavailable, the global network will only undergo partial failure as other distinct resources will be usable.

We consider the tradeoff of having one *vs.* many overlays as a conflict without a cause: having a single global overlay has many obvious advantages and is *de facto* the most natural solution, but it appears unrealistic in the actual setting. In one optimistic case, different overlays are suitable for collaboration by opening their proprietary protocols in order to build an open standard; in many other pessimistic cases, this opening is simply unrealistic for many different reasons (backward compatibility, security, commercial, practical, etc.). With all this said, studying protocols to interconnect collaborative (or competitive) overlay networks appears to be quite an interesting and intriguing research vein.

Contribution. The main contributions of this paper are the introduction of *Synapse*, a scalable protocol for information retrieval over the inter-connection of heterogeneous overlay networks, and the presentation of the results obtained from various simulations and experiments with this protocol. The protocol itself is based on co-located nodes, also called *synapses*, serving as low-cost natural candidates for inter-overlay bridges. In principle, every regular node can become a synapse. In the simplest case (where overlays to be interconnected are ready to adapt their protocols to the requirements of interconnection), every message received by a synapse can be forwarded to other overlays the node belongs to. In other words, upon receipt of a search query, in addition to its forwarding to the next hop in the current overlay (according to their routing policy), the node can possibly start a new search, according to some given strategy, in some or all of the other overlay networks it belongs to. This implies that a Time-To-Live value is provided, and already processed queries can be detected, so as to avoid infinite looping in the network, as in unstructured peer-to-peer systems.

In case of concurrent overlay networks, inter-overlay routing becomes harder, as intra-overlays are provided as black boxes: a *control* overlay-network made of co-located nodes maps one hashed key from one overlay into the original key that, in turn, will be hashed and routed in other overlays, in which the co-located node belongs to. This extra structure is unavoidable, since we would like to route queries along closed overlays and prevent routing loops.

Our experiments and simulations show that a small number of well-connected synapses is sufficient in order to achieve almost exhaustive searches in a "synapsed" network of structured overlay networks. We believe that Synapse can give an answer to circumventing network partitions; the key points being that: (i) several logical links for one node lead to as many alternative physical routes through these overlays, and (ii) a synapse can retrieve keys from overlays that it does not even know simply by forwarding its query to another synapse that, in turn, is better connected. Those features are achieved in Synapse at the cost of some additional data structures and in an orthogonal way to ordinary techniques of caching and replication. Moreover, being a synapse can allow for the retrieval of extra information from many other overlays even if the node isn't directly connected with them. We summarize our contributions as follows: (i) the introduction of *Synapse*, a generic protocol, which is suitable for interconnecting heterogeneous overlay networks without relying on merging in presence of open/collaborative or closed/competitive networks; (ii) extensive simulations in the case of the interconnection of structured overlay networks, in order to capture the real behavior of such platforms in the context of information retrieval, and identify their main advantages and drawbacks; (iii) the deployment of a lightweight prototype of Synapse, called JSynapse on the Grid'5000 platform[1] along with some real deployments showing the viability of such an approach while validating the software itself; (iv) finally, on the basis of the previous item, the description and deployment on the Grid'5000 platform of a open source prototype,

---

[1] http://www.grid5000.fr and http://open-chord.sourceforge.net

called open-Synapse, based on the open-Chord[4] implementation of Chord, inter-connecting an arbitrary number of Chord networks. The final goal is to grasp the complete potential that co-located nodes have to offer, and to deepen the study of overlay networks' inter-connection using these types of nodes.

Outline. The remainder of the paper is organized as follows: In Section 2, we introduce our Synapse protocol at work for open/collaborative overlays (*white box*) viz. closed/competitive (*black box*) overlays, and provide several examples. In Section 3, we present the results of our simulations of the Synapse protocol to capture the behavior of key metrics traditionally used to measure the efficiency of information retrieval. In Section 4, we describe a deployment of a client prototype[2] over the Grid'5000 platform. In Section 5, we summarize the mechanisms proposed in the literature related to the cooperation of overlay networks. Finally, in Section 6, we present our conclusions and discuss ideas for further work. Appendix shows pseudocode: due to lack of space, the protocol is presented in details in a separate web appendix[5].

2   The Synapse Protocol

Architecture and assumptions. We now present our generic *meta*-protocol for information distribution and retrieval over an interconnection of heterogeneous overlay networks. Information is a set of basic (key, value) pairs, as commonly encountered in protocols for information retrieval. The protocol specifies how to insert information (PUT), how to retrieve it through a key (GET), how to invite nodes in a given overlay (INVITE), and how to join a given overlay (JOIN) over a heterogeneous collection of overlay networks linked by co-located nodes. These co-located nodes represent a simple way to aggregate the resources of distinct overlays. We assume each overlay to have its own inner routing algorithm, called by the Synapse protocol to route requests inside each overlay. We assume no knowledge of the logical topology of all the involved overlay networks connected by Synapse. To ensure the usual properties of the underlying network, we assume that communication is both symmetric and transitive. Synapse simply ignores how routing takes place inside the overlays, Synapse only offers a mechanism to route from one overlay to another in a simple, scalable and efficient way.

The inter-overlay network, induced by the Synapse protocol, can be considered as an aggregation of heterogeneous sub-overlay networks (referred to as *intra*-overlay networks henceforth). Each intra-overlay consists of one instance of, *e.g.*, Chord or any structured, unstructured, or hybrid overlay. We recall that an overlay network for information retrieval consists of a set of nodes on which the information on some resources is distributed. Each intra-overlay has its own key/value distribution and retrieval policy, logical topology, search complexity, routing and fault-tolerance mechanisms, so on and so forth. The Synapse protocol can be summarized by the following points:

- *Synapses:* the interconnection of intra-overlay networks is achieved by co-located nodes taking part in several of these intra-overlays, called synapses. Each peer will

---

[2] Code and web appendix at http://www-sop.inria.fr/teams/lognet/synapse

act in accordance with the policies of each of its intra-overlays, but will have the extra-role of forwarding the requests to some other intra-overlay it belongs to. As stated in the introduction, every node can become a synapse.

- *Peer's name:* every peer comes with a proper logical name in each intra-overlay; in particular, synapses have as many logical names as the number of networks they belong to.
- *Keys mapping in peers:* each peer is responsible for a set of resources (key, value) it hosts. Since every intra-overlay has different policies for key distribution, we could say that the inter-overlay induced by Synapse also inherits homogeneous distribution among the intra- and inter-networks. As for peers, every key comes with a proper logical name peculiar to each intra-overlay.
- *Set of resources assigned to set of nodes:* all overlay protocols for information retrieval share the invariant of having a set of peers responsible for a specific set of resources. This invariant allows for routing under structured, unstructured and hybrid networks because, by construction, intra-routing is the one always responsible for its correctness, since Synapse only cares about the overlay's interconnection.
- *Network in dependency and message translation:* intra-network protocols are different by construction: as such, when a message leaves a particular network and enters another network, the first network loses control of the route of that message inside the second one.
- *Topology, exhaustiveness, complexity and scalability:* by construction, the inter-overlay network induced by the Synapse protocol belongs to the category of unstructured overlay networks, with a routing that is not exhaustive, even if Synapse can connect only overlays that guarantee exhaustivity. The same goes for the routing complexity that can be upper-bounded only in the presence of precise and strong hypotheses about the type of intra-overlay networks. The same goes for scalability: a Synapse inter-network is scalable if all of the intra-networks are scalable.
- *Loopy routing avoidance:* to avoid lookup cycles when doing inter-routing, each peer maintains a list of tags of already processed requests, in order to discard previously seen queries, and a TTL value, which is decreased at each hop. These two features prevent the system from generating loops and useless queries, thus reducing the global number of messages in the Synapse inter-network.
- *Replications and Robustness:* to increase robustness and availability, a key can be stored on more than one peer. We introduce a Maximum-Replication-Rate (MRR) value which is decreased each time a PUT message touches a synapse, thus replicating the resource in more than one intra-overlay. This action acts as a special TTL denoting how many overlays can traverse a PUT message.
- *Social primitives:* each peer implements autonomously a good_deal? policy. This is a social-based primitive aimed at making some important choices that may strongly influence the performance and robustness of the Synapse routing. In particular, such a primitive is intended to make easier the choice of whether or not to join another intra-overlay (hence to become a synapse), invite or accept a peer to one of the overlays (hence invite a peer to become a synapse), or even create a new network from scratch. There is no best good deal strategy: for example, if one network wants to increase connectivity with other overlays, it can suggest to

all peers to invite and join all interesting/interested peers: this can be especially useful in case of high churning of the intra-network in order to increase alternative routing-paths through the neighboring intra-networks.

"White box" vs. "black box" synapse protocol. As stated in the introduction, one important issue in interconnecting overlay networks is the ability of one overlay to potentially modify its protocol instead of only accepting that co-located nodes will route packets without any change in the protocol itself. This is a concrete backward compatibility issue, since many overlays already exist, and it is hard to change them at this point for many reasons (security, commercial, technological ...).

As such, we have developed two variants of the synapse protocol; the first *white box* variant, is suitable for interconnecting overlays whose standards are open and collaborative, meaning that the protocol and the software client can be modified accordingly. The second, *black box* variant, is suitable for interconnecting overlays that, for different reasons, are not collaborative at all, in the sense that they only route packets according to their proprietary and immutable protocol. The white box allows for the adding of extra parameters to the current inter-overlay we are connecting, while the black box deals with those extra parameters by means of a *synapse control network*, *i.e.* a distributed overlay that stores all the synapse parameters that cannot be carried on by the overlay we are traversing.

White box synapse. The white box presented here is capable of connecting heterogeneous network topologies, given the assumption that every node is aware of the additions made to existing overlay protocols. The new parameters used to handle the game over strategy and replication need to be embedded into the existing protocols, and so does the unhashed key in order to be rehashed when a synapse is met. One important requirement of the Synapse white box protocol with respect to other protocols using hash functions is that the keys and nodes' addresses circulate *unhashed* from hop to hop. Hash functions have no inverse: once a sought key is hashed, it is impossible to retrieve its initial value, and thus impossible to forward to another overlay having a different hash function, since hash functions may vary (in implementations and key size) from overlay to overlay. Both the hashed and the *clear* key data can be carried within the message, or a fast hash computation can be performed at each step. Standard cryptographic protocols can be used in case of strong confidentiality requirements, without affecting the scalability of the Synapse protocol itself.

Black box synapse. Interconnecting existing overlays made of "blind" peers, who are not aware of any additional parameters, seems to be a natural Synapse evolution and it constitutes a problem worth investigating. The assumption is that an overlay can be populated by blind peers (*e.g.* nodes previously in place) and synapses at the same time. Both interact in the same way in the overlay and exchange the same messages; moreover, those synapses can be members of several overlays independently (thus being able to replicate a request from one overlay to another) and can communicate with each other exclusively through a dedicated *Control Network*. The Control Network is basically a set of DHTs allowing each node to share routing information with other synapses without being aware of the routing of the undergoing message. So far the DHTs implemented are the following: (i) a Key table, responsible for storing unhashed

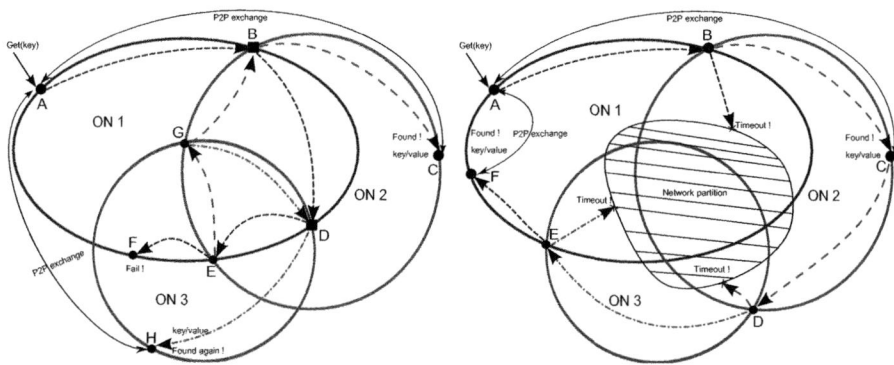

Fig. 1. Routing across different overlays and dealing with a network partition

keys circulating in the underlying overlays. Every synapse accessing this table can easily retrieve the key in clear way using only the information it is aware of; (ii) a Replication table, in which the number of times the key should be replicated across all of the the overlays is stored; (iii) a Cache table, used to implement the replication of GET requests, cache multiple responses and control the flooding of foreign networks. Due to the obvious lack of space, the white and black box models are treated in the web appendix.

Example 1. Routing across different intra-overlays. Figure 1 shows how a value present in one overlay can be retrieved from a GET launched by another overlay. Peer A in the overlay ON1 receives a GET(key) message: the routing goes until synapse B, which triggers a second intra-overlay routing in ON2. The two routings proceed in parallel, and, in particular, the routing in ON2 terminates successfully with a peer-to-peer interaction between peer A and peer C, responsible for the resource. Routing continues on ON1 until synapse D, which triggers a third intra-overlay routing in ON3. The routing proceeds in parallel, and, in particular, routing in ON3 terminates successfully with a second peer-to-peer interaction between A and H, while routing in ON1 proceeds to a failure on peer F via synapse E. Synapse E launches a fourth intra-overlay routing in ON2 that proceeds to a failure on node B (game over strategy) via synapse G. Finally, G launches a fifth intra-overlay routing on ON3, terminating with a failure on D (game over strategy again). Peers playing the game over strategy are depicted as squares.

Example 2. Dealing with network partition. Figure 1 also shows how intra-overlays take advantage of joining each other in order to recover situations where network partitioning occurs (because of the partial failure of nodes or the high churn of peers). Since network partitions affect routing performance and produce routing failures, the possibility of retrieving a value in a failed intra-overlay routing is higher, thanks to alternate inter-overlay paths. More precisely, the figure shows how a value stored in peer E of the overlay ON1 can be retrieved in presence of a generic network partition by routing via ON2 and ON3 through synapses B,C,D, and E. Please refer to the appendix for a description of the protocol pseudocode, in both the white and the black box model.

3   The Simulations

The purpose of the simulations is a better understanding of the behavior of platforms interconnecting structured overlay networks through the Synapse approach. We focus on the key metrics traditionally considered in the distributed information retrieval process, such as exhaustiveness (the extent of existing objects effectively retrieved by the protocol), latency (number of hops required to reach the requested object) and the amount of communications produced (number of messages generated for one request). We want to highlight the behavior of these metrics while varying the topology (the number of synapses and their connectivity, TTL, the number of intra-overlays ...).

Settings. Our simulations have been conducted using Python scripts, and using the *white box* protocol, capturing the essence of the Synapse approach. The topology of the overlay simulated is a set of Chord sub-networks interconnected by some synapses. Information is a set of (key, value) pairs. Each pair is unique and exists once and only once in the network. We study the unstructured interconnection of structured networks. We rely on discrete-time simulation: queries are launched on the first discrete time step, initiating a set of messages in the network, and each message sent at the current step will be received by its destination (next routing hop) at the next time step.

Synapses. Our first set of simulations had the intent of studying how the previously mentioned metrics vary while we add synapses or increase the degree of existing ones (the number of intra-overlays a co-located node belongs to). In this first set of simulations, the topology was built beforehand. The number of nodes was fixed to 10000, uniformly distributed amongst 20 overlays (*i.e.*, approximately 500 nodes within each Chord). Queries are always triggered by one random node, the key sought by a query is also picked uniformly at random from the set of keys stored by the network. A query is said to be *satisfied* if the pair corresponding to the key has been successfully retrieved.

We first studied search latency, *i.e.* the number of hops needed to obtain the first successful response. As illustrated in Figure 2, the first point to notice is that the number of hops remains logarithmic when changing a Chord network into a Synapse network (if the number of nodes is 10000, then the latency never exceeds 14). Other experiments conducted by increasing the number of nodes confirm this. More precisely, (top left) highlights the following behavior: (i) when the network contains only a few synapses, the latency first increases with the degree of synapses: only a few *close* keys are retrieved (keys available in the network of the node that initiated the query); (ii) then, when both parameters (the connectivity and the number of synapses) have reached a certain threshold, the searches can touch more synapses, and the whole network becomes progressively visible, multiple parallel searches become more and more frequent and distant nodes (and keys) are reached faster. As we can see, increasing the number of synapses decreases the latency by only a small constant factor. In other words, synapse topologies do not need a lot of synapses to be efficient. This result fits with random graphs behavior: when the number of neighbors in the graph reaches a (small) threshold, the probability for the graph to be connected tends towards 1. Obviously, multiple searches in parallel lead to an increased number of messages. As illustrated in (top right), this number increases proportionally with the connectivity and the number of synapses.

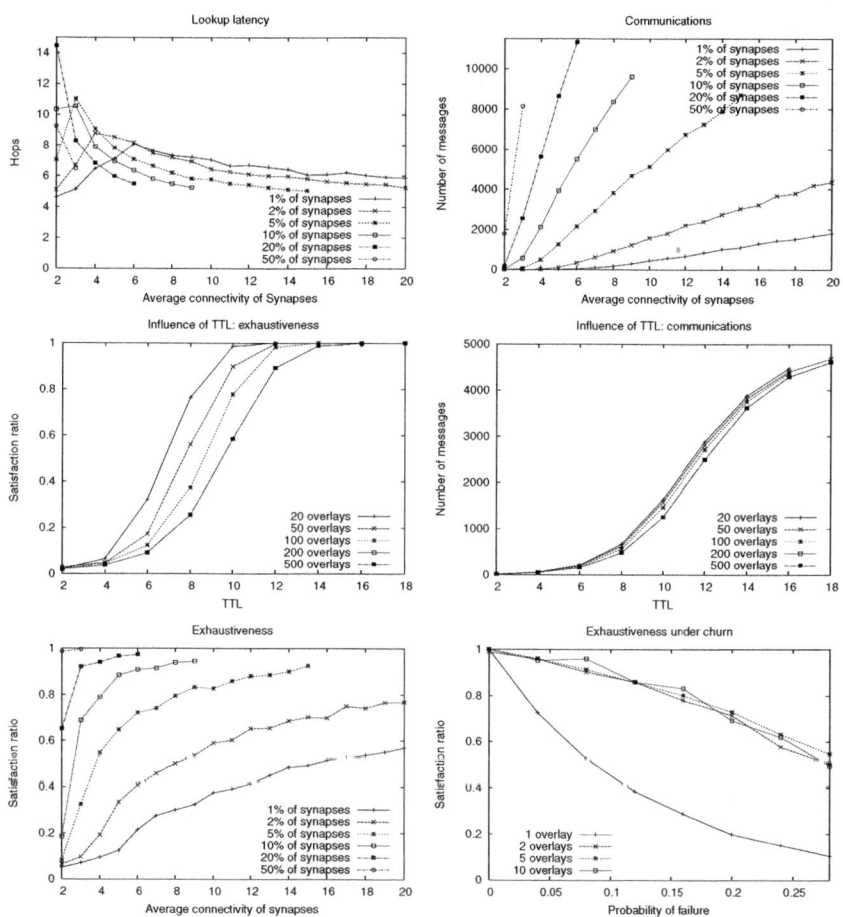

Fig. 2. Latency, communications, exhaustiveness and communications in Synapse

Time-To-Live. As we pointed out, the number of messages can become high when the number of synapses increases. To limit this impact, we introduced a Time-to-Live (TTL) to reduce the overhead while keeping an acceptable level of exhaustiveness. We launched a second set of experiments in order to study the impact of the TTL on the search queries. This TTL is simply decreased every time the query traverses a node.

The purpose here is to preserve significant exhaustiveness, while reducing the amount of communications undergone by the inter-overlay. We made the number of overlays vary, to experiment with the impact of the *granularity* of the network. In other words, a Synapse network made of a few large structured intra-overlays could be called *strongly structured*, while another network with many smaller structured intra-overlays could be called *weakly structured*. The number of nodes was still set to 10000, and every node is a synapse belonging to 2 overlays chosen uniformly at random.

Figure 2 (mid left) confirms that a low synapse degree (2) is enough to achieve quasi-exhaustiveness. Another interesting result is that the TTL can be bounded without any impact on the exhaustiveness (10 or 12 is enough even when the number of overlays interconnected is 500), while, as highlighted (mid right), drastically reducing the amount of communications experienced, with the number of messages being almost divided by 2. To sum up, Synapse architectures can use TTL, leading to a significant exhaustiveness, while drastically reducing the expected overhead. Finally, the *granularity* (defined above) does not significantly influence exhaustiveness and communications when the number and connectivity of the synapses are fixed.

Connectivity and Peers' churn. Figure 2 (bottom left) shows the evolution of the exhaustiveness while increasing the average number of overlays a synapse belongs to. We repeated the experiment for different ratios of synapses (in percentage of the total number of nodes). The exhaustiveness is improved by increasing both factors. We obtain more than 80% of satisfaction with only 5% of nodes belonging to 10 floors, and other nodes belonging to only one intra-overlay. When each node belongs to 2 overlays, the exhaustiveness is, also, almost guaranteed.

Since networks are intended to be deployed in a dynamic settings (nodes joining and leaving the network without giving notice), we conducted a final set of simulations to see the tolerance of Synapse compared to a single Chord overlay network. In other words, the question is: *does an interconnection of small Chords better tolerate transient failures than one large unique Chord?* In this experiment, the number of nodes remained the same, whatever the topology tested. At each step, a constant fraction of nodes was declared temporarily unreachable (the failed nodes would reappear in the next step, thus simulating the churn). As a consequence, the routing of some messages failed. As highlighted (bottom right), improvement on the number of satisfied requests can be obtained through a Synapse network: when the probability of failure/disconnection of a node increases, the global availability of the network is far less reduced with Synapse than with Chord. This shows that such synapse architectures are more robust and, thus, are good candidates for information retrieval on dynamic platforms. In other words, it demonstrates that this promising paradigm (interconnection of independent overlay networks) provides more fault-tolerance than a single global overlay network.

4   The Experimentations

JSynapse. In order to test our protocols on real platforms, we have initially developed JSynapse, a Java software prototype, which uses the Java RMI standard for communication between nodes, and whose purpose is to capture the very essence of our Synapse protocol. It is a flexible and ready-to-be-plugged library, which can interconnect any type of overlay networks. In particular, JSynapse fully implements a Chord-based inter-overlay network. It was designed to be a lightweight, easy-to-extend piece of software. We also provided some practical classes, helping with automating the generation of the inter-overlay network and the testing of specific scenarios. We have experimented with JSynapse on the Grid'5000 platform, connecting more than 20 clusters on 9 different sites. Again, Chord was used as the intra-overlay protocol. Our goal here was to give a proof of concept and show the viability of the Synapse approach,

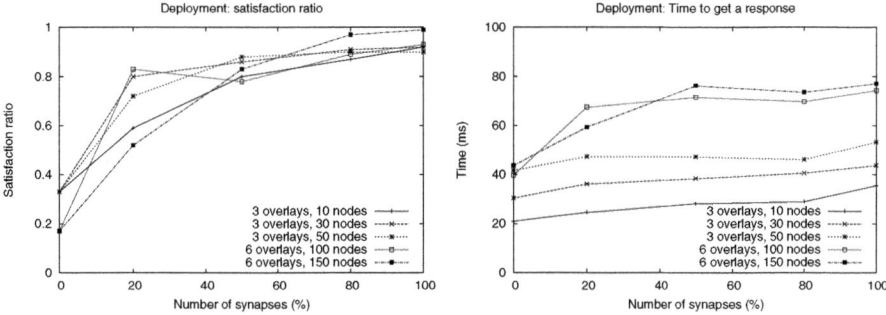

Fig. 3. Deploying Synapse : Exhaustiveness and Latency

while confirming results obtained by simulation about the overlay interconnection paradigm.

We used one cluster located at Sophia Antipolis, France. The `Helios` cluster consists of 56 quad-core AMD Opteron 275 processors linked by a gigabit Ethernet connection. The created Synapse network was first made of up to 50 processors uniformly distributed among 3 Chord intra-overlays. Then, still on the same cluster, as nodes are quad-core, we deployed up to 3 logical nodes by processor, thus creating a 150 nodes overlay network, nodes being dispatched uniformly over 6 overlays. During the deployment, overlays were progressively bridged by synapses (the degree of which was always 2). More precisely, for each number of networks (3 and 6), the number of nodes increases over time steps (10 to 50, and 100 to 150, respectively). One step was as follows: First, a set of queries is sent to the network (and results are collected). Second, some synapses are added for the next step.

We focus on the metrics affecting the user. Once his request was sent, a user waits for 1 second before closing the channels opened to receive responses. If no response was received after 1 second, the query is considered as not satisfied.

Figure 3 (left) shows the satisfaction ratio when increasing the number of synapses, for both the white and the black box version. As expected, the general behavior is comparable to the simulation results, and a quasi-exhaustiveness is achieved, with only a connectivity of 2 for synapses. Figure 3 (right) illustrates the very low latency (a few milliseconds) experienced by the user when launching a request, even when a lot of synapses may generate a lot of messages. Obviously, this result has to be considered keeping the performances of the underlying hardware and network used in mind. However, this suggests the viability of our protocols, the confirmation of simulation results, and the efficiency of the software developed.

Open-Synapse. We have also developed `open-synapse`, based on the stable and widely used `open-chord` implementation, which provides a complete and efficient Chord implementation. Open-Synapse extends the `open-chord` core, thus taking advantage of its robustness and reliability. A preliminary set of tests on `open-synapse` involved 50 nodes and different randomly generated scenarios.

5   Related Work

Cooperation through hierarchy. Several propositions have been made over the years to build topologies based on the coexistence of smaller local overlay networks. The first approach was based on hierarchical systems [6, 18], with some elected super-peers. In a more general view, merging several co-existing structured overlay networks has been shown to be a very costly operation [5, 16].

In the context of mobile ad hoc networks, Ariwheels [1, 12] was designed to provide a large variety of services through a multi-layer overlay network. Ariwheels provides an efficient mapping between physical devices in the wireless underlay network and virtual entities in the overlay network.

Cooperation through gateways. Authors in [4] present two models for two overlays to be (de)composed, known as *absorption* (a sort of merging) and *gatewaying*. Their protocol enables one CAN-network to be completely absorbed into another. They do not specifically take advantage of a simple assumption that nodes can be part of multiple overlays at the same time, thus playing the role of natural bridges.

More recently, authors in [3] propose a novel information retrieval protocol, based on gateways, called *DHT-gatewaying*, which is scalable and efficient across structured overlay networks[3]. They argue that there is not one preferred structured overlay network implementation, and that peers are members of co-existing DHTs. Although the focus is on wireless ad hoc network, they claim that their protocol can be used in wired networks too. Unfortunately, it is unclear how they evaluate their protocol.

Cooperation through co-located nodes. Authors in [11] present Synergy, an architecture which improves routing performance by providing cooperative forwarding of flows between networks. They try to create and maintain long-lived flows (*i.e.* long-lived paths) that can be used to cross overlay boundaries. However they do not go into details as much as we do in this paper regarding the algorithms that enable such overlay inter-connection.

Authors in [10] present algorithms which enable a symbiosis between different overlays networks with a specific application in mind: file sharing. They propose mechanisms for hybrid P2P networks cooperation and investigate the influence of system conditions, such as the numbers of peers and the number of meta-information a peer has to keep. Again, a more comprehensive understanding of this approach is missing. Authors only focus on file sharing whereas our protocol can be applied to any application. Even if the algorithms and the various observations they present are relevant, they fail to provide both any real experiments, and an in-depth analysis of their algorithms.

Authors in [7] consider multiple spaces with some degree of intersection between spaces, *i.e.* with co-located nodes. They focus on different potential strategies to find a path to another overlay from a given overlay, *i.e.* how requests can be efficiently routed from one given overlay to another one. They showed that with some dynamic finger caching, and with multiple gateways, they obtain pretty good performances. Their

---

[3] Ex. Two 160-bit Chord, or two 160/256-bit Chord, or one 160-bit Chord and one 256-CAN.

protocol focuses on the interconnection of DHTs, while we extend it to any kind of overlay.

In our previous preliminary work [13], we introduced BabelChord, a protocol for inter-connecting Chord overlay networks using co-located nodes that are part of multiple Chord "floors". These nodes connect, in an unstructured fashion, several Chord overlays together. The simulations showed that we could achieve pretty high exhaustiveness with a small amount of those co-located nodes. Our current paper, in turn, focuses on the co-located nodes heuristic in far more details than the aforementioned work, by providing not only a more generic protocol which enables inter-overlay routing that can in principle be applied to connect arbitrary heterogeneous overlays, but also more simulations to show the behaviors of such networks as well as a real implementations and live experiments.

Self-Aware Networks (SAN) and Cognitive Packets Networks (CPN). In SAN [9], nodes can discover new paths dynamically when the need to communicate arises; nodes can be able to discover new routes and optimize their QoS. CPN [8] are SAN where *smart packets* are used to discover new paths. A Synapse network is also a SAN but not a CPN. Every overlay keeps its proprietary routing and uses synapses nodes to explore other overlays that couldn't have been reached.

6   Conclusion

We have introduced Synapse, a scalable protocol for information retrieval in hetero-geneous inter-connected overlay networks relying on co-located nodes. Synapse is a generic and flexible *meta*-protocol which provides simple mechanisms and algorithms for easy overlay network interconnection. We have given the set of algorithms behind our protocols and provided a set of simulations allowing to capture the behavior of such networks and show their relevance in the context of information retrieval, using key metrics of distributed information retrieval. We have also developed JSynapse, a lightweight implementation of Synapse, and experimented with it using the Grid'5000 platform, thus confirming the obtained simulation results and giving a proof of concept.

As future work, we first intend to focus on social mechanisms involved in synapses, which can greatly influence the shape of the network. On the practical side, we plan to extend JSynapse and plug in some other overlay network implementations. More intensive tests and deployments of open-synapse, our early prototype based on open-chord will also be considered. Adding CPN capabilities to Synapse networks would also be worth studying.

References

1. Borsetti, D., Casetti, C., Chiasserini, C., Liquori, L.: Content Discovery in Heterogeneous Mobile Networks. In: Heterogeneous Wireless Access Networks: Architectures and Proto-cols, pp. 419–441. Springer, Heidelberg (2009)
2. Castro, M., Druschel, P., Kermarrec, A.-M., Rowstron, A.: SCRIBE: A Large-Scale and Decentralized Application-Level Multicast Infrastructure. IEEE Journal on Selected Areas in Communications (JSAC), 20 (2002)

3. Cheng, L.: Bridging Distributed Hash Tables in Wireless Ad-Hoc Networks. In: Global Telecommunications Conference, GLOBECOM 2007, pp. 5159–5163. IEEE, Los Alamitos (2007)
4. Cheng, L., Ocampo, R., Jean, K., Galis, A., Simon, C., Szabo, R., Kersch, P., Giaffreda, R.: Towards Distributed Hash Tables (De) Composition in Ambient Networks. In: State, R., van der Meer, S., O'Sullivan, D., Pfeifer, T. (eds.) DSOM 2006. LNCS, vol. 4269, pp. 258–268. Springer, Heidelberg (2006)
5. Datta, A., Aberer, K.: The challenges of merging two similar structured overlays: A tale of two networks. In: Proc. of IWSOS (2006)
6. Erice, L.G., Biersack, E.W., Ross, K.W., Felber, P.A., Keller, G.U.: Hierarchical P2P Systems. In: Kosch, H., Böszörményi, L., Hellwagner, H. (eds.) Euro-Par 2003. LNCS, vol. 2790, pp. 1230–1239. Springer, Heidelberg (2003)
7. Furtado, P.: Multiple Dynamic Overlay Communities and Inter-space Routing. In: Moro, G., Bergamaschi, S., Joseph, S., Morin, J.-H., Ouksel, A.M. (eds.) DBISP2P 2005 and DBISP2P 2006. LNCS, vol. 4125, pp. 38–49. Springer, Heidelberg (2007)
8. Gelenbe, E.: Cognitive Packets Networks. U.S. Patent 6,804,201 (October 2004)
9. Gelenbe, E.: Steps Toward Self-Aware Networks. Commun. ACM 52(7), 66–75 (2009)
10. Junjiro, K., Naoki, W., Masayuki, M.: Design and Evaluation of a Cooperative Mechanism for Pure P2P File-Sharing Networks. IEICE Trans. Commun. (Inst. Electron. Inf. Commun. Eng.) E89-B(9), 2319–2326 (2006)
11. Kwon, M., Fahmy, S.: Synergy: an Overlay Internetworking Architecture. In: Proc. of International Conference on Computer Communications and Networks, pp. 401–406 (2005)
12. Liquori, L., Borsetti, D., Casetti, C., Chiasserini, C.: An Overlay Architecture for Vehicular Networks. In: Das, A., Pung, H.K., Lee, F.B.S., Wong, L.W.C. (eds.) NETWORKING 2008. LNCS, vol. 4982, pp. 60–71. Springer, Heidelberg (2008)
13. Liquori, L., Tedeschi, C., Bongiovanni, F.: Babelchord: a social tower of dht-based overlay networks. In: Prof. of IEEE ISCC, pp. 307–312 (2009)
14. Ratnasamy, S., Francis, P., Handley, M., Karp, R., Shenker, S.: A Scalable Content-Adressable Network. In: ACM SIGCOMM, pp. 161–172 (2001)
15. Rowstron, A., Druschel, P.: Pastry: Scalable, Distributed Object Location and Routing for Large-Scale Peer-To-Peer Systems. In: Guerraoui, R. (ed.) Middleware 2001. LNCS, vol. 2218, pp. 329–350. Springer, Heidelberg (2001)
16. Shafaat, T.M., Ghodsi, A., Haridi, S.: Handling network partitions and mergers in structured overlay networks. In: Proc. of P2P, pp. 132–139. IEEE Computer Society, Los Alamitos (2007)
17. Stoica, I., Morris, R., Karger, D., Kaashoek, M., Balakrishnan, H.: Chord: A Scalable Peer-to-Peer Lookup service for Internet Applications. In: Proc. of ACM SIGCOMM, pp. 149–160 (2001)
18. Xu, Z., Min, R., Hu, Y.: HIERAS: A DHT Based Hierarchical P2P Routing Algorithm. In: Proc. of ICPP, p. 187. IEEE Computer Society, Los Alamitos (2003)

A   "White" and "Black" Box Synapse Protocols

Figure 4 presents the pseudo-code of the two implementations of Synapse protocol using the message passing paradigm. For obvious lack of space, the code cannot be commented in great detail; the interested reader can refers to the web appendix.

The White Box protocol (first part) implements the same strategy to inter-route both PUT and GET requests and describes the social primitives JOIN and INVITE which are used to handle the invitation and join of new networks by a Synapse node.

White Box protocol

```
1.01  on receipt of OPE(code,key,value) from ipsend do     receive an opcode, key value from ipsend
1.02  tag = this.new_tag(ipsend);                           create a new unique tag for this lookup
1.03  send FIND(code,ttl,mrr,tag,key,value,ipsend) to this.myip;     send FIND msg to itself
2.01  on receipt of FIND(code,ttl,mrr,tag,key,value,ipdest) from ipsend do     rcv FIND
2.02   if ttl = 0 or this.game_over?(tag)     lookup aborted because of zero ttl or game over strategy
2.03   else this.push_tag(tag);              push the tag of the query as "already processed"
2.04    next_mrr = distrib_mrr(mrr,this.net_list);     split all the maximum replication rate
2.05    for all net ∈ this.net_list do             for all net the synapse belongs do
2.06     if this.isresponsible?(net,key)       the current synapse is responsible for the key in the net
2.07       send FOUND(code,net,mrr,key,value) to ipdest;     send a FOUND msg to ipdest
2.08     else if this.good_deal?(net,ipsend)         the net/ipsend is a "good" net/synapse
2.09       send FIND(code,ttl-1,next_mrr.get(net),tag,key,value,ipdest)
           to this.next_hop(key);                send a FIND msg with ... to ...
3.01  on receipt of FOUND(code,net,mrr,key,value) from ipsend do rcv FOUND msg from ipsend
3.02   this.good_deal_update(net,ipsend);          update my good deal tables with net and ipsend
3.03   match code
3.04    code=GET                                                              GET code
3.05     send READ_TABLE(net,key) to ipsend                  send a READ_TABLE msg to ipsend
3.06    code=PUT                                                              PUT code
3.07     if mrr < 0                              stop replication since no inter PUT is allowed
3.08      else send WRITE_TABLE(net,key,value) to ipsend     send a WRITE_TABLE msg (omitted)
3.09  on receipt of INVITE(net) from ipsend do         receive an invitation to join the net from ipsend
3.10   if this.good_deal?(net,ipsend)     the net/ipsend is a "good" net/synapse (left to peer's strategy)
3.11    send JOIN(net) to ipsend;             send a JOIN message to reach the net to ipsend
4.01  on receipt of JOIN(net) from ipsenddo      receive a JOIN message to reach the net from ipsend
4.02   if this.good_deal?(net,ipsend)   the net/ipsend is a "good" net/synapse (left to synapse's strategy)
4.03    this.insert_net(net,ipsend);           the current synapse insert ipsend in the net
```

Black Box protocol

```
1.04  on receipt of SYN_GET(key,cacheTTL,[targetNetworks]) from ipsend do    GET request
1.05  CacheTable[key].TimeToLive = cacheTTL;                 init the control data
1.06  CacheTable[key].targetedNetworks = [targetNetworks];  specify the net to target if any specific
1.07  for all network ∈ (this.net_list ∩ targetNetworks) do
1.08   KeyTable[network.ID|network.hash(key)] = key;    store the unhashed key in the KeyTable
1.09   result_array += network.get(network.hash(key));   retrieve the result from each network
1.10   result_array += CacheTable[key].cachedResults;   retrieve the cache content results from network
1.11  send SYN_FOUND(key,result_array) to ipsend;
2.01  on receipt of SYN_PUT(key,value,mrr) from ipsend do     a PUT is instantiated by a synapse
2.02  if (mrr > this.networks.size)  if MRR is bigger than the number of connected overlays for the synapse
2.03   mrrInSight = this.network.size;        compute how many replications are done in this synapse
2.04   mrrOutOfSight = mrr-this.networks.size;  compute how many replications are left after this step
2.05   delete ReplicationTable[key];                   refresh the replicationTable entry
2.06   ReplicationTable[key].ReplicasLeft = mrrOutOfSight;       update with new MRR value
2.07  else mrrInSight = mrr;                       else we replicate for only MRR times
2.08  for i = [1:mrrInSight] do
2.09   KeyTable[this.net_list[i].ID|this.net_list[i].hash(key)] = key; store unhashed key
2.10   this.net_list[i].put(this.net_list[i].hash(key),value);  continue forwarding the PUT
3.01  on receipt of GET(hashKey) from this.net_list[i] do    a synapse rcv a GET to be forwarded
3.02  key = KeyTable[network.ID|hashKey];        get the unhashed key from the Key Table
3.03  if (CacheTable[key] exists)             if there is an entry in the Cache Table for the key
3.04   for all net ∈ (CacheTable[key].targetedNetworks ∩ this.net_list) do
                                        compute which of the specified networks we are connected to
3.05    KeyTable[net.ID|net.hash(key)] = key;                    add entry in KeyTable
3.06    results += net.forward_get(replicaNetwork.hash(key));       append the result
3.07   CacheTable[key].cachedResults += results;     store the collected results in the Cache Table
3.08   return results[1];                          return only the first result
3.09  else return net_list[i].get(hashKey);    if there is no entry just forward in the current network
4.01  on receipt of PUT(hashKey,value) from this.net_list[i] do
4.02  key = KeyTable[network.ID|hashKey];        get the unhashed key from the Key Table
4.03  if (ReplicationTable[key] exists)       if there is an entry in the Replication Table for the key
4.04   for all net ∈ this.net_list do
4.05    if (ReplicationTable[key].ReplicasLeft > 0) and    if there are still replicas and the network
4.06       not (ReplicationTable[key].hasNetwork?(replicaNetwork.ID))    hasn't had a PUT yet
4.07     KeyTable[net.ID|net.hash(key)] = key;                    add entry in KeyTable
4.08     ReplicationTable[key].addNetwork(net.ID);               add the target network
4.09     ReplicationTable[key].ReplicasLeft--;       decrement the number of replicas to be done
4.10     net.forward_put();                     forward the message in the foreign network
4.11  else net_list[i].put(hashKey,value);    if there is no entry just forward in the current network
```

Fig. 4. The Synapse white and black box protocol

The Black Box protocol (second part) on the other hand, is a set of messages to an independent Synapse Controller standing between the application layer and the actual overlay networks. The Controller can perform the following operations: (i) it can be notified by the application to initiate a new synapse GET/PUT, via the messages SYN_GET, SYN_PUT; in this cases it will take care of initializing the shared inter-routing data in the Control Network and fire a conventional GET/PUT message on one or more overlays it is connected to; (ii) it can receive a notification from one of the connected overlays about a GET/PUT request going through that could therefore be "synapsed" in the other networks. This is done via the call back methods GET and PUT. The Control Network operations are here represented for clarity as simple associative arrays. The notation KeyTable[network.ID|hashKey] = key is equivalent to send a KEYTABLE_PUT message with arguments network.ID|hashKey and key to the SynapseControlNetwork, while key = KeyTable[network.ID|hashKey] corresponds to send KEYTABLE_GET with argument network.ID|hashKey to SynapseControlNetwork and then to collect the result in the variable key.

A Longitudinal Study of Small-Time Scaling Behavior of Internet Traffic

Himanshu Gupta[1,2], Vinay J. Ribeiro[2], and Anirban Mahanti[3]

[1] IBM Research Laboratory, New Delhi, India
higupta8@in.ibm.com
[2] Department of Computer Science and Engineering,
Indian Institute of Technology, New Delhi, India
vinay@cse.iitd.ac.in
[3] NICTA, Locked Bag 9013, Alexandria, NSW 1435, Australia
anirban.mahanti@nicta.com.au

Abstract. During the last decade, many new Web applications have emerged and become extremely popular. Together, these new "Web 2" applications have changed how people use the Web and the Internet. In light of these changes, we conduct a longitudinal study of the small-time scaling behavior of Internet traffic using network traffic traces, available from the MAWI repository, that span a period of eight years. The MAWI traces are affected by anomalies; these anomalies make correct identification of scaling behavior difficult. To mitigate influence of anomalies, we apply a sketch-based procedure for robust estimation of the scaling exponent. Our longitudinal study finds tiny to moderate correlations at small-time scales, with scaling parameter in the range [0.5, 0.75], across the traces examined. We also find that recent traces show larger correlations at small-time scales than older traces. Our analysis shows that this increased correlation is due to the increase in the fraction of aggregate traffic volume carried by dense flows.

Keywords: Traffic Analysis, Small-time scaling, Dense Flows, Robust Estimation.

1 Introduction

Scaling behavior of Internet traffic has received much attention from networking researchers. It is known that Internet traffic displays two scaling regimes with a transition occurring in the 100ms to 1s time range [3, 7]. Internet traffic when aggregated to large-time scales ($\geq$1s) is quite bursty and is modeled using *long-range dependent* (LRD) processes [11, 12]. The scaling parameter (H) in large-time scales is typically in the (0.8,1) range which is indicative of highly correlated packet arrivals [3, 6, 11, 12].

The focus of this paper is on the small-time scaling behavior of recent Internet traffic. By small-time scale, we refer to timescales smaller than the scaling transition point, which is in the 100 to 1000ms range, wherein factors that cause correlations at large-time scales may not yet be strong or even present. Traffic

M. Crovella et al. (Eds.): NETWORKING 2010, LNCS 6091, pp. 83–95, 2010.

models proposed for small-time scales include simple Poisson processes, independent Gamma interarrivals, and non-Gaussian multifractal processes [5, 9]. Zhang et al. [16] examined traces from backbone networks (collected in 2000-02) and found most of the examined traces to exhibit little or no correlations at small-time scales with a scaling parameter h below 0.6, and often fairly close to 0.5.[1] Only a small fraction of the traces examined exhibited some small-time correlations with scaling parameter within the range of 0.6-0.7. They also introduced the concept of "dense flows", defined as flows with bursts of densely clustered packets, and showed that traces with more dense flows display relatively larger correlations at small-time scales.

In this paper, we present a longitudinal analysis of the small-time scaling phenomena. Our work is driven by the observation that a number of significant changes have occurred since the beginning of 2000 (when prior work in this area was undertaken). In particular, typical backbone network capacities as well as number of Internet connected hosts have increased. Furthermore, composition of traffic has changed with the rapid growth of "Web 2" sites such as YouTube, FaceBook, Flickr, and Twitter, and also with popularity of peer-to-peer file sharing applications [2]. Therefore, we revisit aspects concerning the small-time scaling behavior of Internet traffic. Specific questions that we consider include the following: Do recent traces mostly display no/small/moderate correlations at small-time scales? Do dense flows still drive small-time scaling behavior of Internet traffic? In this paper, we provide answers to these questions by conducting a longitudinal analysis of publicly available Internet traffic traces from the MAWI repository that span 8 years (2001-2009).

The traces in the MAWI repository are known to contain anomalies [4]. In a recent work, Borgnat et al. [3] have shown that anomalies can interfere with the correct identification of scaling behavior. They proposed a robust estimation method based on sketches (random projections) to mitigate the effect of network anomalies and used this estimation method for a longitudinal analysis of the LRD behavior of network traces. We use this robust estimation method for the longitudinal analysis of small-time scaling behavior. We show that in absence of a robust estimation procedure, misleading inferences regarding small-time scaling behavior of Internet traffic may be derived. We believe our work further reinforces the importance of this new method.

Our primary contribution is a study of the evolution of small-time scaling behavior of the MAWI traces throughout this decade. Once the effects of network anomalies have been mitigated, we find the scaling parameter in small-time scales to be consistently between 0.5 and 0.75, thereby, showing the presence of tiny to moderate correlations in small-time scales in the traces examined. We also find that the recent MAWI traces exhibit comparatively larger correlations at small-time scales than the earlier traces. Among the traces examined, traces from 2007 and later years tend to exhibit slightly higher correlation (e.g., h of 0.6 or more) than those prior to 2007. This observation is not in consonance with

---

[1] Scaling parameter in small-time scales is represented as h while in large-time scales is represented as H.

the prediction that Internet traffic will likely be described by simple models (e.g. Poisson) [10]. We further show that dense flows account for a larger fraction of the aggregate traffic in recent MAWI traces vis-a-vis traces from earlier years.

We would like to emphasize here that the results presented in this paper are specific to the MAWI traces. Generalizing the trends regarding Internet traffic by studying packet traces from only one link is difficult as any observations derived from such an analysis run the risk of getting biased by individual events happening at the concerned link. Nonetheless, our results are still valuable as they summarize the evolution of small-time scaling behavior for an Internet backbone for a period of 8 years, and hence can act as a reference for longitudinal studies on other backbone links.

The rest of this paper is organized as follows. Section 2 presents necessary background on the approach used for estimating the scaling parameter, and also summarizes the traces used in this work. Section 3 illustrates the importance of the robust estimation method in the context of scaling at small-time scales. Section 4 focuses on results from our longitudinal analysis. Section 5 analyzes the influence of dense flows on the small-time scaling behavior of the aggregate traffic. Conclusions are presented in Section 6.

2   Background

2.1   The MAWI Dataset

We use publicly available traces from the MAWI repository [1]. This repository provides traces collected from a trans-Pacific backbone. A 15-minute long trace is made public for download every day. We use the traces captured at collection points B and F. The link corresponding to B (100 Mbps) was replaced by F (100 Mbps) in July 2006, and was subsequently upgraded to 150 Mbps in June 2007. For our longitudinal study, we select the traces collected on 1st and 15th of every month, from January 2001 to December 2008. This gives us a set of 180 traces spanning eight years. Each trace is partitioned into two subtraces, one for traffic flowing in each direction referred to as JptoUS and UStoJp. We analyze each subtrace separately.

Overall, the traces exhibit substantial variabilities during the time period we consider. For example, a global increase of throughput from 100 kbps in 2001 to more than 12 Mbps in 2009 is observed. Several long lasting congestion periods are also observed (e.g., UStoJp:2003/04 to 2004/10, UStoJp:2005/09 to 2006/06, JptoUS:2005/09 to 2006/06) [3]. Strong fluctuations in packet number are observed on UStoJp from 2004/07 to 2005/04 due to massive activities of the Sasser worm. Many anomalies such as ping floods and SYN scans have also been observed [3, 4]. Overall, this suite of 180, 15-minute, packet traces is uniformly spread across eight years, and hence likely captures various variations and anomalies one expects to see in Internet traffic traces. We refer the reader to [3, 4] for a detailed statistical characterization of traces available from the MAWI repository.

2.2   Scaling Analysis

The analysis involves studying the data at multiple timescales and hence the term
"scaling". The analysis of the scaling behavior of a process helps to characterize
the way the process will aggregate over different time intervals. We say a process
Y scales with a scaling parameter h if

$$Variance(Y^{(m)}) \approx m^{2h-2}Variance(Y) \tag{1}$$

where $Y^{(m)}$ is the aggregated process of process Y and is defined as follows:

$$Y^{(m)}(k) = \frac{1}{m} \sum_{i=km-m+1}^{i=km} Y(i), k = 1, 2, ...\infty \tag{2}$$

In scaling analysis of network traces, the process Y usually represents the number of bytes or packets sent within a fixed length time interval. The scaling
parameter h measures the strength of correlations present in the data. A value
of 0.5 represents the absence of correlations; a Poisson process displays a scaling
parameter of 0.5. The larger the value of h, the stronger the correlations. We
interpret scaling parameter values lying within the ranges $[0.5, 0.6)$, $[0.6, 0.7)$,
and $[0.7, 0.75]$ as tiny, small, and moderate correlations, respectively.

Of all the methods for estimating the value of scaling parameter (e.g.,
Variance-Time plot, R/S plot etc), Wavelet-based method is the most preferred [15]. Other methods are not reliable in the presence of non-stationary
trends in the data. An implementation of Wavelet-based method is provided at
[13] and we use this software to carry out our analysis. A concise summary of
this method follows.

2.3   The Wavelet Estimator for the Scaling Parameter

This method allows us to observe the scaling behavior of a traffic process over
a certain range of timescales. The working of this estimator is illustrated using
Haar wavelets which are the simplest kind of wavelets. In practice any other kind
of wavelet can be considered.

Consider a reference timescale T_0 and let $T_j = 2^j T_0$ for $j = 1, 2, ...$ be increasingly coarser timescales. To form the process at scale j, we partition the
trace into consecutive and non-overlapping time intervals of size T_j and count
the numbers of bytes (or packets) in these intervals. If t_i^j is the ith time interval
at scale $j > 0$ then t_i^j consists of the intervals t_{2i}^{j-1} and t_{2i+1}^{j-1}. Let X_i^j be the
amount of traffic in t_i^j, with $X_i^j = X_{2i}^{j-1} + X_{2i+1}^{j-1}$. The *Haar wavelet coefficients*
d_i^j at scale j are defined as:

$$d_i^j = 2^{-j/2}(X_{2i}^{j-1} - X_{2i+1}^{j-1}), i = 1, ..., N_j \tag{3}$$

where N_j is the number of wavelet coefficients at scale j. The *energy* ξ_j at scale
j is defined as:

$$\xi_j = E[(d_i^j)^2] \approx \frac{\sum_i (d_i^j)^2}{N_j}. \tag{4}$$

Plotting the logarithm of energy ξ_j as a function of timescale j gives us a *logscale diagram* (LD). The magnitude of ξ_j increases with variability of the traffic process X^{j-1} at scale $j-1$. Variation of energy ξ_j with j captures the scaling behavior of the process. The slope of this energy plot α estimates the *scaling parameter* h through $\alpha = 2h-1$. The wavelet energy can only be estimated at a scale because of limited data. Hence the slope of logscale diagram can itself only give an estimate of the scaling parameter.

If the scaling parameter is constant across all timescales, the process is said to show *global scaling* [8, 14]. If the slope of the energy plot is roughly constant over a range of timescales j to $j+k$, the traffic process is said to exhibit *local scaling* in the timescales T_j to T_{j+k} [8, 14]. This paper is concerned with local scaling at small-time scales (1-100 ms).

2.4 Robust Estimation Technique

Borgnat et al. [3] recently showed that anomalies (e.g., attack, congestion etc.), if present, in traffic traces can radically affect the observed scaling behavior. By looking at such a trace, one can draw misleading inferences regarding the scaling behavior of the traffic. For example, changes in scaling behavior may be incorrectly attributed to network mechanisms at work rather than to network anomalies. Hence, it is important to mitigate the effect of anomalies so that the impact of networking mechanisms can be precisely studied. This is especially important if one is trying to disentangle smooth long-term evolution features from day-to-day fluctuations.

Borgnat et al. [3] presented a sketch-based method for robust estimation of the scaling parameter. Let f denote a hash table of size M. The original collection of packets is split into M sub-collections, each consisting of all packets with identical sketch output $m = f(A)$ where the hashing key A is chosen as one of the packet attributes (IPdst, IPsrc,...). This approach, referred to as "random projections", preserves flow structures as packets belonging to a given flow are assigned to the same sub-collection. Each sub-collection is aggregated and its scaling parameter is computed. A robust estimate of the scaling parameter is the median of the above values [3].

Statistically, robustness in estimation is achieved by performing averages over independent copies of equivalent data. Finding equivalent traces is a complex problem. Random projection using sketches is one way to achieve independent copies of equivalent traces. The resulting logscale diagrams have the same shape as the original, with a variance which is appropriately scaled down[2], consistent with an independent and identically distributed (i.i.d.) superposition model [7]. In the presence of anomalies, sketching the original packet stream reduces their impact, possibly restricting them to only some of the sub-collections. The small-time correlation structure of the traffic in the sub-collections containing anomalies would likely differ from normal traffic as well as traffic in other

---

[2] If one selects flows with probability 0.7, the resulting LD will have a variance that is approximately equal to 70% of the original.

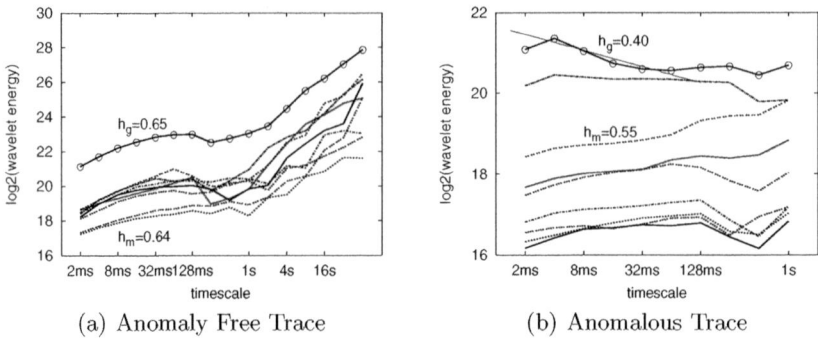

(a) Anomaly Free Trace   (b) Anomalous Trace

Fig. 1. Representative Examples: Sketch based Robust Estimation

sub-collections. Taking the median of h over independent sketches achieves robustness. The median is chosen instead of mean as median is a non-linear statistic that provides robustness against outliers. The estimator can still be fooled if anomalies are a dominant component of the trace. Robustness in such cases can be achieved by maintaining multiple sketches and taking the median over estimates computed from them.

3   Representative Examples

Fig. 1(a) shows the logscale diagram (upper circled plot) for an anomaly-free trace (July 11,2005; UStoJp). The anomaly-free nature of this trace was established (in a prior study) through careful manual inspection and application of an anomaly identification algorithm [3]. The time-series of byte counts every millisecond is used to construct the LD (i.e., reference timescale T_0 is 1 ms and X_i^j represents byte count in interval t_i^j). The scaling parameter estimate is 0.65 which indicates the presence of small correlations at small-time scales.

Next, we estimate the scaling parameter using the sketch-based robust estimation method [3]. We hash the trace into 8 bins using destination IP as the hash-key, and estimate the scaling parameter for each subtrace. Robust estimation is achieved by taking median over this set of 8 estimations of the scaling parameter, one corresponding to each subtrace. Fig. 1(a) plots all the subtrace LDs as well. All sketched subtrace LDs are found to be parallel to the original LD, and therefore, indicate a similar scaling behavior as the original trace. We use notations h_g and h_m to represent global and median estimates, respectively. For this anomaly free trace, both h_g and h_m are found to be identical ($\approx$0.65).

Fig. 1(b) shows the logscale diagram (upper circled plot) for a trace (Oct 11, 2005; JptoUS) which is known to contain a low-intensity long-lasting spoofed flooding anomaly [4]. The anomaly consists of source IP addresses being spoofed (source IP is identical to destination IP) and destination port being 0 (which is

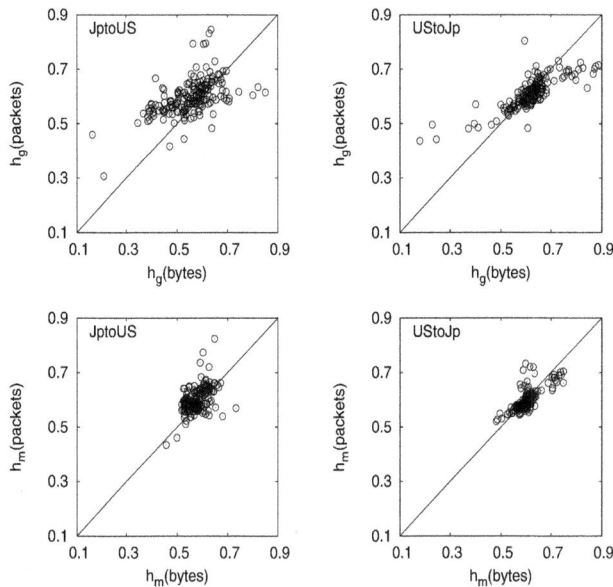

Fig. 2. Global and Median Scaling Parameter: Byte-Count vs Packet-Count

not normally used) [4]. The global scaling estimate h_g is 0.40 which indicates the presence of a small negative correlation (also called *anti-persistent* behavior).

We also estimate the robust (median) value of the scaling parameter. Fig. 1(b) also shows the LDs for all the sketched subtraces. Except for one LD, possibly containing anomaly, all other subtrace LDs have recovered normal behavior, and each of these subtraces now displays tiny positive correlation ($h \approx 0.55$). Note that the scaling behavior of these sketches is markedly different from the scaling behavior shown by the original trace. As a result, the median estimate h_m over 8 sketches, is found to be 0.55 while the global estimate h_g was 0.40. This shows that h_g being 0.40 is an artifact of network anomalies; otherwise all sketched LDs should have had a similar value of scaling parameter.

We next show the value of robust estimation by comparing the values of scaling parameter computed using byte-counts and packet-counts. A debate exists regarding whether scaling analysis should be measured on byte-counts or packet-counts [3]. We hence compute the small-time scaling parameter, both global and median, for all 180 traces in our MAWI tracesuit. Fig. 2 shows scatter plots comparing the values of the scaling parameter using byte-counts and packet-counts. The top row compares global values of the scaling parameter across all traces examined while the bottom row compares the median values. We find that most of the median values of the scaling parameter lie along a 45 degree line indicating that the parameter values as obtained from byte-counts or packet-counts are nearly identical, an observation consistent with prior studies [3, 7, 11] that predict the same scaling parameter for packet and byte counts. In contrast, note that the global values of the scaling parameter show relatively larger variability and dispersion which erroneously suggests inequality of byte-count and packet-count scaling parameter for many traces.

Unless stated otherwise, all results presented in the remainder of this paper are from application of the robust estimation technique with 8 bins and destination IP address as the hashing key. Also, all scaling analysis is done using byte-counts.

4 Longitudinal Analysis of Scaling Behavior

4.1 Importance of Robust Estimation

Fig. 3 plots the scaling parameter values for the traces across eight years, with and without robust estimation. There are multiple traces for which global estimation h_g is found to be less than 0.5. Specifically, from 2005 to mid-2006 the JptoUS subtrace has h_g values consistently less than 0.5, often close to 0.4, suggesting presence of small negative short range correlations. The median values of the scaling parameter h_m, however, computed using the robust estimation procedure, are markedly different with h_m consistently lying close to 0.55. Similarly, for many traces the global scaling parameter h_g is found to be close to 0.8 (e.g., around 2007 UStoJp) suggesting large correlation in small-time scales, in contrast with our current understanding [16]; however, the values of h_m indicate only tiny to moderate correlations. Clearly, robust estimation of the scaling parameter is important, and should not be overlooked when studying small-time scaling behavior of network traffic.

4.2 Evolution of Short Range Correlations

Fig. 3 shows that the median scaling parameter h_m lies between 0.55 and 0.75, thereby indicating that traces across the eight years display tiny to moderate

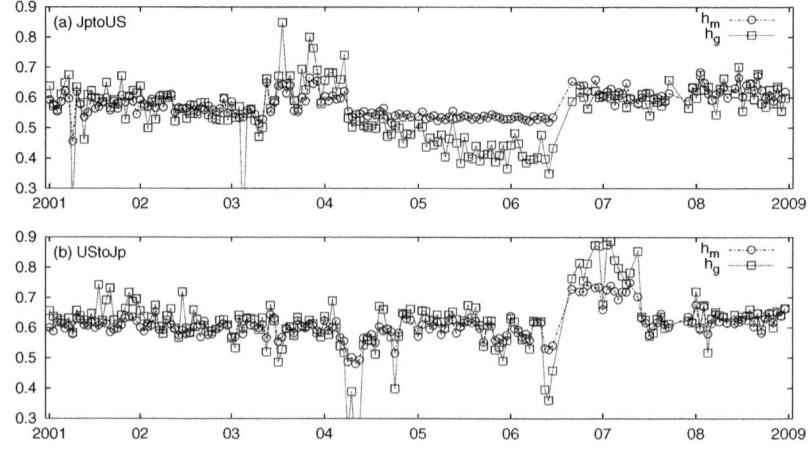

Fig. 3. Scaling behavior of traffic at small-time scales (2001–2009)

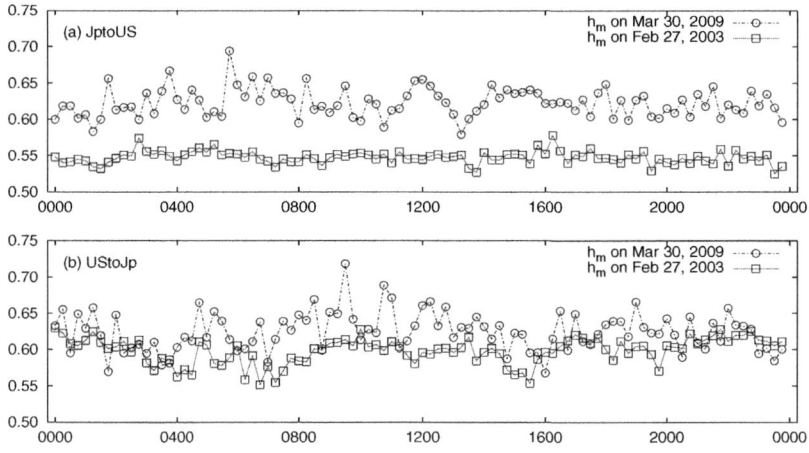

Fig. 4. Results on 24-hour long traces (2003, 2009)

short range correlations. This observation is consistent with that in the original study by Zhang et al [16]. The traces used in their study were collected in 2000–2002, and since then despite many things having changed, the range of the small-time scaling parameter remains unchanged.

A closer look at Fig. 3 reveals that a majority of the post-2007 traces exhibit small to moderate correlations, with corresponding scaling parameter lying between 0.6 and 0.75, whereas a majority of the pre-2007 traces typically have scaling parameters less than 0.6. Our observations for these traces are consistent with those made by Zhang et al. [16] from their traces from 2000-02. We believe this observation provides an evidence against the predictions by Karagiannis et al. [10] that Internet traffic is moving towards simpler to describe models. We further explore this issue in Section 5 wherein dense flows are discussed.

4.3   Analysis of 24-Hour Long Traces

The MAWI repository also makes available some 24-hour long traces. We analyze two such traces, one collected on February 27, 2003 and the other collected on March 30, 2009, to complement our observations on small-time scaling behavior (made using the 15-minute long traces collected at 14:00 hours). As with other MAWI traces, these 24-hour long traces are partitioned into 15-minute long segments. Thus, we have a total of 96 trace segments for each 24-hour trace. We analyzed the small-time scaling behavior for each of these 96, 15-minute, segments.

Fig. 4 compares the median small-time scaling parameter values obtained by application of the sketch-based robust estimation procedure. The results obtained are found to be consistent with observations outlined earlier in Section 4. We find all median values of the scaling parameter to be between 0.5 and 0.75. Moreover, the median values of scaling parameter for recent 2009 trace are typically larger than those for the older 2003 trace. For the recent 2009 trace, the

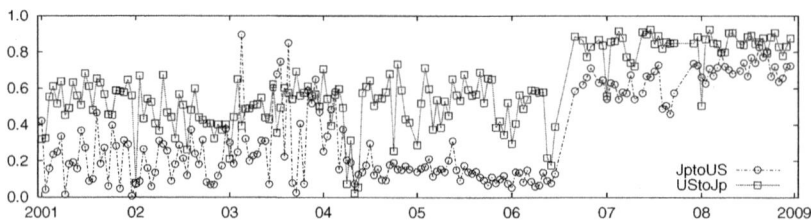

Fig. 5. Fraction of aggregate bytes contributed by dense flows

values of median scaling parameter are found to be more than 0.6 in both directions. The median scaling parameter values for the older 2003 trace are generally less than 0.6. This provides further evidence that recent traces are more likely to display small to moderate correlations in small-time scales as compared to older traces.

5   Longitudinal Analysis of Dense Flows

This section explains some of the observations in the preceding section in terms of dense flows. Zhang et al. [16] defined a flow as *dense* if at least 50% of the packet interarrival times of the flow are below a certain threshold T; otherwise, the flow is called *sparse*. Intuitively, dense flows have bursts of densely clustered packet arrivals. Zhang et al. [16] studied the aggregate of dense and sparse flows which are referred to as the dense and sparse components, respectively. Their study showed that the sparse component has a smaller scaling parameter in small-time scales ($h \leq 0.6$) relative to that of the original trace. Their work further illustrated that flow size alone does not play a significant role in shaping the small-time scaling behavior of network traffic. In this section, we revisit this property in the context of recent traces.

Fig. 5 shows, for the traces considered earlier in Section 4.2, the fraction of traffic carried by dense flows (defined using threshold $T = 2$ ms). Overall, we find that the contribution of dense flows to traffic volume has increased in the recent MAWI traces as compared to earlier traces. This increase in the fraction of aggregate traffic owing to dense flows is one factor that is likely responsible for the increase in the scaling parameter values we noted earlier in Section 4. A detailed analysis of the influence of dense flows follows using the "semi-experiments" based approach developed by Zhang et al. [16].

The process of artificially modifying the packet arrival process is referred to as a semi-experiment in the networking community. A comparison of scaling behavior before and after the semi-experiment leads to conclusions about the importance of the role played by the parameters modified by the semi-experiment. Specifically, for each trace considered, we compute the small-time scaling behavior for the sparse and small components of the trace. The small-time scaling parameter of the sparse component of a trace is computed by analyzing each trace sans the dense flows. Specifically, we removed all dense flows (using threshold

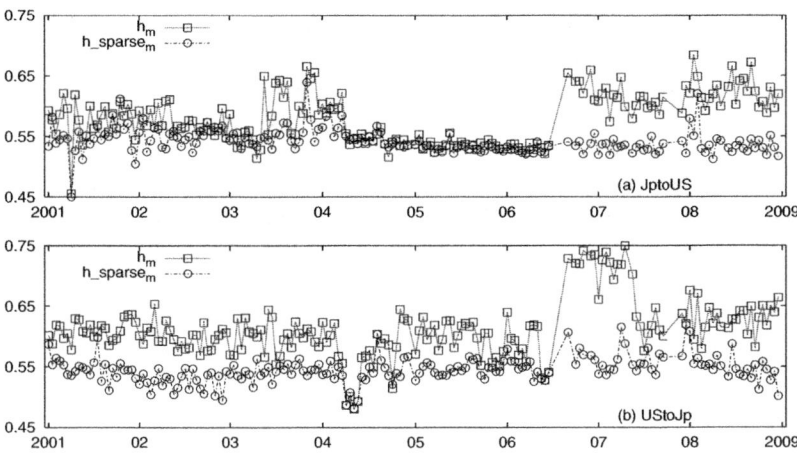

Fig. 6. Scaling behavior of the sparse component

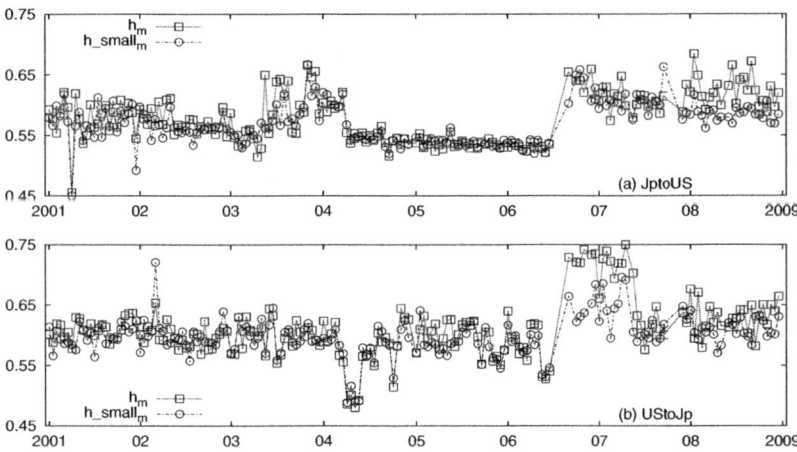

Fig. 7. Scaling behavior of the small component

$T = 2$ ms) from the trace being analyzed and computed the scaling parameter h. To compute the scaling parameter for the small component of a trace, we remove from the trace the largest k flows (in terms of bytes) that contribute as many bytes as the dense flows. The scaling component is estimated using the robust estimation technique discussed earlier in order to weed out the impact of network anomalies.

Fig. 6 shows the median scaling parameter of the original traces and their sparse components. For all the traces exhibiting small to moderate correlations, the removal of dense flows reduces the scaling parameter. The scaling parameter for the sparse component is consistently close to 0.55 for all the traces. For the 2004-2006 JptoUS traces, the aggregate scaling parameter is close to 0.55, and hence no further reduction in scaling parameter is observed for the sparse component. For the 2001-2004 JptoUS traces, the scaling parameter is close to 0.6, and thus only a small decrease in the scaling parameter is observed for the sparse component. The 2006-2009 traces contain small to moderate correlations and in these a clear decrease in the scaling parameter is observed.

Fig. 7 compares the scaling parameters of the original traces and their corresponding small components. We find that, except for a few traces, the scaling parameters for the small component are close to the original traces even though the small component is created by removing large flows that together contribute as many bytes as the dense component with T=2ms. Overall, our analysis reinforces the observation by Zhang et al. [16] that dense flows are one key factor to consider when studying small-time scaling behavior of network traces.

6   Conclusions

This paper presented a longitudinal study of small-time scaling behavior of Internet traffic, using traces spanning a period of eight years obtained from the MAWI trace repository. We believe this study has served multiple purposes. First, this study has re-emphasized the need for robust analysis techniques when studying scaling behavior of network traffic. Second, our study complements previous works on small-time scaling behavior [7, 16]. Specifically, we showed that small-time scaling behavior and many associated properties identified in the literature have more or less remained invariant throughout this decade. Our work is based on traces collected at one trans-pacific link. A longitudinal analysis of traces from other links and a subsequent comparison with our results will aid in understanding the evolution of small-time scaling behavior of Internet traffic.

Acknowledgements

We thank Brighten Godfrey, our shepherd, for his detailed feedback and suggestions. The authors are also grateful to the anonymous reviewers for their constructive suggestions, which helped improve the presentation of the paper. Financial support for this research was provided by NICTA. NICTA is funded by the Australian Government as represented by the Department of Broadband, Communications and the Digital Economy and the Australian Research Council through the ICT Center of Excellence program.

References

1. MAWI working group traffic archive, http://tracer.csl.sony.co.jp/mawi
2. Basher, N., Mahanti, A., Mahanti, A., Williamson, C., Arlitt, M.: A comparative analysis of Web and Peer-to-Peer traffic. In: WWW (2008)
3. Borgnat, P., Dawaele, G., Fukuda, K., Abry, P., Cho, K.: Seven years and one day: Sketching the evolution of Internet traffic. In: IEEE INFOCOM (2009)

4. Dawaele, G., Fukuda, K., Borgnat, P., Abry, P., Cho, K.: Extracting hidden anomalies using sketch and non-gaussian multiresolution statistical detection procedure. In: SIGCOMM LSAD (2007)
5. Feldmann, A., Gilbert, A.C., Willinger, W.: Data networks as cascades: Investigating the multifractal nature of Internet WAN traffic. In: SIGCOMM (1998)
6. Gupta, H., Mahanti, A., Ribeiro, V.: Revisiting coexistence of Poissonity and Self-similarity in Internet traffic. In: MASCOTS (2009)
7. Hohn, N., Veitch, D., Abry, P.: Does fractal scaling at the IP level depend on TCP flow arrival processes? In: IMW (2002)
8. Jiang, H., Dovrolis, C.: Source-level IP packet bursts: Causes and effects. In: IMC (2003)
9. Jiang, H., Dovrolis, C.: Why is the Internet traffic bursty in short time scales? In: SIGMETRICS (2005)
10. Karagiannis, T., Molle, M., Faloutsos, M., Broido, A.: A non-stationary Poisson view of Internet traffic. In: IEEE INFOCOM (2004)
11. Leland, W.E., Taqqu, M., Willinger, W., Wilson, D.: On the self similar nature of Ethernet traffic. IEEE/ACM Transaction on Networking 2(1) (1994)
12. Paxson, V., Floyd, S.: Wide area traffic: the failure of Poisson modelling. IEEE/ACM Transaction on Networking 3(3) (1995)
13. Veitch, D.: Code for estimation of scaling exponents, http://www.cubinlab.ee.mu.oz.au/~darryl
14. Veitch, D., Abry, P.: A Wavelet-based joint estimator of the parameters of long-range dependence. IEEE Transactions on Information Theory 45(3) (1999)
15. Veitch, D., Abry, P.: A statistical test for time constancy of scaling exponents. IEEE Transaction on Signal Processing 49(10) (2001)
16. Zhang, Z.L., Ribeiro, V.J., Moon, S., Diot, C.: Small-time scaling behaviors of Internet backbone traffic: An empirical study. In: IEEE INFOCOM (2003)

Measurement Study of Multi-party Video Conferencing

Yue Lu, Yong Zhao, Fernando Kuipers, and Piet Van Mieghem

Delft University of Technology, P.O. Box 5031, 2600 GA Delft, The Netherlands
{Y.Lu,F.A.Kuipers,P.F.A.VanMieghem}@tudelft.nl

Abstract. More and more free multi-party video conferencing applications are readily available over the Internet and both Server-to-Client (S/C) or Peer-to-Peer (P2P) technologies are used. Investigating their mechanisms, analyzing their system performance, and measuring their quality are important objectives for researchers, developers and end users. In this paper, we take four representative video conferencing applications and reveal their characteristics and different aspects of Quality of Experience. Based on our observations and analysis, we recommend to incorporate the following aspects when designing video conferencing applications: 1) Traffic load control/balancing algorithms to better use the limited bandwidth resources and to have a stable conversation; 2) Use traffic shaping policy or adaptively re-encode streams in real time to limit the overall traffic.

This work is, to our knowledge, the first measurement work to study and compare mechanisms and performance of existing *free multi-party* video conferencing systems.

1   Introduction

The demand for video conferencing (VC) via the Internet is growing fast. VC services are provided in two different ways: (1) either utilizing a high-quality VC room system with professional equipment and dedicated bandwidth or (2) implementing a VC application on personal computers. The first category can guarantee quality, but it is costly and limited to a fixed location, while the second category is often free of charge and easy to install and use, although the quality cannot be guaranteed.

In this paper, we focus on studying *free* applications that provide multi-party (≥ 3 users) VC on the Internet, and focus on the following questions:

- *How do multi-party VC applications work?*
- *How much resources do they need?*
- *What is the Quality of Experience (QoE)?*
- *What is the bottleneck in providing multi-party VC over the Internet?*
- *Which technology and architecture offer the best QoE?*

M. Crovella et al. (Eds.): NETWORKING 2010, LNCS 6091, pp. 96–108, 2010.

In order to answer these questions we have surveyed existing popular VC applications and among them chose and measured four representative applications to investigate, namely *Mebeam*[1], *Qnext*[2], *Vsee*[3], and *Nefsis*[4].

The remainder of this paper is organized as follows. Section 2 presents related work. In Section 3, eighteen popular VC applications will be introduced and classified. Section 4 describes our QoE measurement scenario. Sections 5 and 6 will show the measurement results obtained. Finally, we conclude in Section 7.

2 Related Work

Most research focuses on designing the network architectures, mechanisms and streaming technologies for VC. In this section we only discuss the work on studying and comparing the mechanisms and performance of streaming applications.

Skype supports multi-party audio conferencing and 2-party video chat. Baset and Schulzrinne [1] analyzed key Skype functions such as login, call establishment, media transfer and audio conferencing and showed that Skype uses a centralized P2P network to support audio conferencing service. Cicco *et al.* [2] measured Skype video responsiveness to bandwidth variations. Their results indicated that Skype video calls require a minimum of 40 kbps available bandwidth to start and are able to use as much as 450 kbps. A video flow is made elastic through congestion control and an adaptive codec within that bandwidth interval.

Microsoft Office Live Meeting (Professional User License) uses a S/C architecture and has the ability to schedule and manage meetings with up to 1,250 participants. However, only few participants can be presenters who can upload their videos and the others are non-active attendees.

Spiers and Ventura [3] implemented IP multimedia subsystem (IMS)-based VC systems with two different architectures, S/C and P2P, and measured their signaling and data traffic overhead. The results show that S/C offers better network control together with a reduction in signaling and media overhead, whereas P2P allows flexibility, but at the expense of higher overhead.

Silver [4] discussed that applications built on top of web browsers dominate the world of Internet applications today, but are fundamentally flawed. The problems listed include delays and discontinuities, confusion and errors, clumsy interfacing and limited funtionality.

Trueb and Lammers [5] analyzed the codec performance and security in VC. They tested High Definition (HD) VC and Standard Definition (SD) VC traffic characteristics and their corresponding video quality. In their results, HD provides a better video quality at good and acceptable network conditions, while in poor network conditions HD and SD have similar performance.

Few articles compare the different types of existing free multi-party VC systems or measure their QoE. In this paper, our aim is to provide such a comparison.

---

[1] http://www.mebeam.com/

[2] http://www.qnext.com/

[3] http://www.vsee.com/

[4] http://www.nefsis.com/leads/free-trial.aspx

3   Survey

We have considered eighteen VC applications, for which we list the maximum frame rate they can support (the best video quality they can provide), the maximum number of simultaneous conference participants, and the category (S/C or P2P) they belong to in Table 1.

Table 1. Popular video conferencing applications

	Max. frame rate (frames/second)	Max. # of simultaneous video participants	S/C or P2P
Eedo WebClass		6	web-based S/C
IOMeeting	30	10	web-based S/C
EarthLink	30	24	S/C
VideoLive	30	6	web-based S/C
Himeeting	17	20	S/C
VidSoft	30	10	S/C
MegaMeeting	30	16	web-based S/C
Smartmeeting	15	4	S/C
Webconference	15	10	web-based S/C
Mebeam		16	web-based S/C
Confest	30	15	S/C
CloudMeeting	30	6	S/C
Linktivity WebDemo	30	6	web-based S/C
WebEx	30	6	web-based S/C
Nefsis Free Trial	30	10	S/C
Lava-Lava	15	5	decentralized P2P
Qnext		4	centralized P2P
Vsee	30	8	decentralized P2P

Even though there exist many free VC applications, many of them turn out to be instable once installed. From Table 1, we observe that the maximum frame rate is 30 frames/s which corresponds to regular TV quality. All applications support only a very limited number of participants and the applications that support more than 10 simultaneous participants all use a centralized S/C network structure.

Many other popular online chatting applications (like Skype, MSN, Yahoo messenger, Google talk, etc.) only support multi-party audio conference and 2-party video conference, and therefore are not considered here.

4   Experiments Set-Up

We have chosen four representative applications to study:

- *Mebeam*: web-browser based S/C with a single server center.
- *Qnext (version 4.0.0.46)*: centralized P2P. The node which hosts the meeting is the super node.

- *Vsee (version 9.0.0.612)*: decentralized full-mesh P2P.
- *Nefsis (free trial version)*: S/C, network of distributed computers as servers.

We have chosen these four applications because they each represent one of the four architectures under which all eighteen applications in Table 1 can be classified.

We have performed two types of experiments: (1) local lab experiments, composed of standard personal computers participating in a local video conference, in order to investigate the login and call establishment process, as well as the protocol and packet distribution of the four VC applications; (2) global experiments, to learn more about the network topology, traffic load and QoE, when a more realistic international video conference is carried out.

The global measurements were conducted during weekdays of May, 2009, under similar and stable conditions[5]:

- Client 1: 145.94.40.113; TUDelft, the Netherlands; 10/100 FastEthernet;
  Client 2: 131.180.41.29; Delft, the Netherlands; 10/100 FastEthernet;
  Client 3: 159.226.43.49; Beijing, China; 10/100 FastEthernet;
  Client 4: 124.228.71.177; Hengyang, China; ADSL 1Mbit/s.
- Client 1 always launches the video conference (as the host);
- Clients 1, 3 and 4 are behind a NAT.

To retrieve results, we used the following applications at each participant:

- *Jperf* to monitor the end-to-end available bandwidth during the whole process of each experiment. We observed that usually the network is quite stable and that the available end-to-end bandwidth is large enough for different applications and different participants.
- *e2eSoftVcam* to stream a stored video via a virtual camera at each VC participant. Each virtual camera is broadcasting in a loop a "News" video (.avi file) with a bit rate of 910 Kbit/s, frame rate of 25 frames/s and size 480x270;
- *Camtasia Studio 6.* Because all applications use an embedded media player to display the Webcamera streaming content, we have to use a screen recorder to capture the streaming content. The best software available to us was *Camtasia*, which could only capture at 10 frames/s. In order to have a fair comparison of the original video to the received video, we captured not only the streaming videos from all participants, but also the original streaming video from the local virtual camera[6].
- *Wireshark* to collect the total traffic at each participant.

---

[5] We have repeated the measurements in July, 2009 and obtained similar results to those obtained in May 2009.

[6] We assess the video quality using the full reference batch video quality metric (bVQM) which computes the quality difference of two videos. Capturing at 10 frames/s a video with frame rate of 25 frames/s may lead to a different averaged bVQM score. However, because the video used has a stable content (there are only small changes in the person profile and background), we do not expect a large deviation in bVQM score with that of the 25 frames/s video. The results are accurate for 10 frames/s videos.

5   Measurement Results

5.1   Login and Call Establishment Process

Mebeam: We open the *Mebeam* official website to build a web video-chat room and all participants enter the room. The traces collected with *Wireshark* revealed that two computers located in the US with IP addresses 66.63.191.202 (Login Server) and 66.63.191.211 (Conference Server) are the servers of *Mebeam*. Each client first sends a request to the login server, and after getting a response sets up a connection with the single conferencing server center. When the conference host leaves from the conference room, the meeting can still continue. *Mebeam* uses TCP to transfer the signals, and RTMP[7] to transfer video and audio data.

Qnext: The data captured by *Wireshark* reveals two login severs located in the US. Each client first sends packets to the login servers to join the network. After getting a response, they use SSLv3 to set up a connection with the login servers. In the call establishment process, each client communicates encrypted handshake messages with 3 signaling servers located in the US and Romania and then uses SSLv3 to set up a connection between the client and the signaling server. When client A invites another client B to have a video conference and client B accepts A's request, they use UDP to transfer media data between each other. In a conference, there is only one host and other clients can only communicate with the host. The host is the super node in the network. When the host leaves the meeting, the meeting will end. If another node leaves, the meeting will not be affected. *Qnext* uses TCP for signaling and UDP for video communication among participants.

Vsee: Each client uses UDP and TCP to communicate with the web servers in login process. In the call establishment process, after receiving the invitation of the host, each client uses[8] T.38 to communicate with each other. *Vsee* has many web servers: during our experiment, one in the Netherlands, one in Canada, and 7 located in the US. *Vsee* has a full-meshed P2P topology for video delivery. However, only the host can invite other clients to participant in the conference. When the host leaves the meeting, the meeting cannot continue. Other peers can leave without disrupting the meeting. *Vsee* is a video-conferencing and real-time collaboration service. The communication among users is usually of the P2P type using UDP, with automatic tunneling through a relay if a direct connection is not available.

Nefsis: In the login process, the clients first use TCP and HTTP to connect to the Virtual Conference Servers (with IP addresses 128.121.149.212 in the US and 118.100.76.89 in Malaysia) and receive information about 5 other access points from the Virtual Conference Servers. These 5 access points are also the data server centers owned by *Nefsis*, and they are located in the Netherlands (Rotterdam and Amsterdam), in the UK, India, Australia, and Singapore. Afterwards,

---

[7] Real-Time Messaging Protocol (RTMP) is a protocol for streaming audio, video and data over the Internet, between a Flash player and a server.

[8] T.38 is an ITU recommendation for fax transmission over IP networks in real-time.

the clients choose some access points to set up connections via TCP. After entering the conference room, each client communicates with each other through the access point when firewalls/NAT are present at clients, otherwise clients can set-up an end-to-end connection to communicate with each other directly. *Nefsis* uses TCP for signaling and delivering streaming data.

5.2   Packet Size Distribution and Traffic Load

To differentiate between non-data packets, video and audio packets, we performed three local experiments for each application. The first experiment uses two computers with cameras and microphones to have a video conference. In the second experiment, two computers are only equipped with microphones, but without cameras (no video packets will be received). In the third experiment, two computers set-up a connection, both without microphones and cameras (so only non-data packets will be exchanged).

Based on *Wireshark* traces, we could distill for each VC application the packet size range as shown in Table 2:

Table 2. The packet size distribution of *Mebeam*, *Qnext*, *Vsee* and *Nefsis*

Packet size	*Mebeam*	*Qnext*	*Vsee*	*Nefsis*
Audio packet	> 50 bytes	72 bytes	100 ~ 200 bytes	100 ~ 200 bytes
Video packet	> 200 bytes	50 ~ 1100 bytes	500 ~ 600 bytes	1000 ~ 1600 bytes
Signaling packet	50 ~ 200 bytes	50 ~ 400 bytes	50 ~ 100 bytes	50 ~ 100 bytes

Other interesting observations are: 1) If the person profile or background images in the camera change/move acutely, a traffic peak is observed in our traces. 2) The traffic does not necessarily increase as more users join the conference. Fig. 1 shows the change of the average traffic load at each user when a new participant joins the conference[9]. The decreasing slope after 3 users indicates that *Mebeam*, *Qnext* and *Vsee* either re-encoded the videos or used traffic shaping in order to reduce/control the overall traffic load in the system. We can see from Fig. 1 that only the traffic load at *Nefsis* clients does not decrease when the number of video conferencing participants reaches to 4. Therefore, we introduced more participants into the video conference for *Nefsis*, and we found that the traffic at each *Nefsis* user starts to decrease at 5 participants. Hence, we believe that in order to support more simultaneous conference participants, the overall traffic has to be controlled.

Fig. 1 illustrates that, compared with the traffic generated by *Nefsis* which uses the same coding technology and the same frame rate on the same video, *Qnext* and *Vsee* generate most traffic, especially the host client of *Qnext*. This is because *Qnext* and *Vsee* use P2P architectures where the signaling traffic overhead is much more than the traffic generated by a S/C network with the same number of participants. The host client (super node) of *Qnext* generates 3 times more traffic than

---

[9] We captured the packets after the meeting was set up and became stable.

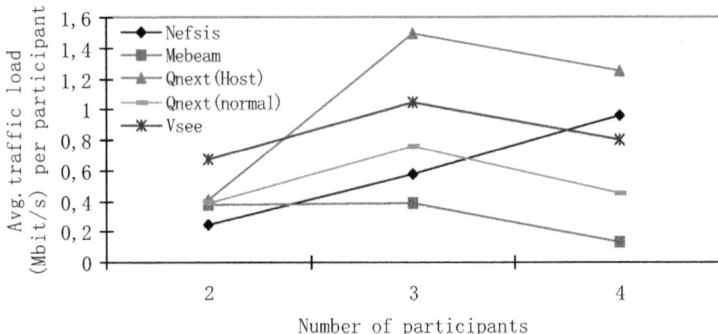

Fig. 1. The average traffic load at an end-user when the number of conference participants increases from 2 to 4 (*Qnext* is limited to 4 participants)

other normal clients. Hence, for this architecture, a super-node selection policy is recommended to choose a suitable peer (with more resources, for example) as the super node.

Fig. 1 also shows that *Mebeam* generates least traffic. Considering that the overall traffic load, which can be supported in a VC network, has an upperbound due to the limited users' bandwidth, and each *Mebeam* client generates much less traffic than the three other applications, it clarifies why *Mebeam* can support 16 simultaneous video users while *Nefsis* can only support 10 users, *Vsee* can support 8 users and *Qnext* can support 4 users.

5.3   Quality of Experience (QoE)

QoE can be measured through objective and subjective measurements. In this section, we assess via global measurements the QoE at the end user with respect to their video quality, audio-video synchronization, and level of interaction.

Video Quality.   In the objective measurements, we use bVQM (Batch Video Quality Metric) to analyze the VC's video quality off-line. bVQM takes the original video and the received video and produces quality scores that reflect the predicted fidelity of the impaired video with reference to its undistorted counterpart. The sampled video needs to be calibrated. The calibration consists of estimating and correcting the spatial and temporal shift of the processed video sequence with respect to the original video sequence. The final score is computed using a linear combination of parameters that describe perceptual changes in video quality by comparing features extracted from the processed video with those extracted from the original video. The bVQM score scales from 0 to approximately[10] 1. The smaller the score, the better the video quality.

---

[10] According to [7], bVQM scores may occasionally exceed 1 for video scenes that are extremely distorted.

We captured at every participant the stream from the imbedded multimedia player of each VC application with *Camtasia Studio 6*, and used[11] *VirtualDub* to cut and synchronize the frames of the compared videos.

Table 3 provides the bVQM scores for VC service per participant.

Table 3. The video quality of *Mebeam*, *Qnext*, *Vsee* and *Nefsis* at 4 clients

VQM score	Client 1	Client 2	Client 3	Client 4	Average
Mebeam (Flash video, MPEG-4)	0.63	0.41	0.94	0.86	0.71
Qnext (MPEG-4, H.263, H.261)	1.05	0.94	0.63	0.83	0.86
Vsee	0.78	0.82	0.80	0.79	0.80
Nefsis (MPEG-4, H.263, H.263+)	0.34	0.61	0.61	0.87	0.61

Table 3 indicates that *Nefsis* features the best video quality among the 4 applications, although with an average bVQM score of 0.61 (its quality is only "fair", which will be explained later with the subjective measurements). The highest bVQM score (the worst video quality) appears at Client 1 (the super node) of *Qnext*. Generally speaking, all four VC applications do not provide good quality[12].

Because no standard has been set for what level of bVQM score corresponds to what level of perceived quality of a VC service, we have also conducted subjective measurements. We use the average Mean Opinion Score (MOS) [8], a measure for user perceived quality, defined on a five-point scale[13]: 5 = *excellent*, 4 = *good*, 3 = *fair*, 2 = *poor*, 1 = *bad*.

We gave 7 different quality videos generated by VC applications to 24 persons who gave a subjective MOS score independently. We also objectively computed their bVQM scores. Fig. 2 shows the correlation between the objective bVQM scores and the subjective MOS values.

We mapped between the bVQM scores and the average MOS scores over 24 persons, and found that they have a linear correlation in the range $0.3 < $bVQM score$\leq 1$. Hence, the VC's video quality is predictable when using the objective metric bVQM.

Compared with the video quality of a global P2PTV distribution service, which has an average MOS value of 4 [9], the video quality of a global VC service is poor (with an average bVQM score of 0.74 and MOS value of around 2.2), because the VC service requires end users to encode and upload their streams in real-time.

---

[11] *VirtualDub* is a video capture and video processing utility for Microsoft Windows.

[12] We also objectively measured the audio quality using metric PESQ-LQ (Perceptual Evaluation of Speech Quality-Listening Quality) [6] [8] and found that the PESQ-LQ average score (scale from 1.0 to 4.5, where 4.5 represents an excellent audio quality) is 2.24, 2.68, 3.08 and 3.15 for *Mebeam*, *Qnext*, *Vsee*, and *Nefsis*, respectively.

[13] The threshold for acceptable TV quality corresponds to the MOS value 3.5.

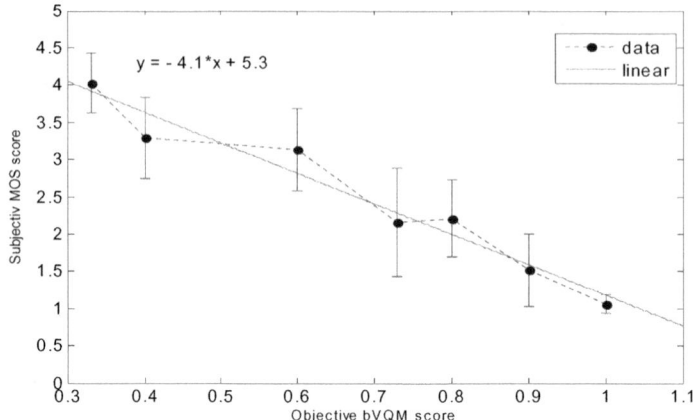

Fig. 2. Relation between bVQM and MOS for video conferencing service

Even the local uploaded video has a largely degraded quality although it is still the best among all participants.

Audio-Video Synchronization. The relative timing of sound and image portions of a streaming content may not be synchronized.

ITU [10] [11] has suggested that the viewer detection thresholds of audio-video lag are about +45 ms to −125 ms, and the acceptance thresholds are about +90 ms to −185 ms, for video broadcasting.

To analyze the A/V synchronization provided by each VC application, we used an "artificially generated" video test sample, in which the video and audio waveforms are temporally synchronized with markers. Similar to the experiments of testing the video quality, we captured at each end user the videos from all other participants. When the audio and video tracks were extracted and compared offline, there was an average difference in time between the two tracks of about 650 ms for *Mebeam*, 470 ms for *Qnext*, 400 ms for *Vsee* and 350 ms for *Nefsis*. Such large audio-video lags are mainly caused by a large amount of frame losses, which lead to the low video quality mentioned already in Section 5.3.

Interactivity (communication delay). During a video conference it is annoying to have large communication delay[14]. Large communication delay implies lack of real-time interactivity in our global multi-party VC experiments. We measured the video delays among participants by injecting in the network another artificial video that mainly reproduced a timer with millisecond granularity.

In the video conference, this artificial "timer" video was uploaded via the virtual camera and transmitted among the participants via the different VC applications.

---

[14] In IP video conferencing scenarios, the maximum communication delay recommended by ITU is 400 ms [12].

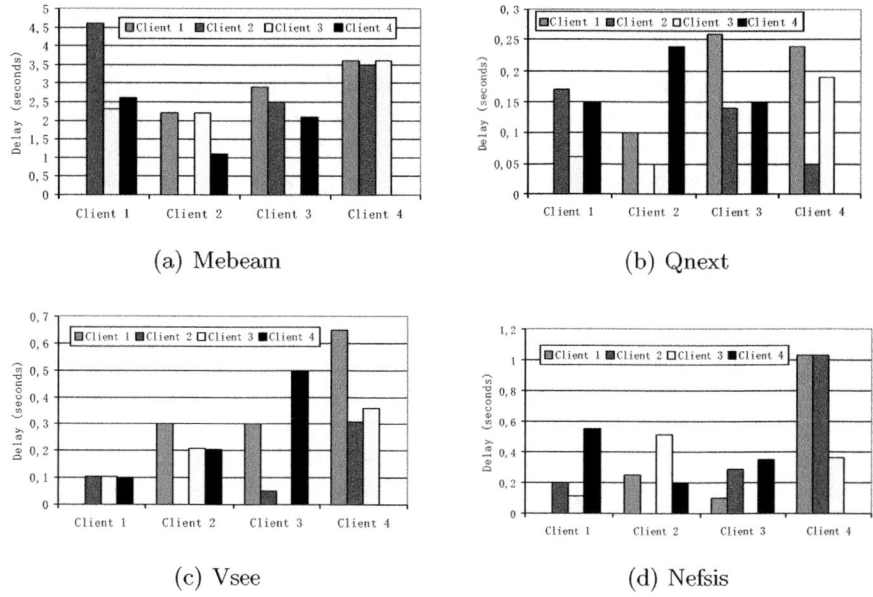

Fig. 3. The video delay between different participants

At each participant, we used a standard universal Internet time as reference[15]. We displayed the "timer" videos of all participants in real time. After a 1-minute long stable video conference, we cut the captured content at each participant with *VirtualDub* to compare the "timers" between any 2 participants. For each application, we took samples at 2 different times to calculate an average delay.

The video delays among participants are shown in Fig. 3. The x axis shows the 4 different clients. The y axis shows the video transmission delay from the participant on the x axis to the participant shown in the legend.

Fig. 3 shows that *Qnext* provides a video that is most synchronized among the clients. *Qnext*, *Vsee*, and *Nefsis* have a comparable level of average video delay, respectively 0.15 s, 0.27 s, and 0.41 s. However, *Mebeam* clients suffer a huge video delay (2.77 s on average), because the processing time at the server is too long.

We also measured the audio delays among participants by injecting in the video an artificial DTMF (Dual-tone multi-frequency) tone. We sent and recorded the audio at Client 1. Other participants kept their speaker and microphone on, but did not produce extra audio. Based on the recorded audio tracks, we compared the time the audio marker was sent from Client 1 and the time the same audio marker was heard again at Client 1 after the transmitted audio was played, recorded, and retransmitted by a client. The time difference is approximately twice the one-way audio delay plus the processing delay at a client. Our results revealed 1 s, 1.4 s,

---

[15] http://www.time.gov/timezone.cgi?Eastern/d/-5/java

0.2 s and almost 0 s on average for *Mebeam*, *Qnext*, *Vsee* and *Nefsis* respectively. *Qnext* in this case provided the least synchronized audio among the users. When we measure the audio delay and the A/V synchronization, the delay is the end-to-end delay including the transmission delay and the delay introduced by the application. In our experiment, the video delay represents the delay of a same video scene that was captured at the application interfaces of the sender and the receiver, which does not include the time used for uploading the video to the sender via applications. Hence, considering the audio delay, video delay, and the A/V synchronization discussed in Section 5.3, we can conclude that the delay introduced by the application, when uploading, is large for *Qnext*.

6   Worst-Case Study

In another set of global experiments in June, 2009, our *Jperf* plots indicated that the end-to-end connections of clients 3 and 4 with the host were very unstable. We found that the two participants in China always passively disconnected from the conference or could not even log into *Mebeam*, *Nefsis* and *Qnext*. *Vsee* could still work, but the quality was awful, with bVQM scores close to 1.

In order to investigate the minimum bandwidth to support a video conference, we repeated many experiments adjusting the upload rate upper-bound (using[16] *Netlimiter*) at each participant for a particular VC application to test the user's upload bandwidth minimally required to launch a video conference.

For *Qnext*, the threshold is 50 Kbit/s. If an end user's available upload bandwidth is < 50 Kbit/s, (s)he cannot launch *Qnext*. For *Vsee*, the threshold is 50 Kbit/s; for *Nefsis* it is 28 Kbit/s; and for *Mebeam* it is 5 Kbit/s, which we believe are the minimally supported streaming bit rates set by the applications.

7   Summary and Conclusions

Through a series of local and global experiments with four representative video conferencing systems, we examined their behavior and mechanisms, and investigated their login process, the call establishment process, the packet size distribution, transfer protocols, traffic load, delivery topology, and different aspects of Quality of Experience.

Our main conclusions from the measurement results on the traffic characteristics of four different video conferencing systems are: (1) The QoE of multi-party video conferencing is very sensitive to bandwidth fluctuations, especially in the uplink. Hence, an adaptive bit rate/frame rate policy should be deployed; (2) When the number of participants increases, the traffic load at each participant does not always increase correspondingly (see Fig. 1), suggesting that re-encoding at the video or a traffic shaping policy take place to control the overall traffic in the system.

---

[16] *NetLimiter* is an Internet traffic control and monitoring tool designed for setting download/upload transfer rate limits for applications.

Our QoE measurement results are summarized as: (1) Compared with the non real-time multimedia services (i.e. P2PTV), existing Internet video conferencing applications in general cannot provide very good quality to their end users (poor video and audio quality, large audio-video lag, and long communication delay in some cases); (2) Only a limited number of multimedia participants are supported and rare high definition webcamera streaming is supported due to the limited available bandwidth or the limited processing capability; (3) The existing systems are not reliable in the worst cases. When the network is unstable or the available upload bandwidth is very limited (thresholds have been found), none of the applications work properly.

It seems that the Server-to-Client architecture with many servers located all over the world is currently the best architecture for providing video conferencing via the Internet, because it introduces the least congestion at both servers and clients. Load balancing and load control algorithms help the overall performance of the system and the codec used is important for the quality that end users perceive. The bottleneck to support video conferencing with more participants and high definition streams is the overhead traffic generated by them. To support more simultaneous participants in a single conferencing session, the traffic load has to be controlled/limited by using traffic shaping policy or re-encoding the video streams.

We have chosen four representative video conferencing systems for our study, but the measurement methodologies mentioned in this paper can also be applied to other video conferencing applications, which could be compared with our study in the future.

Acknowledgements

We would like to thank Rob Kooij for a fruitful discussion on measuring audio delay. This work has been supported by TRANS (http://www.trans-research.nl).

References

1. Baset, S.A., Schulzrinne, H.: An Analysis of the Skype Peer-to-Peer Internet Telephony Protocol. In: INFOCOM '06, Barcelona, Spain (April 2006)
2. De Cicco, L., Mascolo, S., Palmisano, V.: Skype Video Responsiveness to Bandwidth Variations. In: NOSSDAV '08, Braunschweig, Germany (May 2008)
3. Spiers, R., Ventura, N.: An Evaluation of Architectures for IMS Based Video Conferencing, Technical Report of University of Cape Town (2008)
4. Silver, M.S.: Browser-based applications: popular but flawed? Information Systems and E-Business Management 4(4) (October 2006)
5. Trueb, G., Lammers, S., Calyam, P.: High Definition Videoconferencing: Codec Performance, Security, and Collaboration Tools, REU Report, Ohio Supercomputer Center, USA (2007)
6. Rix, A.W.: A new PESQ-LQ scale to assist comparison between P.862 PESQ score and subjective MOS, ITU-T SG12 COM12-D86 (May 2003)
7. Pinson, M.H., Wolf, S.: A New Standardized Method for Objectively Measuring Video Quality. IEEE Transactions on Broadcasting 50(3), 312–322 (2004)

8. ITU-T Rec. P.800, Methods for Subjective Determination of Transmission Quality (1996)
9. Lu, Y., Fallica, B., Kuipers, F., Kooij, R., Van Mieghem, P.: Assessing the Quality of Experience of SopCast. International Journal of Internet Protocol Technology 4(1), 11–23 (2009)
10. ITU BT.1359-1, Relative timing of sound and vision for broadcasting (1998)
11. Lias, J.L.: HDMI's Lip Sync and audio-video synchronization for broadcast and home video, Simplay Labs, LLC (August 2008)
12. Bartoli, I., Iacovoni, G., Ubaldi, F.: A synchronization control scheme for Videoconferencing services. Journal of multimedia 2(4) (August 2007)

Passive Online RTT Estimation
for Flow-Aware Routers Using One-Way Traffic*

Damiano Carra[1,**], Konstantin Avrachenkov[2], Sara Alouf[2],
Alberto Blanc[2], Philippe Nain[2], and Georg Post[3]

[1] University of Verona, Verona, Italy
damiano.carra@univr.it
[2] INRIA, Sophia Antipolis, France
first.last@sophia.inria.fr
[3] Alcatel-Lucent Bell Labs, Nozay, France
georg.post@alcatel-lucent.fr

Abstract. With the introduction of new generation high speed routers, and with the help of "flow-aware" traffic management, it becomes possible to improve the Quality of Service for users as well as the network efficiency for ISPs. An example of the "flow-aware" traffic management is the Alcatel-Lucent "Semantic Networking" framework where short-lived flows are processed with high priority and long-lived flows are controlled on a per flow basis. In order to control efficiently the flows, it is useful to know an estimate of the Round Trip Time (RTT). In the present work, we provide an online RTT estimation algorithm which is passive and needs one-way traffic only. The one-way traffic requirement is essential for the application of the algorithm for "flow-aware" traffic management inside the network. To the best of our knowledge, there was no online one-way traffic RTT estimators. Tests on a controlled testbed and on the Internet demonstrate high accuracy of the proposed estimator.

Keywords: Spectral Analysis; Measurement; Quality of service; Evolution of IP network architecture.

1   Introduction

Customers' increasing demand for better quality of service (QoS) and quality of experience (QoE) [1] have increased the interest in techniques for actively managing and controlling the traffic inside the network. With the introduction of the new generation of high speed routers, it becomes possible to treat individually a significant number of flows. An example of "flow-aware" traffic management is Alcatel-Lucent's framework of "Semantic Networking;" cf. [2].

The framework "Semantic Networking" takes advantage of the *mice-elephants* phenomenon. Many measurement studies have observed that the traffic is approximately composed of two types of connections: short-lived and long-lived

---

* This work was done in the framework of the INRIA and Alcatel-Lucent Bell Labs Joint Research Lab on Self Organized Networks.
** Corresponding author.

M. Crovella et al. (Eds.): NETWORKING 2010, LNCS 6091, pp. 109–121, 2010.
© IFIP International Federation for Information Processing

flows, also known as *mice* and *elephants*. It has been observed that the number of flows of each type and the actual traffic they generate can be summarized by the 80-20 rule: mice account for the majority of the flows (80%), but the volume of the traffic associated to them represents 20% of the total traffic, while elephants (20% of the flows) convey 80% of the total traffic. Recent measurements [1] show that this proportion is shifting to a 90-10 rule. Therefore, by serving the short-lived flows with high priority one can significantly decrease the number of active flows. The efficiency of such approach has been validated by [3, 4]. For the long-lived flows, it has been observed that the instantaneous number of elephants present in the router is small, actually only few hundreds [5]. This means that it is possible to manage long-lived flows on *per flow* basis. Such an approach can give to an ISP a greater flexibility in managing its network, potentially increasing its efficiency and providing customers with high QoE [2].

One of the most important notions for a TCP flow is its aggressiveness that is how fast a TCP connection increases its sending rate. Most deployed TCP versions are based on a congestion window whose value is changed every Round Trip Time (RTT). This implies that the RTT is one of the principal parameters that determine the aggressiveness of a TCP flow, and it needs to be taken into account for the design of new "flow-aware" traffic management schemes.

In this paper, we design a method based on spectral analysis for the *online* estimation of the RTT by passively monitoring, in real-time, only one direction of a TCP flow. Specifically, our algorithm satisfies different constraints. The estimation is passive, as the measuring point (e.g., the router) does not inject packets into an existing flow, nor does it alter their flow. The estimation is done in *real-time*, since the flow rate and its rate growth should be instantaneously available at the measuring point. This constraint implies that the used algorithms must be both computationally efficient and working incrementally as new packets arrive. In other words, we need to find efficient *online* algorithms that provide sufficiently accurate results. The last constraint imposes the use of information on only *one direction* of a TCP flow. In fact, even when forward and reverse traffic do flow through the monitoring point, collecting and real-time processing of two-way traffic may impose excessive load and complexity on the network cards. As we will discuss in detail in the ensuing section on related work, none of the existing methods for RTT estimation satisfies simultaneously the above mentioned criteria.

Our contribution is threefold. First, since typical implementations of spectral analysis tools are for offline use, we reformulate one implementation of the *periodogram* to make it suitable for online use. Second, using a pattern matching technique, we develop an algorithm for extracting the fundamental frequency— whose inverse is the estimated RTT—once the spectrum has been estimated. Third, we validate our RTT estimation algorithm by conducting experiments on a controlled testbed and on the Internet during both peak and off-peak hours.

The experimental results show that our solution is able to accurately estimate the RTT, the estimation error being within ±10% (resp. ±20%) of the true value with probability equal to 75% (resp. 99%). The results given by our methodology show a higher accuracy with respect to the results obtained in previous studies (see Sect. 2 for a review of the literature), using less information, namely packets flowing in only one direction.

2   Related Work

The RTT estimation has been the subject of many studies. The aim of these studies is to understand the characteristics of the TCP connections in the Internet, in order to study different aspects, such as the rate-limiting factors or the non-conforming TCP senders. These works consider methods that can be applied offline, since they are generally computationally intensive. For instance, the work of S. Jaiswal et al. in [6] is based on the reconstruction of the TCP congestion window values, which requires to maintain a "replica" of the TCP sender's state. This work was later extended by J. But et al. in [7], maintaining however the computational complexity.

Another example is the work of Y. Zhang et al. who propose in [8] a method based on time correlation of samples: while the computational complexity is similar to our approach, the main problem is related to robustness, since the results are strongly affected by noise, which impacts the accuracy of the RTT estimation.

In [9], R. Lance et al. make use of spectral analysis—the basic mechanism used by our solution—as one of the possible steps for off-line estimation of the RTT. In particular, the authors apply the spectral analysis to a set of samples, but they do not consider the continuous, real-time update of the spectrum as new samples arrive. Moreover, the post processing of the spectrum for the extraction of the RTT is based on a simple evaluation of the frequency with the maximum power, which not always corresponds to the fundamental frequency.

In [10], B. Veal et al. propose a method based on the TCP timestamp option. Our method does not rely on information provided by the protocols, but it is based only on the inter-arrival times of the packets. In [11], Jiang and Dovrolis estimate the RTT using the first packets of a connection, this method is therefore not generally applicable to the continuous monitoring of a connection.

In [12], Y. Qi et al. propose a method for Bayesian spectrum estimation based on Kalman filtering. Nevertheless, this method is not applied to RTT estimation. The main issue is its complexity: the filter gain computation requires a matrix multiplication, whose complexity is approximately $O(N^{2.8})$ (N is the number of samples used); there exist algorithms with $O(N^{2.376})$, but with much higher constants, which makes them practically unusable. Our methodology has a complexity of $O(N)$ at each sample arrival, since $2N$ is the number of frequencies in the spectrum (see Section 5), thus, if we consider N consecutive samples, the complexity becomes $O(N^2)$.

3   Motivations

The estimation of the RTT is a building block for *actively controlling the traffic* inside the network. We focus on a flow-aware networking approach, where routers

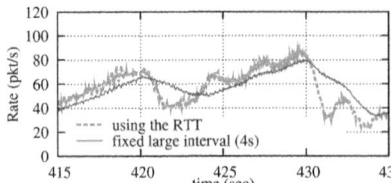 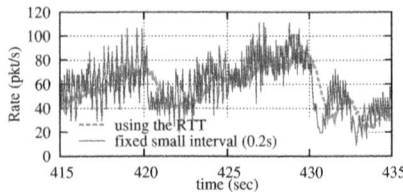

Fig. 1. Rate estimate using the estimated RTT, compared with the same using two fixed averaging intervals: large interval (left) and small interval (right)

are able to manage flows –a connection identified by the classical five-tuple of protocol ID, source and destination addresses and ports– rather than simply packets. For a detailed view of the flow-aware networking approach the interested reader is referred to [2].

It has been observed in [5] that the number of concurrent long-lived flows at a router is of the order of hundreds. This means that it is possible, with the current technology, to individually manage these flows.

Flows for large data transfers, which carry 90% of the total traffic [1], use mainly TCP to compete for bandwidth. More specifically, they use window based congestion control, where each flow increases its sending rate until a packet loss is detected (or an explicit congestion notification is received). Conventional routers that are not flow-aware cannot insure fairness under high traffic loads, especially when the competing flows have different RTTs [13, 14].

With flow-aware routers, there is the possibility of new flow-aware active queue management (AQM) algorithms capable of proactively controlling each flow. Being able to predict the behavior of a flow can be very useful in order to achieve this goal. In the case of a TCP connection, its future behavior (in the absence of a packet drop or congestion signal) is dictated by the increase of the congestion window. The rate of the increase depends on the RTT, as the congestion window is incremented by one or more packets each RTT. Therefore, tracking in real time this parameter is a fundamental step in predicting the evolution of a flow. While the RTT is available at the sender, current TCP versions do not propagate this information to the routers; only experimental protocols like XCP convey this information explicitly.

Furthermore, if the RTT is known, it is possible to better estimate the flow rates: by averaging the number of packets received within an interval equal to RTT it is possible to correctly estimate the TCP throughput. A smaller value of the averaging interval results in a fluctuation of the estimate, while a bigger value does not allow for a prompt detection of rate changes. This is illustrated in Fig. 1, which shows an experiment's rate estimation made with our estimation tool (cf. Section 5), compared with the same using fixed averaging intervals.

If the information about the rate and its growth is made available at the router, it is possible to design novel AQM policies that take into account this information. While the study of such AQM policies is not the focus of this paper, we highlight that the estimation of the RTT represents an important building block for such policies.

4   A Methodology Based on Spectral Analysis

The self-clocking mechanism of TCP introduces *periodic* components into the arrival times of packets, the main one being the RTT. The basic idea of the RTT estimation is to use spectral analysis to extract such periodic components. With spectral analysis, it is possible to build an estimation of the spectrum of a signal starting from a finite sequence of samples. There is a large set of useful methods suitable for different scenarios [15]. In order to choose the best method, the first step in spectral analysis is to identify the characteristics of the *signal* whose spectrum is to be estimated.

In our case, the signal is the packet inter-arrival time of the flow at hand. This signal is sampled at each packet arrival. More precisely, at the arrival of the kth packet of the flow at hand, a new sample of the signal is computed, namely $h_k := t_k - t_{k-1}$, with $k \geq 1$ where we set the time origin at the arrival of the first packet ($t_0 := 0$), and t_k is the arrival time (in seconds) of the kth packet. Since samples are taken at packet arrivals, the sequence h_k is *unevenly* spaced. The signal is said to be *irregularly* sampled (also unevenly sampled or nonuniformly sampled). While the literature on regularly sampled data proposes many efficient methods to estimate the signal's spectrum, the proposed solutions for irregularly sampled data have been found to be impractical [16]: The choice of spectral analysis for irregularly sampled data is essentially limited, for either technical or computational reasons, to the *periodogram*.

In case of irregularly sampled data, the periodogram is computed using the Lomb-Scargle method [17] and it is generally referred to as the Lomb periodogram. Even if the periodogram has some limitations, it remains an efficient solution in many cases [15]. Considering our case, thanks to the characteristics of the signal h_k, the periodogram *does* represent a good estimator (the interested reader is referred to the companion technical report [18] for details).

A problem common to all the methods for spectral analysis (for either regularly or irregularly sampled data, including the periodogram) is that they are based on offline algorithms: they consider a sequence of N samples and they build an estimation of the spectrum. If the signal is composed of more than N samples, it is divided into subsequences of N samples and the methodology is applied to each subsequence separately. This approach is not suitable for the RTT estimation by a flow-aware router, which has to closely follow the variation of the RTT in real-time. It is extremely important to have a spectrum estimation methodology that is able to update the estimate at each packet arrival. The solution should also be computationally efficient. The simplicity of the periodogram helps in designing an *online* version of the method. All the other methods (even the more advanced ones recently proposed for irregular sampled data, see [16]) are based on complex optimized algorithms, and cannot be translated into corresponding online versions.

5   Estimation Process

At each packet arrival a new sample is added to the sequence h_k. The estimation process takes as input the new packet arrival time and updates the current estimation of the RTT. The functional blocks are shown in Fig. 2.

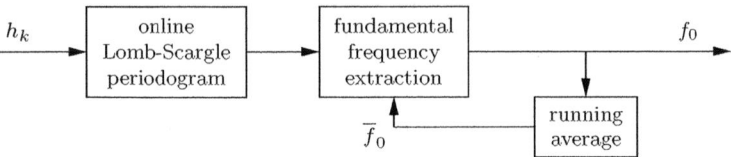

Fig. 2. The RTT estimation process: after the update of the spectrum estimation, the fundamental frequency f_0 is extracted; the RTT estimation is the inverse of f_0

Given a periodic discrete signal, with period T_0 and frequency $f_0 = 1/T_0$, the periodogram of the signal presents peaks at frequencies $f_0, 2f_0, 3f_0, \ldots$, where f_0 is called the *fundamental frequency*. In general, the highest peak in the periodogram is not necessarily at f_0, but it can be at any multiple of f_0; therefore, it is not sufficient to look for the largest peak in the spectrum to find f_0. For this reason, we need another module to extract the fundamental frequency. This module takes as input a list of the spectrum peaks and the average of the previous estimations. The fundamental frequency is extracted with a pattern-matching technique. Basically, after smoothing the periodogram through a low-pass filter, we take the largest W peaks and iteratively search for the frequency in the list that is the least common divisor of the other frequencies in the list.

5.1   Online Lomb Periodogram

The Lomb-Scargle periodogram [17] is built using N samples. Once the initial N samples are collected and the Lomb periodogram is built, every time a new packet arrives, the oldest sample is removed and the new sample is used to update the Lomb periodogram.

The offline version of the Lomb periodogram takes $O(N \log N)$ operations. To the best of our knowledge, no online implementation (i.e., update of the periodogram as the samples arrive) exists, so we had to devise one. For the online computation, we start from the definition of the Lomb periodogram. The power spectrum (Lomb periodogram), at angular frequency $\omega := 2\pi f$, and at the kth sample with $k \geq N$, is

$$P_k^N(\omega) := \frac{1}{2\sigma_k^2} \left\{ \frac{\left[\sum\limits_{j=0}^{N-1} (h_{k-j} - \bar{h}_k) \cos \omega(t_{k-j} - \tau_k) \right]^2}{\sum\limits_{j=0}^{N-1} \cos^2 \omega(t_{k-j} - \tau_k)} + \frac{\left[\sum\limits_{j=0}^{N-1} (h_{k-j} - \bar{h}_k) \sin \omega(t_{k-j} - \tau_k) \right]^2}{\sum\limits_{j=0}^{N-1} \sin^2 \omega(t_{k-j} - \tau_k)} \right\} \quad (1)$$

where $\bar{h}_k$ and σ_k^2 are the mean and variance of the N last samples of h_k:

$$\bar{h}_k := \frac{1}{N} \sum_{i=0}^{N-1} h_{k-i} = \bar{h}_{k-1} + \frac{h_k - h_{k-N}}{N}; \qquad (2)$$

$$\sigma_k^2 := \frac{1}{N-1} \sum_{i=0}^{N-1} h_{k-i}^2 - \frac{N}{N-1} \bar{h}_k^2 = \sigma_{k-1}^2 + \frac{h_k^2 - h_{k-N}^2}{N-1} + \frac{N}{N-1}\left(\bar{h}_{k-1}^2 - \bar{h}_k^2\right), \ (3)$$

and where τ_k is the solution of: $\tan(2\omega\tau_k) = \dfrac{\sum_{i=0}^{N-1} \sin 2\omega t_{k-i}}{\sum_{i=0}^{N-1} \cos 2\omega t_{k-i}}$. The summations in (2) and (3) can be efficiently updated at each packet arrival.

The periodogram (1) is evaluated for a number of frequencies equal to $2N$. In particular, we compute the minimum frequency at arrival of packet k, $f_k^{\min}$, considering the interval of time of the N current samples, i.e., $f_k^{\min} = 1/(t_k - t_{k-N+1})$. The maximum frequency, $f_k^{\max}$, is derived from the Nyquist theorem: since we have N samples, we obtain $f_k^{\max} = \frac{N}{2} f_k^{\min}$. The interval $[f_k^{\min}, f_k^{\max}]$ is divided into $2N$ values, obtaining $\Delta_\omega = 2\pi \frac{f_k^{\max} - f_k^{\min}}{2N}$. Thus, we compute $P_k^N(\omega_i)$ for every $\omega_i = 2\pi f_k^{\min} + i\Delta_\omega, i = 0, \ldots, 2N - 1$.

The power spectrum is to be updated at each packet arrival. Since the values of τ_k and $\bar{h}_k$ change at each packet arrival, the summations in (1) must be always recomputed. This can be avoided by decomposing (1) using basic trigonometric properties. We define

$$\Phi_k := \sum_{j=0}^{N-1} h_{k-j} \cos(\omega t_{k-j}) - \bar{h}_k \sum_{j=0}^{N-1} \cos(\omega t_{k-j}); \qquad \Phi_k^* := \sum_{j=0}^{N-1} \cos(2\omega t_{k-j});$$

$$\Gamma_k := \sum_{j=0}^{N-1} h_{k-j} \sin(\omega t_{k-j}) - \bar{h}_k \sum_{j=0}^{N-1} \sin(\omega t_{k-j}); \qquad \Gamma_k^* := \sum_{j=0}^{N-1} \sin(2\omega t_{k-j}).$$

Note that Φ_k, Γ_k, Φ_k^* and Γ_k^* are simple summations that can be updated at each packet arrival in a similar way as done for (2)-(3). We can then rewrite the Lomb periodogram as follows:

$$P_k^N(\omega) = \frac{1}{\sigma_k^2} \left\{ \frac{[\Phi_k \cos(\omega\tau_k) + \Gamma_k \sin(\omega\tau_k)]^2}{N + \Phi_k^* \cos(2\omega\tau_k) + \Gamma_k^* \sin(2\omega\tau_k)} + \frac{[\Gamma_k \cos(\omega\tau_k) - \Phi_k \sin(\omega\tau_k)]^2}{N - \Phi_k^* \cos(2\omega\tau_k) - \Gamma_k^* \sin(2\omega\tau_k)} \right\}. \ (4)$$

When a new packet k arrives, the values of $\bar{h}_k$, τ_k, σ_k^2, Φ_k, Γ_k, Φ_k^* and Γ_k^* are immediately updated, and the periodogram is recomputed accordingly. The number of operations done at each packet arrival is $O(N)$, since updating $\bar{h}_k$, τ_k, σ_k^2, Φ_k, Γ_k, Φ_k^* and Γ_k^* takes $O(1)$ and we have $2N$ angular frequencies ω.

5.2 Fundamental Frequency Extraction

The outcome of the computation of the Lomb periodogram is usually *noisy*, with many local maxima and a global maximum that not always corresponds to the fundamental frequency f_0: sometimes the global maximum corresponds to a frequency multiple of the fundamental one (see for instance Fig. 5 in [18]). The operations performed *at each packet arrival* by the Fundamental Frequency Extraction module on the updated periodogram can be summarized as follows:

Periodogram smoothing. A basic low-pass FIR filter is applied to the sequence that composes the Lomb Periodogram: in particular, the filter is a moving average filter of order three. We have tested different orders for the filter obtaining similar results, as long as the order is not too high (e.g., greater than 8), since the smoothing effect decreases the dynamic range.

Peak detection. We consider the value at frequency f_k to be a peak if it is greater than the values at frequencies f_{k+1} and f_{k-1}; the detected peaks are put in a "peak list."

Peak pruning and ordering. We remove from the list the entries whose values are not the top W largest peaks, with $W = 10$. The parameter W is configurable: in all our experiments, the number of peaks in the spectrum was between 15 and 25, but only a subset of them were sufficiently strong (one order of magnitude greater than the white noise). By considering the average characteristics of the TCP flows observed in [6], we can conclude that this value can be applied in general. From the peak list, we remove also some frequency values considering the following boundary conditions: the RTT should be greater than 2 ms and less than 500 ms. In [6], it has been observed that 70% of the flows have RTT smaller than 500 ms. The list is then ordered by frequency, smallest first.

Multiple frequency search. Starting from the smallest frequency in the list, we evaluate if the other frequencies in the list can be a multiple of the considered frequency. If we find at least two multiples, this step ends and returns the estimated fundamental frequency f_0. Otherwise it goes on with the following value in the list. If no fundamental frequency is found, this step returns a *NULL* value.

Comparison. We compare the output with the average of the previous estimations, $\overline{f}_0$. If the current estimation is not *NULL*, we consider the ratio between the current estimation and the average of the previous estimations, $r = f_0/\overline{f}_0$. In [6], the authors show that, in 95% of the flows, the ratio between the maximum and the minimum RTT is less than $3/2$. Thus, if $2/3 < r < 3/2$, the algorithm returns f_0 and terminates. Otherwise, it returns $\overline{f}_0$.

The output of this module is then the estimated fundamental frequency whose inverse is the estimated RTT.

6   Numerical Results

We validated our methodology with experiments both in a controlled environment and over the Internet. We focus on a single long-lived *scp* transfer between a source and a destination and we collect traces with *tcpdump* at two points: (i) at the source, where we capture the traffic in both directions, in order to have the packets and their acknowledgments; (ii) at a measuring point between the source and the destination, where we record the packets in one direction. In addition to the observed long-lived *scp* transfer, there are different flows on the path between the source and the destination: In the controlled testbed, the cross-traffic flows are generated by *scp* transfers.

The traces collected at the source are post-processed computing the instantaneous RTT and the smoothed RTT using an exponentially weighted moving average (EWMA) algorithm whose parameter α is set to $1/8$ (see RFC 2988). Note that the RTT is updated for every packet without considering retransmitted packets.

The traces collected at the measuring point are analyzed using our solution. For the spectral analysis, the length of the sequence considered is $N = 256$. The spectral analysis, along with the fundamental frequency extraction, gives the estimation of the instantaneous RTT. Similarly to TCP, we compute a smoothed estimated RTT through an EWMA algorithm with parameter $\alpha = 1/8$. Hereinafter, we will use the general term "RTT" to refer to the smoothed value of the RTT (at the source and estimated).

The file size used for the testbed and for the Internet experiments is 120 MBytes. From the samples of the RTTs we create the empirical Cumulative Distribution Function (CDF). From this empirical CDF we derive any performance index of interest, such as the mean RTT and the variance of the RTT.

In order to evaluate the accuracy of the estimation process over time, we consider the error between the estimate and the real RTT, these being the average values observed over small, non overlapping, intervals (of 5 seconds, in practice). We compute the error of the estimate as follows:

$$\text{error} = \frac{\text{mean estimated RTT} - \text{mean RTT at the source}}{\text{mean RTT at the source}}$$

and build the corresponding empirical CDF.

6.1 Controlled Environment

We first considered a testbed using Dell Precision 380 workstations running Fedora Linux version 10 (kernel 2.6.27). The topology is shown in Fig. 3: all machines are directly connected as shown (using Ethernet cables) and there is no outside traffic. Given that the propagation and processing delay are very small, we introduce random delays between the machines. The link between the source of the traffic and the machine *cross* has a delay uniformly distributed on $[50 - d_1, 50 + d_1]$ ms, where d_1 is set to either 5 or 40 ms. The link between the machines *cross* and *btlnk* has a delay uniformly distributed on $[25 - d_2, 25 + d_2]$ ms, where d_2 is set to either 5 or 20 ms.

All the links have a capacity of 100 Mbps (fast Ethernet) except the link between the machines *btlnk* and *dest* which is capped to 10 Mbps in order to act as the bottleneck link. The random delays and the capacity limitations are implemented using the Netem kernel module. Throughout the experiments we explicitly configured the sender to use Reno TCP and not Cubic (the default option for kernel 2.6.27).

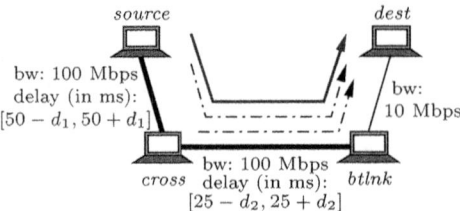

experim. ID	# cross flows source → dest	# cross flows cross → dest	d_1	d_2
exp1	0	12	5	5
exp2	2	6	5	5
exp3	3	9	5	5
exp4	3	9	40	20

Fig. 3. Scheme of the testbed: the continuous line represents the observed flow, dashed lines represent the cross traffic

Fig. 4. Configuration of the testbed experiments, with the different cross flows and delays

The main flow we consider is between machines *source* and *dest*. We add additional cross traffic directed to *dest*, starting from *source* and from *cross*. We have a measuring point of the packets' arrivals at the machine *btlnk*. Figure 4 summarizes the experiments done.

Figure 5 shows an example of the output of one experiment (exp2). For each packet, we compute the RTT at the source and its estimation at the measuring point. We have found that the estimated RTT is of the same order of magnitude as the real RTT, even though it seems to be less variable than the RTT at the source. This is due to the averaging effect of the spectral analysis that uses the last observed 256 packets (which include 5-8 flights) for the spectrum estimation.

While this representation shows that the estimation and the real RTT are close, we need to characterize in detail the performance of the algorithm. In order to minimize the effects of the lag time between the RTT samples collected at the source and at the measuring point, we consider the empirical cumulative distribution function (CDF) of the RTTs built from the samples and report the corresponding mean and variance in Table 1 (cf. columns 3–6). While the real RTT has more variability, the estimated RTT is more stable due to the averaging effect of the spectral analysis.

Another performance index we are interested in is the error in the estimation: as Fig. 6 shows, in experiments exp1–exp3, the estimate lies within 10% of the true value with probability 95%. In the worst case, our solution is able to

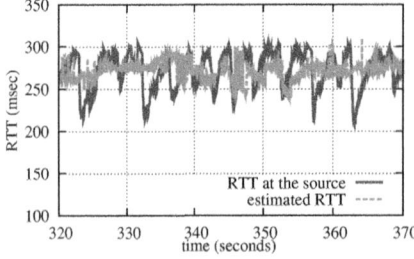

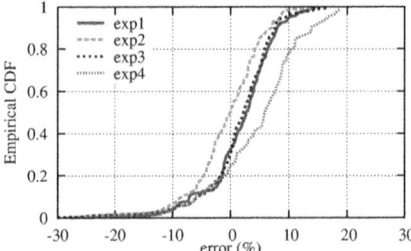

Fig. 5. Example of the RTT measured at the source and the estimated RTT (exp2)

Fig. 6. Empirical CDF of the error between true and estimated RTTs (testbed)

Table 1. Mean and variance of the RTT (real and estimated)

		Testbed experiments				Internet experiments	
		exp1	exp2	exp3	exp4	peak	off-peak
Real RTT	mean	0.2795	0.2760	0.2782	0.2752	0.0389	0.0448
(in s)	variance	$1.85\ 10^{-3}$	$1.84\ 10^{-3}$	$1.62\ 10^{-3}$	$2.70\ 10^{-3}$	$1.72\ 10^{-5}$	$2.40\ 10^{-6}$
Estimated	mean	0.2814	0.2706	0.2792	0.2875	0.0422	0.0446
RTT (in s)	variance	$9.46\ 10^{-5}$	$1.41\ 10^{-4}$	$3.57\ 10^{-6}$	$5.43\ 10^{-4}$	$2.40\ 10^{-6}$	$2.94\ 10^{-6}$

accurately estimate the RTT with an error in $[-10\%, 10\%]$ with probability equal to 75%, and with an error in $[-20\%, 20\%]$ with probability equal to 99%.

The results given by our methodology show a higher accuracy with respect to the results obtained in previous studies (e.g. [6]): our results have been obtained online observing the traffic in one direction, unlike the results of [6] which require to collect traffic in both directions and analyze it offline.

6.2   Internet Experiments

In this section, we show the results of experiments using three machines connected through Internet (Fig. 7). The source is at INRIA Sophia Antipolis (France), the intermediate machine (*hop1*) is at University of Trento (Italy) while the destination is at Eurecom (France). We created two SSH tunnels, one between *source* and *hop1*, and one between *hop1* and the destination *dest*. The traffic generated from the source passes through the tunnel at *hop1*, where packets interarrivals are measured, and reaches the destination.

Columns 7–8 of Table 1 show the mean and the variance of the RTTs (real and estimated) for two different experiments: the first, during a peak hour (Monday, May 4th, 2009, 10 AM, CEST) and the second, during an off-peak hour (Sunday, May 3rd, 2009, 12 PM, CEST). The high variability of the peak hour results in a higher error, that is not smoothed by considering the overall mean. The reason can be understood looking at Fig. 5: the estimation algorithm often does not detect sudden decreases of the RTT. When the RTT drops, we obtain large

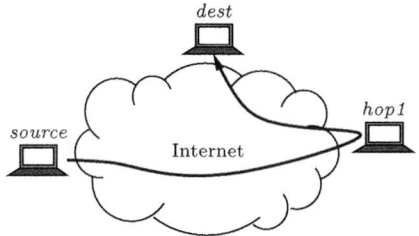

Fig. 7. Scheme of the Internet tests

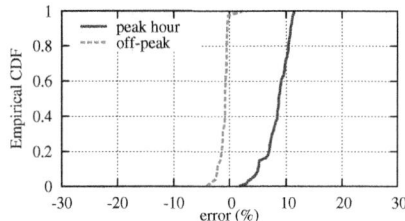

Fig. 8. Empirical CDF of the error between the RTT measured at the source and estimated (Internet experiments)

positive error values, while there are no large negative error values to compensate for the large positive ones.

As seen in Fig. 8 which shows the empirical CDF of the error, the estimate lies within 10% of the true value with probability equal to 99%.

7 Open Issues and Conclusions

In this paper we have presented a methodology, based on spectrum analysis, for the passive online estimation of the RTT of a long-lived TCP connection, using one-way traffic. The estimation of the RTT represents a basic building block in the framework of "Semantic Networking," where flow-aware routers can implement novel AQM techniques in order to control the traffic.

We validate our solution through measurements in a controlled testbed and over the Internet showing a good accuracy. We plan to extend the evaluation using different scenarios: for instance, it is interesting to understand the variation in estimation accuracy as the RTT varies due to changes in the routing or in the number of flows sharing a link.

Since the method is based only on the arrival pattern, it is clear that if the arrival pattern is strongly modified inside the network, the method may not be accurate. For instance, if we observe a flow after it has passed through a bottleneck, the arrival pattern may become a continuous stream of packets. Should this happen, the method may not be applicable, however one would still be able to estimate the rate of a connection through a moving average estimator for instance. In future evaluations, we plan to investigate in detail the scenarios where the methodology may not be accurate, in order to provide a comprehensive view of the benefits of our solution.

References

1. Collange, D., Costeux, J.: Passive estimation of quality of experience. Journal of Universal Computer Science 14(5), 625–641 (2008)
2. Noirie, L., Dotaro, E., Carofiglio, G., Dupas, A., Pecci, P., Popa, D., Post, G.: Self-* features for semantic networking. In: Proc. of FITraMEn (December 2008)
3. Avrachenkov, K., Ayesta, U., Brown, P., Nyberg, E.: Differentiation between short and long TCP flows: predictability of the response time. In: Proc. of IEEE INFOCOM, Hong Kong (March 2004)
4. Rai, I., Biersack, E., Urvoy-Keller, G.: Size-based scheduling to improve the performance of short TCP flows. IEEE Network 19(1), 12–17 (2005)
5. Kortebi, A., Muscariello, L., Oueslati, S., Roberts, J.: Evaluating the number of active flows in a scheduler realizing fair statistical bandwidth sharing. In: Proc. of ACM SIGMETRICS, Banff, Alberta, Canada (June 2005)
6. Jaiswal, S., Iannaccone, G., Diot, C., Kurose, J., Towsley, D.: Inferring TCP connection characteristics through passive measurements. In: Proc. of IEEE INFOCOM, Hong Kong (March 2004)
7. But, J., Keller, U., Kennedy, D., Armitage, G.: Passive TCP stream estimation of RTT and jitter parameters. In: Proc. of IEEE CLCN (November 2005)

8. Zhang, Y., Breslau, L., Paxson, V., Shenker, S.: On the characteristics and origins of Internet flow rates. In: Proc. of ACM SIGCOMM (August 2002)
9. Lance, R., Frommer, I., Hunt, B., Ott, E., Yorke, J., Harder, E.: Round-trip time inference via passive monitoring. ACM SIGMETRICS PER 33, 32–38 (2005)
10. Veal, B., Li, K., Lowenthal, D.: New methods for passive estimation of TCP round-trip times. In: Proc. of PAM, Boston, MA, USA (April 2005)
11. Jiang, H., Dovrolis, C.: Passive estimation of TCP round-trip times. Computer Communication Review 32(3), 75–88 (2002)
12. Qi, Y., Minka, T., Picard, R.: Bayesian spectrum estimation of unevenly sampled nonstationary data. In: Proc. of ICASSP, Orlando, FL, USA (May 2002)
13. Brown, P.: Resource sharing of TCP connections with different round trip times. In: Proc. of IEEE INFOCOM, Tel Aviv, Israel (March 2000)
14. Altman, E., Jimenez, T., Nunez Queija, R.: Analysis of two competing TCP/IP connections. Performance Evaluation 49(1-4), 43–55 (2002)
15. Stoica, P., Moses, R.: Introduction to spectral analysis. Prentice-Hall, Englewood Cliffs (1997)
16. Stoica, P., Sandgren, N.: Spectral analysis of irregularly-sampled data: Paralleling the regularly-sampled data approaches. Digital Signal Processing 16(6), 712–734 (2006)
17. Scargle, J.: Statistical aspects of spectral analysis of unevenly spaced data. Journal of Astrophysics 263, 835–853 (1982)
18. Carra, D., Avrachenkov, K., Alouf, S., Blanc, A., Nain, P., Post, G.: Passive online RTT estimation for flow-aware routers using one-way traffic. Research Report RR-7124, INRIA (November 2009)

A Flow Scheduler Architecture

Dinil Mon Divakaran[1,*], Giovanna Carofiglio[2], Eitan Altman[3,**],
and Pascale Vicat-Blanc Primet[1]

[1] INRIA / Université de Lyon / ENS Lyon,
LIP, ENS Lyon, Lyon - 69007, France
{Dinil.Mon.Divakaran,Pascale.Primet}@ens-lyon.fr
[2] Alcatel-Lucent Bell Labs
Giovanna.Carofiglio@alcatel-lucent.com
[3] INRIA
Eitan.Altman@sophia.inria.fr

Abstract. Scheduling flows in the Internet has sprouted much interest
in the research community leading to the development of many queueing
models, capitalizing on the heavy-tail property of flow size distribution.
Theoretical studies have shown that 'size-based' schedulers improve the
delay of small flows without almost no performance degradation to large
flows. On the practical side, the issues in taking such schedulers to implementation have hardly been studied. This work looks into practical
aspects of making size-based scheduling feasible in future Internet. In this
context, we propose a flow scheduler architecture comprising three modules — Size-based scheduling, Threshold-based sampling and Knockout
buffer policy — for improving the performance of flows in the Internet.
Unlike earlier works, we analyze the performance using five different performance metrics, and through extensive simulations show the goodness
of this architecture.

Keywords: Scheduling, Sampling, QoS, Future Internet, Architecture.

1 Introduction

Recent works have advocated the importance of networks being 'flow-aware'.
Bonald *et al.* have listed the need for having a flow-aware architecture [1]. In a
flow-aware network, the performance is measured at flow level. This is in line
with the utility of end-users, where e.g., the delay of small flows, throughput
of large flows, instantaneous rate of streaming traffic etc. are most often more
important than packet-level QoS metrics. In this context, our goal is to come up
with a flow scheduler architecture for improving the delay performance of small
(and middle size) flows. The current Internet architecture has a FCFS scheduler
and Droptail buffer at each of its nodes. These, along with the fact that most

---

* Corresponding author.
** This work was done in the framework of the INRIA and Alcatel-Lucent Bell Labs
Joint Research Lab on Self Organized Networks.

M. Crovella et al. (Eds.): NETWORKING 2010, LNCS 6091, pp. 122–134, 2010.

of the flows in the Internet are carried by TCP, makes this current architecture biased against small TCP flows for the following reasons. (i) A packet loss to a small flow most often results in a timeout due to the small 'congestion window' (*cwnd*) size; whereas, a large flow is most probably in the congestion avoidance phase, and hence has large *cwnd* size. Therefore, for a large flow, packet losses are usually detected using duplicate ACKs, instead of timeouts, thus avoiding slow-start. (ii) The increase in round trip time (RTT) due to large queueing delays hurts the small flows more than the large flows. Again, for the large flows, the large *cwnd* makes up for the increase in RTT; whereas, this is not the case for small flows.

These problems being well-known, researchers have explored scheduling algorithms that give priority to small flows. They range from SRPT (Shortest Remaining Processing Time) [2] to LAS (Least Attained Service) [3] to MLPS (Multi-level Processor Sharing) scheduling policies [4]. While scheduling algorithms give priority in time, buffer management policies give priority in space. Guo *et al.* showed the gain in performance attained by giving space priority to small flows [5]; but it is a stand-alone concept that does not consider giving time-priority to small flows. We argue that it is important to give priority in both time and space to small flows, in order to reduce the delay as well as timeouts faced. To the best of our knowledge, LAS scheduling policy is the only policy that gives space priority to packets of small flows [6], thereby giving priority in both time and space. It does so by inserting incoming packet in the appropriate position and dropping from the tail whenever the buffer is full. But, it has been observed that LAS is unfair to very large flows [7]. Moreover, it is challenging to perform a strict ordering of packets of each flow at high line rates.

This work proposes a flow scheduler architecture, that gives priority in time as well as space to small flows, and uses sampling for performing size-based scheduling. To be precise, our flow scheduling architecture combines three essential modules that help in improving the delay performance of flows:

1. Generalized size-based scheduling;
2. Threshold-based sampling;
3. Knockout buffer policy.

The motivation for such an architecture is given in Section 2, where we also summarize related works. The architecture is detailed in Section 3. We perform extensive simulations and compare different performance metrics to show how each of these three strategies contributes in improving the performance of small flows, without affecting the performance of large flows. Unlike most previous works, where the performance was analyzed using just one metric (usually the conditional mean response time), we consider five different metrics. They are:

1. Conditional mean completion time of small flows;
2. Number of timeouts encountered by small flows;
3. Mean completion time for range of flow sizes;
4. Mean completion time for small flows, large flows and all flows;
5. Maximum completion time of small flows.

The goal of simulations and the setting are described in Section 4. The benefits of using the Knockout policy are analyzed in Section 5. In Section 6, we evaluate the proposed flow scheduler architecture and compare with other schemes.

2   Related Works and Motivation

The literature in this research area being vast, we limit the references to a small but important subset. A large number of researchers have considered giving priority in time to small flows. These have given rise to the study of scheduling disciplines like SRPT, LAS and MLPS disciplines in the context of Internet flows. While SRPT requires the knowledge of flow size in advance [2], LAS is a 'blind' scheduling policy — requires no in-advance information on flow size [3]. These differentiating policies perform better in terms of delay, when compared to the naive PS (processor sharing) system[1]. The MLPS scheduling discipline is a generalized version with high flexibility, having N different priority levels distinguished by $N-1$ thresholds, and strict priority among these levels [7,8,9].

The drawbacks of LAS policy, such as unfairness and scalability issue, have motivated researchers to explore other means of giving priority to small flows, one such being the strict $PS + PS$ model proposed in [7]. The $PS + PS$ model, as the name indicates, uses two PS queues, with priority between them. The first θ packets of every flow are served in the higher priority queue (Q_1), and the remaining in the lower priority queue (Q_2). The service discipline is such that, Q_2 is served only when Q_1 is empty. Therefore, it is a strict $PS+PS$ model. This work also takes a step forward in performance analysis of size-based scheduling systems, by analyzing another metric — maximum response time — other than the usual conditional mean response time. In addition, the authors proposed an implementation of this model; but it relies on TCP sequence numbers, requiring them to start from a set of possible initial numbers. This not only makes the scheme TCP-dependent, but also reduces the randomness of initial sequence numbers. Again, this is another work which does not account for space priority for small flows.

Authors of [5] considered prioritizing small flows in space. This is achieved by preferentially treating small flows inside the bottleneck queue which implement RIO (RED with In and Out). Small and large flows were assigned different drop functions. To facilitate this, they proposed an architecture where the edge routers mark packets as belonging to small or large flow, using a threshold-based classification. With priority given only in space, the performance gains in terms of average response times (apart from analyzing the fairness) is not complete.

We observe that most of the works dealing with giving preferential treatment based on size (or age) assume that the router keeps per-flow information. In fact, this assumption is challenged by the scalability factor, as the number of flows in progress is in the order of hundreds of thousands under a high load. One solution is to use sampling to detect large flows (thus classifying them), and

---

[1] At the flow level, the queues in the Internet are generally modelled as an $M/G/1-PS$ system, even when the queue is served using a FCFS policy at the packet level.

use this information to perform size-based scheduling. Since the requirement here is only to differentiate between small and large flows, the sampling strategy need not necessarily track the exact flow size. A simple way to achieve this is to probabilistically sample every arriving packet, and store the information of sampled flows along with the sampled packets of each flow [10]. SIFT, proposed in [11], uses such a sampling scheme along with the $PS + PS$ scheduler. A flow is 'small' as long as it is not sampled. All such undetected flows go to the higher priority queue until they are sampled. The authors analyzed the system using the 'average delay' (average of the delay of all small flows, and all large flows) for varying load, as a performance metric. Though it is an important metric, it does not reveal the worst-case behaviour in the presence of sampling. This is more important here, as the sampling strategy can have false positives; small flows if sampled will be sent to the lower priority queue. In such scenarios, it is necessary to compare other performance metrics, which we listed earlier.

3   Architecture

This section describes the modules of the architecture, and the cost of implementing this architecture.

3.1   The Modules

The flow scheduler architecture consists of three functional modules: a size-based scheduler, a threshold-based sampling technique to detect large flows, and a Knockout buffer policy giving space priority to small flows.

Size-Based scheduling: Since sampling introduces errors in the detection of large flows, thereby permitting misclassification, using a strict priority-scheduling strategy is not advisable. Therefore, we take a generalized model of the strict $PS + PS$ scheduling, called generalized size-based scheduling, or simply SB scheduling. As before, packets of all flows are served in Q_1, as long as the ongoing size is less than θ packets. Once the ongoing flow size crosses θ, it is queued in Q_2. But instead of giving the whole capacity to Q_1, only a fraction of the capacity is assigned to Q_1. That is, the high priority queue is assigned a weight $0 \leq w \leq 1$. If C is the link capacity, Q_1 and Q_2 are serviced at rates wC and $(1 - w)C$ respectively, whenever the queues are not empty. If Q_1 is empty, Q_2 is served at full capacity C. We assume that the scenario of Q_2 being empty and Q_1 being non-empty is a rare possibility. Note that, if $w = 1$, this becomes the $PS + PS$ scheduling policy. The scheduling module in the figure is shown as deciding which queue to dequeue based on the parameter w.

Threshold-Based sampling: For the sampling part, we use the well-studied 'Sample and hold' strategy proposed for detecting large flows [12]. It works as follows. For every sampled packet, a flow entry is created in the flow table if it does not exist. A packet of s bytes is sampled with a probability p, which is expressed in terms of byte-sampling probability β. We have, $p = 1 - (1 - \beta)^s$. When a packet arrives, a flow table lookup is performed. If the arriving packet is found to be part of an existing flow, the flow-size counter in the flow table is updated. Thus, for each sampled flow, there is a counter that maintains the estimated size. This process is performed during every measurement interval. Thresholds are used to reduce false positives, and to preserve continuing large flows across intervals. Observe that the flow-table lookup is done for every arriving packet, and size update is performed for every detected flow. This is costly in terms of processing, but reduces the flow table's size considerably (to a few thousands of entries). Therefore, it is possible to use SRAM to store the flow table for efficient lookups. A useful property of this sampling strategy is that, since the estimated size is never greater than the actual flow size, by choosing an appropriate threshold, false positives can be completely avoided.

Knockout buffer policy: The third part is the Knockout buffer policy for giving space priority to small flows. Though there is only one single physical queue, it is shared by two virtual queues, one for enqueueing packets of flows classified as small, the other for enqueueing packets of flows classified as large. These correspond to the two queues Q_1 and Q_2 described earlier. The policy is different from Droptail only during packet discard instants [13]. Upon the arrival of a packet when the physical buffer is full, the Knockout policy operates thus: if the packet is for Q_2 (i.e., the system has classified it as belonging to a large flow), it is dropped. If the arriving packet is for Q_1 (i.e., the system has classified it as belonging to a small flow), the last packet from Q_2 is 'knocked out' making space for this new packet. In the scenario of Q_2 being empty (i.e., the physical buffer has packets of only flows classified as small), the arriving packet is dropped. Assuming most large flows are carried by TCP, dropping a packet from a large flow is meaningful as it will be retransmitted by the TCP source.

Fig. 1 gives a pictorial representation of the architecture. An arriving packet first goes to the sampling module, which does a flow-table lookup. Packet sampling and flow-table update are performed if necessary. The queueing module decides to queue the packet in Q_1 or Q_2 based on the flow-size estimate

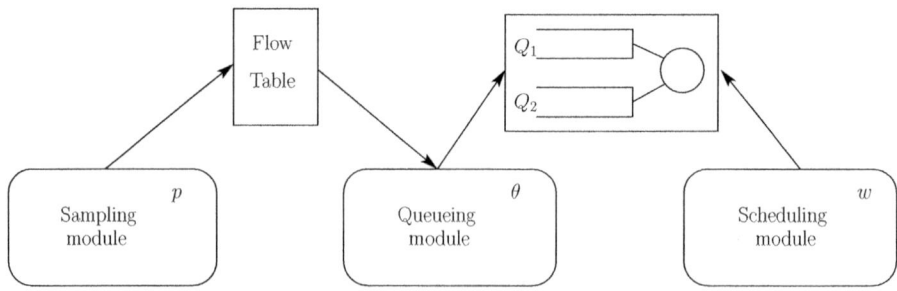

Fig. 1. Flow scheduler architecture

available from the flow table and the parameter θ. If the physical buffer is full, the Knockout policy is used to select the packet to be dropped. The scheduling module uses the weight parameter w to perform SB scheduling.

3.2 Implementation Cost

SB scheduling requires two queues. These can be implemented as virtual queues, on top of the physical queue. The scheduling of packets as such can then be implemented by assigning weights to these queues.

For the sampling, an SRAM with sufficient size to hold the flow table is required. There is extra processing for updating the flow size of detected flows. The flow-table lookup can be combined along with route-table lookup.

Knockout policy uses the two virtual queues, Q_1 and Q_2, with Q_1 being the higher priority queue. Observe that a virtual queue can grow to the actual size of the physical queue with the other virtual queue being empty. To be able to knock out an already queued packet from Q_2, the tail of Q_2 needs to be tracked. All these can be achieved if the physical queue is implemented as a linked list, and pointers to the head and tail of the two virtual queues are maintained. This deviates from the simplest way of implementing a queue as a circular buffer, thus adding extra overhead in maintaining the linked list.

4   Simulation

4.1   Goal

The goal of the simulations is to evaluate the performance of the flow scheduler architecture. As described earlier, small flows are biased against large flows when it comes to timeouts. Hence, we are interested in analyzing not only the improvement in delay performance, but also the reduction in number of timeouts faced by small flows. On the other hand, since prioritizing small flows should not adversely affect large flows, the mean completion times of large flows conditioned on their flow sizes are also analyzed. To see the improvement over today's Internet architecture, we compare results with the FCFS scheduler. Along with the FCFS scheduler, the buffer policy used in all simulations here is Droptail (as is the case in Internet nodes), though not stated explicitly in figures.

4.2   Settings

Simulations are performed using NS-2. A dumbbell topology, representing a single bottleneck link connecting source-destination pairs, was used throughout. The bottleneck link capacity was set to 1 Gbps, and the capacities of the source nodes were all set to 100 Mbps. The delays on the links were set such that the base RTT (consisting of only propagation delays) is equal to 100 ms. The size of the bottleneck queue is set in bytes, as the bandwidth delay product (BDP) for 100 ms base RTT. There were 100 node pairs, with the source nodes generating flows according to a Poisson traffic. The flow arrival rate is adapted to have a packet loss rate of around 1.25% with FCFS scheduler and Droptail buffer. Flow sizes are taken from a Pareto distribution with shape $\alpha = 1.1$, and mean flow size set to 500 KB.

All flows are carried by TCP, in particular, using the SACK version. Packet size is kept constant and is equal to 1000 B. For simplicity, we keep the threshold in packets; θ is set to 25 packets in all the scenarios, unless explicitly stated otherwise. For post-simulation analysis, we define 'small flow' as a flow with size less than or equal to 20 KB, and 'large flow' as one with size greater than 20 KB. Here the flow size is the size of data generated by the application, not including any header or TCP/IP information. Also note that, a small flow of 20 KB can take more than 25 packets to transfer the data, as it includes control packets (like *SYN*, *FIN* etc.) and retransmitted data packets.

5   Performance Analysis of the Scheduler Using Knockout

Here we analyze SB scheduler using Knockout buffer policy, but without sampling. In this case, flows are accurately classified as small and large by tracking the ongoing size of the flow. The focus of this section is to show the importance of having the Knockout buffer policy.

Before choosing the weights, we present an observation. First, it should be noted that, by giving priority to small flows, a policy essentially tries to keep the corresponding buffer for small flows almost empty. With this in mind, we conducted simulations to analyze the average occupancy of Q_1 for different weights. The result is shown in Fig. 2. The number of packets of small flows in queue is almost constant for weights $w \geq 0.6$. Hence, any $w \geq 0.6$ should give close performance for small flows. Dynamically adapting w according to the buffer occupancy being outside the scope of this work, we set w to 0.8 for SB scheduler in our simulations. The other scenario considered is with w set to 1.0; thus we also analyze the strict $PS + PS$ system. Even when there is no sampling involved, we see that there is no notable gain in using a strict SB scheduler.

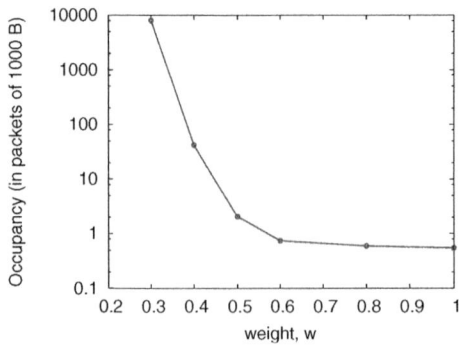

Fig. 2. Average occupancy for Q_1 for different weights

5.1   Results

Fig. 3(a) shows the mean completion time conditioned on the flow sizes, for small flows. The naive packet-level FCFS scheduling policy is shown as a comparison. The other curves correspond to SB scheduling with different weights, and with and without the Knockout policy. A value of '0' for KO implies that the Knockout policy is not in use, and '1' implies the contrary. The figure shows the goodness of size-based scheduling compared to the FCFS scheduling. Knockout buffer policy is seen to complement the SB schedulers. Observe also that, with this metric, there is no notable difference using a weight of 0.8 or 1.0.

   Fig. 3(b) indicates that the large flows are not affected by giving priority to small flows (both in space and time). In fact, it can be seen in Fig. 4(a) that the SB scheduler with $w = 0.8$ and $KO = 1$, gives the same mean delay for very large flows, as does the FCFS scheduler. Fig. 4(a) plots the mean completion time of flows within different size ranges (e.g., 0-20 packets, 21-200 packets etc.). The mean values show that, in general, the SB scheduler also performs better for medium flows (those with a size around 2000 packets).

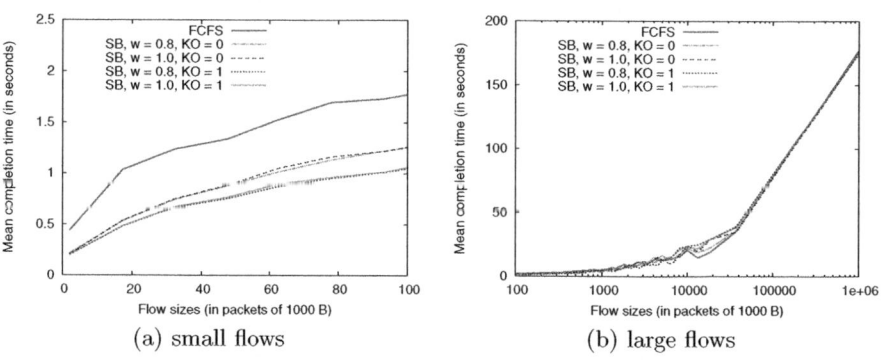

(a) small flows   (b) large flows

Fig. 3. Conditional mean completion time

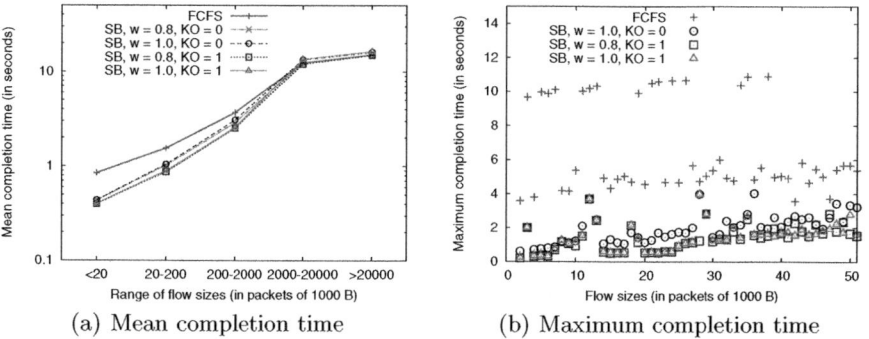

(a) Mean completion time   (b) Maximum completion time

Fig. 4. Other metrics

Table 1. Comparison of different metrics

Metrics	small TOs	small $\overline{CT}$	large $\overline{CT}$	all $\overline{CT}$
FCFS	579	0.8432	2.3294	1.9022
$w = 0.8, KO = 0$	386	0.4325	1.7532	1.3736
$w = 1.0, KO = 0$	449	0.4375	1.8540	1.4468
$w = 0.8, KO = 1$	6	0.3996	1.5715	1.2347
$w = 1.0, KO = 1$	5	0.3997	1.6219	1.2706

For each scheduling, Table 1 lists the number of timeouts faced by small flows, along with the mean completion time (indicated by $\overline{CT}$) for small, large and all flows. We see that the number of timeouts encountered by small flows is highest for FCFS, followed by the schedulers without Knockout. This happens as some of the flows in Q_1, after being served with priority for the first θ packets, come back with more packets (due to a larger $cwnd$) and join Q_2, thereby increasing the total buffer occupancy. Without space priority, the packets of small flows are dropped when the buffer is full. With the Knockout policy, the timeouts are brought down tremendously as the packets of small flows are the last to experience drops. Fig. 4(b), which plots the worst-case completion time per flow size for small flows, also supports the necessity of giving space priority in addition to time priority (the figure does not plot the scenario of $\{w = 0.8, KO = 0\}$ for better clarity).

Comparing the mean CTs, it can be noted that the Knockout policy gives better results for all the means, compared to those without Knockout policy. Note that the prioritized service enjoyed by the first θ packets of a large flow helps in having a 'quicker' slow-start phase when compared to the FCFS-Droptail system. Similarly, non-strict schedulers give better performance (in terms of means) for large flows, compared to the strict counterparts (both with $KO = 0$, and $KO = 1$). At the same time, the mean CT of small flows remain almost the same. With these comparisons, it becomes clear that a non-strict scheduler with Knockout buffer policy performs better than strict scheduler (strict $PS + PS$) without Knockout buffer. In general, these results also confirm a well-known result — SB scheduling outperforms FCFS scheduling in improving the delay performance.

6 Performance Analysis of the Scheduler with Sampling

This section analyzes the performance of the flow scheduler architecture, which combines the SB scheduler, the threshold-based sampling strategy and the Knockout buffer policy. For the scheduler, we set the weight w to 0.8. The results are compared to the SIFT scheme [11]. Note that SIFT does not use the Knockout policy; nor does it use a threshold to classify large flows. Instead, a sampled flow is considered 'large' and sent to Q_2; all other undetected flows go to Q_1. To see the degradation due to sampling, we also compare these schemes with the basic

SB scheduling scheme (with no sampling). The packet-sampling probability is set to 1/100 in the sampling schemes of both SIFT and our flow scheduler architecture. Here, we analyze the system under two traffic scenarios which differ in flow size distribution. Scenario 1 corresponds to the one considered before, where flow sizes were taken from a Pareto distribution. In Scenario 2, 85% of flows are generated using an Exponential distribution with a mean 20 KB; the remaining 15% are contributed by large flows using Pareto distribution with shape $\alpha = 1.1$, and mean flow size set to 1 MB.

In the figures below, the name 'SB-SH' represents our flow scheduler architecture, coming from Size-Based scheduling using 'Sample and Hold'.

6.1   Results with Traffic Scenario 1

Figures 5(a), 5(b), 6(a) and 6(b) show the results. The conditional mean completion time curves for small flows in Fig. 5(a) reveal that sampling-cum-scheduling strategies (including SIFT) give improved performance for small flows in comparison to FCFS scheduling. This is anticipated, as most small flows go undetected and get prioritized. Even the maximum delay experienced by small flows using

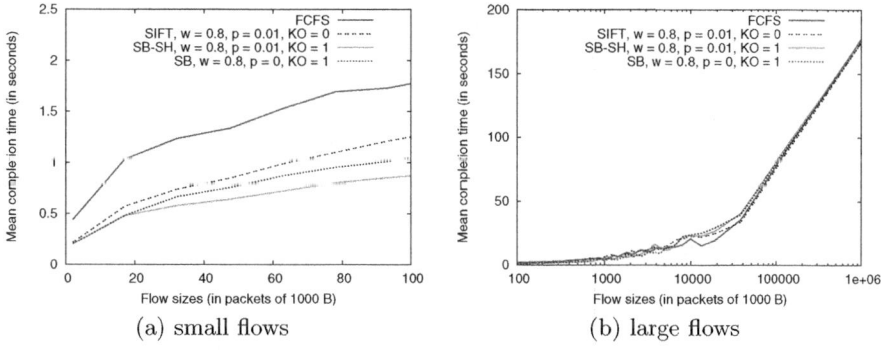

(a) small flows   (b) large flows

Fig. 5. Conditional mean completion time

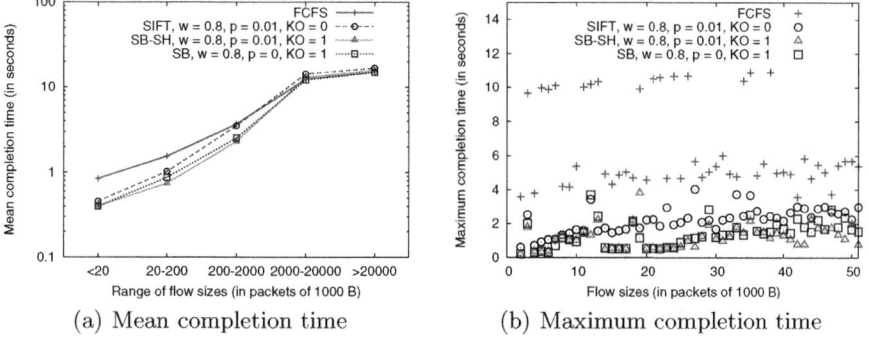

(a) Mean completion time   (b) Maximum completion time

Fig. 6. Other metrics

Table 2. Comparison of different metrics

Metrics	small TOs	small $\overline{CT}$	large $\overline{CT}$	all $\overline{CT}$
FCFS	579	0.8432	2.3294	1.9022
$p = 0.01, KO = 0$, SIFT	502	0.4585	1.9483	1.5200
$p = 0.01, KO = 1$, SB-SH	5	0.3998	1.4406	1.1414
$p = 0.0, KO = 1$, SB	6	0.3996	1.5715	1.2347

sampling-cum-scheduling is lesser as seen in Fig. 6(b). In the figure, the completion time in SIFT is sometimes close to that of FCFS. These are cases when SIFT samples small flows and de-prioritizes them. Between the sampling-cum-scheduling strategies, SB-SH scheme is seen to give smaller delay to small flows than SIFT, both in the mean case and in the worst case.

Additional metrics are compared in the Table 2. In all SB schedulers, the weights are the same ($w = 0.8$), and hence not made explicit in the table. The SIFT scheme induces large number of timeouts for small flows, as it gives no space priority to packets of small flows. In addition, a small flow that is sampled, gets de-prioritized in SIFT, leaving it to compete with the large flows. This is also clear from Fig. 6(b), which plots the maximum delay per size for small flows. From Fig. 6(a) and Table 2, it is seen that the delay for large flows is higher in SIFT than in the SB-SH scheme. Observe that we have the same sampling probability for both schemes. This means, a flow once sampled is de-prioritized immediately in SIFT; whereas a sampled flow still enjoys priority (both in time and space) for the next θ packets in SB-SH scheme. This helps the large flows to attain a large TCP *cwnd* faster (than in FCFS and SIFT).

Comparison of SB-SH scheme with the naive SB scheduling (without sampling), which shows that the former performs better than the latter, might appear to be surprising. But in fact, it is not — recall, that we have not tried to find the optimal threshold, θ, in our study here. The false negatives that results from the sampling strategy increase the mean number of packets being served at Q_1, which is similar to increasing the threshold θ. Increasing the threshold, increases the rate at which TCP *cwnd* increases (due to negligible queueing delay and very few losses). To confirm, we performed SB scheduling (with Knockout policy, and without sampling) where θ was set to 100 packets. It was found that number of timeouts for small flows was 5, and the mean CTs for small, large and all flows were 0.3997, 1.3740 and 1.0939 respectively. Except for the mean CT for small flow, which is almost the same for all, these values are better than all the results shown in Table 2.

6.2   Results with Traffic Scenario 2

Similar graphs were obtained for the second traffic scenario. We show only two plots here — figures 7(a) and 7(b), and refer to an internal report for other figures [14]. Table 3 compares other interested metrics. Comparing the values in

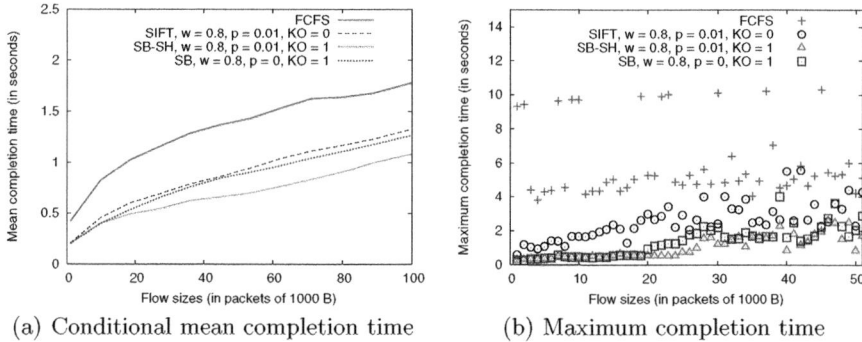

(a) Conditional mean completion time   (b) Maximum completion time

Fig. 7. Metrics for small flows

Table 3. Comparison of different metrics

Metrics	small TOs	small $\overline{CT}$	large $\overline{CT}$	all $\overline{CT}$
FCFS	792	0.7603	1.7491	1.2168
$p = 0.01, KO = 0$, SIFT	778	0.4204	1.3195	0.8354
$p - 0.01, KO - 1$, SB-SH	0	0.3671	1.2327	0.7666
$p = 0.0, KO = 1$, SB	14	0.3698	1.3101	0.8039

the table reveals that the results are similar to that with the first traffic scenario. Note that, as the number of small flows is higher in this scenario, SIFT gives a worse performance for the maximum completion time of small flows in this scenario (Fig. 7(b)) in comparison to the previous scenario (Fig. 6(b)).

7   Conclusions

In this paper, we proposed a new flow scheduler architecture to improve the performance of flows in the Internet. Through arguments and simulations we have emphasized the importance of each of the modules in the architecture. The architecture is shown to improve the performance of flows in comparison to the naive FCFS scheduler. Besides, in comparison to SIFT, the flow scheduler architecture brings in better performance in terms of conditional mean completion time and timeouts for small flows, and mean CTs (for small, large, and all flows). Apart from these, the worst-case delay performance is also appealing.

In general, our study confirms previous observation that size-based scheduling induces negligible degradation to large flows. While sampling is known to be a practical solution in tracking large flows, here we also see that it does not affect the performance of small flows. This work opens different directions for future work. The parameters such as threshold θ, weight w and sampling probability p were kept constant here. Finding the right values for each of these so as to obtain the optimal delay performance is dependent on the other two parameters.

134 D.M. Divakaran et al.

All of them have an influence on the mean queue length. A larger value of θ will result in a larger number of packets sent to Q_1, a smaller value of w indicates a reduction in the service rate at Q_1, and decreasing the sampling probability will also increase the average number of packets of large flows served at Q_1. So, the variation in the average queue length (for a given load) can be used to decide the optimal values for these parameters.

References

1. Bonald, T., Oueslati-Boulahia, S., Roberts, J.: IP traffic and QoS control: the need for a flow-aware architecture. In: World Telecommunications Congress (September 2002)
2. Schrage, L.: A proof of the optimality of the Shortest Remaining Processing Time Discipline. Operations Research (16), 687–690 (1968)
3. Rai, I.A., Urvoy-Keller, G., Vernon, M.K., Biersack, E.W.: Performance analysis of las-based scheduling disciplines in a packet switched network. SIGMETRICS Perform. Eval. Rev. 32(1), 106–117 (2004)
4. Kleinrock, L.: Queueing Systems. Computer Applications, vol. II. Wiley Interscience, Hoboken (1976)
5. Guo, L., Matta, L.I.: The war between mice and elephants. In: ICNP 2001, pp. 180–188 (November 2001)
6. Rai, I.A., Biersack, E.W., Urvoy-Keller, G.: Size-based scheduling to improve the performance of short TCP flows. IEEE Network 19(1), 12–17 (2005)
7. Avrachenkov, K., Ayesta, U., Brown, P., Nyberg, E.: Differentiation Between Short and Long TCP Flows: Predictability of the Response Time. In: INFOCOM (2004)
8. Aalto, S., Ayesta, U., Nyberg-Oksanen, E.: Two-level processor-sharing scheduling disciplines: mean delay analysis. SIGMETRICS Perform. Eval. Rev. 32(1), 97–105 (2004)
9. Aalto, S., Ayesta, U.: Mean delay analysis of multi level processor sharing disciplines. In: INFOCOM (2006)
10. Zseby, T., et al.: RFC 5475: Techniques for IP Packet Selection (March 2009), http://www.rfc-editor.org/rfc/rfc5475.txt (Network Working Group)
11. Psounis, K., Ghosh, A., Prabhakar, B., Wang, G.: SIFT: A simple algorithm for tracking elephant flows, and taking advantage of power laws. In: 43rd Annual Allerton Conference on Control, Communication and Computing (2005)
12. Estan, C., Varghese, G.: New directions in traffic measurement and accounting. SIGCOMM Comput. Commun. Rev. 32(4), 323–336 (2002)
13. Chang, C.G., Tan, H.H.: Queueing analysis of explicit policy assignment pushout buffer sharing schemes for atm networks. In: Proceedings of the 13th IEEE Networking for Global Communications, vol. 2, pp. 500–509 (June 1994)
14. Divakaran, D.M., Carofiglio, G., Altman, E., Primet, P.V.B.: A flow scheduler architecture. Research Report 7133, INRIA (December 2009)

Stateless RD Network Services

Maxim Podlesny[1] and Sergey Gorinsky[2]

[1] University of Calgary, Calgary, Canada
[2] Madrid Institute for Advanced Studies in Networks (IMDEA Networks),
Madrid, Spain
mpodlesn@ucalgary.ca, sergey.gorinsky@imdea.org

Abstract. Rate-Delay (RD) Network Services constitute a promising differentiated-services architecture for multi-provider networks, by offering users a choice between high throughput or low queuing delay at bottleneck links. An RD router provides service differentiation via transmission scheduling and by managing two FIFO queues. To ensure strict delay bounds, an RD router tracks arrival times of packets in the D service queue, and discards late packets at the queue head. However, maintaining the per-packet state is undesirable for complexity and cost reasons. In this paper, we present a Stateless RD (S-RD) router design that provides low queuing delay to the D service exclusively via buffer dimensioning, without requiring any per-packet state. After proving analytically that the S-RD design meets the delay guarantees, we use simulation to evaluate the performance of the stateless design, confirming that S-RD routers preserve the delay bounds of RD network services. As case studies, we consider Voice-over-IP (VoIP) and Web browsing as particular examples of Internet applications. The extensive simulation results demonstrate that S-RD Network Services significantly improve VoIP quality and increase the goodput of short-lived Web flows, without degrading the throughput of long-lived flows.

Keywords: Quality of service, resource allocation, Internet applications.

1 Introduction

The RD (Rate-Delay) Network Services [1] constitute a promising recent design for service differentiation in multi-provider network environments. In particular, the D (Delay) service assures low queuing delay at bottleneck links; it is suitable for Internet telephony and other applications that require low end-to-end packet delays. While delay is the most important consideration for delay-sensitive applications, the loss rate tends to be higher for the D service. In turn, the higher loss rate increases the occurrence of TCP (Transmission Control Protocol) [2] retransmission timeouts, which can negatively affect the performance of web and other TCP-based applications.

In this paper, we propose a modification to the RD router design. The original design involves recording packet arrival times in order to schedule transmissions and guarantee low queuing delay. This state information requires additional

M. Crovella et al. (Eds.): NETWORKING 2010, LNCS 6091, pp. 135–147, 2010.
© IFIP International Federation for Information Processing

memory and processing, which increases the router cost (e.g., expensive Static Random Access Memory (SRAM) [3]), makes the router more complex, and can affect router performance. Our modification avoids the per-packet state information required for arrival-time tracking. Our theoretical analysis and simulations confirm that the proposed S-RD (Stateless RD) Network Services still assure bounded queuing delays.

To understand the effects on application-perceived performance, we perform two case studies, in which we assess the performance of typical Internet applications (VoIP and web browsing) with S-RD Network Services. Our extensive simulations examine a wide variety of network topologies and traffic scenarios. The simulation results show that S-RD Network Services significantly improve the quality of Internet telephony. More surprisingly, web browsing also benefits from the S-RD Network Services.

The rest of the paper is organized as follows. Section 2 describes the S-RD design. In Section 3, we present the theoretical analysis of this stateless version. Section 4 evaluates the S-RD Network Services using ns-2 simulations. Sections 5 and 6 report our assessment of application-perceived performance for VoIP and web browsing, respectively. Finally, Section 7 concludes the paper with a summary of its contributions.

2   RD Network Services

2.1   Overview

The key idea in RD Network Services is to separate delay-oriented traffic and throughput-oriented traffic into two classes and serve them using separate queues. Class D provides low queueing delay, while class R provides high(er) per-flow throughput. The two main parameters of the RD Network Services are the delay constraint d, which is the maximum queuing delay for flows from class D, and the desired ratio k between per-flow rates for classes D and R.

The queues are scheduled such that the ratio of traffic volume serviced from the D and R queues is maintained close to $\alpha = \frac{n_D}{kn_R}$, where n_D and n_R represent the number of class D and class R flows, respectively. For a link of capacity C, the effective service rates for the D and R queues are:

$$R_D = \frac{n_D C}{n_D + kn_R}, \quad R_R = \frac{kn_R C}{n_D + kn_R} \tag{1}$$

We will refer to the traffic sent from the R and D queues during the recalculation period as L_D and L_R, respectively. The size of buffer for the D queue is configured so that the draining time of the D queue is close to d. Thus, the control rules employed to allocate the buffers for each class are the following:

$$B_D = \frac{n_D C d}{n_D + kn_R}, \quad B_R = \min\left\{B_{max}; \ B - \frac{n_D C d}{n_D + kn_R}\right\} \tag{2}$$

where B is the total buffer dedicated to a link, B_{max} is the size of a buffer required for effective support of throughput-greedy TCP traffic [4].

Routers periodically reallocate the buffers to queues for the D and R queues according to Equation (2). The numbers of flows in the classes are estimated through the time-stamp vector algorithm [5]. Packets from each queue are served in a FIFO (First-In First-Out) manner for scheduling the departures of packets. The selection of a queue to be served is based on the values of L_D and L_R since the last reset of L_D, L_R. If L_D is no more than αL_R, then the D queue is chosen for transmission. Otherwise, the next transmission is from the R queue. To enforce the delay constraint, the router tracks the arrival time of each packet in the D queue, and drops a packet from the head of queue D if the queueing delay of the packet exceeds d. In the event of queue overflow, the DropTail policy is used for dropping packets from either class.

2.2   Stateless Design

In the stateless version of the RD Network Services, we remove the requirement to explicitly track packet arrival times. Instead, we carefully limit the buffer size for class D traffic to ensure that queueing delay does not exceed d. In particular, we specify the size of the D buffer as follows:

$$B_D = \lfloor (d - w)R_D \rfloor^+, \quad w = \frac{2}{C}\left(\frac{S_D^{max}}{\alpha} + S_R^{max}\right) \tag{3}$$

where S_D^{max} and S_R^{max} are the maximum sizes of packets from classes D and R, respectively, and w is a delay adjustment. Through the subsequent analysis, we derive the foregoing expression for w, and prove that configuring the buffer size of queue D of the S-RD link via Equation (3) provides strict support for the delay constraint.

3   Analysis

In this section, we formally analyze the worst-case delay bounds for a backlogged S-RD router. There are in fact two different cases: one queue is backlogged, or both queues are backlogged. For space reasons, we ignore the simple case[1] of a single backlogged queue, and examine the more interesting case when both the D and R queues are backlogged.

The dual-queue backlog scenario also has two cases, based on the S-RD buffer configuration. The first (trivial) one corresponds to a buffer of size zero, which causes no queuing delay. The second case involves a non-zero buffering delay, i.e., $d - w > 0$, where d is the delay constraint.

Our goal is to consider the latter case, and derive the minimum value of w required for supporting queuing constraint if $B_D = \lfloor (d - w)R_D \rfloor^+$. Let us consider an arbitrary packet p from the D queue. We assume that p arrives at

---

[1] For this case, it suffices to add one more counter, beyond L_D and L_R, to indicate the traffic that must depart from the D queue in order to avoid exceeding the delay constraint.

the D queue at time t_a and departs from the D queue at time t_d. Suppose that at times t_a, t_d:

$$\frac{L_D(t_a)}{L_R(t_a)} = \alpha + \delta(t_a), \qquad \frac{L_D(t_d)}{L_R(t_d)} = \alpha + \delta(t_d) \qquad (4)$$

where $L_R(t_a) > 0$ and $L_R(t_d) > 0$. Thus, if $\delta(t) < 0$ then the packet sent at time t is from the D queue, otherwise - from the R queue. We consider the scenario where both the D and R queues are backlogged during time period $[t_a; t_d]$. We will refer to the traffic sent from the D and R queues during time period $[t_a; t_d]$ as ΔL_D and ΔL_R, $\Delta L_R > 0$, respectively. We prove the following theorem:

Theorem 1. *For any packet p in any traffic pattern, the maximum queuing delay is d-w provided that:*

$$\frac{\Delta L_D}{\Delta L_R} \geq \alpha \qquad (5)$$

Proof. Indeed, if inequality (5) holds, then $\frac{R'_D}{R'_R} \geq \alpha$, where R'_D, R'_R are actual serving rates for the D and R queues during $[t_a; t_d]$, respectively, and $R'_D \geq R_D$. Since $B_D = (d - w)R_D$, the maximum packet delay does not exceed $d - w$. ∎

Since $\Delta L_D = L_D(t_d) - L_D(t_a)$, $\Delta L_R = L_R(t_d) - L_R(t_a)$, we can rewrite inequality (5) as follows:

$$\frac{L_D(t_d) - L_D(t_a)}{L_R(t_d) - L_R(t_a)} \geq \alpha \qquad (6)$$

Let us denote the left side of inequality (6) as γ. Then, using expressions in Equation (4) and performing a simple transformation, we establish that:

$$\gamma = \alpha + \left(\delta(t_d) + \frac{L_R(t_a)}{\Delta L_R}(\delta(t_d) - \delta(t_a)) \right) \qquad (7)$$

Therefore, inequality (5) holds if and only if:

$$\delta(t_d) + \frac{L_R(t_a)}{\Delta L_R}(\delta(t_d) - \delta(t_a)) \geq 0 \qquad (8)$$

Let us now prove the following:

Theorem 2. *For any packet p in any traffic pattern, $\frac{\Delta L_D}{\Delta L_R} \geq \alpha$ if and only if:*

$$\delta(t_a) \leq 0, \quad \delta(t_d) \geq 0 \qquad (9)$$

Proof. First, let us prove this is a sufficient condition. Indeed, if $\delta(t_a) \leq 0$, $\delta(t_d) \geq 0$, then inequality (8) holds for any values of $L_R(t_a)$ and ΔL_R, i.e., for any traffic pattern and any packet p. Second, let us prove it is a required condition. Suppose that it is not true. We need to consider three possible cases:

Case 1: $\delta(t_a) \leq 0$, $\delta(t_d) < 0$. Then from inequality (8) we have that:

$$\frac{\Delta L_R}{L_R(t_a)} + 1 \leq \frac{\delta(t_a)}{\delta(t_d)} \qquad (10)$$

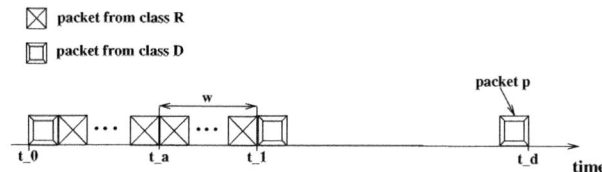

Fig. 1. Schedule of packet departures when $\delta(t_a) > 0$, $\delta(t_d) \geq 0$

Since the left side of inequality (10) exceeds 1, and there exist traffic patterns and packets p such that the right side is less than 1, we have a contradiction.

Case 2: $\delta(t_a) > 0$, $\delta(t_d) \geq 0$. Then from inequality (8) we derive that:

$$\frac{\Delta L_R}{L_R(t_a)} + 1 \geq \frac{\delta(t_a)}{\delta(t_d)} \tag{11}$$

Since there exist traffic patterns and packets p such that the left side of inequality (11) is smaller than 2, whereas the right side of inequality (11) exceeds 2, we have a contradiction.

Case 3: $\delta(t_a) > 0$, $\delta(t_d) < 0$. Inequality (8) leads us to:

$$\frac{\Delta L_R}{L_R(t_a)} + 1 \leq \frac{\delta(t_a)}{\delta(t_d)} \tag{12}$$

Since the left part of inequality (12) is positive, and its right side is negative, we have a contradiction. Thus, we have shown that our assumption cannot be true, which means that (9) is a required condition.   ∎

From Theorem 1 and Theorem 2 we conclude that (9) expresses a sufficient condition for supporting queueing delay of at most $d - w$ for any packet with an arbitrary traffic pattern at the S-RD link.

Theorem 3. *For any packet p in any traffic pattern, the maximum queuing delay is d-w if and only if $\frac{\Delta L_D}{\Delta L_R} \geq \alpha$.*

Proof. The sufficiency of this condition follows from Theorem 1. Let us now prove the necessity. Let us consider an arriving packet p that completely fills the buffer of the D queue, i.e., the enqueing of that packet causes $q_D = B_D$, where q_D is the size of queue D. Indeed, if $\frac{\Delta L_D}{\Delta L_R} < \alpha$, then $\frac{R'_D}{R'_R} < \alpha$, and $R'_D < R_D$. From the fact that $B_D = (d - w)R_D$ we conclude that the maximum packet delay of a packet p exceeds $d - w$.   ∎

Let us now consider all possible cases when the packet delay exceeds $d - w$ for a packet p.

Case 1: $\delta(t_a) > 0$, $\delta(t_d) \geq 0$. In Figure 1, we show the schedule of packet departures in the considered case. According to Theorem 2 and Theorem 3, if packet p arrived at time t_1 and departed at t_d, then its queuing delay would

140   M. Podlesny and S. Gorinsky

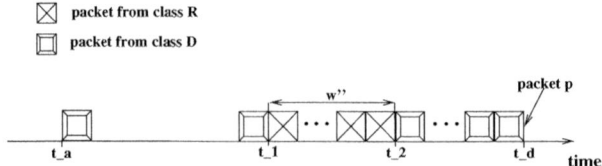

Fig. 2. Schedule of packet departures when $\delta(t_a) \leq 0$, $\delta(t_d) < 0$

not exceed $d - w$, since $\delta(t_1) < 0$, $\delta(t_d) \geq 0$. In the interval $[t_a; t_1]$, there is no potential arrival time t'_a of packet p at which $\delta(t'_a) < 0$, so the queuing delay of packet p can exceed the delay constraint by the length of the interval $[t_a; t_1]$. We will refer to the length of that interval as w', and to the amount of R traffic sent during this interval as X. Suppose that a D packet departing at time t_0 has size S_D, then the following inequalities must hold:

$$L_D(t_0) \leq \alpha L_R(t_0), \quad L_D(t_0) + S_D > \alpha L_R(t_0), \quad L_D(t_0) + S_D \leq \alpha L_R(t_0) + X\alpha \tag{13}$$

Lemma 1. *The value of X in the worst case scenario that satisfies the inequalities in (13) is:*

$$X = \frac{S_D^{max}}{\alpha} + S_R^{max}. \tag{14}$$

Proof. By inspection, it is clear that the X defined by equation (14) satisfies 13. Next, we need to show that there is no smaller solution. Let us suppose that there exists a smaller solution X':

$$X' = X - \Delta X \tag{15}$$

where X is defined by equation (14), $\Delta X > 0$, $\Delta X < X$, ΔX is an integer, i.e., there exists ΔX such that X' satisfies inequalities in (13). Let us assume that the traffic scenario is such that the first inequality in (13) achieves equality. Then from the third inequality in (13) and $L_D(t_0) = \alpha L_R(t_0)$ we derive that:

$$S_D \leq S_D^{max} + \alpha S_R^{max} - \alpha \Delta X \tag{16}$$

Assuming that $S_D = S_D^{max}$ and $S_R^{max} < \Delta X$, we have that inequality (16) is not valid. Since there exists a traffic scenario such that the third inequality in (13) is not valid, we have a contradiction. Finally, we mention that S_R^{max} in (14) reflects that traffic is in packets, i.e., not fluid. ∎

From Lemma 1 we conclude that the maximum queuing delay in excess of $d - w$ in the considered case is as follows:

$$w' = \frac{1}{C}\left(\frac{S_D^{max}}{\alpha} + S_R^{max}\right) \tag{17}$$

Case 2: $\delta(t_a) \leq 0$, $\delta(t_d) < 0$. In Figure 2 we demonstrate how the packets are scheduled for this case. If during time interval $[t_1; t_2]$, the link continued to serve the D queue up to packet p instead of packets from class R, then, according to Theorems 2 and 3, queuing delay of packet p would not exceed $d - w$, since $\delta(t_a) < 0$, $\delta(t'_d) \geq 0$, where t'_d would be its departure time. Therefore, queuing delay of packet p can exceed the delay constraint by the length of the interval $[t_1; t_2]$. We refer to the length of that interval as w'', and to the amount of D traffic sent during this time interval as Y. As in Case 1, Y is defined by the right side of equation (14). Therefore, w'' is the same as w' defined by equation (17).

Case 3: $\delta(t_a) > 0$, $\delta(t_d) < 0$. Since this case is a combination of the two previous ones, the maximum queuing delay in excess of $d - w$ is the sum of w' and w'':

$$w = \frac{2}{C}\left(\frac{S_D^{max}}{\alpha} + S_R^{max}\right) \tag{18}$$

This expression completes the derivation of 3. Since we did not use the information that p fills the buffer of queue D while considering the three possible cases of exceeding the $d - w$ delay, we have in fact proved the following:

Theorem 4. *Sizing the D buffer according to Equation (3) ensures that the S-RD router algorithm supports maximum queuing delay d for class D.*

4   Simulation Evaluation of S-RD Network Services

In this section, we evaluate our modification to the original design of the RD Network Services through simulations using version 2.29 of ns-2 [6]. We run the experiments in a dumbbell topology where the bottleneck and access links have capacities 100 Mbps and 200 Mbps, respectively. The bottleneck link carries 100 long-lived D flows and 100 long-lived R flows in both directions and has propagation delay 50 ms. We choose propagation delays for the access links so that the propagation RTT (Round-Trip Time) for the flows is uniformly distributed between 104 ms and 300 ms. In addition, there is one web server and one web traffic receiver connected to the bottleneck link. The long-lived flows from the both classes join the network during the initial 1 s of an experiment. The size of web flows is described by the Pareto distribution with the average of 30 packets and shape index 1.3, and the web flows arrive at the network according to a Poisson process. All flows employ TCP NewReno [7] and data packets of size 1 KB. We configure link buffer to $B = B_{max} = C \cdot 250$ ms, where C is the capacity of the link. Every experiment lasts 60 s, and we repeat it five times for each of the considered parameter settings. We apply the same settings for the design and the measurements as in [1]. We average the utilization and loss rate over the whole experiment with exclusion of its first five seconds. We report the maximum packet delay observed over all runs of each experiment. All our simulation results show that the S-RD Network Services support the required throughput differentiation between the classes.

Capacity scalability. To explore the impact of the link capacity, we vary the bottleneck link speed from 10 Mbps to 1 Gbps. The arrival rate of the web-like flows is 50 flows per second (fps). Figure 3 shows that there is no violation of

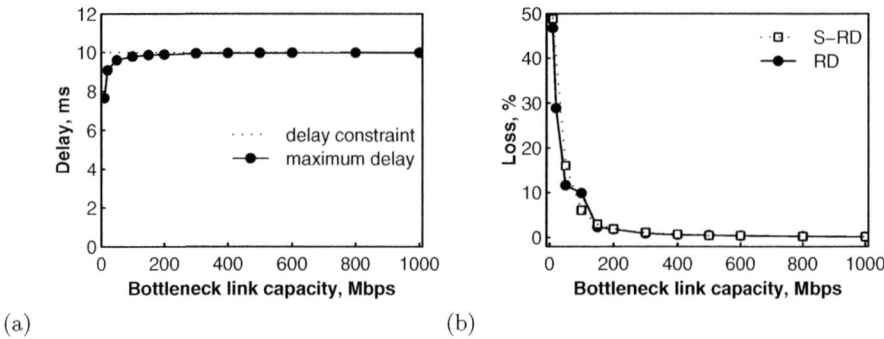

Fig. 3. Capacity scalability of the S-RD algorithm: (a) maximum queuing delay of class D; (b) comparison of the loss rate for class D to original version of the RD link

the delay constraint across the range of link speeds considered. In addition, the loss rate of class D shows only a marginal increase compared to the original RD design. The worst-case delay bound is tight for high link speeds, but not as tight for lower link speeds. The maximum queuing delay observed for low bottleneck link capacities close to 10 Mbps is due to the increased significance of the term w in Equation (3) specifying the size of the D buffer.

We also studied the impact of the intensity of web-like traffic, the value of delay constraint, and the maximum packet sizes for classes D and R by varying the appropriate parameter. The results indicate that the delay constraint is strictly supported over the whole range of the varied parameter. For space reasons, we do not provide these results here.

5   VoIP Traffic Results

To evaluate the quality of the delivered service for VoIP, we use Mean Opinion Score (MOS) [8], a subjective score for voice quality ranging from 1 (Unacceptable) to 5 (Excellent). To estimate a MOS score through network characteristics, we employ the E-Model [9], which assesses VoIP quality by accounting for network characteristics like loss and delay. The E-Model uses the R-factor, which is computed as a function of all of the impairments occurring with the voice signal. The R-factor ranges from 0 to 100, with 100 being the best, which is reflected on Table 1.

Table 1. Categories of voice transmission quality

R-factor range	MOS	Quality category	User satisfaction
90 - 100	4.34 - 4.50	Best	Very satisfied
80 - 90	4.03 - 4.34	High	Satisfied
70 - 80	3.60 - 4.03	Medium	Some users dissatisfied
60 - 70	3.10 - 3.60	Low	Many users dissatisfied
50 - 60	2.58 - 3.10	Poor	Nearly all users dissatisfied

5.1   Evaluation Methodology

To generate VoIP traffic and perform measurements of voice quality, we use the tool developed in [10], an additional module of the network simulator ns-2. We use the same network topology as in Section 4 with the same traffic from R class and web-like traffic from both classes, but the bottleneck link delay is 10 ms. Instead of long-lived D flows, there are 100 VoIP flows with the same propagation RTTs of 150 ms. The value of d is 50 ms. Web flows arrive with the intensity of 50 fps. We perform five experiments for each settings, and each experiment lasts for 70 sec. To encode the speech, we employ AMR (Adaptive Multi-Rate) Audio Codec [11] operating at audio bit rate of 12.2 kbps. The parameters we measure are average MOS and the average utilization of class R. While measuring MOS, first ten seconds of the experiment are neglected. All flows join the network during the initial 1 s. We compare the performance of the S-RD Network Services with the performance of DropTail.

5.2   Simulation Results

Influence of the web-like traffic. To study the influence of the web-like flows, we change the intensity of the web-like flows from 1 fps to 150 fps. In Figure 4, we observe that the S-RD Network Services demonstrate better performance for VoIP over almost the whole range of the varied parameter, while the R traffic achieves the same bottleneck link utilization as the DropTail link.

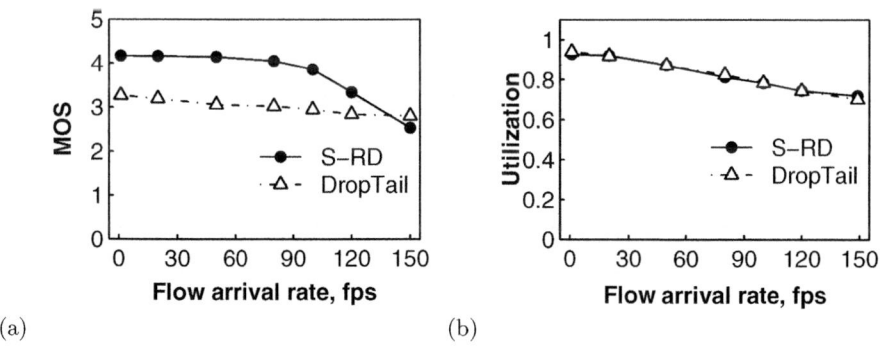

(a)   (b)

Fig. 4. Influence of the intensity of the web-like flows: (a) Average MOS; (b) average utilization of class R

Transient behavior. In this experiment, VoIP flows join the network during the whole experiment lasting for 600 s. There are 500 VoIP flows that start arriving from the beginning. The arrival process is Poisson with an average rate of 1 fps. Whereas the average MOS with the DropTail link is 2.97, MOS with the S-RD Network Services is 4.16. The utilization of class D is 84.45% and 83.85% with the

S-RD Network Services and DropTail link, respectively. Thus, the S-RD Network Services deliver better service for VoIP in the considered dynamic scenario.

Impact of VoIP population size. To examine the scalability of the design concerning the population of VoIP flows, we vary the number of them from 100 to 500. The results show that the number of VoIP flows does not affect the quality of VoIP. In particular, MOS with the S-RD Network Services is between 4.14 and 4.16 whereas MOS with the DropTail link is between 3.01 and 3.13. The consistent performance of VoIP over the whole range of the varied parameter is because a VoIP flow requires a relatively small connection throughput. The R flows achieve 82-87% utilization of the bottleneck link.

Partial deployment. In this experiment, we explore the situation when VoIP flows use the service provided by two different ISPs. There is one bottleneck within each ISP, 50 VoIP flows, 50 R flows going through both the ISPs, and two groups of 50 R flows each traversing a single ISP. In particular, we can consider two deployment scenarios, which reflect the deployment of the S-RD Network Services by only one ISP and by both the ISPs. The propagation RTT of VoIP flows is varied between 64 ms and 500 ms. In Figure 5, we observe that even under the partial deployment of the S-RD Network Services, which is labeled as "partial" in the graph, VoIP flows get better service. Moreover, the full deployment of the design (labeled as "full") further improves the VoIP quality. More importantly, the improvements of VoIP quality do not affect the service delivered to the R class concerning the flow rates.

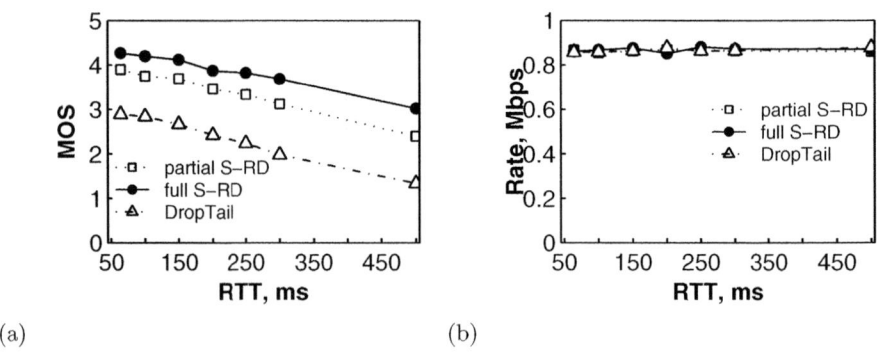

(a)   (b)

Fig. 5. Performance under the partial deployment for different propagation RTTs: (a) Average MOS; (b) average per-flow throughput of class R

6   Web Traffic Results

To evaluate the performance of a web application, which generates flows with different sizes, we calculate the average goodput of the web-like flows as the average of the goodput of each web-like flow. The goodput of a web-like flow is the ratio between the flow size and its FCT (Flow Completion Time).

6.1   Evaluation Methodology

In the experiments, we employ a dumbbell topology with the same experimental settings as in Section 5. To compare the S-RD Network Services design, we also run the experiments under the same settings for the DropTail link. There are 100 long-lived flows in the forward and reverse directions that are served as class R. The R flows join the network during the first second of an experiment. The value of d is 50 ms. In addition, there is one web server and one web traffic receiver connected to the bottleneck link. The web server generates flows with the same parameters as in Section 4, which are served as class D.

6.2   Simulation Results

Influence of the web-like traffic. We study the influence of the intensity of the web-like flows by varying their arrival rate in the interval between 5 fps and 400 fps. In Figure 6, we see that S-RD improves the performance of the web-like flows over the whole range of the varied parameter. On the other hand, the performance of the long-lived flows deteriorates when the intensity of the web-like flows exceeds 100 flows per second. The decreased performance of web-like flows beyond intensities of 250 fps is attributed to the increased packet loss rate.

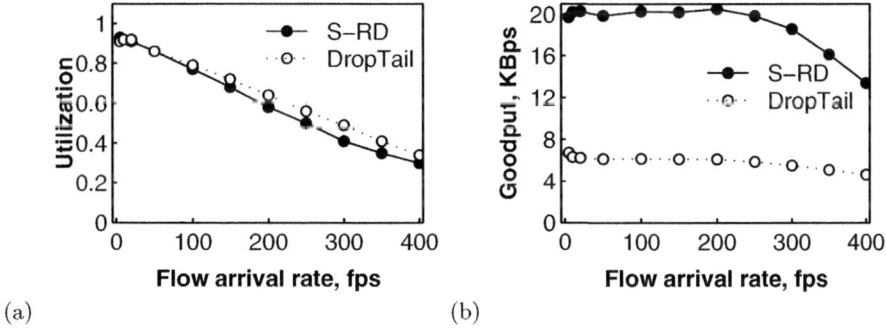

(a)   (b)

Fig. 6. Influence of the intensity of the web-like flows: (a) average utilization of the long-lived flows; (b) average goodput of the web-like flows

Influence of the long-lived flows. To explore the population scalability, we vary the number of long-lived flows between 50 and 600. The intensity of the web-like flows is 100 fps. Figure 7 shows that the S-RD Network Services provide consistently better performance for the web-like flows over the whole range of the varied parameter, whereas the long-lived flows have the same goodput with the S-RD Network Services and DropTail link. In particular, the former improves the goodput of the web-like flows by 50%-200%.

Influence of propagation RTT of the web-like traffic. In this experiments, we vary the propagation RTT of the web-like flows in the range between 30 ms

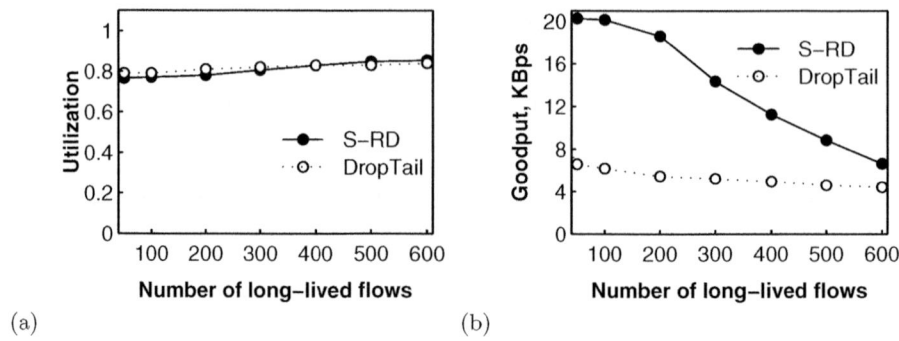

(a)   (b)

Fig. 7. Influence of the number of the long-lived flows: (a) average utilization of the long-lived flows; (b) average goodput of the web-like flows

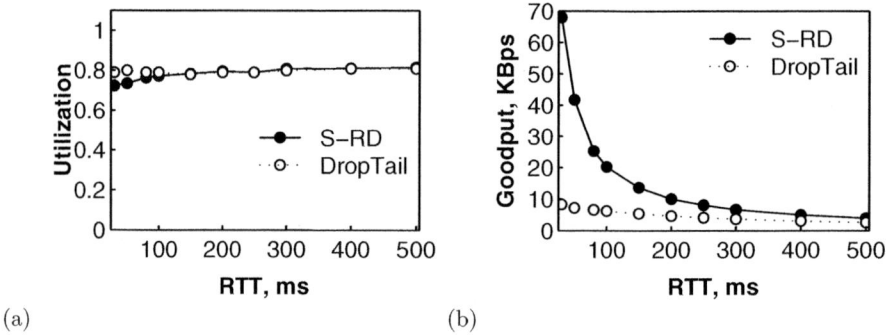

(a)   (b)

Fig. 8. Influence of propagation RTT of the web-like flows: (a) average utilization of the long-lived flows; (b) average goodput of the web-like flows

and 500 ms. The number of long-lived flows is 100. In Figure 8, we observe that the S-RD Network Services significantly improve the goodput of the web-like flows for small RTTs. Besides, the performance of the S-RD Network Services and DropTail link for the long-lived flows is similar except for RTTs less than 50 ms, for which the S-RD Network Services show slightly lower bottleneck link utilization than the DropTail scheme.

7   Conclusion

In this paper, we proposed a modification to the original router design of the RD Network Services. The new simpler design does not track packet arrival times. Our theoretical analysis and simulations showed that the proposed S-RD Network Services guarantee the low bounded queuing delay of the original design. As case studies, we explored the application-perceived performance of VoIP and web browsing with the S-RD Network Services. Through the extensive simulations, we determined that the architecture improves the application performance.

Acknowledgement

Financial support for the work was provided in part by iCORE (Informatics Circle of Research Excellence) in the Province of Alberta. The authors are thankful to Carey Williamson for his invaluable comments about this paper.

References

1. Podlesny, M., Gorinsky, S.: RD Network Services: Differentiation through Performance Incentives. In: Proceedings ACM SIGCOMM 2008 (August 2008)
2. Postel, J.: Transmission Control Protocol - DARPA Internet Program Protocol Specification. RFC 793 (September 1981)
3. Iyer, S., Kompella, R., McKeown, N.: Designing Packet Buffers for Router Line Cards. Technical Report, TR-02-HPNG-031001 (October 2002)
4. Villamizar, C., Song, C.: High performance TCP in ANSNET. ACM SIGCOMM Computer Communication Review 24(5), 45–60 (1994)
5. Kim, H., O'Halloran, D.: Counting Network Flows in Real Time. In: Proceedings IEEE GLOBECOM 2003 (December 2003)
6. McCanne, S., Floyd, S.: ns Network Simulator, http://www.isi.edu/nsnam/ns/
7. Floyd, S., Henderson, T.: The NewReno Modification to TCP's Fast Recovery Algorithm. RFC 2582 (April 1999)
8. ITU-T: Methods for subjective determination of transmission quality. Recommendation P.800 (August 1996)
9. Bergstra, J., Middelburg, C.: The E-model, a Computational Model for Use in Transmission Planning. ITU-T Recommendation G.107 (June 2006)
10. Bacioccola, A., Cicconetti, C., Stea, G.: User-level Performance Evaluation of VoIP Using ns-2. In: Proceedings NSTools 2007 (October 2007)
11. Sjoberg, J., Westerlund, M., Lakaniemi, A., Xie, Q.: Real-Time Transport Protocol (RTP) Payload Format and File Storage Format for the Adaptive Multi-Rate (AMR) and Adaptive Multi-Rate Wideband (AMR-WB) Audio Codecs. IETF RFC 3267 (June 2002)

Multicast in Multi-channel Wireless Mesh Networks

Ouldooz Baghban Karimi[1], Jiangchuan Liu[1], and Zongpeng Li[2],[*]

[1] School of Computing Science, Simon Fraser University
[2] Department of Computer Science, University of Calgary
{oba2,jcliu}@cs.sfu.ca, zongpeng@ucalgary.ca

Abstract. We study high-throughput multicast solutions for wireless mesh networks (WMN). Two techniques in WMN design are considered for combating wireless bandwidth limitations and wireless interference, respectively: introducing multiple mesh gateways and exploiting the diversity of wireless channels. We target a cross-layer solution that jointly (a) selects appropriate channels for each mesh node to use, at judiciously tuned power, and (b) computes the optimal multicast flows associated with the channel assignment. Our solution is obtained by first formulating the WMN multicast problem into a mathematical program, and then designing an iterative primal-dual optimization framework for it based on Lagrange relaxation and primal problem decomposition. Solution algorithms for the decomposed sub-problems are designed to complete the solution. In particular, a progressive channel assignment heuristic is introduced at the MAC/PHY layer. Through extensive simulations, we demonstrate the effectiveness of the proposed solution framework and the sub-problem heuristics. In particular, a throughput improvement of up to 100% is observed when compared to straightforward approaches of utilizing multiple wireless channels for mutlicast routing.

Keywords: Wireless Mesh Networks, Multicast, Multi-Channel Communication, Primal-Dual Optimization.

1 Introduction

Wireless mesh networks (WMN) are emerging as a promising solution for broadband connectivity, due to its flexibility and cost-effectiveness in bringing a large number of users online, in comparison to competing solutions that depend on a wireline infrastructure [1, 3]. In a WMN, Internet gateways, mesh routers and client nodes are organized into a mesh topology. Data flows are routed between the clients and the gateways through wireless links, in a multi-hop fashion. A notable challenge in WMN is to provide support for multicast applications that surged on the Internet during the past decade, such as file dissemination, video conferencing and live media streaming. Such applications usually serve a large number of users, and consume high network bandwidth.

We consider two techniques for addressing the high-throughput requirement of multicast applications in WMNs. The first is to use multi-gateways. A gateway is directly

---

[*] This research is supported by an NSERC Discovery Grant, an NSERC Strategic Project Supplemental Competition Grant, an AI-TF New Faculty Award, and a MITACS Project Grant.

M. Crovella et al. (Eds.): NETWORKING 2010, LNCS 6091, pp. 148–159, 2010.

connected to the Internet, and hence serves as the data source for users in a WMN. A single gateway design makes the gateway node a bottleneck, and is prone to congestion during high network activities. Having multiple gateways can dramatically improve the network performance at a reasonable cost. The second is to exploit the diversity in wireless channels, and provide a multi-channel multicast solution. Wireless interference is a critical limitation on throughput of WMN applications [8]. Utilizing distinct channels at neighboring nodes for transmission can help reduce interference to minimum. For example, the IEEE 802.11b/g protocol defines 13 channels within a 2.4GHz frequency band [5]. The further apart two channels are, the less interference exists between them; in particular, channels 1, 6 and 13 are totally orthogonal.

We first formulate the multi-gateway multi-channel multicast problem in WMNs as a mathematical programming problem, which jointly considers channel assignment and transmission power tuning at the MAC/PHY layer, as well as multicast routing at the network layer. Two important regions that the formulation is based on, the channel capacity region and the routing region, are both convex. Furthermore, the objective function that models the utility of multicast throughput is strictly concave. Therefore, the entire optimization model we obtain is a convex program, if we can freely select the frequency band for a channels. However, with pre-defined channels such as in IEEE 802.11, the optimization model contains discrete variables, which complicates the solution design.

In order to provide an efficient and practical solution the the optimization model, we apply the classic Lagrange relaxation technique [4, 20], and derive an iterative primal-dual optimization algorithm that leads to a cross-layer multicast solution. Towards this direction, we first relax the link capacity constraints that couple the channel region and the routing region, and decompose the overall optimization into two smaller sub-problems, one for channel assignment at the MAC/PHY layer, and one for multicast routing at the network layer. Our primal-dual solution framework then iteratively refines the primal solution, with help of the Lagrange dual that signalizes capacity demand at each wireless link. The dual is updated during each iteration based on the latest primal solutions.

To complete the solution defined by the primal-dual framework, we need to precisely define the channel region and the routing region, and design a solution algorithm for each of the channel assignment and routing sub-problems. We formulate the channel assignment problem as a mathematical program, in which channel capacities are computed from their signal-to-noise-and-interference ratio (SINR), and the computation of SINR in turn appropriately takes into account the separation between different wireless channels used at neighboring mesh nodes. The main challenge in solving this mathematical program lies in the presence of discrete channel assignment variables. We design an efficient heuristic, *progressive channel assignment*, for overcoming this difficulty. Finally, we discuss both multicast tree based and network coding based solutions for the multicast routing sub-problem. Extensive simulations, with various network sizes, were conducted for evaluating the effectiveness of both the overall primal-dual optimization framework and the sub-problem solutions. Throughput improvement of up to 100% were observed, when the proposed solution is compared to straightforward channel assignment schemes such as orthogonal channel assignment and consecutive channel assignment.

The rest of the paper is organized as follows. We review related research in Sec. 2. Sec. 3 presents the optimization problem formulation for multicast in WMN. Sec. 4 introduces the problem decomposition and the overall primal-dual solution framework. Sec. 5 presents solutions for the sub-problems. Sec. 6 is simulation results and Sec. 7 concludes the paper.

2   Related Work

There have been significant research on wireless mesh networking in recent years. Channel assignment has consistently been a focus with diverse static and dynamic solutions being proposed [15] [7] [6] [12] [13]. Given the tight coupling of different layers in such networks, joint optimization across layers have attracted great interest [5] [10] [13]. Alicherry et al. [3] presented a joint orthogonal channel assignment and unicast throughput maximization framework. Rad et al. [13] investigated channel allocation, interface assignment and MAC design altogether. Merlin et al. [11] further provided a joint optimization framework for congestion control, channel allocation, interface binding and scheduling to enhance the throughput of multi-hop wireless meshes. Their framework accommodates different channel assignments, but neighboring channel interference has yet to be addressed. Recently, Chiu et al. [5] proposed a joint channel assignment and routing protocol for 802.11-based multi-channel mobile ad hoc networks. While sharing many similarities with wireless meshes, the mobility concern and associated overheads are not critical in mesh networks given that the mesh routers and gateways are generally static.

Our work was motivated by these pioneer studies; yet our focus is mainly on throughput maximization in the multicast context. For multicast routing, Nguyen and Xu [14] systematically compared the conventional minimum spanning trees or shortest path trees in wireless meshes. Novel approaches customized for wireless meshes have also been proposed [16] [21] [20]. Our work is closely related to the latter two. In [21], two heuristics for multicast channel assignment were proposed, which also applies to a multi-gateway configuration. They however did not explicitly address route optimization. In [20], routing and wireless medium contention were jointly considered. The impact of link interferences and power amplitude variations on each link were also closely examined, but were limited to single channel usage. Our work differs from them in that we examine both multicast routing and channel assignment in a coherent cross-layer framework, and present effective solutions. We also explicitly explore the potentials of multi-gateway configurations.

3   The Multi-Channel Multicast Problem Formulation

We first construct mathematical programming formulations of the optimal multicast problem in WMNs, with multi-gateways and multi-channels. Envisioning two different physical layer technologies for selecting a frequency band for a channel, we present two corresponding optimization models. The first one is based on flexible frequency bands enabled by variable frequency oscillators, such as assumed in software-defined radios. This ideal radio model leads to optimal multicast throughput that can be computed precisely, through the classic primal-dual optimization framework. The second model is rather similar, but makes a more realistic assumption on frequency bands based on the state-of-the-art IEEE 802.11 standard: each transmission has to use one of the 13 pre-defined channels.

3.1 Network Model and Notations

We model a WMN as a graph $G = (V, E)$, with nodes V and links E. Assume $T \subseteq V$ is the set of gateways. Each gateway has a high-bandwidth connection to the Internet, and can be viewed as a data source. Let S be the set of data transmission sessions. We define five vectors of variables. The first four are: the vector of data flows $\mathbf{f} =< f_e^i | i \in S, e \in E >$; the vector of multicast throughput $\mathbf{r} =< r^i | i \in S >$; the vector of link capacities $\mathbf{c} =< c_e | e \in E >$; and the power assignment vector $\mathbf{p} =< p_u \leq p_{u,max} | u \in V >$. The last one is on channel assignment. We assume that each node is equipped with one radio with capacity b, which can transmit at different frequencies with adjustable power. In the flexible channel model, we have the vector of centre frequencies $\mu =< \mu_u | u \in V >$. The frequency band of the channel used by u is then $[\mu_u - b/2, \mu_u + b/2]$. In the case of fixed channels, we have the vector $\gamma =< \gamma_u \in \Gamma | u \in V >$ to represent the channel assignment at each node. Here Γ represents the set of pre-defined channels, such as the 13 in the IEEE 802.11 standard.

3.2 The Flexible Channel Model

Two capacity regions are fundamental to our multicast problem formulation: the *channel region* and the *routing region* , at the MAC/PHY layer and the network layer, respectively. The channel region H defines a set of $(\mathbf{c},\mathbf{h})$ such that channel assignment in $\mathbf{h}$ can support link capacity vector $\mathbf{c}$. The routing region R defines a set of $(\mathbf{r}, \mathbf{f})$ such that the throughput vector $\mathbf{r}$ can be supported by flow rates in $\mathbf{f}$. Detailed characterization of the two regions are not immediately relevant to the overall optimization structure, and are postponed to Sec 5.1 and Sec 5.3 respectively, where we select optimal solutions from each region.

The multicast throughput for each session is measured as the data receiving rates at the receivers, which are equal for receivers across the same session. A basic physical rule that establishes a connection between the routing region and the channel region is that the aggregated data flow rates have to be bounded by the corresponding link capacities. Furthermore, we follow the convention [20] in modeling throughput utility, and adopt the concave utility function $log(1 + r_i)$ for session throughput r_i. Then, the throughput maximization problem can be formulated as:

$$\text{Maximize } U(r) = \sum_{i \in S} U(r_i) = \sum_{i \in S} log(1 + r_i)$$

$$\text{Subject to } (c, \mu, p) \in H$$

$$(r, f) \in R$$

$$\sum_{i \in S} f_e^i \leq c_e, \forall e \in E \tag{1}$$

The first constraint $(\mathbf{c}, \mu, \mathbf{p}) \in H$ models the dependence of effective channel bandwidth on channel assignment and power assignment at each node. The second constraint $(\mathbf{r}, \mathbf{f}) \in R$ models the dependence of multicast throughput $\mathbf{r}$ on the routing scheme $\mathbf{f}$. $\sum_{i \in S} f_e^i \le c_e, \forall e \in E$ model link capacity constraints. The objective function $U(r)$ is concave, and both the routing and channel regions are convex regions. Therefore, convex optimization methods [4] can be used to compute the optimal solution (μ^*, p^*, f^*). In Sec. 4 and Sec. 5, we present a primal-dual solution based on based on Lagrange relaxation and iterative primal-dual optimization.

If nodes can transmit using pre-defined channels only, we can modify the mathematical program in (1), by replacing the frequency vector μ with the channel assignment vector γ. Since γ is an integer vector, the mathematical program can not be directly solved to optimal using conventional convex optimization methods, in polynomial time. Nonetheless, the solutions in Sec. 4 and Sec. 5 will be flexible enough to compute approximate solutions, based on a heuristic channel assignment method.

4   The Primal-Dual Solution Framework

The overall solution framework we propose for solving (1) is an iterative primal-dual schema, which switches between solving primal sub-problems and updating dual variables. We describe in Sec. 4.1 how to decompose the primal problem while introducing dual variables, and then present the primal-dual solution framework in Sec.4.2.

4.1   The Routing vs. Channel Assignment Decomposition

A critical observation of the optimization problem (1) is that, the channel region H and the routing region R characterize variables from the MAC/PHY layer and the network layer respectively, and are relatively independent. The only coupling constraint between them is $\mathbf{f} \le \mathbf{c}$. We can apply the Lagrange relaxation technique [4,9] to remove $\mathbf{f} \le \mathbf{c}$ from the constraint set, and add a corresponding price term into the objective function: $L = U(r) + \sum_{e \in E} \alpha_e [c_e - \sum_{i \in S} f_e^i]$. Here α is a vector of Lagrange multipliers, which can be viewed as prices governing the link capacity supply — the larger α_e is, the tighter bandwidth supply at link e is.

After the relaxation, the resulting optimization problem is naturally decomposed into two smaller, easier-to-solve sub-problems, including the Channel Assignment Sub-problem at the MAC/PHY layer:

$$Maximize \sum_{e \in E} \alpha_e c_e \qquad (2)$$
$$Subject\ to\ (c, \gamma, p) \in H$$

and the Routing Sub-problem at the network layer:

$$Maximize\ U(r) - \sum_{e \in E}(\alpha_e \sum_{i \in S} f_e^i) \qquad (3)$$
$$Subject\ to\ (r, f) \in R$$

It is interesting to observe that, given a link e with high price α_e, the routing sub-problem will automatically attempt to reduce the amount of flow f_e through e during the next round, since its objective function implies minimizing $\sum_{e \in E} \alpha_e \sum_{i \in S} f_e^i$. On the other hand, the channel assignment sub-problem will automatically attempt to create more capacity for e, since its objective function is to maximize $\sum_{e \in E} \alpha_e c_e$.

4.2 The Primal-Dual Solution Schema

The primal-dual approach iteratively updates the primal $(\mathbf{f}, \mu, \mathbf{p})$ and dual (α) solutions. During each iteration, we solve the two primal sub-problems given the current dual vector α, and subsequently update α with the newly computed primal vectors as below. Here t is the round number, and β is the step size vector.

i. Set $t = 1$; initialize $\alpha(0)$, e.g., set $\alpha_e(0) = 0, \forall e \in E$
ii. Solve primal sub-problems (2) and (3).
iii. Update the dual domain variables as below:

$$\alpha(t) = max(0, [\alpha(t-1) + \beta(t)(\sum_{s \in G} \sum_{t \in T_s} f_e^t)])$$

iv. Set $t = t + 1$ and return to step ii, until convergence.

Theorem 1. *The primal-dual algorithm above converges to an optimum primal solution ($\mathbf{f}^*$, μ^*, $\mathbf{p}^*$). of the optimization problem (1), as long as the regions R and H are convex and the step sizes $\beta(t)$ are appropriately chosen.*

Proof. The constraint $\mathbf{f} < \mathbf{c}$ is linear, the objective function in (1) is strictly concave. The convexity of the capacity regions R and H then ensures that the update in the dual domain (iii) is a sub-gradient for the dual variables in α. Therefore as long as the step sizes are appropriate chosen, the dual update converges [9, 20]. Strong duality further assures that the convergence point of the primal-dual algorithm corresponds to a global optimum of the network optimization problem in (1).$\beta[t] \geq 0, \lim_{t \to \infty} \beta[t] = 0$, and $\sum_{t=1}^{\infty} \beta[t] = \infty$. A simple sequence that satisfies the conditions above, is $\beta[k] = a/(mk + n)$, for some positive constants a, m and n. $\square$

5 Solving Channel Assignment and Routing Sub-problems

In order to obtain a complete solution under the primal-dual schema, we need to design algorithms for solving each of the two primal sub-problems. We next discuss how to solve the channel assignment sub-problem in Sec. 5.1 and Sec. 5.2, and the routing sub-problem in Sec. 5.3

5.1 The Channel Assignment Sub-problem

We now construct a detailed model for the channel capacity region H, and discuss how the resulting channel assignment problem from (2) can be solved. The effective link bandwidth are determined by the signal-to-noise-and-interference ratio (SINR) of the transmission; following the Gaussian channel capacity model [20]:

$$c_e = b\log_2(1 + SINR_e), \quad SINR_e = \frac{G_{ee}P_e}{(\sum_{l\neq e} I_{le}.P_l.G_{le}) + \sigma^2}$$

Here G_{ee}, P_e and σ^2 are gain, power and noise associated with a link respectively. G_{le} and σ^2 denote the interference coefficient and noise from link l to link e respectively. I_{le} is the *channel correlation coefficient*, which depends on the *separation* between channels used by l and e, e.g., the separation between channels 1 and 4 is 3.

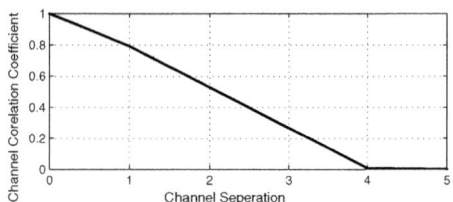

Fig. 1. Power leakage for neighbouring 802.11g channels

The correlation $I_{\gamma\gamma'}$ between two channels γ and γ' are known for all possible channel separations, as shown in Fig. 1 [2]. From this figure, the correlation between any two channels, either flexibly selected or pre-defined, can be found. For example, for the 13 IEEE 802.11 channels, $I_{\gamma_1\gamma_1} = 1.0$, $I_{\gamma_1\gamma_2} = 0.7906$, $I_{\gamma_1\gamma_3} = 0.5267$, and $I_{\gamma_1\gamma_7} = 0$. Furthermore, we assume the total budget at each node v is $p_{v,max}$ and $O(v)$ is the set of outgoing links from node v. Then, the channel assignment problem for capacity maximization can be formulated as:

$$\text{Maximize } \sum_{e\in E} \alpha_e c_e \tag{4}$$

$$\text{subject to } c_e = b\log_2(1 + SINR_e), \forall e \in E$$

$$SINR_e = \frac{G_{ee}p_e}{(\sum_{l\neq e} I_{\gamma_l\gamma_e}.p_l.G_{le}) + \sigma^2}, \forall e \in E$$

$$\sum_{e\in O(v)} p_e \leq p_{v,max}, \forall v \in V$$

$$\gamma_e \in \Gamma, \forall e \in E$$

Without the discrete variables in γ, (4) can be solved using known techniques, such as using geometric programming or through a power control game [20]. The new challenge in (4) is to compute good channels for vector γ. In Sec. 5.2, we present a heuristic solution for efficiently solving γ, and evaluate its performance later in Sec. 6.

5.2   Heuristic Channel Assignment Algorithm

We design a heuristic channel assignment algorithm based on the *interference factor* $\varphi_\gamma = \sum_{\gamma'\in\Gamma} I_{\gamma\gamma'}/d'_\gamma$, for a given candidate channel γ. Here $d_{\gamma'}$ is the distance to the nearest node transmitting at channel γ'. In measurement-based systems, $d_{\gamma\gamma'}$ is not necessary because the node can sense if a channel is in use in the range of this node and signal strength could be used instead. In coordination-based systems, $d_{\gamma\gamma'}$ could be found based on the coordination information. We assume $d_{\gamma'} = \infty$ if channel γ' is not in use in the network.

Our heuristic solution, Algorithm 1, performs a breadth-first-traversal of the WMN. At each node, candidate channels are sorted by the interference factor to already assigned channels at other nodes. Different options are possible in selecting the channel. A greedy algorithm selects the channel γ with the smallest φ_γ value, *i.e.*, as apart from neighboring channels in use as possible. We propose a progressive channel assignment approach instead, and select a channel γ with the highest φ_γ below an acceptable threshold φ_{th}. The rational here is to look beyond channel assignment at the current node, and to leave good candidate channels for neighbor nodes.

Algorithm 1. Progressive Channel Assignment

Initialization:
$\Gamma(v) := \emptyset, \forall v \in V$;
$\varphi Set := \emptyset$;
forall $v \in V$ **do**
    forall $\gamma \in \Gamma$ **do**
        Compute φ_γ^v;
        if $\varphi_\gamma^v \le \varphi_{th}$ **then**
            $\varphi Set := \varphi Set \cup \varphi_\gamma^v$;

    if $\varphi Set = \emptyset$ **then**
        Choose γ_v with smallest ϕ_{γ_v} and activate it on node v;
    else
        Choose γ_v from φSet with largest ϕ_{γ_v} and activate it on node v;
    $\varphi Set := \emptyset$

Algorithm 1 consists of a double loop. The outer loop iterates through nodes in the network, and the inner loop iterates through all possible channels. The number of channels is 13 and therefore the total number of iterations is $13|V|$.

5.3 The Routing Sub-problem

The multicast flow routing problem at the network layer has been extensively studied in the literature during the past decade. Two classes of solutions have been proposed. The first class includes multicast tree based solutions. Since achieving optimal multicast throughput using multicast trees corresponds to the NP-hard problem of Steiner tree packing, one needs to resort to efficient approximation algorithms, such as the KMB algorithm. The second class includes network coding based solutions. By assuming information coding capabilities for nodes in the network, the complexity of the optimal multicast problem decreases from NP-hard to polynomial time solvable [9]. In particular, *conceptual flow* based linear programming models have been successfully developed for multicast in various network models [9, 20]. In this section, we apply similar techniques and formulate our routing sub-problem into a convex program with all-linear constraints, which can be solved using general convex optimization algorithms such as the interior-point algorithm [4], or tailored subgradient algorithms [9].

We model flows from each of the gateways to different destinations as conceptual flows that do not compete for link bandwidth. $e_l^{i,j}$ denotes the conceptual flow rate on link l in ith multicast session to its jth destination. $I(v)$ is set of incoming links to node v and $O(v)$ is the set of outgoing links from node v. The multi-gateway multicast routing sub-problem with network coding can be stated as below. For compact LP formulation, the convention of assuming a virtual feedback link from multicast receivers to sources is followed [9].

$$\text{Max } U(r) - \sum_{e \in E} \alpha_e \sum_{s \in G} \sum_{t \in T_s} f_e^t$$

$$\text{s.t. } r^t \leq \sum_{t \in T_s} \sum_{l \in I(D_j^{i,t})} e_l^{i,j} \, , \forall i, \forall j, \forall D_j^{i,t} \in V, \forall t \in T_s$$

$$e_l^{i,j} \leq f_l^i, \forall i, \forall j, \forall l \in E$$

$$\sum_{l \in O(v)} e_l^{i,j} = \sum_{l' \in I(v)} e_{l'}^{i,j} \, , \forall v \in T_s, \forall i, j$$

$$f_l^i \geq 0, e_l^{i,j} \geq 0, r^t \geq 0$$

6   Simulation Results

We have implemented the overall primal-dual solution framework, the sub-problem solutions, to examine their performance. For comparison purposes, we have also implemented two other solutions: (a) *orthogonal channel assignment* and (b) *consecutive channel assignment*. In (a), the 13 802.11 channels are assigned to mesh nodes in a

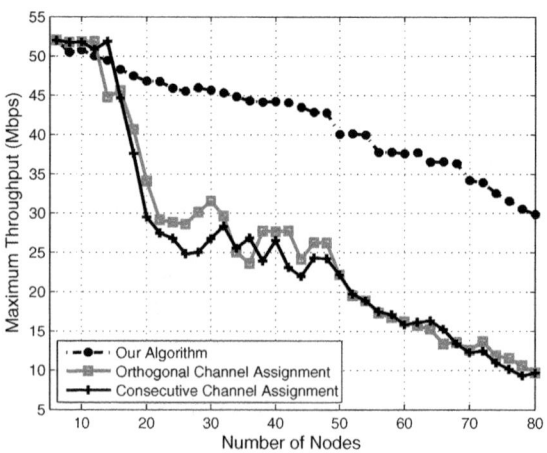

Fig. 2. Maximum throughput (Mbps) vs. number of nodes

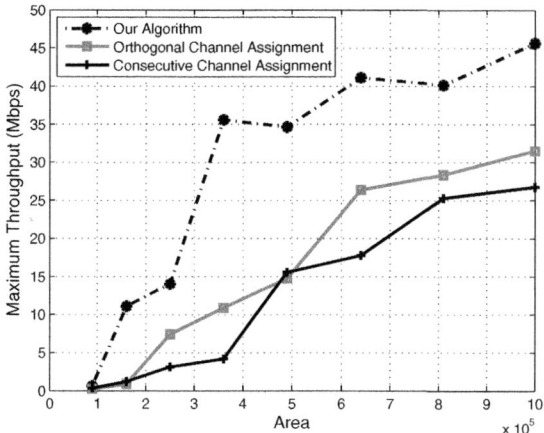

Fig. 3. Maximum throughput (Mbps) vs. area

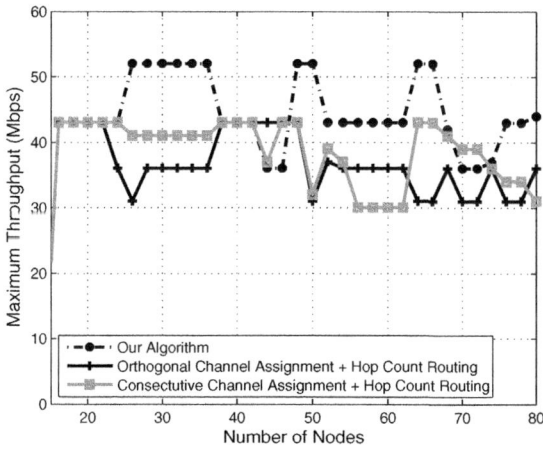

Fig. 4. Maximum multicast throughput(Mbps) vs. number of nodes

consecutive fashion (from channel 1 to 13, then back to 1), during a BFS traversal. In (b), the greedy approach of selecting a channel with maximum separation is adopted.

For results shown in Fig. 2 and Fig. 3, the network area varies from 10^5 to 10^6 m^2, with 6 to 60 nodes equipped with radios running IEEE 802.11 protocol and transmission power of $1mW$. The area is square-shaped and the nodes are placed randomly. Half of the nodes act as receivers in a multicast session. The link capacities are computed using $c_e = b \log_2(1 + SINR_e)$. The variance of the number of nodes is intended for observing the performance of the solutions with different levels of interference. First we observe the channel assignment in different methods, our proposed channel assignment algorithm, orthogonal channel assignment (BFS on neighbours of a node then assign channels with maximum difference) and consecutive channel assignment (BFS

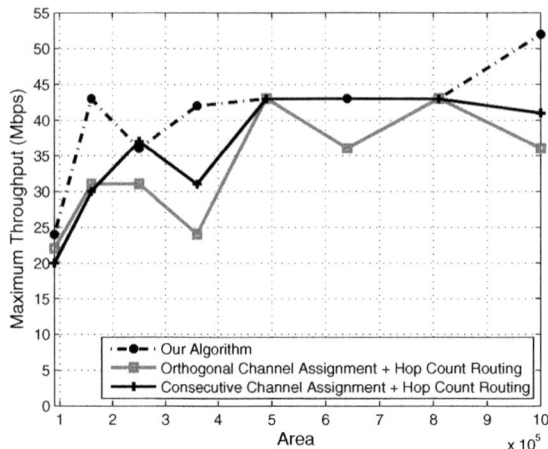

Fig. 5. Maximum multicast throughput(Mbps) vs. area

on neighbours of a node then assign channels consecutively). An overall observation in Fig. 2 and Fig. 3 is that our solution leads both orthogonal and consecutive channel assignment methods, with a largest margin of up to 100%. The improvement increases as the network size increases. Note that the throughput of orthogonal channel assignment and our proposed algorithm could be improved by scheduling mechanisms.

Fig. 4 and 5 show results similar to those in Fig. 2 and 3, but with less number of data sessions, and therefore lower level of interference. The primal-dual solution still performs better overall, but the leading margin is less obvious. We also note the throughput measured is more or less unstable. That is due to the randomness in the network topologies used in the simulations. Intuitively, when interference is low and not a serious concern, a judiciously designed multi-channel transmission scheme becomes less important. We conclude that our proposed solution is more beneficial when applied in networks with high transmission activities and high interference.

7   Conclusion

Multicast applications that require high throughput have recently gained popularity. We studied in this paper the challenges in achieving high multicast throughput in wireless mesh networks. Two techniques in system design were assumed: introducing multiple mesh gateways for mitigating the gateway bottleneck problem and utilizing multiple wireless channels for combating wireless interference. Our overall solution framework is a primal-dual schema based on a mathematical programming formulation of the optimal multicast problem. The framework iteratively switches between solving primal sub-problems for channel allocation and routing, and dual variable update, and gradually progresses towards optimal or approximately optimal solutions. We further presented precise models for each primal sub-problem, and discussed solutions for each of them. Simulation results confirmed the proposed solutions, in considerate throughput gains that were observed over straightforward approaches of multi-channel multicast.

References

1. Akyildiz, F., Wang, X.: A survey on wireless mesh networks. IEEE Communications Magazine 43(9), 23–30 (2005)
2. Angelakis, V., Traganitis, A., Siris, V.: Adjacent channel interference in a multi-radio wireless mesh node with 802. 11a/g interfaces IEEE INFOCOM poster session (2007)
3. Alicherry, M., Bhatia, R., Li, L.: Joint channel assignment and routing for throughput optimization in multi-radio wireless mesh networks. In: ACM MobiCom, pp. 58–72 (2005)
4. Bertsimas, B., Tsitsiklis, J.N.: Introduction to Linear Optimization. Athena Scientific, Belmont (1997)
5. Chiu, H.S., Yueng, K.L., Lui, K.: J-CAR: An efficient joint channel assignment and routing protocol for IEEE 802.11-based multi-channel multi-interface mobile ad-hoc networks. IEEE Transactions on Wireless Communications 8(4), 1706–1714 (2009)
6. Gong, M.X., Midkiff, S.F., Mao, S.: Design principles for distributed channel assignment in wireless ad-hoc networks. In: IEEE ICC, pp. 3401–3406 (2005)
7. Ko, B., Misra, V., Padhye, J., Rubenstein, D.: Distributed channel assignment in multi-radio 802.11 mesh networks. In: IEEE WCNC, pp. 3978–3983 (2007)
8. Li, J., Blake, C., De Couto, D.S.G., Lee, H.I., Morris, R.: Capacity of ad-hoc wireless networks. In: ACM MobiCom, pp. 61–69 (2001)
9. Li, Z., Li, B., Lau, L.C.: On Achieving Maximum Multicast Throughput in Undirected Networks. IEEE Transactions on Information Theory 52(6), 2467–2485 (2006)
10. Ling, X., Yeung, K.L.: Joint access point placement and channel assignment for 802.11 wireless LANs. IEEE Transactions on Wireless Communications 5(10), 2705–2711 (2006)
11. Merlin, S., Vaidya, N., Zorzi, M.: Resource allocation in multi-radio multi-channel multi-hop wireless networks. In: IEEE INFOCOM, pp. 610–618 (2008)
12. Mishra, A., Banerjee, S., Arbaugh, W.: Weighted coloring-based channel assignment for WLANs. In: ACM SIGMOBILE, pp. 19 31 (2005)
13. Mohsenian Rad, A.H., Wong, V.W.S.: Joint channel allocation, interface assignment and MAC design for multi-channel wireless mesh networks. In: IEEE INFOCOM, pp. 1469–1477 (2007)
14. Nguyen, U.T., Xu, J.: Multicast routing in wireless mesh networks: minimum cost trees or shortest path trees. IEEE Communications Magazine 45, 72–77 (2007)
15. Ramachandran, K.N., Belding, E.M., Almerot, K.C., Buddhikot, M.M.: Interference-aware channel assignment in multi-radio wireless mesh networks. In: IEEE INFOCOM, pp. 1–12 (2006)
16. Rong, B., Qian, Y., Lu, K., Qingyang, R.: Enhanced QoS multicast routing in wireless mesh networks. IEEE Transactions on Wireless Communications 7(6), 2119–2130 (2008)
17. Approximation Algorithms. Springer, Heidelberg (2001)
18. Villegas, E.G., Lopez-Aguilera, E., Vidal, R., Paradells, J.: Effect of adjacent-channel interference in IEEE 802.11 WLANs. In: Crowncom, pp. 118–125 (2007)
19. Wu, Y., Chou, P.A., Jain, K.: A comparison of network coding and tree packing. IEEE ISIT, 143 (2004)
20. Yuan, J., Li, Z., Yu, W., Li, B.: A cross-layer optimization framework for multihop multicast in wireless mesh networks. IEEE Journal on Selected Areas in Communication 24(11), 2092–2103 (2006)
21. Zeng, G., Wang, B., Ding, Y., Xiao, L., Mutka, M.W.: Efficient multicast algorithms for multi-channel wireless mesh networks. IEEE Transactions on Parallel and Distributed Systems 21(1), 86–99 (2010)

Ambient Interference Effects in Wi-Fi Networks

Aniket Mahanti[1], Niklas Carlsson[1], Carey Williamson[1], and Martin Arlitt[1,2]

[1] Department of Computer Science, University of Calgary, Calgary, Canada
{amahanti,ncarlsso,carey}@cpsc.ucalgary.ca
[2] Sustainable IT Ecosystem Lab, HP Labs, Palo Alto, U.S.A.
martin.arlitt@hp.com

Abstract. This paper presents a measurement study of interference from six common devices that use the same 2.4 GHz ISM band as the IEEE 802.11 protocol. Using both controlled experiments and production environment measurements, we quantify the impact of these devices on the performance of 802.11 Wi-Fi networks. In our controlled experiments, we characterize the interference properties of these devices, as well as measure and discuss implications of interference on data, video, and voice traffic. Finally, we use measurements from a campus network to understand the impact of interference on the operational performance of the network. Overall, we find that the campus network is exposed to a large variety of non-Wi-Fi devices, and that these devices can have a significant impact on the interference level in the network.

Keywords: Wi-Fi, Interference, Spectrogram, Duty Cycle, Data, Video, Voice.

1 Introduction

Wireless Fidelity (Wi-Fi) networks allow users with wireless-capable devices to access the Internet without being physically tied to a specific location. Today's Wi-Fi networks are not only being used for Web surfing, but also for viewing live video content (e.g., live sports and news events) and voice communication (e.g., voice over IP or VoIP). Many of these services require high quality of service and reliability, which can be degraded by wireless interference.

Wi-Fi networks employ the IEEE 802.11 protocol, which uses the unlicensed 2.4 GHz Industrial, Scientific, and Medical (ISM) Radio Frequency (RF) band [6]. Since the ISM band is unlicensed, it is available for use by multiple devices (both Wi-Fi and non-Wi-Fi), inherently causing interference for one another. The 802.11 protocol is considered to be a polite protocol in that an 802.11 device will transmit only if it senses that the RF channel is free. Non-Wi-Fi devices such as microwave ovens are oblivious to this protocol. These devices transmit regardless of whether the channel is free or not. This makes the interference problem challenging in Wi-Fi networks.

Network practitioners often deploy Wi-Fi networks without knowledge of the ambient usage of the ISM band by non-Wi-Fi devices. For example, trace capture programs, employed for site surveys used to select RF channels for problem-free functioning of Wi-Fi networks, only recognize devices using the 802.11 protocol. Since this process concentrates on the link layer and up, any activity from non-Wi-Fi devices is ignored.

M. Crovella et al. (Eds.): NETWORKING 2010, LNCS 6091, pp. 160–173, 2010.

Unfortunately, since several non-Wi-Fi devices may be operating on the same channel as a Wi-Fi network, and cause severe interference, it is difficult to identify interference sources using this trace capture technique.

For a more complete picture of the activity on the 2.4 GHz ISM band, one needs to consider the physical layer. In this paper, we use an off-the-shelf wireless spectrum analyzer to understand how non-Wi-Fi devices impact the functioning of Wi-Fi networks. Using controlled experiments, we characterize the interference properties of six non-802.11 devices. Five are *unintentional* interferers: a microwave oven, two cordless phones (one analog and one digital), an analog wireless camera, and a Bluetooth headset. We also evaluate one *intentional* interferer, a wireless jammer, for comparison purposes. In addition to capturing the basic characteristics of these devices, we measure, quantify, and discuss implications of their interference on data, video, and voice traffic. Finally, using passive measurements from an operational campus network, we try to understand the impact of interference on the network.

Our results show that among the unintentional interferers, microwave ovens, analog cordless phones, and wireless cameras have the most adverse impact on Wi-Fi networks. Because of its wideband interference, microwave signals affect several Wi-Fi channels at close range, however, their impact is still felt at longer distances. Analog cordless phones and wireless cameras are continuous narrowband interferers that completely obliterate Wi-Fi service on any channels they are using. Digital cordless phones and Bluetooth headsets have minimal impact on Wi-Fi because of their frequency hopping nature. From our production network measurements, we find that the campus network is exposed to a large variety of non-Wi-Fi devices, and that these devices can have a significant impact on the interference level in the network. For example, during certain times of the day, almost 80% of a channel may be occupied by interferers, and it is common to see some interference device active (in the background) almost all the time.

The rest of the paper is organized as follows. Section 2 discusses the interferers we studied and our experimental setup. Section 3 describes the organization of the Wi-Fi channels and characterizes the physical layer properties of the six interferers. Section 4 studies the impact of non-802.11 interference on data, video, and voice traffic over Wi-Fi. Section 5 studies the channel utilization of Wi-Fi and various interferers in a production network. Section 6 describes related work. Section 7 concludes the paper.

2 Methodology

2.1 Interferers

Microwave Oven: We used a Panasonic NNS615W microwave oven. This device had a maximum output power of 1,200 W. During our experiments, the oven was operating at maximum power and had a container of food inside the oven.

Analog Wireless Video Camera: We used a Lorex SG8840F analog wireless video camera operating at 2.4 GHz. These cameras provide long-range surveillance using analog signals, typically use directional antennas, and can have a range up to 1.5 km.

Analog Cordless Phone: We used a Vtech GZ2456 analog cordless phone operating at 2.4 GHz. The phone consists of a base and a handset, and as with other analog phones,

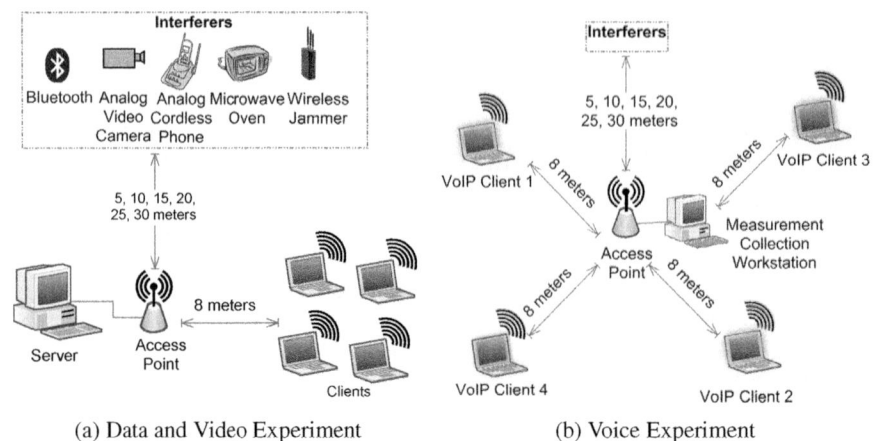

(a) Data and Video Experiment   (b) Voice Experiment

Fig. 1. Experimental setup

works by converting voice into electronic pulses and sending them on a designated channel between the base and handset.

Digital Cordless Phone: We used a Uniden DCT648 digital phone (base and handset) operating at 2.4 GHz. This particular phone uses Digital Spread Spectrum (DSS) to change channels frequently for enhanced voice quality and security.

Bluetooth Headset: We used a Plantronics Pulsar 590A headset. These are short range devices and have a maximum operating range of 10 m.

Wireless Jammer: Wireless jammers may be used for RF Denial of Service attacks. The power levels of these jammers vary from 1 mW to 30 mW.

2.2   Experimental Setup

We conducted two sets of measurement experiments to study the physical layer characteristics of the above-mentioned interferers and their impact on Wi-Fi networks.

The first set of experiments examines the physical-layer characteristics of these interferers in an isolated environment. We used an off-the-shelf spectrum analyzer called AirMagnet Spectrum Analyzer[1] for this purpose. The spectrum analyzer is a hardware/software unit that consists of a radio for detecting RF energy in the 2.4 GHz ISM band, and a software engine that performs Fast Fourier Transforms (FFTs). The spectrum analyzer uses these FFTs to classify known interferers. The physical layer measurements were taken in an interference-neutral environment such that we only captured RF energy from the specific device under study.

The second set of experiments quantifies the impact of interferers on the performance of a Wi-Fi network. We measured the performance degradation of the network in the

---

[1] http://www.airmagnet.com/products/spectrum_analyzer/

presence of the interferers for three types of traffic workloads: data, video, and voice. Our Wi-Fi network consisted of a single D-Link 2100 access point (AP) running in IEEE 802.11g mode. We chose the 802.11g standard because it offers higher transmission rates than 802.11b and is a popular choice for deployment in enterprise, academic, home, and hot spot networks. Additionally, most modern laptops are equipped with 802.11g capable Network Interface Cards (NICs).

We used the setup shown in Figure 1(a) for data and video experiments. We connected a server workstation using an Ethernet cable. We placed four Lenovo T61 laptop clients with built-in 802.11abg NICs at a distance of 8 m from the AP. We placed a single interferer at successive distances of 5, 10, 15, 20, 25, and 30 m from the AP to record the performance degradation of the interferer on the network. For each interferer, we repeated this scenario four times to record our readings.

The experimental setup for the voice experiment is shown in Figure 1(b). The AP was placed in the centre and was connected to a measurement workstation using an Ethernet cable. We placed two pairs of laptops facing each other diagonally at a distance of 8 m from the AP. We configured Ekiga[2] VoIP software on the laptops to allow direct IP calling without using a SIP server. VoIP client 1 was communicating with VoIP client 2 and VoIP client 3 was communicating with VoIP client 4. Interferer placement was the same as in the case of the data and video experiments.

3   Physical-Layer Characteristics

3.1   Channel Structure in Wi-Fi

Wi-Fi channels are organized into 14 overlapping channels each having a spectral bandwidth of 22 MHz. Figure 2 shows a graphical representation of the Wi-Fi channels in the 2.4 GHz band. The figure shows the centre frequencies of each Wi-Fi channel. Adjacent channels are separated by 5 MHz, except for channel 14 whose centre frequency is separated from channel 13 by 12 MHz. A single channel can handle 50 or more simultaneous users [6]. Usage of Wi-Fi channels are governed by national regulatory agencies of the respective countries. In North America, only the first 11 channels are available for use. In the rest of the world the first 13 channels are available for use. Japan allows the use of channel 14 as well, however, it is only available to 802.11b using Direct Sequence Spread Spectrum (DSSS) modulation. We restrict our attention to the 11 channels in use in North America.

To avoid interference, wireless radios are expected to operate on non-overlapping channels, i.e., channels separated by at least 22 MHz. For example, if two APs are operating on the same channel in a wireless cell, then their signals will interfere with each other. The same applies to any other radiating device, such as a microwave oven or cordless phone. From Figure 2, we observe that the following channel combinations do not overlap with each other: {1,6,11}, {2,7}, {3,8}, {4,9}, and {5,10}. Channels 1, 6, and 11 are the most commonly used non-overlapping channels in Wi-Fi deployments.

---

[2] http://ekiga.org/

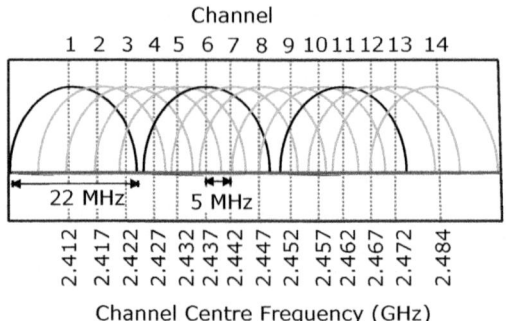

Fig. 2. Structure of Wi-Fi Channels in the 2.4 GHz band

3.2 Metrics

We use spectrograms and duty cycles to characterize interferers at the physical layer. The spectrogram is a representation of the RF power levels over time in the spectrum. Each vertical line in the spectrogram shows the RF power as a function of frequency measured over a time interval of 1 second. Spectrograms offer a temporal perspective of RF power in the frequency domain. The duty cycle measures the RF power in the spectrum. In this work, duty cycle is calculated by measuring the percentage of time the RF signal is 20 dBm above the noise floor. Duty cycle is an indicator of the impact of RF power on network performance. We next describe the physical-layer characteristic of each interferer in isolation.

3.3 Measurement Results

Figure 3 shows the spectrograms for the interferers. The X-axis represents the time period of the measurements. The Y-axis tic marks represent the centre frequencies of the even numbered Wi-Fi channels; i.e., channels 2, 4, 6, 8, 10, 12, and 14. The colour contour lines represent the power levels of the signal, where red indicates the strongest and blue the weakest power levels. When interpreting these graphs, it is important to note that each device may use a different range of RF power levels and the same colour therefore may refer to a different power level in the different graphs. Figure 4 shows the duty cycle FFT measurements for the interferers. The bottom X-axis tic marks represent the centre frequencies of the even numbered Wi-Fi channels, while the top X-axis tic marks show the channel numbers corresponding to those centre frequencies.

Microwave Oven: Figure 3(a) shows that the microwave oven affects about half the available channels (6-12) in the 2.4 GHz band with the highest energy concentrated on channel 9. Note that the microwave oven is operating at high power levels of -80 to -60 dBm. RF power levels above -80 dBm are enough to cause interference with Wi-Fi networks. Figure 4(a) shows that the average duty cycle of the microwave is 50%. Since the microwave signal sweeps through a wideband of the spectrum and has a high average duty cycle, it is likely to affect nearby Wi-Fi devices. Microwave ovens are created with a shield such that all radiation is restricted to the oven cavity, however, with use over time these ovens can leak some radiation.

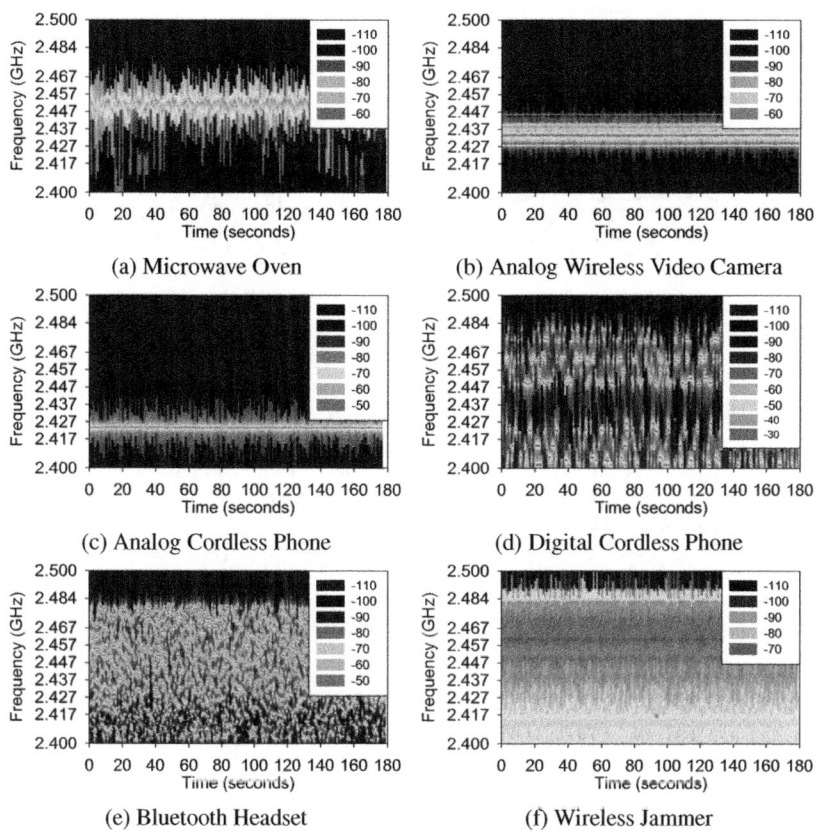

Fig. 3. Spectrograms for six non-Wi-Fi devices

Analog Wireless Video Camera: Figure 3(b) illustrates the narrowband continuous transmitting nature of an analog video camera. The transmit power is similar to that of the
microwave oven, however, in this case this energy is concentrated on a very small portion of the spectrum (channels 4-8). Figure 4(b) shows that the duty cycle of the analog
video camera reaches 100% indicating that no Wi-Fi device in the vicinity will be able
to operate on channels 4-8. Because of its continuous transmission nature, this device
can cause prolonged periods of service disruption.

Analog Cordless Phone: The analog phone concentrates most of its energy on channel 3
in Figure 3(c). The figure also highlights the narrowband fixed-frequency transmission
nature of the analog phone. Figure 4(c) shows that the analog phone has duty cycle as
high as 85%. The high duty cycle of the analog phone indicates that it will severely
impact Wi-Fi operation. Analog phones are quickly becoming out-dated, however, they
are still available for purchase and are widely used around the world.

Digital Cordless Phone: The frequency hopping feature of the digital phone is illustrated in Figure 3(d). This phone utilizes DSS, a newer technology than the one used

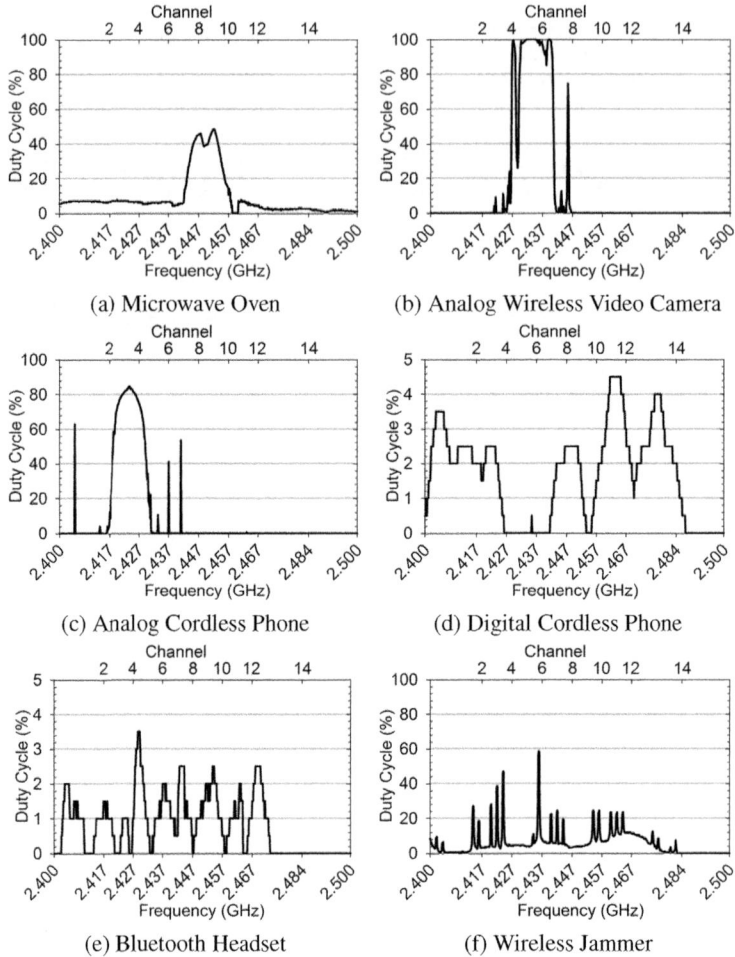

Fig. 4. Duty cycle analysis for six non-Wi-Fi devices

by analog phones. The phone continuously changes channels, only staying on a portion of the spectrum for a small amount of time, reducing its interference. In Figure 4(d) we observe that the maximum duty cycle is 4.5%, indicating that this device may not severely interfere with Wi-Fi devices.[3]

Bluetooth Headset: The Frequency Hopping Spread Spectrum (FHSS) nature of Bluetooth is highlighted in Figure 3(e). The Bluetooth device hops across all the channels. Although the energy emitted by the Bluetooth device may appear high, its duty cycle values are much lower. Furthermore, FHSS technology is limited to 2 Mbps and does

---

[3] More recently, Digital Enhanced Cordless Telecommunications (DECT) phones have become popular. These phones operate on the 1.9 and 5.8 GHz band and do not interfere with 802.11 devices.

not consume much of the available bandwidth. Figure 4(e) shows that the maximum duty cycle attained is 3.5%, which may not affect Wi-Fi devices seriously.

Wireless Jammer: Figure 3(f) shows the wideband characteristics of a wireless jammer. The jammer emits signals on all channels at high power levels in quick succession. Figure 4(f) shows the duty cycle varies between 10% and 60%. Wi-Fi devices use Clear Channel Assessment to sense when the channel is clear for transmission. The wideband jammer ensures that the RF medium is never clear, thus preventing Wi-Fi devices from functioning properly.

4   Impact of Interferers on Wi-Fi Traffic

In this section, we study the impact of interferers on Wi-Fi traffic. In particular, we study the impact of interferers on data traffic, video traffic, and voice traffic. The first experiment for each workload was conducted in an interference-neutral environment. Next, we repeated the experiments where the Wi-Fi link was subjected to interference from one specific interferer at a certain distance. The distances used were 5, 10, 15, 20, 25, and 30 m. We did not consider the digital phone since its physical layer characteristics are similar to that of Bluetooth. Each experiment was performed four times, and we report the average percentage difference between the baseline performance (no interference) and that when the experiment was subjected to interference. For the data experiment, we used the throughput Quality of Service (QoS) metric. For the video and voice experiments, we used a Quality of Experience[4] (QoE) metric called Mean Opinion Score (MOS).

4.1   Experimental Workload Traffic

Data Traffic: We used the Iperf[5] tool to measure throughput of the Wi-Fi link. We ran Iperf in server mode at the server and in client mode at the four wireless client laptops. Iperf can run throughput tests using TCP or UDP packets. We used the TCP option for our experiment. For the TCP tests, Iperf requires the user to set an appropriate TCP window size. If the window size is set too low, the throughput measurements may be incorrect. We found that a window size of 148 KB was sufficient to properly measure the throughput of the Wi-Fi link. The workload consisted of creating bidirectional TCP traffic between the server and the four clients for a period of 3 minutes.

Video Traffic: Video (or Voice) quality can be quantified using subjective methods such as MOS. For our experiments, we used a hybrid subjective assessment scheme called Pseudo Subjective Quality Assessment (PSQA) [7,8]. PSQA uses random neural networks to automatically calculate MOS of a video or voice sample. We set up the server workstation to stream a 3-minute video using the VLC[6] media player. The (normal quality standard-definition) video had a resolution of 624×352, frame rate of 29 frames per second, and was encoded using the Xvid[7] codec at a bit rate of 656 Kbps. The VLC

---

[4] http://en.wikipedia.org/wiki/Quality_of_experience
[5] http://iperf.sourceforge.net/
[6] http://www.videolan.org/vlc/
[7] http://www.xvid.org/

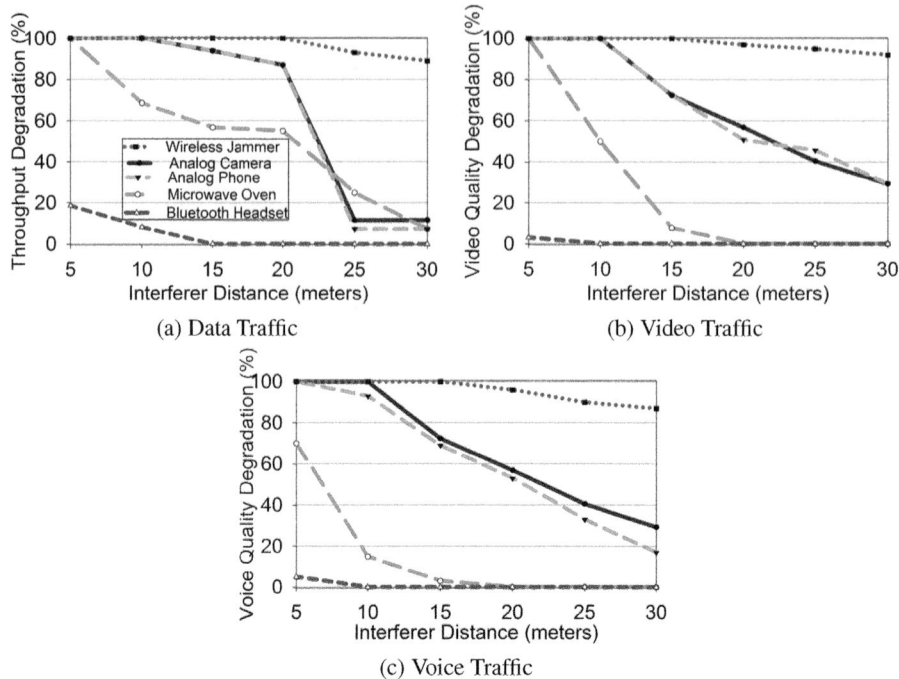

Fig. 5. Impact of interference on different workloads

players on the four clients were configured to receive this video stream. We collected the MOS measurements using PSQA from the client-side.

Voice Traffic: We recorded a 3-minute VoIP conversation between two people. We next separated the voices of the two people and created two different audio files. Recall our voice experiment setup had two pairs of communicating laptops. After establishing a direct connection via the Wi-Fi link for a pair of laptops, we played one audio file on one end and the second audio file on the other end. The two audio files were synchronized such that at no time the two audio files played a human voice simultaneously. The (telephone quality) audio files were encoded using the FLAC[8] codec at a sampling rate of 16 kHz and a bit rate of 56 Kbps. We used a workstation connected to the AP to collect the MOS measurements using PSQA of the voice communication.

4.2 Experimental Results

Figures 5(a), (b) and (c) show the degradation in data throughput, video quality, and voice quality, respectively, in the presence of interference.

As the diverse nature of these workloads require different quality measures, direct comparisons of the degradation of the different services require some care. Voice traffic, which typically uses smaller packets, handles interference the best (with the metrics

---

[8] http://flac.sourceforge.net/

used here). As expected, we consistently notice close to 100% degradation for our experiments using the wireless jammer. This is consistent up to 20 m, beyond which we observe a slight decline in its impact. In the following, we discuss the impact of the unintentional interferers on the different workloads.

Data Traffic: The Bluetooth headset reduced the throughput by 20% at close distances. The degradation may be surprising, since Bluetooth devices have low duty cycle and are designed to accommodate Wi-Fi devices; however, it is less serious compared to other interferers. For example, the microwave oven resulted in zero throughput at close distance. While its impact declined gradually as it was moved away from the Wi-Fi network, it caused a 25% degradation in throughput even at a distance of 25 m. The analog phone and video camera are both continuous transmitters. Hence, their impact on throughput is similar. Their interference is significant at close distance, however, there is a sharp decrease at distances beyond 20 m.

Video Traffic: Although Bluetooth had some impact on data traffic, there was minimal impact on video traffic. The microwave oven at close distance severely disrupted the video stream, while at longer distances it reduced the video quality by only 10%. The analog camera and analog phone had similar impacts on the video stream. Even at longer distances, they reduced the video quality by 50%.

Voice Traffic: Bluetooth had minor impact at short distance and no impact at longer distances. The microwave oven caused approximately 75% degradation at close range, and this plummeted as the distance increased. The analog phone and video camera had severe impact at close range, but this consistently decreased with longer distances. They still caused about 30% degradation at a distance of 30 m. The impact of interference is comparatively lower for voice traffic.

5   Interference in Campus Network

We used the Spectrum Analyzer to study the channel utilization of our campus network. We took physical-layer measurements for 8 hours during a weekday (Tuesday, January 20, 2009). The passive measurements were taken at a popular location in the campus frequented by students and faculty members. We report our results using the channel utilization metric. Channel utilization is the percentage of time a transmission is present from a known RF source in a given channel. It helps us understand how much of the channel is available for use at a given time.

Figure 6(a) shows the channel utilization of the three channels used in our campus network. The figures record the channel utilization by all devices (Wi-Fi and Interferers) transmitting in the 2.4 GHz ISM band.[9] We observe that channels 1 and 11 had heavier usage in comparison to channel 6. In case of channels 1 and 6, utilization peaked near 60%, while for channel 11 it was over 90%. After close inspection, we found that most of the spikes were caused by interferers.

On channel 11, the spikes are the most alarming. We focus on three spikes and try to understand the interferers involved. These spikes are observed at times 12:45, 15:38,

---

[9] The Spectrum Analyzer cannot separate the Wi-Fi and non-Wi-Fi channel utilization over time, but it did allow to us to investigate the utilization at particular points in time.

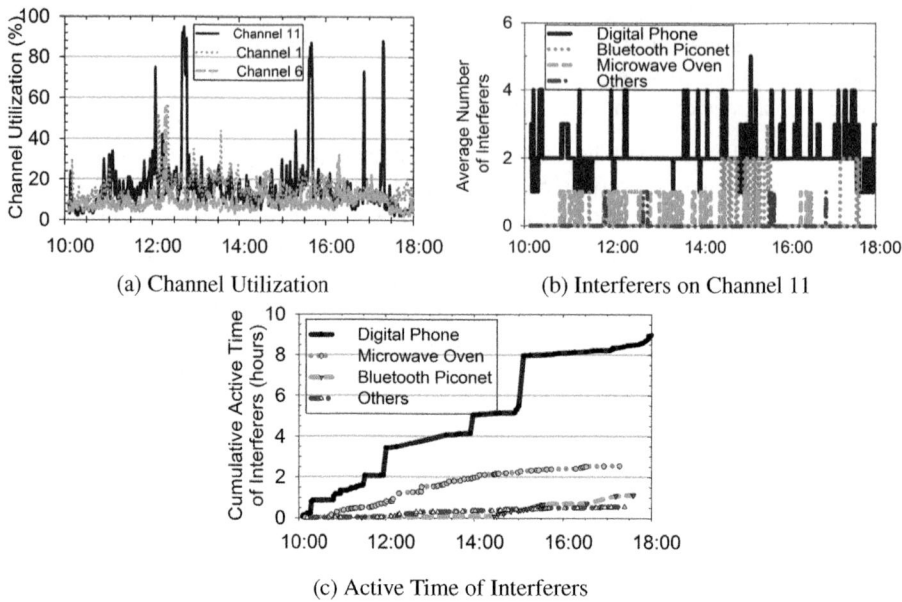

(a) Channel Utilization   (b) Interferers on Channel 11

(c) Active Time of Interferers

Fig. 6. Passive measurements from campus Wi-Fi network

and 17:17. We found that each of these spikes was mainly caused due to microwave ovens and cordless phones operating simultaneously in the area of the Wi-Fi network. For the first spike, we also noticed some other fixed-frequency interferers, whereas the second and third spike were highly influenced by Bluetooth communication. In total, the interferers consumed 80% of the channel during the first spike. The second and third spike used much less of the channel.

Figure 6(b) shows the average number of interferers observed on channel 11 as a function of time. We observed 100s of interference occurrences involving non-Wi-Fi devices on channel 11. These include 22 instances of Bluetooth device pairs communicating (6 unique piconets), 3 instances of analog cordless phones (base and handset), 71 instances of digital cordless phones (30 bidirectional base and handset instances and 41 cases where only the base station was observed), as well as 46 instances involving other unclassified interferers.

Overall, we found that microwave ovens and cordless phones were surprisingly active throughout the trace period, while Bluetooth piconet instances were temporally clustered. Furthermore, we note that most active times for the various interferers were relatively short, but with a few long instances. For example, the average active period for microwave ovens (2:12), Bluetooth (3:05), digital cordless phones (7:32) and other interferers (0:42) were relatively short compared to the corresponding devices' longest observed active periods (16:37, 22:27, 2:28:01, 5:02).[10]

Figure 6(c) shows the cumulative active time (i.e., the time the interferers were transmitting) of interferers over the trace period. Microwave ovens were active 31% of the

---
[10] The active period length is reported in *hours:minutes:seconds*.

total trace period and had an average (and maximum) duty cycle of 9% (40%). The average signal strength was approximately -80 dBm, which is enough to cause some interference. Bluetooth devices were active 12.5% of the time and had an average (low) duty cycle of roughly 5%. With digital base stations always transmitting (at a duty cycle < 7%) when not connected with the handset, it was not surprising that we found that cordless phones were cumulatively active for more than the entire trace period. We note, however, that the average (maximum) duty cycle increased to 15% (20%) when the base and handset were connected. Other interferers were observed active only 6% of the total trace period, but had the highest average duty cycle (15%), maximum duty cycle (60%), and signal strength (-43 dBm) of all the interferers observed in the campus Wi-Fi network.

The exact channel usage of these observed non-Wi-Fi devices may differ from that of the particular devices studied in our controlled experiments, however, the general characteristics are similar. With a small number of heterogeneous non-Wi-Fi devices active at each point in time, it is not surprising that we do not find strong correlation between the average number of active devices and the peaks observed above (e.g., comparing Figures 6(a) and 6(b)). Instead, individual device characteristics and the distance of devices from the network are likely to have the most impact. Referring back to the impact of device and distance (Section 4), it is clear that a few non-Wi-Fi devices can have a huge effect on the performance of the Wi-Fi traffic, especially if located close to the network. Furthermore, while our measurements were taken (in a student/faculty lounge) away from the campus food court and office areas, the interference from microwave ovens and cordless phones would likely be even more significant in such areas. We leave comparisons of the non-Wi-Fi interference in different areas as future work.

6 Related Work

Interference in wireless networks has been investigated mostly in the context of coexisting Wi-Fi networks, Bluetooth networks, and microwave ovens.

Taher et al. [14] used laboratory measurements to develop an analytical model of microwave oven signals. Karhima et al. [12] performed measurements on an ad-hoc wireless LAN under narrowband and wideband jamming. They found that in case of wideband jamming, 802.11g can offer higher transmission rates than 802.11b, when the packet error rate is the same. They also found that 802.11g was more prone to complete jamming, while 802.11b could still operate at lower transmission rates due to its DSSS modulation scheme. Golmie et al. [9] explored the mutual impact of interference on a closed loop environment consisting of Wi-Fi and Bluetooth networks. They found that even by sufficiently increasing the transmission power levels of the Wi-Fi network to that of the Bluetooth network could not reduce packet loss. Our work complements these studies. We use measurements to characterize a wide range of common non-Wi-Fi devices (including microwave ovens and Bluetooth) at the physical layer and use experiments to quantify their impact on different traffic workloads.

Vogeler et al. [15] presented methods for detection and suppression of interference due to Bluetooth in 802.11g networks. Ho et al. [11] studied the performance impact on a Wi-Fi network due to Bluetooth and HomeRF devices using simulations and described design challenges for deployment of 5 GHz wireless networks. Our work is orthogonal to these papers. We focus on passive measurements to study interference from common non-Wi-Fi devices.

Gummadi *et al.* [10] studied the impact of malicious and unintentional interferers on 802.11 networks. They found that some of these interferers could cause considerable performance degradation for commercial 802.11 NICs. Using an SINR model, they noted that changing 802.11 parameters was not helpful, and proposed a rapid channel hopping scheme that improved interference tolerance for these NICs. Farpoint Group studied the effect of interference on general, video, and voice traffic in a Wi-Fi network from two vantage points: short-range (25 feet) and long-range (50 feet) [3,4,5]. In contrast, we present a comprehensive measurement study of impact of interference from non-Wi-Fi devices. We also used QoS and QoE metrics to quantify the degradation of video and voice quality in a Wi-Fi network due to interference. We observed that operational Wi-Fi networks may be subject to ambient interference effects at anytime.

7   Concluding Remarks

In this paper, we characterized the RF behaviour of non-Wi-Fi devices, analyzed the impact of interference on data, video, and voice traffic, and examined interference in a live campus network. Overall, we found that the campus network is exposed to a large variety of non-Wi-Fi devices, and that these devices can have a significant impact on the interference level in the network. Our controlled experiments showed that, even at distances up to 30 m, some of these non-Wi-Fi devices can have a significant negative impact on data, video, and voice traffic. While microwave ovens, wireless analog video cameras, and analog cordless phones typically have the most significant negative impact on Wi-Fi networks, the performance degradation due to a digital cordless phone or a Bluetooth device (which tries to be Wi-Fi friendly) can be noticeable (e.g., 20% at close distances). With the campus network (and likely many other home and enterprise networks) being highly exposed to different types of unintentional interferers, it is important to find (new) ways to identify and mitigate non-Wi-Fi interference.

Network practitioners often use link-layer and transport-layer statistics to investigate interference in a Wi-Fi network. We note that these alone should not be the means for troubleshooting a network. We observe that physical-layer characteristics may be used as primary indicators to identify and mitigate interferers [2]. In general, we believe that interference can be mitigated by identifying and removing the interfering device (if possible) or shielding interferers instead. Careful channel selection at APs may be helpful (e.g., in both our controlled and real world measurements we found that microwave ovens primarily affected channels 6 and higher). APs may be fitted with sensors to detect interference and switch channels automatically [1]. Using multi-sector antennas [13] and controlling data rates to avoid false backoffs can make the network more interference-resilient (although this is a trade-off since the lower data rates allow more noise-immune communication). We conclude by noting that there are many different non-Wi-Fi devices that may cause interference in Wi-Fi networks, and it is therefore important to understand and quickly adapt to the devices affecting the performance of the Wi-Fi channels.

Acknowledgements

Financial support for this research was provided by Informatics Circle of Research Excellence (*i*CORE) in the Province of Alberta, as well as by Canada's Natural Sciences and Engineering Research Council (NSERC). The authors also thank the anonymous reviewers for their constructive comments on an earlier version of this paper.

References

1. Aruba Networks: Advanced RF Management for Wireless Grids. Technical Brief (2007), `http://tinyurl.com/aruba-report`
2. Cisco: 20 Myths of Wi-Fi Interference: Dispel Myths to Gain High-Performing and Reliable Wireless. White Paper (2007), `http://tinyurl.com/cisco-report`
3. Farpoint Group: The Effects of Interference on General WLAN Traffic. Techical Note FPG 2006-328.3 (January 2008), `http://tinyurl.com/FarpointGeneral`
4. Farpoint Group: The Effects of Interference on Video Over Wi-Fi. Technical Note FPG 2006-329.3 (January 2008), `http://tinyurl.com/FarpointVideo`
5. Farpoint Group: The Effects of Interference on VoFi Traffic. Techical Note FPG 2006-330.3 (January 2008), `http://tinyurl.com/FarpointVoFi`
6. Flickenger, R.: Building Wireless Community Networks. O'Reilly, Sebastopol (2003)
7. Ghareeb, M., Viho, C.: Performance Evaluations of a QoE-Based Multipath Video Streaming Mechanism over Video Distribution Network. In: Proc. of International Workshop on Future of Multimedia Networking (FMN), Coimbra, Portugal, June 2009, pp. 236–241 (2009)
8. Ghareeb, M., Viho, C., Ksentini, A.: An Adaptive Mechanism For Multipath Video Streaming Over Video Distribution Network. In: Proc. of International Conference on Advances in Multimedia (MMEDIA), Colmar, France, July 2009, pp. 6–11 (2009)
9. Golmie, N., VanDyck, R., Soltanian, A., Tonnerre, A., Rebala, O.: Interference evaluation of Bluetooth and IEEE 802. 11b Systems. Wireless Networks 9(3), 201–211 (2003)
10. Gummadi, R., Wetherall, D., Greenstein, B., Seshan, S.: Understanding and Mitigating the Impact of RF Interference on 802.11 Networks. In: Proc. of ACM SIGCOMM International Conference, Koyoto, Japan, August 2007, pp. 385–396 (2007)
11. Ho, M.J., Rawles, M., Vrijkorte, M., Fei, L.: RF Challenges for 2.4 and 5 GHz WLAN Deployment and Design. In: Proc. of IEEE Wireless Communications and Networking Conference (WCNC), Orlando, USA, March 2002, pp. 783–788 (2002)
12. Karhima, T., Silvennoinen, A., Hall, M., Haggman, S.: IEEE 802.11 b/g WLAN Tolerance to Jamming. In: Proc. of Military Communications Conference (MILCOM), Monterey, USA, October 2004, pp. 1364–1370 (2004)
13. Lundgren, H., Subramanian, A., Salonidis, T., Carrera, M., Guyadec, P.L.: Interference Mitigation in WiFi Networks using Multi-sector Antennas. In: Proc. of ACM international Workshop on Experimental Evaluation and Characterization (WiNTECH), Beijing, China, September 2009, pp. 87–88 (2009)
14. Taher, T., Al-Banna, A., Ucci, D., LoCicero, J.: Characterization of an Unintentional Wi-Fi Interference Device - The Residential Microwave Oven. In: Proc. of Military Communications Conference (MILCOM), Washinton DC, USA (October 2006)
15. Vogeler, S., Brotje, L., Kammeyer, K.D., Ruckriem, R., Fechtel, S.: Suppression of Bluetooth Interference on OFDM in the 2.4 GHz ISM Band. In: Proc. of International OFDM-Workshop (InOWo), Hamburg, Germany (September 2003)

A Zone Assignment Algorithm for Fractional Frequency Reuse in Mobile WiMAX Networks

Michael Einhaus, Andreas Mäder, and Xavier Pérez-Costa

NEC Laboratories Europe
Kurfürstenanlage 36, 69121 Heidelberg, Germany
{michael.einhaus,andreas.maeder,xavier.perez-costa}@nw.neclab.eu
http://www.nw.neclab.eu

Abstract. Fractional Frequency Reuse (FFR) is a key concept for increasing the spectral efficiency of OFDMA-based broadband wireless access systems such as Mobile WiMAX. One of the main challenges when using FFR is to decide which mobile stations are assigned to which FFR time zone within a WiMAX MAC frame, so that the frame resource utilization efficiency is maximized while meeting the QoS demands of all users. In this paper we propose an efficient algorithm to address this problem with low computational complexity. The performance of the algorithm is compared with the optimum zone assignment. The evaluation has been conducted by means of a comprehensive simulation study of a typical Mobile WiMAX deployment scenario with various numbers of VoIP users.

Keywords: Broadband Wireless Access, WiMAX, OFDMA, FFR, QoS Traffic, Resource Scheduling.

1 Introduction

Fractional Frequency Reuse (FFR) is one of the key concepts proposed for the deployment of $3.9\,G$ and $4\,G$ cellular mobile radio systems. Next to other techniques such as power control and beamforming, FFR is an important means to mitigate interference in future radio networks [1]. The use of Orthogonal Frequency Division Multiple Access (OFDMA) at the air interface of broadband radio systems such as Mobile WiMAX [2] provides efficient means to support this concept. Furthermore, currently developed and extended standards such as 3GPP LTE/LTE-Advanced [3] and IEEE 802.16m [4] consider the FFR concept in order increase total cell data rates as required for ITU IMT-Advanced [5].

In conventional cellular deployment scenarios, orthogonal groups of resources (frequency channels) are used in neighboring cells, and these resource groups are reused in cells with a distance large enough from each other so that the interference is within tolerable limits [6]. The disadvantage of this concept is that the actual interference experienced by individual users from neighboring cells is not considered. Users that do not experience large inter-cell interference could use the whole frequency band instead of just a fraction orthogonal to neighboring cells, and thus increase the throughput.

M. Crovella et al. (Eds.): NETWORKING 2010, LNCS 6091, pp. 174–185, 2010.

The basic idea behind FFR, also known as reuse partitioning [7], combines frequency reuse with full utilization of the frequency band in each cell. A mobile station (MS) that is close to the base station (BS) is allowed to utilize the whole frequency channel (reuse 1), while an MS at the cell edge uses only a fraction of the spectrum which is orthogonal to adjacent cells (reuse 3 in case of three orthogonal fractions of the spectrum[1]). Since inter-cell interference for an MS is less severe in the first case, this allows for high data rates for users close to the BS, and still sufficient data rates for users at cell edge. The throughput for each MS is thereby determined by the SINR (signal-to-noise-plus-interference ratio). The described concept can easily be extended to a sectorized cellular deployment as it is considered in this paper.

In Mobile WiMAX, the concept of FFR is supported by subdividing the frequency channel in the time domain into zones for frequency reuse 1 (Reuse 1 Zone) and zones for reuse 3 (Reuse 3 Zone). A fundamental problem in this context is how to assign service flows, corresponding to MSs, to the different zones. Service flows have to be assigned dynamically to the zones in such a way that QoS (quality of service) requirements are met while the amount of required resources should be minimized. Furthermore, the allocation algorithm has to exhibit low computational complexity in order to reduce costs.

Although the application of FFR in Mobile WiMAX has lately been addressed in several publications, proposals for practical zone assignment algorithms are scarce. The performance with different reuse patterns and subcarrier permutations is evaluated in [8] by means of simulations, showing that the coverage outage probability significantly increases with the number of users using full spectrum. However, the evaluation is based on the assumption that the zone assignment is performed according to the cell geometry and not based on SINR levels. A similar approach is followed in [9], although here an analytical model is used for the evaluation. Furthermore, results of an FFR simulation study with different Reuse Zone sizes and deployment configurations are presented in [10]. In that publication, Reuse Zone assignment is performed with a threshold-based approach: all users with downlink preamble SINR below a certain value are assigned to Reuse 3 Zone, while the others are assigned to the Reuse 1 Zone. The problem of minimizing resource utilization is for example mentioned in [11], but not explicitly evaluated.

In this paper, we extend the previously mentioned related work by proposing a new efficient algorithm for assigning QoS service flow (VoIP) with given throughput requirements to downlink FFR zones. This is done in a way so that the resource usage efficiency for these flows is maximized while using an algorithm with very low computational complexity. The purpose of maximizing the resource usage efficiency for the QoS service flows is to have as much resources as possible available for additional best effort (BE) service flows.

To the best of the authors knowledge there is no related work which addresses these objectives simultaneously. The performance of the proposed algorithm is

---

[1] Another notation is the reuse factor give by $1/N$ (1 and 1/3 in our case) , describing the channel fraction assigned to each cell [6].

compared with an optimum zone assignment in terms of minimizing the amount of required resources while satisfying a given number of QoS service flows. The performance evaluation has been conducted by means of a comprehensive simulation study of a typical Mobile WiMAX deployment.

The next section provides a detailed description of how FFR is applied in Mobile WiMAX. The problem statement of FFR zone assignment and the proposed algorithm are presented in Section 3. The results of a comprehensive performance evaluation by means of a simulation study is given in Section 4. The paper ends with some concluding remarks and an outlook.

2   Fractional Frequency Reuse in Mobile WiMAX

In Mobile WiMAX, the frequency channel is divided into successive MAC frames in the time domain. In TDD mode, the MAC frame is further subdivided in downlink (DL) and uplink (UL) subframes. A detailed description can be found in the underlying IEEE 802.16-2009 standard [12]. In the following we focus on the downlink but a similar reasoning could be applied to the uplink direction.

FFR is realized by time zones in each WiMAX MAC frame [2]. In the conventional approach, one part of the frame is reserved for reuse 1 (Reuse 1 Zone), and another part is reserved for frequency reuse 3 (Reuse 3 Zone), as shown in Fig. 1 for the downlink in a scenario with sectorized cells. The switching point between both zones can be adapted in each MAC frame. Since in Mobile WiMAX all resource is performed by the BS, it also has to decide which user data packets are assigned to which zone. All user data packets are transmitted within so-called service flows. A service flow is defined by the connection endpoints within BS and MS, and a set of QoS parameters such as throughput and delay.

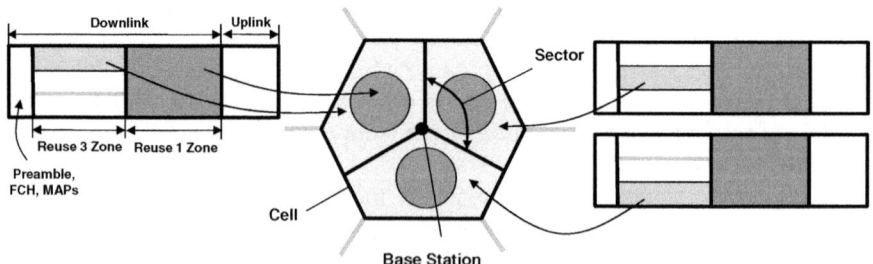

Fig. 1. FFR in a Mobile WiMAX deployment with sectorized cells

Mobile WiMAX provides different OFDMA subcarrier permutations for the formation of subchannels. For the use of FFR, PUSC (partial use of subcarriers) should be used since it provides orthogonal subchannel sets which coincide in different cells which is an inevitable prerequisite for the application of FFR. PUSC subchannels are formed by pseudo randomly distributed subcarrier sets in the frequency domain. As illustrated in [8], the same subcarrier permutation base has to be used by each BS in the whole network in order to provide orthogonal subchannels.

Resources are assigned to service flows in form of slots. A DL PUSC slot has the dimensions of one subchannels and two symbols. The slot capacity in terms of bits/slot is determined by the modulation and coding scheme, which is dynamically selected based on the SINR level by means of adaptive modulation and coding (AMC). A more detailed description of the system profile used in this paper is given in the performance evaluation (Section 4.2).

3   Zone Assignment for Fractional Frequency Reuse

3.1   Problem Statement

The problem that has to be solved is the assignment of a number of QoS service flows with given throughput requirements to the different Reuse Zone of the DL MAC frame in a way so that the amount of required resources for that service flows is minimized.

We define the number of available DL PUSC slots in Reuse 1 and Reuse 3 Zone as S_1 and S_3, respectively. Each service flow is assigned to and scheduled in one of these two zones of the DL MAC frame, depending on the corresponding MS position and channel conditions. Each service flow k requires the transmission of a certain number of bits T_k per MAC frame, which corresponds to the assignment of $R_{z,k}$ number of slots in zone z. Due to the use of AMC in Mobile WiMAX, this number of slots is determined by the slot capacity $C_{z,k}$ depending on the SINR level of the MS in each zone.

Based on knowledge of the SINR level of each MS (and corresponding service flows) in both zones, the BS has to decide which service flow is served in which Reuse Zone. Mathematically, the problem of zone assignment with minimum resource usage can be formulated as the following Integer program:

$$
\begin{aligned}
\text{min.} \quad & \sum_{k=0}^{N-1} R_{1,k} + \sum_{k=0}^{N-1} R_{3,k} \\
\text{s.t.} \quad & R_{1,k} \cdot C_{1,k} + R_{3,k} \cdot C_{3,k} \geq T_k \quad \forall k \\
& \sum_{k=0}^{N-1} R_{z,k} \leq S_z \quad , \quad z \in \{1,3\} \\
& R_{1,k} \cdot R_{3,k} = 0 \quad \forall k
\end{aligned}
\tag{1}
$$

The given conditions state that the QoS requirements of all service flows have to be satisfied in terms of bits T_k per MAC frame, that the sum of all assigned slots $R_{z,k}$ in each zone shall not exceed the corresponding number of available slots S_z, and that each service flows is just assigned to one zone. Since resources cannot be shared by different service flows, the number of assigned slots $R_{z,k}$ is either a positive Integer value or zero.

The complexity of the assignment problem necessitates the development of efficient heuristics with low computational complexity, which in the best case can be customized to meet operator requirements. Furthermore, the algorithm should preferably be easy to implement.

3.2   Proposed Algorithm

To find the solution for the problem described in Eqs. (1), a complete search comparing all possible slot assignment patterns with complexity $O(2^N)$ has to be performed. To minimize the computational complexity, we propose a zone assignment algorithm that is based on the sequential assignment of service flows to preferred zone after first sorting all flows by a certain metric. Thus the complexity is determined by the sorting of N service flows, which is $O(N \log N)$.

The key idea behind the proposed sorting metric is that it should not just consider zone preference from a single service flow's point of view (based on SINR levels), it should rather consider how the assignment of each flow to a certain zone would affect the total resource utilization.

In this context, the relative resource usage efficiency within a zone should be considered because the preference for a certain zone should be high for a service flow with high efficiency in that zone compared to other flows. The reason for this is demonstrated by the following simple example. Consider a service flow which has in Reuse 1 Zone an SINR level which is much larger than the average SINR level of all flows in that zone, but in the Reuse 3 Zone its SINR level is only slightly above the corresponding average. Hence, the overall system performance will benefit more from assigning that flow to the Reuse 1 Zone instead of the Reuse 3 Zone. The important feature here is that this applies even if the absolute SINR level of that flow is higher in the Reuse 3 Zone than in the Reuse 1 Zone, which is an essential difference compared to algorithms where the zone assignment is based just on the absolute SINR level.

In addition to that, the relative zone size should also be considered. In general, the average number of service flows assigned to a certain zone increases with the zone size, independently of the applied allocation algorithm. That means that the efficient utilization of zone resources in a large zone can amount to a large number of resources that are conserved compared to an inefficient allocation. The amount of resources that can be saved by efficient allocation in a small zone is compared to that normally rather small, in the worst case not even enough to bear an additional service flow.

The proposed sorting metric Φ_k of service flow k which considers both factors (relative SINR level and relative zone size) in a multiplicative manner is determined by the following equations:

$$\Phi_{z,k} = \frac{\gamma_{z,k} \cdot N}{\sum_{k=0}^{N-1} \gamma_{z,k}} \times \frac{S_z}{S_1 + S_3} \quad , \quad z \in \{1, 3\} \tag{2}$$

$$\Phi_k = \max\left(\Phi_{1,k}, \Phi_{3,k} \cdot \alpha\right) \quad , \quad \alpha \geq 0 \tag{3}$$

In these equations, N is the number of service flows, $\gamma_{z,k}$ is the absolute SINR level of service flow k in FFR zone z, and S_z is the number of available resources in the Reuse Zones (z can be 1 or 3), corresponding to Eqs. 1. The algorithm comprises one degree of freedom, represented by the tuning factor α, for adjustment to a certain environment and user distribution.

In the sorted list, each service flow has a preferred FFR zone: if $\Phi_k = \Phi_{1,k}$, the preferred zone is the Reuse 1 Zone, otherwise it is the Reuse 3 Zone.

The tuning factor α is used to adapt the algorithm to the number of service flows. It will be shown in the performance evaluation why this additional factor has a significant impact on the performance. The impact of extreme settings of α is given by:

$$\begin{aligned}
\Phi_k|_{S_1>0,\ \alpha=0} &= \Phi_{1_k} \\
\Phi_k|_{S_3>0,\ \alpha\to\infty} &= \Phi_{3_k} \cdot \alpha
\end{aligned} \tag{4}$$

In the first case ($\alpha = 0$), all service flows are sorted according to the quality (SINR level) in the Reuse 1 Zone, and this zone is preferred by all flows. In the second case ($\alpha \to \infty$), all flows prefer the Reuse 3 Zone and are sorted by the quality in that zone.

After sorting all flows in descending order of Φ_k, the actual zone assignment is then done in the following way: beginning with the first flow in the sorted list, each flow is assigned to the preferred zone if there are enough resources available in that zone, otherwise it has to be assigned to the other zone. If not all service flows can be served, this is considered as an outage in the following.

4   Performance Evaluation

4.1   Methodology

The efficiency of the algorithm is evaluated by the total resource utilization $U_T(x)$ of the QoS service flows in the DL MAC frame, where x is the relative zone switching point position within the frame, ranging from 0 to 1:

$$U_T(x) = \frac{\sum_{k=0}^{N-1} R_{1,k}(x) + \sum_{k=0}^{N-1} R_{3,k}(x)}{S_1(x) + S_3(x)} \tag{5}$$

Corresponding to the description in Eq. 1, $R_{z,k}$ represents the amount of resources assigned to service flow k in Reuse Zones z, and S_z is available number of resources in the zones. With $x = 0$, the whole DL MAC frame is used as Reuse 1 Zone, and with $x = 1$ it is used as Reuse 3 Zone.

Furthermore, the outage probability $P_o(x)$ is evaluated. We define it here as the probability that at least one of the given number of the QoS service flows cannot be provided with the required data rate. This means that the amount of available resources in the DL MAC frame is not sufficient.

For the comparison of the proposed FFR zone assignment algorithm with the optimum assignment, the difference between both curves describing the expectation of $U_T(x)$ is evaluated by the mean squared error (MSE):

$$E(\alpha) = \frac{1}{|\mathbf{X}|} \sum_{x \in \mathbf{X}} \left(O(x) - A(\alpha, x) \right)^2 \tag{6}$$

In this equation, $O(x)$ and $A(\alpha, x)$ are the curves for the optimum given by the solution of Eqs. (1) and the proposed algorithm, respectively. $\mathbf{X}$ is the set of all evaluated zone switching points.

4.2 Simulation Scenario

The simulated scenario consists of 19 hexagonal cells in a wrap-around scenario with three sectors per cell, as described in the evaluation methodology for IMT-Advanced [5]. The SINR is calculated based on pathloss and shadowing. The impact of frequency-selective small scale fading is not considered in the simulation study since the performance is evaluated with distributed subcarrier permutations (PUSC in WiMAX) which in combination with channel coding minimizes the impact of frequency-selective fading due to averaging effects.

We use the ITU Suburban Macro model for pathloss and shadowing, as defined in [5]. The inter-site distance in this model is 1299 m, and the carrier frequency is 2.0 GHz. We assume a fixed downlink transmission power of 43 dBm and a thermal noise density of -174 dBm/Hz for the 10 MHz channel.

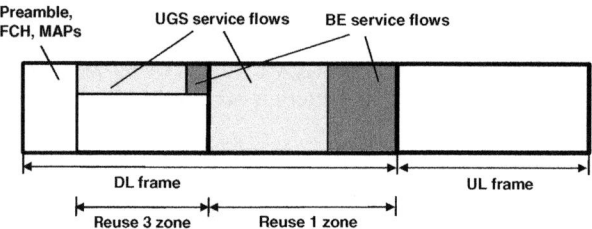

Fig. 2. Mobile WiMAX MAC frame model with DL FFR

The MAC frame has a duration of 5 ms, and we consider a bandwidth of 10 MHz and an FFT size of 1024 which results in 47 symbols within a frame. It is assumed that 35 OFDM symbols are reserved for downlink and 12 symbols for uplink transmissions. The first downlink symbol is used for the preamble, and four symbols are reserved for signalling in form of WiMAX FCH and MAPs [2]. The remaining 30 symbols are partitioned into DL Reuse Zones 1 and 3. In the frequency domain the subcarriers are grouped to 30 DL subchannels (and 35 UL subchannels). For the application of FFR in downlink direction this means that in the Reuse 1 Zone all 30 subchannels are used, and in the Reuse 3 Zone just 10 subchannels.

As resource (slot) allocation strategy we assume that the DL MAC frame is filled up column-wise starting from the upper left corner. This corresponds to the allocation strategy which is mandatory for use of HARQ [12] and enables an one-dimensional representation of the available resource per frame.

For the QoS service flows, we assume CBR data streams (UGS service flows) of 80 kbit/s. This corresponds to 64 kibt/s VoIP traffic (G.711) with a packet generation interval of 20 ms plus 16 kbit/s overhead due to IP, UPD and RTP packet headers (40 bytes per data packet each 20 ms). The resources of the DL MAC frame that are not used for the transmission of CBR traffic are assumed to be completely occupied by BE traffic as shown in Fig. 2. Due to this, all resources are always used for DL data transmissions, which means that the SINR levels of the reuse zones do not depend on the allocation decision.

The mapping of SINR level to MCS given in Tab. 1 is based on the performance evaluation results presented in [13]. Due to the frame duration of 5 ms, the 80 kbit/s per VoIP service flow results in a required average transmission of 200 bits per frame.

Table 1. SINR to MCS mapping

MCS	min SINR	bits/slot
QPSK 1/2	3.5 dB	48
16QAM 1/2	10.0 dB	96
64QAM 1/2	15.5 dB	144
64QAM 2/3	21.0 dB	192
64QAM 3/4	24.5 dB	216

For the performance evaluation, 10000 random user (VoIP service flow) placements are generated for each zone switching point setting. In a each user placement, there is a fixed uniformly distributed number of users generated per sector. All presented results show the average performance of all user placements.

4.3 Simulation Results

Optimum FFR Zone Assignment. Fig. 3 shows the expectation of the resource utilization $U_T(x)$ and the outage probability $P_o(x)$ depending on the zone switching point position for different numbers of uniformly distributed VoIP service flows per sector. The results reveal that there exists an optimum Reuse Zone switching point concerning the minimum average resource utilization. The larger the number of VoIP service flows, the larger is the optimum size of the Reuse 3 Zone in the DL MAC frame, represented by relative zone size x.

The results also show that the optimum switching point for the minimum resource utilization does not automatically minimize the outage probability. If beginning from the optimum zone switching point the Reuse 3 Zone size is further increased, this results in a further reduced outage probability at the cost of reduced efficiency in resource utilization for the VoIP service flows.

The very high outage probabilities and resource utilization values with a very small Reuse 3 Zone ($x \to 0$) is based in the fact that in general the use of frequency reuse 1 cannot be applied due to very low SINR values at the cell edge. The resource utilization for $x = 0$ corresponds to a conventional cellular deployment with frequency reuse 3. Therefore the results clearly prove the advantage of FFR since the resource utilization can be significantly reduced, i.e. with the optimum frame partition x_{opt} more than 20 % in case of 16 VoIP flows. It can

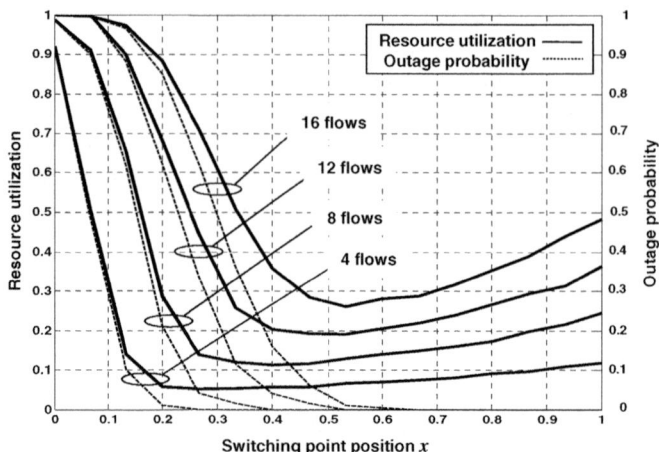

Fig. 3. Resource utilization and outage probability

be seen that FFR utilization gain increases with the number of flows which is based on exploitation of multiuser diversity.

Proposed FFR Zone Assignment Algorithm. As explained in Section 3, the proposed algorithm features a degree of freedom represented by the tuning factor α. It should be set in a way so that the algorithm provides close to optimum performance, meaning that the mean squared error $E(\alpha)$ in Eq. (6) has to be minimized.

The impact of α on the performance for 8 and 14 VoIP service flows is exemplary shown in Fig. 4. The results reveal that the performance of the proposed algorithm converges to the optimum with an appropriate setting of α. This optimum α depends on the number of flows; when the number of flows is increased, the magnitude of α has to be reduced. With 8 flows, $\alpha = 8.0$ shows very good performance; and with 14 flows, the setting of $\alpha = 4.0$ is more appropriate.

The results show that the proposed algorithm meets the optimum performance quite well if one FFR zone is significantly larger than the other. According to the sorting metric of the sequential assignment given by Eqs. (2) and (3), the larger FFR zone is filled first with the best service flows first in these cases. Actually, if the whole DL MAC frame is used as Reuse 1 Zone or as Reuse 3 Zone ($x = 0.0$ and $x = 1.0$, respectively), the performance curves coincide since anyway all flows have to be assigned to a single zone. The largest differences between optimum and α-dependent algorithm appears in the range of medium FFR zone sizes with $0.2 \leq x \leq 0.8$. In general the differences increase with the number of flows, implying more possible assignment combinations from which the best one can be selected with the complete search approach. However, the optimum switching point in terms of minimum resource utilization almost coincides for optimum zone assignment and proposed algorithm with optimum α selection.

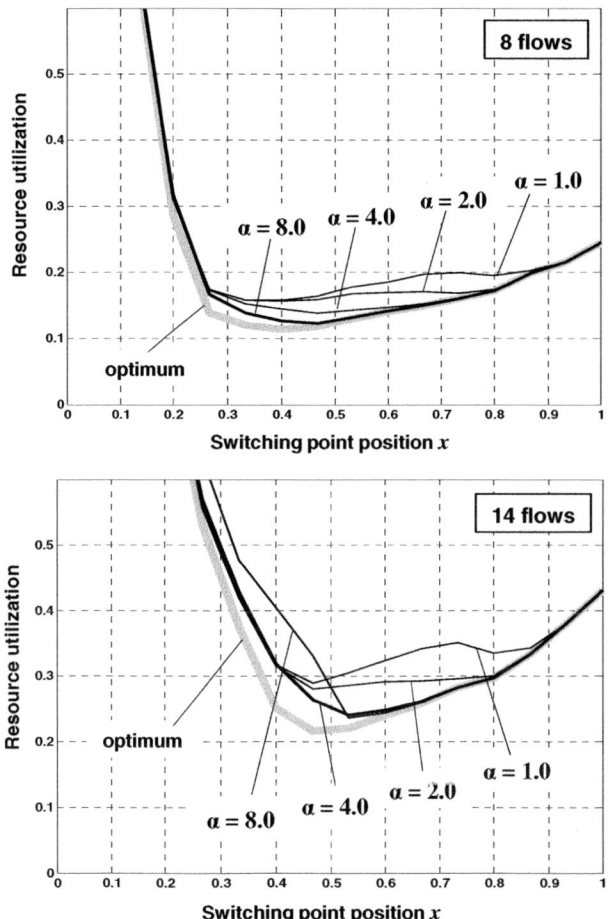

Fig. 4. Resource utilization of proposed algorithm with 8 and 14 VoIP flows

Concerning the tuning factor α, another effect is revealed in the evaluation results (especially in case of 14 flows). If α has a large value, the performance of the proposed algorithm meets the optimum very good for large x values; and if α is small, the performance is very good for small x values, correspondingly.

For an in-depth analysis of the impact of α, the difference between the optimum and the performance curve of the proposed algorithm is quantified in Fig. 5. It shows the mean squared error $E(\alpha)$, as defined in Eq. (6), for different numbers of service flows. The results exhibit that the error is determined by a convex function of tuning factor α, which forms a fundamental finding.

Concerning the dependency on the number of service flows, the results show that the dependency of the performance on α increases with the number of flows. Another important finding is that the minimum achievable error $E(\alpha_{opt})$ increases with the number of flows. Furthermore it can be seen that the α

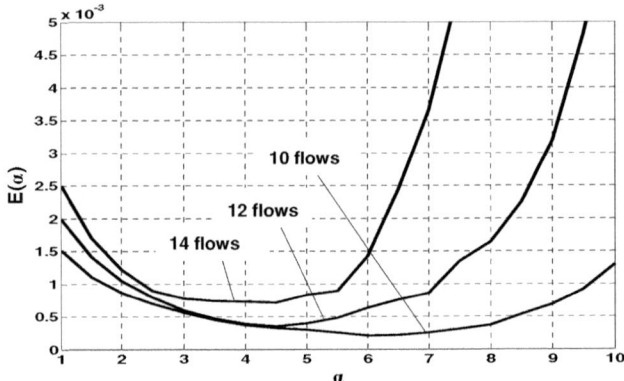

Fig. 5. Comparison of proposed algorithm with optimum

Table 2. Optimum α values depending on number of VoIP service flows

flows	4	6	8	10	12	14
α_{opt}	10.0	9.0	8.0	6.0	4.5	4.5

value providing this minimum error decreases when the number of flows is increased.

The value of this α_{opt} depending on the number of VoIP service flows is given in Tab. 2. For the evaluation, the tuning factor α has been varied in steps of 0.5. The results show that the optimum value decreases in an approximately linear manner with the increasing number of service flows until reach saturation at 4.5. The major findings in this context are that, corresponding to Eq. (3) in the algorithms description, the mean preference for the Reuse 3 Zone should be always higher than the preference for the Reuse 1 Zone ($\alpha > 1.0$). Furthermore, the degree of the average preference for the Reuse 3 Zone in the algorithm should be reduced when the number of VoIP service flows is increased.

5   Conclusions and Outlook

In this paper we presented an efficient algorithm for Mobile WiMAX to assign service flows to FFR zones considering QoS needs in terms of throughput. The proposed assignment scheme provides low complexity, yet achieves close to optimum performance in terms of resource utilization efficiency. This has been proven by means of a detailed simulation study of a typical Mobile WiMAX deployment with various numbers of VoIP service flows.

The main conclusions drawn from our results are: i) an FFR algorithm exhibiting close to optimum performance in terms of resource utilization is feasible at low complexity costs, ii) the performance difference compared to the optimum is a convex function of our algorithm tuning factor α, and iii) the optimum α

value providing the minimum difference to the optimum decreases as the number of flows increases.

As future work we plan to extend the FFR algorithm proposed in this paper so that α does not need to be configured by the operator but is found dynamically in a self-organizing way, automatically adapting to variations in the number of service flows and changing channel conditions. Furthermore, the algorithm has to be extended to support different QoS classes.

References

1. Boudreau, G., Panicker, J., Guo, N., Chang, R., Wang, N., Vrzic, S.: Interference Coordination and Cancellation for 4G Networks. IEEE Communications Magazine 47(4), 74–81 (2009)
2. WiMAX Forum, Mobile WiMAX - Part I: A Technical Overview and Performance Evaluation (August 2006)
3. 3GPP, TS 36.201 Evolved Universal Terrestrial Radio Access (E-UTRA); Long Term Evolution (LTE) physical layer; General description, 3GPP, Tech. Rep. (March 2009)
4. IEEE 802.16m D1, DRAFT Amendment to IEEE Standard for Local and metropolitan area networks, Part 16: Air Interface for Broadband Wireless Access Systems, Advanced Air Interface (July 2009)
5. ITU-R, REPORT ITU-R M.2134, Requirements related to technical performance for IMT-Advanced radio interface(s), ITU-R, Tech. Rep. (2008)
6. Rappaport, T.S.: Wireless Communications - Principles and Practice. Prentice-Hall, Englewood Cliffs (2002)
7. Zander, J.: Generalized Reuse Parititioning in Cellular Mobile Radio. In: Proc. of 43rd IEEE VTC, May 1993, pp. 181–184 (1993)
8. Jia, H., Zhang, Z., Yu, G., Cheng, P., Li, S.: On the Performance of IEEE 802.16 OFDMA System Under Different Frequency Reuse and Subcarrier Permutation Patterns. In: Proc. of IEEE ICC '07, June 2007, pp. 5720–5725 (2007)
9. Godlewski, P., Maqbool, M., Klif, J.-M., Coupechoux, M.: Analytical Evaluation of Various Frequency Reuse Schemes in Cellular OFDMA Networks. In: Proc. of ACM ValueTools, October 2008. ICST, Athens (2008)
10. Zhou, Y., Zein, N.: Simulation Study of Fractional Frequency Reuse for Mobile WiMAX. In: Proc. of IEEE VTC Spring 2008, Singapore, May 2008, p. 2595 (2008)
11. Sarperi, L., Hunukumbure, M., Vadgama, S.: Simulation Study of Fractional Frequency Reuse in WiMAX Networks. Fujitsu Scientific and Technical Journal 44(3), 318–324 (2008)
12. IEEE 802.16 Rev 2, Part 16: Air Interface for Broadband Wireless Access Systems (May 2009)
13. Balachandran, K., Calin, D., Cheng, F.-C., Joshi, N., Kang, J.H., Kogiantis, A., Rausch, K., Rudrapatna, A., Seymour, J.P., Sun, J.: Design and Analysis of an IEEE 802.16e-Based OFDMA Communication System. Bell Labs Technical Journal 11(4), 53–73 (2007)

Handling Transient Link Failures Using Alternate Next Hop Counters

Suksant Sae Lor, Raul Landa, Redouane Ali, and Miguel Rio

Network & Services Research Laboratory
Department of Electronic & Electrical Engineering
University College London, UK
{s.lor,r.landa,r.ali,m.rio}@ee.ucl.ac.uk

Abstract. In this paper, we propose a routing technique to alleviate packet loss due to transient link failures, which are major causes of disruption in the Internet. The proposed technique based on Alternate Next Hop Counters (ANHC) allows routers to calculate backup paths and re-route packets accordingly, thereby bypassing transient failures. This technique guarantees full repair coverage for single link failures, without significantly changing the way traditional routing works and with minimal impact on the computation and memory requirements for routers. We evaluate the performance of our proposed ANHC approach through extensive simulations and show that the stretch of its pre-computed alternate paths, its failure-state link load increase, and its computational and memory overheads are minimal.

Keywords: Internet routing, transient failures, fast re-routing.

1 Introduction

The development of error sensitive and critical applications over the Internet, such as health services and military has demanded a rethinking of the traditional best effort service approach. Link failures, and in particular transient link failures, represent major obstacles in guaranteeing reliability. There have been many proposals that handle failures. A well-known framework for minimising packet loss rates due to link and/or node failures is IP Fast Re-Route (IPFRR) [1]. In [2], analysis shows that the duration of forwarding disruptions must incorporate the times taken to detect the failure, react to it, propagate the failure notification, re-compute forwarding tables, and the time taken to actually install the changes into the Forwarding Information Base (FIB). IPFRR eliminates the time required by all of these except for the failure detection time, which can be shortened by adjusting protocol parameters [3]. This significantly reduces disruption time, as it allows packets to be re-routed via pre-computed backup paths as soon as the failure is detected. In this paper, we use the terms "backup" and "alternate" interchangeably to refer to these paths. The idea of using fast re-route to handle failures is not new. Several techniques such as Loop-Free Alternates (LFAs) [4] and not-via addresses [5] have been proposed and implemented in

M. Crovella et al. (Eds.): NETWORKING 2010, LNCS 6091, pp. 186–197, 2010.
© IFIP International Federation for Information Processing

real networks. However, these techniques are not sufficient, since LFAs cannot guarantee full protection against recoverable failures and the use of not-via addresses can lead to router performance degradation due to IP-in-IP tunnelling. On the other hand, a technique such as Failure-Insensitive Routing (FIR) [6] does not require a mechanism such as tunnelling to provide a 100% repair coverage against single link failures. Nevertheless, a router requires an interface-specific forwarding table which makes it more difficult to implement in hardware.

Other proposed approaches that deal with link and/or node failures include Multiple Routing Configurations (MRC) [7], Resilient Overlay Network (RON) [8] and Failure-Carrying Packets (FCP) [9]. MRC computes a number of configurations for various failure scenarios such that for a given failure, one or more configurations can be used for bypassing. However, this technique requires excessive computational and memory overheads. RON identifies a subset of nodes called the RON nodes that monitor the network and decide whether or not to send packets directly or between themselves, thereby bypassing failures if detected. A drawback of RON is that it requires continual probing of the alternate paths, hence adding overhead. FCP embeds the failed link information in the packet header. Based on this information, a router receiving FCP can calculate the latest shortest path available. Nonetheless, this increases the packet overhead and requires a very expensive dynamic computation.

In this paper, we propose a combination of an algorithm for computing backup paths and a fast packet re-routing mechanism based on Alternate Next Hop Counters (ANHC) which guarantees that packets can be fully recovered in the presence of single link failures without any of the disadvantages of the existing solutions.

The rest of the paper is organised as follows: we introduce our resilient routing technique as an alternative to IPFRR mechanisms in Sect 2. To guarantee full repair coverage and loop-free forwarding, we prove the properties of our routing strategy. In Sect. 3, we evaluate the performance of the routing technique with respect to various characteristics of the alternate paths, its impacts on network traffic, and the overheads and conclude the paper in Sect. 4.

2   Alternate Next Hop Counting

If source routing is available, the use of backup paths is greatly simplified. In contrast, it has been proven difficult to design hop-by-hop resilient routing protocols that can recover from failure quickly while keeping complexity to a minimum. The main cause of this difficulty lies in topology inconsistency among routers during re-convergence, which often leads to packet loss and forwarding loops.

Alternate next hop counting is a novel mechanism used in conjunction with specific backup path computation to provide fast re-route in traditional IP networks. Re-routing in this technique relies on the Alternate Next Hop Counter (ANHC) in the packet header to ensure correct forwarding. After the normal shortest paths are calculated, a router first determines the sum of all link weights W_t. For each node pair (s, d), a router adds W_t to the weights of the links present in the shortest path between s and d and re-calculates the shortest path from s to d. This *secondary path* is used to calculate the *alternate next hop* that s will use to send packets to d if the shortest path is unavailable. To allow consistent routing only on the basis of alternate next hops, each router sets the ANHC value for every destination equal to the number of times the packet needs to be forwarded using the alternate next hop. If the ANHC holds a positive number, a router decreases its value by 1 and forwards the packet to the alternate next hop corresponding to its own secondary path to d. If the ANHC value equals to 0, the router forwards the packet via its normal path to d (the next hop according to the shortest path route).

2.1   Computing Backup Paths

Let $G = (V, E)$ be the graph with vertices $V = \{v_1, v_2, ...\}$ and edges $E \in V \times V$ representing the network topology. Given an edge (i, j), we assign it a weight $w(i, j) \in \mathbb{R} > 0$. We define W_t as the total weight of all links in E:

$$W_t := \sum_{(i,j) \in E} w(i, j) \tag{1}$$

We seek to assign to each destination v_i a *maximally disjoint* secondary path, so that the backup path will have as few links in common with the normal path as possible.

Let $E_p(s, d)$ be the set of links used in a normal path from s to d. The primary next hop and its alternate are denoted as $n_p(s, d)$ and $n_s(s, d)$. Using W_t as a link weight re-calculation factor, Algorithm 1 is run on each router for each source-destination pair. In Algorithm 1, the output of *ShortestPath* is a shortest path tree T_s rooted at s, with $T_s(d)$ being the shortest path from s to d excluding s, and $\phi(T_s(d))$ the first node in $T_s(d)$.

Algorithm 1. Computing the alternate next hop

Input: s, d, G, $E_p(s, d)$
Output: $n_s(s, d)$
1: $G' \leftarrow G$
2: **for all** $(i, j) \in E_p(s, d)$ **do**
3:     $w'(i, j) \leftarrow w(i, j) + W_t$
4: **end for**
5: $T'_s = ShortestPath(G', s)$
6: $H_s(s, d) \leftarrow T'_s(d)$
7: $n_s(s, d) \leftarrow \phi(T'_s(d))$
8: **return** $n_s(s, d)$

The outputs of Algorithm 1 construct $S(d) = \{n_s(v_1, d), n_s(v_2, d), ...\}$, the alternate next hop nodes that every node will use to route packets to d under failure conditions, and $H_s(s, d) = \{h_1, h_2, ...\}$, where $H_s(s, d)$ is the backup path from s to d excluding s, and h_i represents the i-th next hop in the backup path from s to d. Thus, Algorithm 1 calculates an alternate next hop for each source-destination pair, which is the first hop of an alternate path that is maximally link disjoint from its corresponding normal path. Of course, each node $s \in G$ will calculate a different $n_s(s, d)$ for a given destination d: alternate paths for a destination d are not consistent throughout the network. Our technique is different from some other resilient mechanisms precisely because the alternate path locally known to each router does not have to be consistent: there is no guarantee that a path formed by concatenating the alternate next hops to a particular destination will be loop-free. Thus, we need to employ a mechanism that uses these pre-computed alternate next hops to route packets to their destinations via loop-free paths.

We propose a mechanism called *alternate next hop counting* to eliminate this path inconsistency problem.

2.2   Computing ANHC Values

As each router in the network may have different backup paths for the same destination, forwarding must be aided with an additional mechanism that allows correct operation under failure conditions. Our technique uses alternate next hop counting to ensure that no forwarding loop is possible in the presence of a single link failure. We now illustrate how, with inconsistent information on the local alternate paths, packets can be forwarded consistently under our routing scheme.

Alternate next hop counting makes use of an Alternate Next Hop Counter (ANHC) stored in the packet header. A few bits in the Type of Service (ToS) field of IPv4 or the Traffic Class field of IPv6 are sufficient to store the ANHC.

If we recall $S(d) = \{n_s(v_1, d), n_s(v_2, d), ...\}$ and $H_s(s, d) = \{h_1, h_2, ...\}$, the $AHNC(s, d)$ value can be obtained using Algorithm 2.

Algorithm 2. Computing the ANHC value

Input: s, d, $H_s(s, d)$, $S(d)$
Output: $ANHC(s, d)$
 1: $ANHC \leftarrow 0$
 2: $current\_node \leftarrow s$
 3: $i \leftarrow 1$
 4: **while** $h_i \neq d$ **do**
 5:     **if** $h_i == n_s(current\_node, d)$ **then**
 6:         $ANHC \leftarrow ANHC + 1$
 7:         $current\_node \leftarrow h_i$
 8:         $i \leftarrow i + 1$
 9:     **else**
10:         **break**
11:     **end if**
12: **end while**
13: **return** $ANHC$

The algorithm first considers the *current_node* which is initialised with s. The first hop in the alternate path from s to d is then compared with the alternate next hop of *current_node* to d. If they are the same node, $ANHC(s, d)$ is incremented by 1 and the alternate next hop of *current_node* to d becomes *current_node*. After that, the second hop in the alternate path from s to d is used for comparison. This process iterates until either it reaches d or the condition fails, implying that it is no longer necessary to route via the alternate next hop. From this point, the packet can be forwarded normally. When the algorithm is terminated, $ANHC(s, d)$ is obtained.

2.3   Packet Forwarding

Since the alternate next hop counting mechanism does not affect normal route calculation, packets can be forwarded to all destinations via the shortest paths in the absence of failures. When a node v detects a failure in one of its outgoing links, it marks those packets which would be forwarded through the affected link with the ANHC value corresponding to the destination router of the packet. If an alternate path to that node exists, it decrements the ANHC value in the packet and forwards it to its alternate next hop. When a node receives a marked packet, it determines the value of ANHC. If ANHC holds a positive number, its value is decremented by 1 and the packet is forwarded to the node's alternate next hop. However, if the ANHC value is 0, the node forwards the packet to its normal next hop until it reaches the destination. Algorithm 3 summarises the operations when a packet arrives at each node.

Algorithm 3. Packet processing at node s

Input: in_pkt
Output: out_pkt
1: **if** in_pkt.$ANHC == 0$ **then**
2:      **if** $(s, n_p(s, \text{in\_pkt}.d)) ==$ failed **then**
3:          in_pkt.$ANHC \leftarrow ANHC(s, \text{in\_pkt}.d) - 1$
4:          **return** out_pkt $\leftarrow$ in_pkt
5:      **else**
6:          **return** out_pkt $\leftarrow$ in_pkt
7:      **end if**
8: **else**
9:      **if** $(s, n_s(s, \text{in\_pkt}.d)) ==$ failed **then**
10:         Drop(in_pkt)
11:         **return** null
12:     **else**
13:         in_pkt.$ANHC \leftarrow$ in_pkt.$ANHC - 1$
14:         **return** out_pkt $\leftarrow$ in_pkt
15:     **end if**
16: **end if**

It is important to note that, our routing technique handles only **single** link failures. Certain cases of multiple failures can lead to forwarding loops. However, this problem can be trivially corrected. We propose the use of an extra bit to indicate a re-routed packet. That is, if a packet encounters a failure, the detecting node also marks it using this bit, in addition to the ANHC. If a marked packet experiences a failure again, it will be dropped immediately as the routing technique does not handle multiple failures.

Intuitively, the resilient routing technique described will, in case of failure, route packets along the first hops of maximally disjoint paths terminating on their destination node until, after a number of hops equal to the ANHC, a node is reached where they can be routed using the normal shortest path tree of the network. The shortest path route from this node need not be equal to the backup path calculated by the failure-detecting node - in fact, it is frequently shorter, as it does not need to be maximally disjoint. Thus, the length of the actual path that the packets will traverse under failure scenarios is no greater than the length of the backup path to their destination starting from their failure-detecting node.

Furthermore, since all failure-avoiding packets will traverse a number of hops equal to their ANHC before they are shortest-path routed, the length of the actual path that the packets will traverse under failure scenarios is no shorter than the ANHC to their destination starting from their source.

Evidently, the computation of alternate next hops and their corresponding ANHC values implies additional computations for network elements. Since these only need to be performed for stable topology configurations to pre-compute and cache relevant values (as opposed to be carried out constantly), these can be "amortised" over longer time periods. Thus, it is feasible to perform the algorithm at practical speeds, even using commodity hardware.

If additional efficiency is required, optimised shortest path algorithms can be used. One such algorithm is the incremental shortest path first algorithm (iSPF) [10], which avoids the calculation of the whole shortest path tree and instead terminates the computation once the shortest path between the source and destination has been found. This significantly reduces the computation time of the alternate next hops.

2.4 Properties

The two key properties of our routing technique are *a)* full repair coverage for recoverable single link failures, and *b)* loop-free forwarding. These properties are guaranteed if the routing scheme is *complete* and *correct* in the presence of *recoverable* failures. Definitions for these concepts are now given. For the remainder, we assume that equal-cost paths can be distinguished, so that all paths are essentially cost-unique, and all algorithms choose from between equal cost paths following a deterministic algorithm. Typical ways of achieving this involve differentiating by the number of hops on each path, or choosing on the basis of the interface ID for the first link.

Definition 1. *A single link failure is* recoverable *for a source-destination node pair if there is at least one alternate path from the source to the destination which does not traverse the failed link.*

Definition 2. *The routing technique is* complete *if the combination of local alternate next hops and the packet forwarding mechanism guarantees a successful packet delivery in case of any single link recoverable failures.*

Definition 3. *The routing technique is* correct *if the combination of local alternate next hops and the packet forwarding mechanism can forward packets to the destination in case of any single link recoverable failures without traversing through the failed links.*

First, we show that our routing strategy is complete in the presence of any recoverable single link failures.

Theorem 1. *If G is not disconnected after the removal of link $(s, n_p(s, d))$, then there exists a path from s to d via $(s, n_s(s, d))$ that does not traverse link $(s, n_p(s, d))$ under fast re-route with ANHC.*

Proof. We call $\mathcal{D}$ the set of paths from s to d that do not include $(s, n_p(s, d))$. If $\mathcal{D} = \emptyset$, the failure is non-recoverable by definition. Thus, we proceed to prove that if $\mathcal{D} \neq \emptyset$, the ANHC algorithm will always find an alternate path from s to d in $\mathcal{D}$.

We proceed by contradiction, assuming that $\mathcal{D} \neq \emptyset$ and nonetheless the algorithm has found that $n_p(s, d) = n_s(s, d)$. This implies that the weight of all paths in $\mathcal{D}$ is strictly higher than W_t, since the algorithm adds W_t to the weight of each one of the links of the primary shortest path in order to find the alternate path from s to d, and $w(i, j) > 0$; $\forall (i, j)$. However, the longest path from s to d over G would be an *Eulerian path*, whose weight could be of at most W_t, and thus the weight of all paths in $\mathcal{D}$ could be of at most W_t. Hence, we have a contradiction, and $\mathcal{D} \neq \emptyset$ implies that $n_p(s, d) \neq n_s(s, d)$. ☐

Second, we also show that incorporating the pre-computed alternate next hops with the alternate next hop counting mechanism can forward re-routed packets to the destination correctly under failure scenarios. Note that, the packet forwarding in normal case is based on the shortest path tree and hence, it is correct.

Theorem 2. *If there exists a path from s to d without link $(s, n_p(s, d))$, fast re-route using ANHC can forward packets from s to d without traversing $(s, n_p(s, d))$.*

Proof. Let $T_p(d)$ be the shortest path tree rooted at d and $H_s(s, d) = \{h_1, h_2, ...\}$ be the hop sequence of the alternate path from s to d excluding s, which is locally known to s. We denote $T_p(d_s)$ as the subgraph of $T_p(d)$ below s, which includes s with a set of vertices N.

Given E is a set of vertices in N that are employed by the alternate path from s to d. Each node e_i in E has the alternate next hop, $n_s(e_i, d)$. As each node e_i shares some links in the $T_p(d)$ with s, $H_s(s, d)$ must involve $n_s(e_i, d)$.

A re-routed packet can encounter a failed link $(s, n_p(s, d))$ if and only if it traverses along $T_p(d)$ starting from any node in E. However, a node will forward a re-routed packet through $T_p(d)$ only if the ANHC value is 0 - after this the packet will no longer be routed by using alternate next hops.

Since all nodes in E have alternate next hops that coincide with the alternate path from s to d, no re-routed packets arriving at e_i will have a zero value ANHC. Thus, packets will not be routed along $T_p(d)$ starting from e_i. Furthermore, packets will not be routed via $T_p(d)$ starting from a node in N that does not belong to E either, since $N - E$ and $H_s(s, d)$ are disjoint sets.

Finally, routing via $T_p(d)$ from any node outside N will not cause packets to traverse $(s, n_p(s, d))$, because these nodes are not elements of $T_p(d_s)$.

Therefore, we conclude that a path for re-routing packets from s to d does not involve the failed link $(s, n_p(s, d))$.   □

It is important to note that, alternate path locally known to each router is used for ANHC value calculation only and the routing technique does not hinder the network operator from knowing the actual alternate path used for packet re-routing.

3   Performance Evaluation

This section presents our evaluation of fast re-route resiliency using ANHC. The broad areas we investigate include protocol overhead, characteristics of the alternate paths, the impact of failures and associated recovery mechanisms on link load.

3.1   Method

We create our own software model to compute alternate next hops and their corresponding ANHC values. We run our simulations on a machine with a 2.16 GHz Intel Core 2 Duo processor and 2 GB memory. We use the Abilene [11] and GEANT [12] topologies, with corresponding real traffic matrices [13]. To illustrate that our technique can perform well for arbitrary network topologies, we also use the Abovenet and Sprintlink topologies inferred from Rocketfuel data [14], and synthetic topologies generated by BRITE [15] based on Waxman and Barabasi-Albert (BA) models.

We use the gravity models [16] to generate traffic matrices composed of edge-to-edge flows. To use realistic inputs to the gravity models, we use data from the U.S. Census Bureau [17] and the United Nations Statistics Division [18] to calculate the city population of each node in the network. Furthermore, we use a technique based on the Breadth-First Search (BFS) algorithm to assign link capacity [19]. To examine the traffic characteristics, we scale the traffic matrices such that the maximum link utilisation does not exceed 100% under normal routing or after re-convergence.

We use the normal shortest path routing as a benchmark comparison. The path used upon completion of the re-convergence process is denoted as *OSPF re-route*.

Of course, ANHC achieves the performance level reported immediately after a failure is detected, whereas OSPF re-route normally requires several seconds of re-convergence stabilisation before it can achieve its reported performance.

3.2 Overheads

We evaluate the computational overhead of our routing strategy by reporting the time required for computation of the alternate next hops and their corresponding ANHC values as evaluation metrics. The largest topology (*i.e.* Sprintlink) requires less than 100 ms for alternate next hop computation. ANHC value computation time is negligible for all topologies. Table 1 summarises the simulation results based on different topologies. Overall, results show that the algorithms used to compute essential parameters (*i.e.* alternate next hops and ANHC values) of our routing scheme do not incur any significant computational overhead.

Table 1. Summary of topologies used in simulations and the corresponding computational overhead introduced by fast re-route using ANHC

Topology	Type	Number of nodes	Number of links	In-/Out-degree	Computation time (ms) ANHs	ANHC	Total
Abilene	Real	11	28	1.273	0.108	0.006	0.114
GEANT	Real	23	74	1.609	0.120	0.008	0.128
Abovenet	Inferred	138	744	2.906	11.920	0.020	11.940
Sprintlink	Inferred	315	1944	3.086	77.091	0.051	77.142
Waxman	Random	100	400	2.000	3.091	0.015	3.106
Barabasi-Albert	Random	100	394	1.970	3.383	0.016	3.339

In order to enable re-routing with alternate next hop counting, a router must store additional information for each existing destination. That is, apart from the normal next hop, its alternate and the corresponding ANHC value must be maintained. However, no additional routing table entries are required and, hence, our routing technique does not entail excessive memory overhead.

Our forwarding scheme requires a few bits to store the ANHC value. We suggest that a part of the ToS field can be used for IPv4, and a part of the Traffic Class field can be used for IPv6. From simulation results, more than 90% of the alternate paths have an ANHC value of less than 3. In practice, a maximum ANHC of 7 is needed in the packet, as the failure-detecting node will decrement the ANHC value before it forwards the packet to its alternate next hop. To prevent forwarding loops in the presence of multiple failures, we propose the use of an extra bit to indicate a re-routed packet. With this in mind, our routing technique needs no more than 4 bits for an optimal packet re-routing in practical topologies.

3.3 Path Stretch

We measure the excess latency or weight required for delivery via alternate paths using *stretch*. Although the main objective of the IPFRR framework is to prevent packets from being dropped in case of failures, it is important

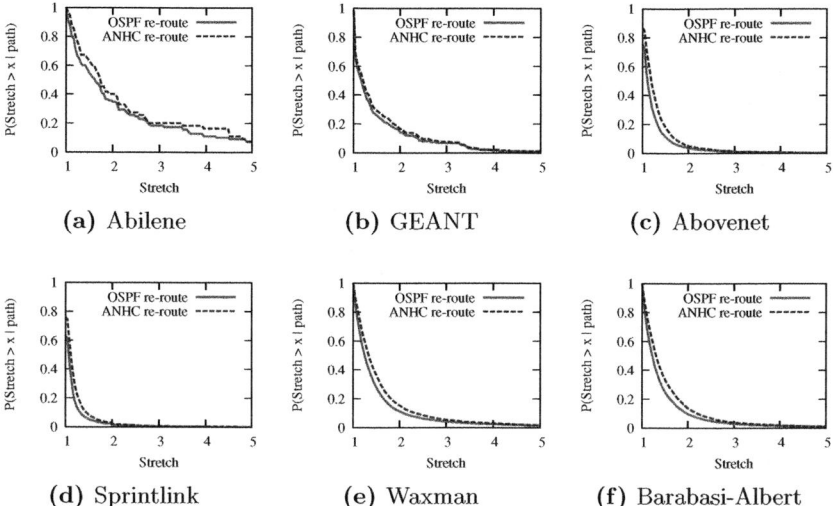

(a) Abilene   (b) GEANT   (c) Abovenet

(d) Sprintlink   (e) Waxman   (f) Barabasi-Albert

Fig. 1. Stretch comparison between OSPF re-route and IPFRR using ANHC

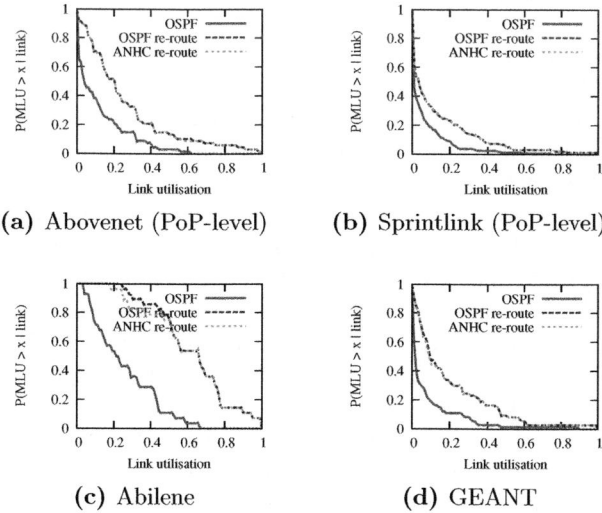

(a) Abovenet (PoP-level)   (b) Sprintlink (PoP-level)

(c) Abilene   (d) GEANT

Fig. 2. Maximum link utilisation before and after single link failures of different topologies

that packet delivery via alternate paths does not cause too high delay so that end users of sensitive applications can perceive a poorer Quality of Service (QoS).

As can be seen in Fig. 1, our fast re-route technique offers recovery via alternate paths that are near optimal regardless of the underlying network topology. The average of optimal stretch across all topologies is 1.221 while the average of stretch provided by our proposed technique is 1.305. In most cases, the cost of an alternate path is close to that of the shortest path.

3.4   Maximum Link Utilisation

We analyse the maximum link utilisation over different failure scenarios of each topology. Figure 2 illustrates the fraction of links in the network with maximum link utilisation exceeding the value in x-axes. Interestingly, unlike other IPFRR techniques [20], none of the links in all topologies are overloaded under our routing strategy. Fast re-route with ANHC also performs as well as normal re-convergence in any arbitrary network.

4   Conclusion

Packet dropping due to transient failures is a major cause of disruption in today's Internet and will become more critical as more demanding applications are developed. Several approaches such as IPFRR, multi-homing and multi-path routing, and overlay networks have been proposed to alleviate the problem. Although IPFRR provides better modularity, none of its techniques can guarantee full protection against single failures without using mechanisms (*e.g.* tunnelling) that degrade the performance of a router.

We proposed a new IPFRR technique based on the Alternate Next Hop Counter (ANHC) to handle transient link failures. In the normal scenario, packets are forwarded along the shortest path calculated similarly as in traditional IP routing. When a link fails, the detecting router is responsible for setting the pre-computed ANHC value in the packet header. This value is used by intermediate routers to determine the packet's next hop. We presented two algorithms for computing the alternate paths and their corresponding ANHC values. Furthermore, we proved that our technique is complete and correct.

As the path provided by OSPF re-route is the shortest path after a failure, we used it as a benchmark throughout our evaluation. From simulation results, we concluded that fast re-route using ANHC requires no significant overheads and can be easily deployed without major modifications. Moreover, it can fully protects single link failures using low stretch alternate paths and has minimal impact on network traffic. We greatly believe that fast re-route using ANHC can greatly enhance the network reliability without any expensive requirements.

References

1. Shand, M., Bryant, S.: IP fast reroute framework. RFC 5714 (January 2010),
   http://tools.ietf.org/html/rfc5714
2. Li, A., Francois, P., Yang, X.: On improving the efficiency and manageability of
   notvia. In: Proc. ACM CoNEXT, New York, December 2007, pp. 1–12 (2007)
3. Goyal, M.R., Feng, K.K.W.: Achieving faster failure detection in OSPF networks.
   In: Proc. IEEE ICC, Anchorage, AK, May 2003, pp. 296–300 (2003)
4. Atlas, A., Zinin, A.: Basic specification for IP fast reroute Loop-Free Alternates.
   RFC 5286 (September 2008), http://tools.ietf.org/html/rfc5286
5. Bryant, S., Shand, M., Previdi, S.: IP fast reroute using not-via addresses. IETF
   Internet draft (July 2009),
   http://tools.ietf.org/html/draft-ietf-rtgwg-ipfrr-notvia-addresses-04
6. Nelakuditi, S., Lee, S., Yu, Y., Zhang, Z.L., Chuah, C.N.: Fast local rerouting for
   handling transient link failures. IEEE/ACM Transactions on Networking 15(2),
   359–372 (2007)
7. Kvalbein, A., Hansen, A.F., Cicic, T., Gjessing, S., Lysne, O.: Fast IP network recovery using multiple routing configurations. In: Proc. IEEE INFOCOM,
   Barcelona, Spain, April 2006, pp. 23–29 (2006)
8. Andersen, D.G., Balakrishnan, H., Kaashoek, M.F., Morris, R.: Resilient Overlay
   Network. In: Proc. ACM SOSP, Banff, Canada, October 2001, pp. 131–145 (2001)
9. Lakshminarayanan, K., Caesar, M., Rangan, M., Anderson, T., Shenker, S., Stoica, I.: Achieving convergence-free routing using failure-carrying packets. In: Proc.
   ACM SIGCOMM, Kyoto, Japan, August 2007, pp. 241–252 (2007)
10. McQuillan, J.M., Richer, I., Rosen, E.C.: The new routing algorithm for the
    ARPANET. IEEE Transactions on Communications 28(5), 711–719 (1980)
11. Zhang, Y.: The Abilene topology and traffic matrices (December 2004),
    http://www.cs.utexas.edu/~yzhang/research/AbileneTM/
12. GEANT: The GEANT topology (December 2004),
    http://www.geant.net/upload/pdf/GEANT_Topology_12-2004.pdf
13. Uhlig, S., Quoitin, B., Balon, S., Lepropre, J.: Providing public intradomain traffic
    matrices to the research community. ACM SIGCOMM Computer Communication
    Review 36(1), 83–86 (2006)
14. Spring, N., Mahajan, R., Wetherall, D., Anderson, T.: Measuring ISP topologies
    with Rocketfuel. IEEE/ACM Transactions on Networking 12(1), 2–16 (2004)
15. Medina, A., Lakhina, A., Matta, I., Byers, J.: BRITE: an approach to universal
    topology generation. In: Proc. IEEE MASCOTS, Cincinnati, OH, August 2001,
    pp. 346–353 (2001)
16. Medina, A., Taft, N., Salamatian, K., Bhattacharyya, S., Diot, C.: Traffic matrix
    estimation: Existing techniques and new directions. In: Proc. ACM SIGCOMM,
    Pittsburgh, PA, August 2002, pp. 161–174 (2002)
17. U.S. Census Bureau: Census 2000 gateway (April 2000),
    http://www.census.gov/main/www/cen2000.html
18. United Nations Statistics Division: Demographic and social statistics (August
    2008), http://unstats.un.org/unsd/demographic/
19. Liu, W., Karaoglu, H.T., Gupta, A., Yuksel, M., Kar, K.: Edge-to-edge bailout
    forward contracts for single-domain Internet services. In: Proc. IEEE IWQoS, En-
    schede, The Netherlands, June 2008, pp. 259–268 (2008)
20. Menth, M., Hartman, M., Martin, R., Cicic, T., Kvalbein, A.: Loop-free alternates
    and not-via addresses: A proper combination for IP fast reroute? Computer Net-
    works (2009)

Efficient Recovery from False State in Distributed Routing Algorithms

Daniel Gyllstrom, Sudarshan Vasudevan, Jim Kurose, and Gerome Miklau

Department of Computer Science, University of Massachusetts Amherst,
140 Governors Drive Amherst, MA 01003
{dpg,svasu,kurose,miklau}@cs.umass.edu

Abstract. Malicious and misconfigured nodes can inject incorrect state into a distributed system, which can then be propagated system-wide as a result of normal network operation. Such false state can degrade the performance of a distributed system or render it unusable. For example, in the case of network routing algorithms, false state corresponding to a node incorrectly declaring a cost of 0 to all destinations (maliciously or due to misconfiguration) can quickly spread through the network. This causes other nodes to (incorrectly) route via the misconfigured node, resulting in suboptimal routing and network congestion. We propose three algorithms for efficient recovery in such scenarios and prove the correctness of each of these algorithms. Through simulation, we evaluate our algorithms – in terms of message and time overhead – when applied to removing false state in distance vector routing. Our analysis shows that over topologies where link costs remain fixed and for the same topologies where link costs change, a recovery algorithm based on system-wide checkpoints and a rollback mechanism yields superior performance when using the poison reverse optimization.

Keywords: Routing, Security, Recovery, Checkpointing, Fault Tolerance.

1 Introduction

Malicious and misconfigured nodes can degrade the performance of a distributed system by injecting incorrect state information. Such false state can then further propagate through the system either directly in its original form or indirectly, e.g., by diffusing computations initially using this false state. In this paper, we consider the problem of removing such false state from a distributed system.

In order to make the false-state-removal problem concrete, we investigate distance vector routing as an instance of this problem. Distance vector forms the basis for many routing algorithms widely used in the Internet (e.g., BGP, a path-vector algorithm) and in multi-hop wireless networks (e.g., AODV, diffusion routing). However, distance vector is vulnerable to compromised nodes that can potentially flood a network with false routing information, resulting in erroneous least cost paths, packet loss, and congestion. Such scenarios have occurred in

M. Crovella et al. (Eds.): NETWORKING 2010, LNCS 6091, pp. 198–212, 2010.
© IFIP International Federation for Information Processing

practice. For example, recently a routing error forced Google to redirect its traffic through Asia, causing congestion that left many Google services unreachable [1]. Distance vector currently has no mechanism to recover from such scenarios. Instead, human operators are left to manually reconfigure routers. It is in this context that we propose and evaluate automated solutions for recovery.

In this paper, we design, implement, and evaluate three different approaches for correctly recovering from the injection of false routing state (e.g., a compromised node incorrectly claiming a distance of 0 to all destinations). Such false state, in turn, may propagate to other routers through the normal execution of distance vector routing, making this a network-wide problem. Recovery is correct if the routing tables in all nodes have converged to a global state in which all nodes have removed each compromised node as a destination, and no node has a least cost path to any destination that routes through a compromised node.

Specifically, we develop three new distributed recovery algorithms: 2nd best, purge, and cpr. 2nd best performs localized state invalidation, followed by network-wide recovery. Nodes directly adjacent to a compromised node locally select alternate paths that avoid the compromised node; the traditional distributed distance vector algorithm is then executed to remove remaining false state using these new distance vectors. The purge algorithm performs global false state invalidation by using diffusing computations to invalidate distance vector entries (network-wide) that routed through a compromised node. As in 2nd best, traditional distance vector routing is then used to recompute distance vectors. cpr uses local snapshots and a rollback mechanism to implement recovery.

We prove the correctness of each algorithm and use simulations to evaluate the efficiency of each algorithm in terms of message overhead and convergence time. Our simulations show that cpr using poison reverse outperforms 2nd best and purge (with and without poison reverse) – at the cost of checkpoint memory – over topologies with fixed and changing link costs. This is because cpr efficiently removes all false state by rolling back to a checkpoint immediately preceding the injection of false routing state. In scenarios where link costs can change, purge using poison reverse yields performance close to cpr with poison reverse. purge makes use of computations subsequent to the injection of false routing state that do not depend on false routing state, while cpr must process all valid link cost changes that occurred since false routing state was injected. Finally, our simulations show that poison reverse significantly improves performance for all three algorithms, especially for topologies with changing link costs.

Recovery from false routing state has similarities to the problem of recovering from malicious transactions [12] in distributed databases. Our problem is also similar to rollback in optimistic parallel simulation [11]. We are unaware of existing solutions to the problem of recovering from false routing state. However, a related problem is that of discovering misbehaving nodes. In Section 2, we discuss existing solutions to this problem. In fact, the output of these algorithms serve as input to the recovery algorithms proposed in this paper.

This paper has five sections. In Section 2 we define the problem and state our assumptions. We present our three recovery algorithms in Section 3. Section 4 describes our simulation study. We detail related work in Section 5 and finally we conclude and comment on directions for future work in Section 6.

2   Problem Formulation

We consider distance vector routing [4] over arbitrary network topologies.[1] We model a network as an undirected graph, $G = (V, E)$, with a link weight function $w : E \to \mathbb{N}$. Each node, v, maintains the following state as part of distance vector: a vector of all adjacent nodes $(adj(v))$, a vector of least cost distances to all nodes in G ($\overrightarrow{min_v}$), and a *distance matrix* that contains distances to every node in the network via each adjacent node ($dmatrix_v$).

For simplicity, we present our recovery algorithms in the case of a single compromised node. We describe the necessary extensions to handle multiple compromised nodes in our technical report [9].

We assume that the identity of the compromised node is provided by a different algorithm, and thus do not consider this problem in this paper. Examples of such algorithms include [6,7,13,16,18]. Specifically, we assume that at time t, this algorithm is used to notify all neighbors of the compromised node. Let t' be the time the node was compromised.

For each of our algorithms, the goal is for all nodes to recover correctly: all nodes should remove the compromised node as a destination and find new least costs that do not use the compromised node. If the network becomes disconnected as a result of removing the compromised node, all nodes need only compute new least costs to the nodes in their connected component. For simplicity, let $\bar{v}$ denote the compromised node, let $\overrightarrow{old}$ refer to $\overrightarrow{min_{\bar{v}}}$ before the first $\bar{v}$ was compromised, and let $\overrightarrow{bad}$ denote $\overrightarrow{min_{\bar{v}}}$ after $\bar{v}$ has been compromised. Intuitively, $\overrightarrow{old}$ and $\overrightarrow{bad}$ are snapshots of the compromised node's least cost vector taken at two different timesteps: $\overrightarrow{old}$ marks the snapshot taken before $\bar{v}$ was compromised and $\overrightarrow{bad}$ represents a snapshot taken after $\bar{v}$ was compromised.

3   Recovery Algorithms

In this section we propose three new recovery algorithms: 2nd best, purge, and cpr. With one exception, the input and output of each algorithm is the same.[2]

Input: Undirected graph, $G = (V, E)$, with weight function $w : E \to \mathbb{N}$. $\forall v \in V$, $\overrightarrow{min_v}$ and $dmatrix_v$ are computed (using distance vector). Also, each $v \in adj(\bar{v})$ is notified that $\bar{v}$ was compromised.

Output: Undirected graph, $G' = (V', E')$, where $V' = V - \{\bar{v}\}$, $E' = E - \{(\bar{v}, v_i) \mid v_i \in adj(\bar{v})\}$, and link weight function $w : E \to \mathbb{N}$. $\overrightarrow{min_v}$ and $dmatrix_v$ are computed via the algorithms discussed below $\forall v \in V'$.

---

[1] Recovery is simple with link state routing: each node uses its complete topology map to compute new least cost paths that avoid all compromised nodes. Thus we do not consider link state routing in this paper.

[2] cpr requires each $v \in adj(\bar{v})$ be notified of the time, t', in which $\bar{v}$ was compromised.

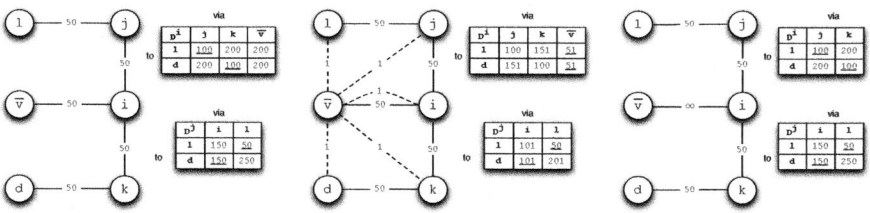

(a) Before $\bar{v}$ is compromised. (b) After $\bar{v}$ is compromised. (c) After recovery.

Fig. 1. Three snapshots of a graph, G, where $\bar{v}$ is the compromised node: (a) G before $\bar{v}$ is compromised, (b) G after $\overrightarrow{bad}$ has finished propagating but before recovery has started, and (c) G after recovery. The dashed lines in (b) mark false paths used by $\overrightarrow{bad}$. Portions of $dmatrix_i$ and $dmatrix_j$ are displayed to the right of each sub-figure. The least cost values are underlined.

First we describe a preprocessing procedure common to all three recovery algorithms. Then we describe each recovery algorithm. Due to space constraints, the proof of correctness and pseudo code for each algorithm can be found in our technical report [9].

3.1 Preprocessing

All three recovery algorithms share a common preprocessing procedure. The procedure removes $\bar{v}$ as a destination and finds the node IDs in each connected component. This is implemented using diffusing computations [5] initiated at each $v \in adj(\bar{v})$. A diffusing computation is a distributed algorithm started at a source node which grows by sending queries along a spanning tree, constructed simultaneously as the queries propagate through the network. When the computation reaches the leaves of the spanning tree, replies travel back along the tree towards the source, causing the tree to shrink. The computation eventually terminates when the source receives replies from each of its children in the tree. In our case, each diffusing computation message contains a vector of node IDs. When a node receives a diffusing computation message, the node adds its ID to the vector and removes $\bar{v}$ as a destination. At the end of the diffusing computation, each $v \in adj(\bar{v})$ has a vector that includes all nodes in v's connected component. Finally, each $v \in adj(\bar{v})$ broadcasts the vector of node IDs to all nodes in their connected component. In the case where removing $\bar{v}$ partitions the network, each node only computes shortest paths to nodes in the vector.

3.2 The 2nd best Algorithm

2nd best invalidates state locally and then uses distance vector to implement network-wide recovery. Following the preprocessing described in Section 3.1, each neighbor of the compromised node locally invalidates state by selecting the least cost pre-existing alternate path that does not use the compromised node as the first hop. The resulting distance vectors trigger the execution of traditional distance vector to remove the remaining false state.

We trace the execution of 2nd best using the example in Figure 1. In Figure 1(b), i uses $\bar{v}$ to reach nodes l and d. j uses i to reach all nodes except l. Notice that when j uses i to reach d, it transitively uses $\overrightarrow{bad}$ (e.g., uses path $j - i - \bar{v} - d$ to d). After the preprocessing completes, i selects a new next-hop node to reach l and d by finding its new smallest distance in $dmatrix_i$ to these destinations: i selects the routes via j to l with a cost of 100 and i picks the route via k to reach d with cost of 100. (No changes are required to route to j and k because i uses its direct link to these two nodes). Then, using traditional distance vector i sends $\overrightarrow{min_i}$ to j and k. When j receives $\overrightarrow{min_i}$, j must modify its distance to d because $\overrightarrow{min_i}$ indicates that i's least cost to d is now 100. j's new distance value to d becomes 150, using the path $j - i - k - l$. j then sends a message sharing $\overrightarrow{min_j}$ with its neighbors. From this point, recovery proceeds according by using traditional distance vector.

2nd best is simple and makes no synchronization assumptions. However, 2nd best is vulnerable to the count-to-∞ problem: because each node only has local information, the new shortest paths may continue to use $\bar{v}$.

3.3   The purge Algorithm

purge globally invalidates all false state using diffusing computations and then uses distance vector to compute new distance values that avoid all invalidated paths. Recall that diffusing computations preserve the decentralized nature of distance vector. The diffusing computations are initiated at the neighbors of $\bar{v}$ and spread to the network edge, invalidating false state at each node along the way. Then ACKs travel back from the network edge to the neighbors of $\bar{v}$, indicating that the diffusing computation is complete. Next, purge uses distance vector to recompute least cost paths invalidated by the diffusing computations.

In Figure 1, the diffusing computation executes as follows. First, i sets its distance to l and d to ∞ (thereby invalidating i's path to l and d) because i uses $\bar{v}$ to route these nodes. Then, i sends a message to j and k containing l and d as invalidated destinations. When j receives i's message, j checks if it routes via i to reach l or d. Because j uses i to reach d, j sets its distance estimate to d to ∞. j does not modify its least cost to l because j does not route via i to reach l. Next, j sends a message that includes d as an invalidated destination. l performs the same steps as j. After this point, the diffusing computation ACKs travel back towards i. When i receives an ACK from j and k, the diffusing computation is complete. At this point, i needs to compute new least costs to node l and d because i's distance estimates to these destinations are ∞. i uses $dmatrix_i$ to select its new route to l (which is via j) and to find i's new route to d (which is via k). Finally, i sends $\overrightarrow{min_i}$ to its neighbors, triggering the execution of distance vector to recompute the remaining distance vectors.

An advantage of purge is that it makes no synchronization assumptions. Also, the diffusing computations ensure that the count-to-∞ problem does not occur by removing false state from the entire network. However, globally invalidating false state can be wasteful if valid alternate paths are locally available.

3.4   The cpr Algorithm

cpr[3] is our third and final recovery algorithm. Unlike 2nd best and purge, cpr requires that clocks across different nodes be loosely synchronized. We assume a maximum clock offset between any two nodes. For ease of explanation, we describe cpr as if clocks are perfectly synchronized. Extensions to handle loosely synchronized clocks should be clear. Accordingly, we assume that all neighbors of $\overline{v}$, are notified of the time, t', at which $\overline{v}$ was compromised.

For each node, $i \in V$, cpr adds a time dimension to $\overrightarrow{min}_i$ and $dmatrix_i$, which cpr then uses to locally archive a complete history of values. Once the compromised node is discovered, the archive allows the system to rollback to a system snapshot from a time before $\overline{v}$ was compromised. From this point, cpr needs to remove $\overline{v}$, $\overrightarrow{old}$, and update stale distance values resulting from link cost changes. We describe each algorithm step in detail below.

Step 1: Create a $\overrightarrow{min}$ and $dmatrix$ archive. We define a *snapshot* of a data structure to be a copy of all current distance values along with a timestamp. [4] The timestamp marks the time at which that set of distance values start being used. $\overrightarrow{min}$ and $dmatrix$ are the only data structures that need to be archived.

Our distributed archive algorithm is quite simple. Each node can archive at a given frequency (e.g., every m timesteps) or after some number of distance value changes (e.g., each time a distance value changes). Each node must choose the same option, which is specified as an input parameter to cpr. A node archives independently of all other nodes. A side effect of independent archiving, is that even with perfectly synchronized clocks, the union of all snapshots may not constitute a globally consistent snapshot.[5]

Step 2: Rolling back to a valid snapshot. Rollback is implemented using diffusing computations. Neighbors of the compromised node independently select a snapshot to roll back to, such that the snapshot is the most recent one taken before t'. Each such node, i, rolls back to this snapshot by restoring the $\overrightarrow{min}_i$ and $dmatrix_i$ values from the snapshot. Then, i initiates a diffusing computation to inform all other nodes to do the same.

Step 3: Steps after rollback. After Step 2, the algorithm in Section 3.1 is executed. When the diffusing computations complete, there are two issues to address. First, nodes may be using $\overrightarrow{old}$. Second, nodes may have stale state as a result of link cost changes that occurred during $[t',t]$ and consequently are not reflected in the snapshot. To resolve these issues, each $i \in adj(\overline{v})$ sets its distance to $\overline{v}$ to ∞ and then selects new least cost values that avoid the compromised node, triggering the execution of distance vector to update the remaining distance vectors.

---

[3] The name is an abbreviation for **C**heck**P**oint and **R**ollback.
[4] In practice, we only archive distance values that have changed. Thus each distance value is associated with its own timestamp.
[5] A globally consistent snapshot is not required for correctness [9].

In the example from Figure 1, the global state after rolling back is nearly the same as the snapshot depicted in Figure 1(c): the only difference between the actual system state and that depicted in Figure 1(c) is that in the former $(i,\overline{v}) = 50$ rather than ∞. Step 3 of cpr makes this change. Because no nodes use $\overrightarrow{old}$, no other changes take place.

In summary, rather than using an iterative process to remove false state (like in 2nd best and purge), cpr does so in one diffusing computation. However, cpr incurs storage overhead resulting from periodic snapshots of $\overrightarrow{min}$ and $dmatrix$. Also, after rolling back, stale state may exist if link cost changes occur during $[t', t]$. This can be expensive to update. Finally, unlike purge and 2nd best, cpr requires loosely synchronized clocks because without a bound on the clock offset, nodes may rollback to highly inconsistent local snapshots. Although correct, this would severely degrade cpr performance.

4   Evaluation

In this section, we use simulations to characterize the performance of each of our three recovery algorithms in terms of message and time overhead. Our goal is to illustrate the relative performance of our recovery algorithms over different topology types (e.g., Erdös-Rényi graphs, Internet-like graphs) and different network conditions (e.g., fixed link costs, changing link costs). We evaluate recovery after a single compromised node has distributed false routing state. An evaluation of our algorithms in the case of multiple compromised nodes can be found in our technical report [9].

We build a custom simulator with a synchronous communication model: nodes send and receive messages at fixed epochs. In each epoch, a node receives a message from all its neighbors and performs its local computation. In the next epoch, the node sends a message (if needed). All algorithms are deterministic under this communication model. The synchronous communication model, although simple, yields interesting insights into the performance of each of the recovery algorithms. Evaluation of our algorithms using a more general asynchronous communication model is currently under investigation. However, we believe an asynchronous implementation will demonstrate similar trends.

We simulate the following scenario:

1. Before t', $\forall v \in V$ $\overrightarrow{min}_v$ and $dmatrix_v$ are correctly computed.
2. At time t', $\overline{v}$ is compromised and advertises $\overrightarrow{bad}$ (a vector with a cost of 1 to *every* node in the network) to its neighboring nodes.
3. $\overrightarrow{bad}$ spreads for a specified number of hops (this varies by experiment). Variable k refers to the number of hops $\overrightarrow{bad}$ spreads.
4. At time t, some $v \in V$ notifies all $i \in adj(\overline{v})$ that $\overline{v}$ was compromised.[6]

The message and time overhead are measured in step (4) above. The precomputation (Section 3.1) is not counted towards message and time overhead.

---

[6] For cpr this node also indicates the time, t', $\overline{v}$ was compromised.

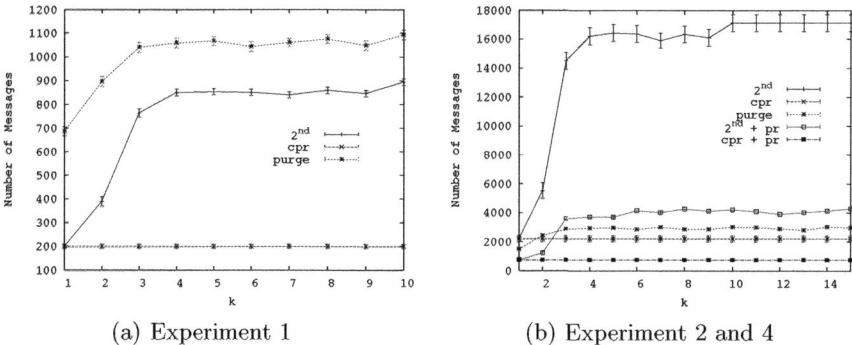

(a) Experiment 1   (b) Experiment 2 and 4

Fig. 2. Experiment 1, 2, and 4 plots. (a) Experiment 1 - message overhead for Erdös-Rényi Graphs with fixed unit link weights, where $n = 100$, $p = 0.05$, and diameter= 6.14. (b) Experiment 2 and 4 - message overhead for Erdös-Rényi graph with random link weights, $n = 100$, $p = .05$, and average diameter=6.14. The 2nd best, purge, and cpr curves correspond to Experiment 2. Experiment 4 additionally includes 2nd best + pr (2nd best using poison reverse) and cpr + pr (cpr using poison reverse).

4.1   Fixed Link Weight Experiments

In the next four experiments, we evaluate our recovery algorithms over different topology types in the case of fixed link costs.

Experiment 1. We start with a simplified setting and consider Erdös-Rényi graphs with parameters n and p. n is the number of graph nodes and p is the probability that link (i, j) exists where $i, j \in V$. The link weight of each edge in the graph is set to 50. We iterate over different values of k. For each k, we generate an Erdös-Rényi graph, $G = (V, E)$, with parameters n and p. Then we select a $\bar{v} \in V$ uniformly at random and simulate the scenario described above, using $\bar{v}$ as the compromised node. In total we sample 20 unique nodes for each G. We set $n = 100$, $p = \{0.05, 0.15, 0.25, 0.25\}$, and let $k = \{1, 2, ...10\}$. Each data point is an average over 600 runs (20 runs over 30 topologies). We then plot the 90% confidence interval.

The results for this experiment are shown in Figure 2(a). We omit figures for $p = \{0.15, 0.25, 0.50\}$ because the results follow the same trends as $p = 0.05$ [9]. cpr outperforms purge and 2nd best because $\overrightarrow{bad}$ is removed using a single diffusing computation, while the other algorithms remove $\overrightarrow{bad}$ state through distance vector's iterative process.

With 2nd best, distance values increase from their initial value until they reach their final (correct) value. Any intermediate, non-final, distance value uses $\overrightarrow{bad}$ or $\overrightarrow{old}$. Because $\overrightarrow{bad}$ and $\overrightarrow{old}$ no longer exist during recovery, these intermediate values must correspond to routing loops.[7] Our profiling numbers indicate that

---

[7] We formally prove these properties in our technical report [9].

there are few (if any) pairwise routing loops during 2nd best recovery, indicating that nodes quickly count up to their final least costs.

Although no pairwise routing loops exist during purge recovery, purge incurs overhead in its purge phase. Roughly, 50% of purge's messages come from the purge phase. This accounts for purge's high message overhead.

purge and 2nd best message overhead increases with larger k. Larger k implies that false state has propagated further in the network, resulting in more paths to repair, and therefore increased messaging. For values of k greater than a graph's diameter, the message overhead remains constant, as expected.

The trends for time overhead match those for message overhead. The interested reader can refer to our technical report [9] for these figures.

Experiment 2. The experimental setup is identical to Experiment 1 with one exception: link weights are selected uniformly at random between $[1, n]$ (rather than using fixed link weight of 50). Figure 2(b) shows the message overhead for $n = 100$ and $p = .05$. We omit the figures for the other p values because they follow the same trend as $p = .05$ [9].

In striking contrast to Experiment 1, purge outperforms 2nd best for all values of k. 2nd best performs poorly because the count-to-∞ problem: when $k < 4$, there are 1K pairwise routing loops (a strong indicator of the occurrence of the count-to-∞ problem) and over 10K routing loops occur for each $k \geq 4$. No routing loops are found with purge because they are removed by purge's diffusing computations. cpr performs well because $\overrightarrow{bad}$ is removed using a single diffusing computation, while the other algorithms remove $\overrightarrow{bad}$ state through distance vector's iterative process.

In addition, we count the number of epochs in which at least one pairwise routing loop exists. For 2nd best (across all topologies), on average, all but the last three timesteps have at least one routing loop. This suggests that the count-to-∞ problem dominates the cost for 2nd best.

Experiment 3. In this experiment, we simulate our algorithms over Internet-like topologies downloaded from the Rocketfuel website [3] and generated using GT-ITM [2]. We show the results for one Rocketfuel graph in Figure 3(a). The results follow the same pattern as in Experiment 2.

Experiment 4. We repeat Experiments 2 and 3 using poison reverse for 2nd best and cpr. We do not apply poison reverse to purge because no routing loops (resulting from the removal of $\overline{v}$) exist during purge's recovery. Additionally, we do not repeat Experiment 1 using poison reverse because we observed few routing loops in that experiment. The results are shown for one representative topology in Figure 2(b), where 2nd best + pr and cpr + pr refer to each respective algorithm using poison reverse.

cpr + pr has modest gains over standard cpr because few routing loops occur with cpr. On other hand, 2nd best + pr sees a significant decrease in message overhead when compared to the standard 2nd best algorithm because poison reverse removes the many pairwise routing loops that occur during 2nd best recovery. However, 2nd best + pr still performs worse than cpr + pr and

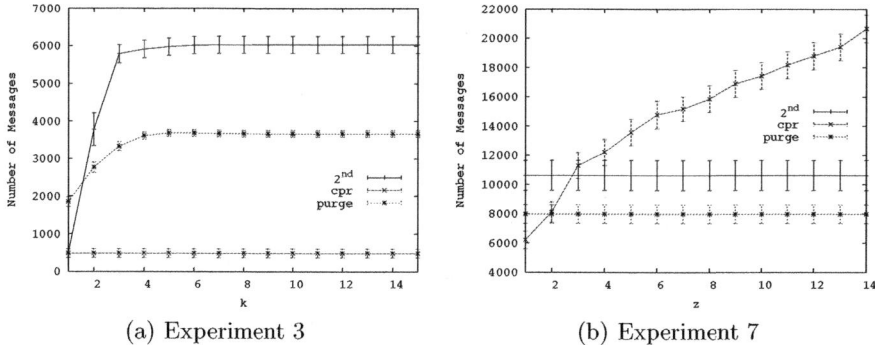

(a) Experiment 3   (b) Experiment 7

Fig. 3. Plots for Experiment 3 and 7. (a) Experiment 3 - Rocketfuel graph number 6461 with $n = 141$ and diameter=8. (b) Experiment 7 - message overhead for Erdös-Rényi with random link weights, $n = 100$, $p = 0.05$, $k = 2$, and $\lambda = 4$. z refers to the number of timesteps cpr must rollback.

purge. When compared to cpr + pr, the same reasons described in Experiment 2 account for 2nd best + pr's poor performance. Comparing purge and 2nd best + pr yields interesting insights to the two different approaches for eliminating routing loops: purge prevents routing loops using diffusing computations and 2nd best + pr uses poison reverse. Because purge has lower message complexity than 2nd best + pr and poison reverse only eliminates pairwise routing loops, it suggests that purge removes routing loops larger than 2. We are currently investigating this claim.

4.2 Link Weight Change Experiments

In the next three experiments we evaluate our algorithms over graphs with changing link costs. We introduce link cost changes between the time $\overline{v}$ is compromised and when $\overline{v}$ is discovered (e.g. during $[t', t]$). In particular, let there be λ link cost changes per timestep, where λ is deterministic. To create a link cost change event, we choose a link equiprobably among all links (except all $(v, \overline{v})$ links) and change its cost. The new link cost is selected uniformly at random from $[1, n]$.

Experiment 5. Except for λ, our experimental setup is identical to the one in Experiment 2. We let $\lambda = \{1, 4, 8\}$. In order to isolate the effects of link costs changes, we assume that cpr checkpoints at each timestep.

Due to space constraints, we only show results for $p = .05$ and $\lambda = \{4, 8\}$ in Figure 4.[8] purge yields the lowest message overhead, but only slightly lower than cpr. cpr's message overhead increases with larger k because there are more link cost change events to process. After cpr rolls back, it must process all link cost changes that occurred in $[t', t]$. In contrast, 2nd best and purge process some of the link cost change events during $[t', t]$ as part of normal distance vector

---

[8] Our experiments for different λ and p values, yield the same trends [9].

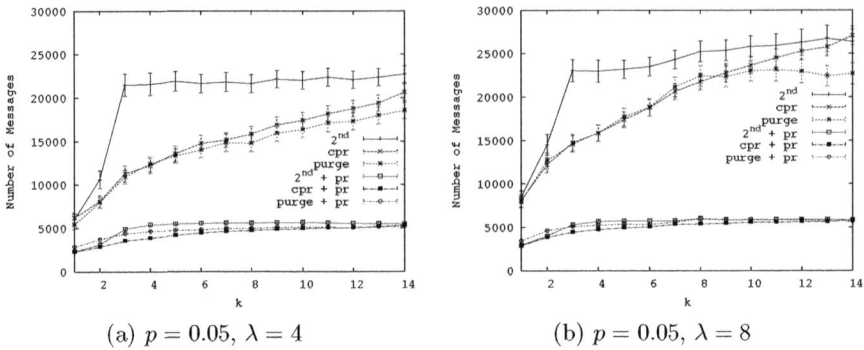

(a) $p = 0.05$, $\lambda = 4$ (b) $p = 0.05$, $\lambda = 8$

Fig. 4. Plots for Experiment 5 and 6. Each figure shows message overhead for Erdös-Rényi graphs with link weights selected uniformly at random, $p = 0.05$, average diameter is 6.14, and $\lambda = \{4, 8\}$. Experiment 5 includes the 2nd best, purge, and cpr curves. The curves for 2nd best + pr, purge + pr, and cpr + pr correspond to Experiment 6.

execution. In our experimental setup, these messages are not counted because they do not occur in Step 4 of our simulation scenario described in Section 4.

We also find that 2nd best performance suffers because of the count-to-∞ problem. The gap between 2nd best and the other algorithms shrinks as λ increases because link cost changes have a larger effect on message overhead with increasing λ.

Experiment 6. In this experiment, we apply poison reverse to each algorithm and repeat Experiment 5. Because purge's diffusing computations only eliminate routing loops corresponding to $\overrightarrow{bad}$ state, purge is vulnerable to routing loops stemming from link cost changes. Thus, contrary to Experiment 4, poison reverse improves purge performance. We include the three new curves in both Figure 4 plots (each new curve has label "algorithm-name" + pr). Results for different λ and p values yield the same trends and so we omit the corresponding plots.

All three algorithms using poison reverse show remarkable performance gains. As confirmed by our profiling numbers, the improvements are significant because routing loops are more pervasive when link costs change. Accordingly, the poison reverse optimization yields greater benefits as λ increases.

As in Experiment 4, we believe that for $\overrightarrow{bad}$ state *only*, purge + pr removes routing loops larger than 2 while 2nd best + pr does not. For this reason, we believe that purge + pr performs better than 2nd best + pr. We are currently investigating this claim. cpr + pr has the lowest message complexity. In this experiment, the benefits of rolling back to a global snapshot taken before $\overline{v}$ was compromised outweigh the message overhead required to update stale state pertaining to link cost changes that occurred during $[t', t]$. As λ increases, the performance gap decreases because cpr + pr must process all link cost changes that occurred in $[t', t]$ while 2nd best + pr and purge + pr process some link cost change events during $[t', t]$ as part of normal distance vector execution.

However, cpr + pr only achieves such strong results by making two optimistic assumptions: we assume perfectly synchronized clocks and checkpointing occurs at each timestep. In the next experiment we relax, the checkpoint assumption.

Experiment 7. Here we study the trade-off between message overhead and storage overhead for cpr. To this end, we vary the frequency at which cpr checkpoints and fix the interval $[t', t]$. Otherwise, our experimental setup is the same as Experiment 5.

Due to space constraints, we only display a single plot. Figure 3(b) shows the results for an Erdös-Rényi graph with link weights selected uniformly at random between $[1, n]$, $n = 100$, $p = .05$, $\lambda = 4$ and $k = 2$. We plot message overhead against the number of timesteps cpr must rollback, z. The trends are consistent when using the poison reverse optimization for each algorithm. cpr's message overhead increases with larger z because as z increases there are more link cost change events to process. 2nd best and purge have constant message overhead because they operate independent of z.

We conclude that as the frequency of cpr snapshots decreases, cpr incurs higher message overhead. Therefore, when choosing the frequency of checkpoints, the trade-off between storage and message overhead must be carefully considered.

4.3 Summary

Our results show cpr using poison reverse yields the lowest message and time overhead in all scenarios. cpr benefits from removing false state with a single diffusing computation. Also, applying poison reverse significantly reduces cpr message complexity by eliminating pairwise routing loops resulting from link cost changes. However, cpr has storage overhead, requires loosely synchronized clocks, and requires the time $\bar{v}$ was compromised be identified.

2nd best's performance is determined by the count-to-∞ problem. In the case of Erdös-Rényi graphs with fixed unit link weights, the count-to-∞ problem was minimal, helping 2nd best perform better than purge. For all other topologies, poison reverse significantly improves 2nd best performance because routing loops are pervasive. Still, 2nd best using poison reverse is not as efficient as cpr and purge using poison reverse.

In cases where link costs change, we found that purge using poison reverse is only slightly worse than cpr + pr. Unlike cpr, purge makes use of computations that follow the injection of false state, that do not depend on false routing state. Because purge does not make the assumptions that cpr requires, purge using poison reverse is a suitable alternative for topologies with link cost changes.

Finally, we found that an additional challenge with cpr is setting the parameter which determines checkpoint frequency. Frequent checkpointing yields lower message and time overhead at the cost of more storage overhead. Ultimately, application-specific factors must be considered when setting this parameter.

5 Related Work

There is a rich body of research in securing routing protocols [10,17,20]. Unfortunately, preventative measures sometimes fail, requiring automated techniques for recovery. Previous approaches to recovery from router faults [15,19] focus on allowing a router to continue forwarding packets while new routes are computed. We focus on a different problem: recomputing new paths following the detection of a malicious node that may have injected false routing state into the network.

Our problem is similar to that of recovering from malicious but committed database transactions. Liu et al. [12] develop algorithms to restore a database to a valid state after a malicious transaction has been identified. purge's algorithm to globally invalidate false state can be interpreted as a distributed implementation of the dependency graph approach in [12].

Database crash recovery [14] and message passing systems [6] both use snapshots to restore the system in the event of a failure. In both problem domains, the snapshot algorithms are careful to ensure snapshots are globally consistent. In our setting, consistent global snapshots are not required for cpr, since distance vector routing only requires that all initial least costs are non-negative.

Garcia-Lunes-Aceves's DUAL algorithm [8] uses diffusing computations to coordinate least cost updates in order to prevent routing loops. In our case, cpr and the prepossessing procedure (Section 3.1) use diffusing computations for purposes other than updating least costs (e.g., rollback to a checkpoint in the case of cpr and remove $\bar{v}$ as a destination during preprocessing). Like DUAL, the purpose of purge's diffusing computations is to prevent routing loops. However, purge's diffusing computations do not verify that new least costs preserve loop free routing (as with DUAL) but instead globally invalidate false routing state.

Jefferson [11] proposes a solution to synchronize distributed systems called Time Warp. Time Warp is a form of optimistic concurrency control and, as such, occasionally requires rolling back to a checkpoint. Time Warp does so by "unsending" each message sent after the time the checkpoint was taken. With cpr, a node does not need to explicitly "unsend" messages after rolling back. Instead, each node sends its $\overrightarrow{min}$ taken at the time of the snapshot, which implicitly undoes the effects of any messages sent after the snapshot timestamp.

6 Conclusions and Future Work

In this paper, we developed methods for recovery in scenarios where malicious nodes inject false state into a distributed system. We studied an instance of this problem in distance vector routing. We presented and evaluated three new algorithms for recovery in such scenarios. In the case of topologies with changing link costs, we found that poison reverse yields dramatic reductions in message complexity for all three algorithms. Among our three algorithms, our results showed that cpr – a checkpoint-rollback based algorithm – using poison reverse yields the lowest message and time overhead in all scenarios. However, cpr has storage overhead and requires loosely synchronized clocks. purge does not have these restrictions and we showed that purge using poison reverse is only slightly worse than cpr with poison reverse. Unlike cpr, purge has no stale state to update because purge does not use checkpoints and rollbacks.

As future work, we are interested in finding the worst possible false state a compromised node can inject (e.g., state that maximizes the effect of the count-to-∞ problem). We have also started a theoretical analysis of our algorithms.

Acknowledgments

The authors greatly appreciate discussions with Dr. Brian DeCleene of BAE Systems, who initially suggested this problem area.

References

1. Google Embarrassed and Apologetic After Crash,
   http://www.computerweekly.com/Articles/2009/05/15/236060/
   google-embarrassed-and-apologetic-after-crash.htm
2. GT-ITM, http://www.cc.gatech.edu/projects/gtitm/
3. Rocketfuel, http://www.cs.washington.edu/research/networking/rocketfuel/
   maps/weights/weights-dist.tar.gz
4. Bertsekas, D., Gallager, R.: Data Networks. Prentice-Hall, Inc., Upper Saddle River (1987)
5. Dijkstra, E., Scholten, C.: Termination Detection for Diffusing Computations. Information Processing Letters (11) (1980)
6. El-Arini, K., Killourhy, K.: Bayesian Detection of Router Configuration Anomalies. In: MineNet '05: Proceedings of the 2005 ACM SIGCOMM workshop on Mining network data, pp. 221–222. ACM, New York (2005)
7. Feamster, N., Balakrishnan, H.: Detecting BGP Configuration Faults with Static Analysis. In: 2nd Symp. on Networked Systems Design and Implementation (NSDI), Boston, MA (May 2005)
8. Garcia-Lunes-Aceves, J.J.: Loop-free Routing using Diffusing Computations. IEEE/ACM Trans. Netw. 1(1), 130–141 (1993)
9. Gyllstrom, D., Vasudevan, S., Kurose, J., Miklau, G.: Recovery from False State in Distributed Routing Algorithms. Technical Report UM-CS-2010-017
10. Hu, Y.C., Johnson, D.B., Perrig, A.: SEAD: Secure Efficient Distance Vector Routing for Mobile Wireless Ad Hoc Networks. In: Proceedings Fourth IEEE Workshop on Mobile Computing Systems and Applications, pp. 3–13 (2002)
11. Jefferson, D.: Virtual Time. ACM Trans. Program. Lang. Syst. 7(3), 404–425 (1985)
12. Liu, P., Ammann, P., Jajodia, S.: Rewriting Histories: Recovering from Malicious Transactions. Distributed and Parallel Databases 8(1), 7–40 (2000)
13. Mittal, V., Vigna, G.: Sensor-Based Intrusion Detection for Intra-domain Distance-vector Routing. In: CCS 2002: Proceedings of the 9th ACM Conf. on Comp. and Communications Security, pp. 127–137. ACM, New York (2002)
14. Mohan, C., Haderle, D., Lindsay, B., Pirahesh, H., Schwarz, P.: ARIES: A Transaction Recovery Method Supporting Fine-Granularity Locking and Partial Rollbacks Using Write-Ahead Logging. ACM Trans. Database Syst. 17(1), 94–162 (1992)
15. Moy, J.: Hitless OSPF Restart. In: Work in progress, Internet Draft (2001)

16. Padmanabhan, V., Simon, D.: Secure Traceroute to Detect Faulty or Malicious Routing. SIGCOMM Comput. Commun. Rev. 33(1), 77–82 (2003)
17. Pei, D., Massey, D., Zhang, L.: Detection of Invalid Routing Announcements in RIP Protocol. In: Global Telecommunications Conference, GLOBECOM '03, December 2003, vol. 3, pp. 1450–1455. IEEE, Los Alamitos (2003)
18. School, K., Westhoff, D.: Context Aware Detection of Selfish Nodes in DSR based Ad-hoc Networks. In: Proc. of IEEE GLOBECOM, pp. 178–182 (2002)
19. Shaikh, A., Dube, R., Varma, A.: Avoiding Instability During Graceful Shutdown of OSPF. Technical report. In: Proc. IEEE INFOCOM (2002)
20. Smith, B., Murthy, S., Garcia-Luna-Aceves, J.J.: Securing Distance-vector Routing Protocols. In: Symposium on Network and Distributed System Security, p. 85 (1997)

IP Fast Reroute in Networks with Shared Risk Links

Yan Li[1] and Mohamed G. Gouda[2]

[1] Department of Computer Science,
The University of Texas at Austin,
1 University Station (C0500),
Austin, TX 78712-0233
[2] The National Science Foundation, and
The University of Texas at Austin,
1 University Station (C0500),
Austin, TX 78712-0233
{yanli,gouda}@cs.utexas.edu

Abstract. IP fast reroute is a mechanism that is used to reroute packets around a failed link as soon as the link fails. Most of the IP fast reroute mechanisms, that have been proposed so far, focus on single or dual link failures but can not handle Shared Risk Link Group (SRLG) failures when several links fail at the same time because of some common underlying component failure. Furthermore, most of current work is based on the assumption that each node in the network has access to some global topology information of the network. In this paper, we present the first IP fast reroute mechanism for SRLG failures that is not based on the assumption that the nodes in the network have global topology information of the network. In our mechanism, nodes in the network use "relay bits" to identify themselves as "relay nodes" for a reroute link in a fully distributed mannner. Through simulation, we show that our mechanism succeeds in rerouting around SRLG failures alomst 100% of the time, with average length of a reroute path about 1.5 times the re-converged shortest path.

Keywords: IP Fast Reroute, Shared Risk Link Group Failure, Distance Vector, Failure Recovery, Reliability.

1 Introduction

Demands on reliability and availability of the Internet are becoming more and more stringent, especially with the development of more real-time applications like VoIP and Video on demand [5]. Unfortunately, failures are very common in daily operations of a network and what makes things worse is that most failures are not predictable. It is reported in [22] that 80% of all failures are unexpected. Among these unexpected failures, besides the most common single link failures, another significant part is Shared Risk Link Group (SRLG) failures. Links that

M. Crovella et al. (Eds.): NETWORKING 2010, LNCS 6091, pp. 213–226, 2010.

belong to the same SRLG share some underlying component either in the optical infrastructure like a fiber or at a router like a line card.

The convergence process for failure recovery in traditional routing protocols, link state and distance vector, is time consuming and may result in instability in case of frequent transient link failures. Although much work has been dedicated to reduce the convergence time of routing to even under a second [12], it is still quite far from the 50 milliseconds target for mission critical applications [25]. Recently, IP fast reroute [10, 13, 26] has been proposed to proactively compute backup paths before a failure happens. And as soon as a failure is detected, the backup path can be invoked immediately to reroute around the failure during the convergence period. Thus the routing disruption time can be limited to only the failure detection time. Although several mechanisms have been developed within the IP fast reroute realm, most of them focus on single or dual link failures and can not handle SRLG failures [3, 4, 6, 14, 16, 20, 23, 24]. Also, most of existing work relies on the existence of some global topology information to precompute backup paths [3, 4, 6, 8, 14, 16, 17, 18, 23, 24, 27].

When global topology information is not available (for example, in distance vector like routing protocols), to recover from SRLG failures, IP fast reroute faces more challenges: how to get necessary information to compute alternative backup paths to avoid all the links in the same SRLG, without changing the original routing tables? To address this challenge, we design an IP fast reroute mechanism for handling SRLG failures with the following goals:

- **No Global Topology Information**: Our mechanism does not assume nodes in the network have access to any global topology information: neither the connectivity information of the network nor any additional IP addresses associated with each node. Each node only has access to its local topology, i.e., links associated with the node itself. Also, each node only knows to which SRLG its associated links belong but does not know other links in the same SRLG.

- **Distributed Computation of Backup Paths**: Each backup path for a reroute link is designated by a node called *relay node*. In order to find relay nodes in a fully distributed way, we introduce two *relay bits* for each reroute link. Using the two relay bits, a node in the network can automatically decide if itself can serve as a relay node for a reroute link or not, in a distributed manner.

- **Reroute Only When You Want**: Rerouting information is propagated only for links that are currently under protection using IP fast reroute, which are called *reroute links*. Reroute links can be changed at any time. So the cost of our mechanism is dynamically associated with the number of links currently protected in the network.

We propose a tunneling scheme to ensure that loops are never formed. We also propose an algorithm to suppress redundant relay notification messages. Finally we show, through extensive simulations on a variety of networks of different sizes and varying SRLG size, that the coverage of our mechanism is close to 100%. Suppression can effectively cut about 80% notification messages when the network has at least one hundred nodes. Also, the average length of a reroute path is around 1.5 times the average length of the re-converged shortest path.

The structure and organization of this paper follows from our technical report [20] which focuses on how to handle single link failures using IP fast reroute. Section 2 presents the concept of reroute links. Then in Section 3, we introduce the concept of a relay node for a reroute link that is a member of a SRLG and how to use the relay node in rerouting. Section 4 presents how a node learns a relay node, without access to the global topology information, for a reroute link which belongs to a SRLG. Section 5 describes the suppression mode to suppress redundant notification messages. We show the efficiency and overhead of our rerouting algorithms in Section 6. Related work is reviewed in Section 7. Finally we conclude in Section 8.

2   Reroute Links in Shared Risk Link Group

We model a network as an undirected graph where each node represents a router and each (undirected) edge between two nodes represents two links. For each link $a{\rightarrow}b$, node a is called *the source* of link $a{\rightarrow}b$ and node b is called *the sink* of link $a{\rightarrow}b$. We assume that packets are routed between different nodes in the network using distance vector routing protocols [2, 21].

A Shared Risk Link Group (SLRG) is a set of links that share the same underlying physical point of failure such as a fiber cut or a line card failure. We assume that some links in the network are partitioned into SRLGs with unique identifiers. Thus, each link belongs to at most one SLRG. We also assume that the source node a of any link $a{\rightarrow}b$ in the network knows the identifier of the SRLG to which link $a{\rightarrow}b$ belongs, if any. (But node a does not need to know the other links in the same SRLG to which link $a{\rightarrow}b$ belongs, if any.)

As an example, Figure 1 shows a network N_1 in which each edge is labeled with a distance. The routing table $RT.a$ of node a in network N_1 is shown in Table 1. Also, there are three SRLGs in N_1 with SRLG number g_1, g_2 and g_3. Link $a{\rightarrow}b$, $c{\rightarrow}d$ and $e{\rightarrow}f$ belong to SRLG g_1; link $b{\rightarrow}f$ and $f{\rightarrow}g$ belong to SRLG g_2; link $g{\rightarrow}h$ and $h{\rightarrow}i$ belong to SRLG g_3.

Now consider the situation where node a has a packet whose ultimate destination is node d. But then node a notices that link $a{\rightarrow}b$ used to reach destination

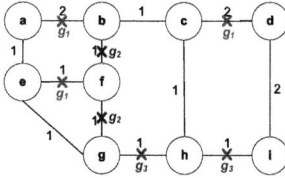

Fig. 1. An example network N_1

Table 1. Routing table $RT.a$ of node a

dest.	next hop	dist.
a	-	0
b	b	2
c	b	3
d	b	5
e	e	1
f	e	2
g	e	2
h	e	3
i	e	4

d has failed. The question now is "what does node a do with this packet that need to be transmitted over the failed link $a{\rightarrow}b$ before the entries of the routing table $RT.a$ have converged to their new correct values?"

Node a has two options in this situation. The first option is that node a drops every packet. This option is attractive if link $a{\rightarrow}b$ is very reliable or if the rate of packets that need to be transmitted over link $a{\rightarrow}b$ is very small. The second option is that node a anticipates the failure of link $a{\rightarrow}b$ and maintains alternative routes that can be used to reroute around link $a{\rightarrow}b$ when this link fails. To do so, node a needs to advertise to every node in the network that link $a{\rightarrow}b$ has been designated (by node a) to be a *reroute link*. Thus every node in the network can proceed to help node a identify and maintain alternative routes that can be used to reroute around link $a{\rightarrow}b$ when it fails.

Each node a in the network is provided with a *rerouting table RR.a* that has four columns: *(rlink, srlg, rbits, relay)*. The first column, *rlink*, in every rerouting table lists all the links that have been designated, by their source nodes, as reroute links. The second column, *srlg*, lists the id of the shared risk link group that this reroute link belongs to. The other two columns, *rbits* and *relay*, are discussed below in Section 4.

Initially, the rerouting table $RR.a$ of each node a is empty except that node a adds one entry for each link $a{\rightarrow}s_n$ that a wants to designate as a reroute link. Whenever node a sends a copy of its routing table $RT.a$ to each neighboring node, node a also sends a copy of its rerouting table $RR.a$ (excluding the "relay" column) to the neighboring node. This periodic exchange of routing and rerouting tables between neighboring nodes in the network eventually causes every link that has been designated as a reroute link to have one entry in every rerouting table in the network. At any time, each node a can change the set of links that it has designated as reroute links by adding new links to this set or by removing old links from this set.

3   Relay Nodes for SRLG Failures

In this section, we introduce the concept of a relay node for a reroute link that is a member of a SRLG in a network. (Note that the relay node defined in this section also works for reroute links that does not belong to any SRLG in the network.) We then discuss how a relay node for a reroute link can be used in rerouting around its reroute link when all the links that belong to the same SRLG fail.

Let s and d be two nodes in a network N, and let $R(s, d)$ denote the shortest route from node s to node d as determined by the routing tables. A node r is called a *relay node* for a reroute link $a{\rightarrow}b$ in N iff neither the route $R(a, r)$ nor the route $R(r, b)$ contains any link which belongs to the same SRLG as link $a{\rightarrow}b$ (including $a{\rightarrow}b$ itself).

As an example, consider network N_1 in Figure 1. Link $a{\rightarrow}b$, $c{\rightarrow}d$ and $e{\rightarrow}f$ belong to the same SRLG g_1. If link $a{\rightarrow}b$ in network N_1 is designated, by node a, as a reroute link, then by definition, node g, h and i are relay nodes for link

$a{\rightarrow}b$. However, if link $b{\rightarrow}c$ and $c{\rightarrow}h$ belong to the same SRLG (not shown in N_1), and link $b{\rightarrow}c$ is designated, by node b, as a reroute link, then no node in network N_1 is a relay node for link $b{\rightarrow}c$. These two examples simply show that a reroute link in a SRLG can have one or more relay nodes or even no relay nodes. Fortunately, we show by extensive simulations below in Section 6 that reroute links in a SRLG that have no relay nodes are extremely rare.

Next we describe the procedure for rerouting around a reroute link $a{\rightarrow}b$, when this link fails, assuming that node a knows the identity of a relay node r for link $a{\rightarrow}b$:

1. Assume that a packet is to be sent from a node s to a node d along the route $R(s, d)$ which contains the reroute link $a{\rightarrow}b$. In this case, the IP header of the packet can be represented as *(from s, to d)*.
2. Assume also that when this packet reaches node a, node a discovers that the reroute link $a{\rightarrow}b$ has failed and so decides to reroute the packet towards the relay node r for link $a{\rightarrow}b$. In this case, node a encapsulates the packet in two outer IP headers *(from a, to b)* and then *(from a, to r)* and forwards the encapsulated packet towards r.
3. When the encapsulated packet reaches the relay node r, node r removes the outermost IP header *(from a, to r)* and discovers that the packet has an inner IP header *(from a, to b)*. Thus node r forwards the encapsulated packet towards b.
4. When the encapsulated packet reaches node b, node b removes the outer IP header *(from a, to b)* and discovers that the packet has an inner IP header *(from s, to d)*. Thus node b forwards the packet towards d. Note that now the packet is not encapsulated any more.
5. Assume that, while the packet is traversing the route $R(b, d)$ (non-encapsulated now like a normal packet), the packet reaches a node x that needs to transmit the packet over a reroute link $x{\rightarrow}y$ except that it discovers that link $x{\rightarrow}y$ has failed. Node x can use the same procedure, step 1 through 4 as describe above, to reroute the packet around the failed link $x{\rightarrow}y$, no matter whether link $x{\rightarrow}y$ belongs to the same SRLG as link $a{\rightarrow}b$ or not.
6. However, assume that while the encapsulated packet is traversing the route $R(a, r)$ or the route $R(r, b)$, the packet reaches a link $x{\rightarrow}y$ that has failed. Recognizing that the packet is being rerouted because it is an encapsulated packet, node x drops the packet and does not attempt to reroute it a second time.

The fact that no routing loops are created due to the repeated rerouting of the same packet is established in the following lemma and theorem. The proofs can be found in [19].

Lemma 1. *Let* r *be a relay node for a reroute link* a$\rightarrow$b *in a network* N, *and let* d *be any node in network* N *such that the route* R(a, d) *contains the reroute link* a$\rightarrow$b. *If link* a$\rightarrow$b *fails, and node* a *reroutes a packet, whose ultimate destination is node* d, *to node* r, *then this packet will not traverse any loop in the network before reaching node* b *no matter whether link* a$\rightarrow$b *belongs to a SRLG or not.*

Theorem 1. *No routing loops are created due to repeated rerouting of the same packet to its ultimate destination using the rerouting procedure, no matter whether all encountered failed links belong to the same SRLG or not.*

4   Relay Bits to Identify Relay Nodes

In the previous section we presented a procedure by which a node a can reroute packets around a reroute link $a{\rightarrow}b$ when all the links that belong to the same SRLG fail. This procedure is based on the assumption that node a knows a relay node for the reroute link $a{\rightarrow}b$. So the question now is "How does node a know a relay node for link $a{\rightarrow}b$ which belongs to a SRLG without access to the global topology information?" In this section we present a fully distributed procedure by which node a learns all the relay nodes for link $a{\rightarrow}b$ although node a does not know any other links that belong to the same SRLG as link $a{\rightarrow}b$. This procedure consists of the following three parts.

The First Part: For node a to announce that it has designated link $a{\rightarrow}b$ as a reroute link, node a adds the entry $(a{\rightarrow}b,\ g_1,\ 00,\ \text{-})$ to its rerouting table $RR.a$. Recall that each entry in a rerouting table consists of four components *(rlink, srlg, rbits, relay)*, where *rlink* is a reroute link; *srlg* lists the id of the SRLG that this link belongs to; *rbits* are two relay bits (to be discussed shortly) for the reroute link; and *relay* is the set of all known relay nodes for the reroute link. The initial value of *relay* is "-" which indicates that node a does not know yet this value.

Because the rerouting table of every node is sent periodically to every neighbor of this node, the fact that link $a{\rightarrow}b$ has been designated a reroute link, as well as the SRLG g_1 it blongs to, is eventually recorded in every rerouting table in the network according to the following rule. If a node x receives a rerouting table $RR.y$ from a neighbor y, and the next hop for reaching node a in the routing table $RT.x$ of node x is node y, and if $RR.y$ has an $a{\rightarrow}b$ entry but $RR.x$ does not have $a{\rightarrow}b$ entry, then node x adds an entry $(a{\rightarrow}b,\ g_1,\ 00,\ \text{-})$ to its rerouting table $RR.x$. Conversely, if $RR.y$ has no $a{\rightarrow}b$ entry but $RR.x$ has an $a{\rightarrow}b$ entry, then node x removes the $a{\rightarrow}b$ entry from its rerouting table $RR.x$.

The Second Part: For each reroute link $a{\rightarrow}b$, node x maintains two bits, named the *relay bits* of link $a{\rightarrow}b$, in its rerouting table $RR.x$. These two bits are denoted *rbits.x[a→b]* and each of the two bits has anyone of two values. The value "0" in the first bit indicates two cases: either node x does not know yet the correct value of the bit (i.e., initial value of the bit), or node x has checked that some link that belongs to the same SRLG g_1 as link $a{\rightarrow}b$ occurs in the route $R(x, a)$. The value "1" in the first bit indicates that x has checked that no link that belongs to the same SRLG g_1 as link $a{\rightarrow}b$ occurs in the route $R(x, a)$. Similarly, the value in the second bit indicates the same meaning except that node x checks whether there is any link that belongs to the same SRLG g_1 as link $a{\rightarrow}b$ occurs in the route $R(x, b)$. Only when the two bits are both "1"s, i.e., no link that belongs to the same SRLG g_1 as link $a{\rightarrow}b$ occurs in route $R(x, a)$ and route $R(x, b)$, node x is a relay node for reroute link $a{\rightarrow}b$.

Next, we describe how to set up the two relay bits for a reroute link. Initially, the value of $rbits.x[a{\rightarrow}b]$ is "00" in the rerouting table $RR.x$ in every node x in the network, meaning that every node does not know the correct value of the bits yet. The source node a of link $a{\rightarrow}b$ assigns the relay bits $rbits.a[a{\rightarrow}b]$ in its rerouting table $RR.a$ the value 10. The first bit "1" means that no link that belongs to the same SRLG g_1 as link $a{\rightarrow}b$ occurs in the route $R(a, a)$, and the second bit "0" means that link $a{\rightarrow}b$ occurs in the route $R(a, b)$. Then the sink node b of link $a{\rightarrow}b$ assigns the bits $rbits.b[a{\rightarrow}b]$ in its rerouting table $RR.b$ the value 01. The first bit "0" means that link $b{\rightarrow}a$ that belongs to the same SRLG g_1 as link $a{\rightarrow}b$ occurs in the route $R(b, a)$, and the second bit "1" means that no link that belongs to the same SRLG g_1 as link $a{\rightarrow}b$ occurs in the route $R(b, b)$.

Then every other node x in the network assigns each of the two relay bits $rbits.x[a{\rightarrow}b]$ in its rerouting table $RR.x$ the value val, where val is either 0 or 1, according to the following rule: If x receives $RR.y$ from neighbor y, and if the next hop for reaching node a in the routing table $RT.x$ of node x is node y, then node x checks whether the SRLG of link $x{\rightarrow}y$ is the same as the SRLG of link $a{\rightarrow}b$, if yes, then node x assigns the first bit $rbits.x[a{\rightarrow}b][0]$ in its $RR.x$ the value 0. Otherwise, node x assigns the first bit $rbits.x[a{\rightarrow}b][0]$ in its $RR.x$ the value of the first bit in rbits.y[a→b][0]. Similarly, node x assigns the second relay bit.

The first and second parts outlined above are part of updating the rerouting table $RR.x$ after node x receives the rerouting table $RR.y$ from the neighboring node y shown in Algorithm 1.

Algorithm 1. Update rerouting table **$RR.x$** after x receives rerouting table **$RR.y$** from neighbor y

```
 1: for (s_r→s_n ∈ rlink.x) and (s_r→s_n ∉ rlink.y) do
 2:     if nexthop.x[s_r] == y then
 3:         remove s_r→s_n entry from RR.x;
 4: for (s_r→s_n ∉ rlink.x) and (s_r→s_n ∈ rlink.y) do
 5:     if nexthop.x[s_r] == y then
 6:         add s_r→s_n entry to RR.x;
 7: for s_r→s_n ∈ rlink.x do
 8:     if x == s_r then
 9:         rbits.x[s_r→s_n] := 10;
10:     else if x == s_n then
11:         rbits.x[s_r→s_n] := 01;
12:     else
13:         if nexthop.x[s_r] == y then
14:             if srlg.x[x→y] == srlg.x[s_r→s_n] then
15:                 rbits.x[s_r→s_n][0] := 0;
16:             else
17:                 rbits.x[s_r→s_n][0] := rbits.y[s_r→s_n][0];
18:         if nexthop.x[s_n] == y then
19:             if srlg.x[x→y] == srlg.x[s_r→s_n] then
20:                 rbits.x[s_r→s_n][1] := 0;
21:             else
22:                 rbits.x[s_r→s_n][1] := rbits.y[s_r→s_n][1];
```

Table 2. Rerouting table **$RR.a$** of node a in network N_1 with the relay nodes for the reroute links $a{\rightarrow}b$ whose source node is a

rlink	srlg	rbits	relay
a→b	g_1	10	g, h, i
a→e	-	10	b, c, d
c→d	g_1	00	-
e→f	g_1	10	-
b→f	g_2	11	-
f→g	g_2	11	-
g→h	g_3	10	-
h→i	g_3	10	-
d→i	-	11	-

The Third Part: When node a receives a *notify(x, a→b)* message, Node a then adds x to the set of *relay.a[a→b]* in the rerouting table $RR.a$ of node a.

In Figure 1, assume that all the links in three SRLGs g_1, g_2 and g_3 have been designated by their respective source nodes to be reroute links. Also, assume that link $a→e$ and link $d→i$ have been designated as reroute links too. But these two links do not belong to any SRLG. Then the rerouting table $RR.a$ of node a, after these links have been designated as reroute links, is shown in Table 2.

Correctness of the procedure for updating the relay bits follows from the next theorem. The proof can be found in [19].

Theorem 2. *For any node* x *in a network N, if the relay bits in node* x *for a reroute link* a→b *are both ones, i.e.,* rbits.x[a→b]=11, *then neither route* R(a,x) *nor route* R(x,b) *contains any link that belongs to the same SRLG as reroute link* a→b.

5 Suppression Mode

There is one problem concerning the second and third part of the procedure discussed in the previous section: for some reroute links many nodes in the network qualify to be relay nodes and so these many nodes start to send notify messages to the source node of the link, and the source node has to process all the notify messages even though only one relay node is enough to reroute packets around the link when it fails.

In order to minimize the notification messages sent in the network, we introduce a *suppression mode*. In the suppression mode, when the relay bits *rbits.x* *[a→b]* in the rerouting table $RR.x$ have the value 11, node x recognizes that it is a relay node for link $a→b$ and so it sends a *notify(x, a→b)* message to its next hop for reaching node a, which either drops the message (as explained below) or forwards the message to its next hop for reaching node a. Thus, if the *notify(x, a→b)* message reaches, along the route $R(x, a)$, a node y where the relay bits *rbits.y[a→b]* in the rerouting table $RR.y$ have the value 11, then node y

---

Algorithm 2. Actions of node x on sending and receiving relay notify messages

```
      /* ---------sending action---------- */
1   for s_r→s_n ∈ rlink.x do
2       if (rbits.x[s_r→s_n] == 11) then
3           send notify(x, s_r→s_n) to nexthop.x[s_r];

      /* ---------receiving action-------- */
4   rcv notify(z, s_r→s_n) from a neighbor y do:
5       if x == s_r then
6           add z to relay.x[s_r→s_n]
7       else if (rbits.x[s_r→s_n] == 11) then
8           suppress notify(z, s_r→s_n)
9       else
10          forward notify(z, s_r→s_n) to nexthop.x[s_r]
```

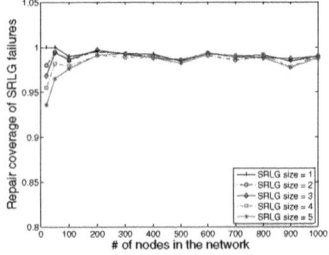

Fig. 2. Repair coverage of SRLG failures for Barabasi-Albert networks

drops the *notify(x, a→b)* message knowing that its own *notify(y, a→b)* message is sufficient for node *a* to have one relay node for link *a→b*.

If the suppression mode is used in network N_1 in Figure 1, then for reroute link *a→b*, the notify message from node *i* is dropped by node *h*, and the notify messages from *h* is dropped by *g*.

The actions of a node *x* concerning the sending and receiving of notify messages are shown in Algorithm 2.

6   Simulation Results

We now evaluate the performance of our IP fast reroute mechanism for various size of SRLGs (i.e., the number of links that are members of a SRLG) using simulations. Through simulation, we intend to answer the following questions: 1) What is the repair coverage for various size of SRLGs? 2) what is the efficiency of suppression under different size of SRLGs? 3) What is the chance that a node can have multiple relay nodes to choose for various size of SRLGs? Will the suppression affect this? 4) What is the overhead of using a relay path, which may include several relays for links in the same SRLG, instead of using the re-converged shortest path? How does the suppression affect this?

We conduct our simulations using two general networks, generated using the BRITE tool [1]. The first network satisfies the power law distribution based on the Barabasi-Albert (BA) model. The second is a random network based on the Waxman model. For each toplogy with E edges, we randomly select S edges, $1 \leq S \leq 5$ that are close to each other to form a SRLG (the number of hops between the first selected edge and any other selected edge is no larger than 0.6 the maximum number of hops in the network). For S = 1, we count every single link failure. For S > 1, we generate up to 1000 different SRLG failures and make sure each SRLG is not a cut of the topology graph.

Let *repair coverage* be the percentage of source-destination pairs in which, when the link *source→sink* of a SRLG used to traverse packets from the source to the destination fails, the source can reroute around any failed link in the same SRLG which appears along the path to reach the destination. Figure 2 and 3 show the repair coverage for SRLG failures for BA and Waxman network respectively. For both BA and Waxman network, no matter what is the size of the network, the repair coverage for smaller SRLG size is greater than the repair coverage for larger SRLG size. However, when the network size is at least 100 nodes, the SRLG size does not have much effect on the repair coverage and our IP fast reroute mechanism can achieve close to 100% repair coverage in these cases.

We measure the efficiency of suppression using *suppress ratio*, which is defined as the percentage of suppressed relay notify messages. As shown in Figure 4 and 5, in both BA and Waxman networks, the size of the SRLG does not affect the suppress ratio much. If the network size is larger than one hundred nodes, then the suppress ratio is about 80%. This demonstrates that suppression will effectively save the processing overhead for the source node and the bandwidth in the network.

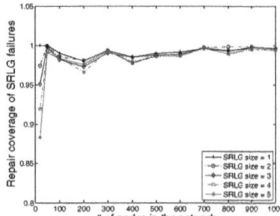

Fig. 3. Repair coverage of SRLG failures for Waxman networks

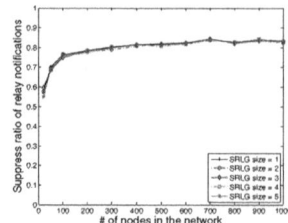

Fig. 4. Suppress ratio of relay notify messages for BA networks

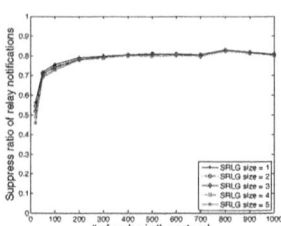

Fig. 5. Suppress ratio of relay notify messages for Waxman networks

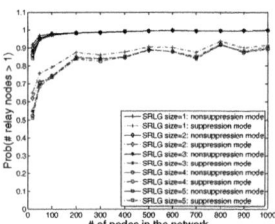

Fig. 6. Probability of more than one relay node for SRLG failures in BA networks

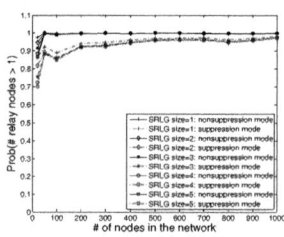

Fig. 7. Probability of more than one relay node for SRLG failures in Waxman networks

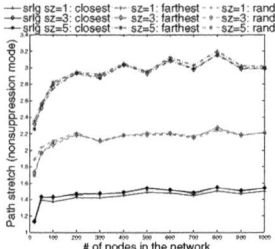

Fig. 8. The average path stretch when choosing different relay nodes for SRLG failures in non-suppression mode for BA networks

In both BA and Waxman networks, no matter what's the size for the SRLG, when the network size is over one hundred nodes and there is no suppression, the chance that a source node can find multiple relay nodes to choose from instead of only one relay node is over 97%, shown in Figure 6 and 7. While in the suppression mode, since some relay notify messages are suppressed, the chance that a source node can find multiple relay nodes drops to over 88% in Waxman networks and to about 80% in BA networks. However, we will show that the suppression mode will not affect the best relay node in terms of reroute path length and it also gives a source node better choices in terms of reroute path length.

For a reroute link, the pre-computed alternative path through a relay node is not necessarily the shortest path. This is because only the source node of the reroute link is aware of the failure and no other nodes are. So compared to the globally re-converged shortest path, IP fast reroute gains the lossless forwarding with a possible longer path penalty. However, we show that the penalty is not significant. Let *path stretch* be the ratio of the length of the pre-computed alternative path going through the relay node(s) divided by the length of the shortest path after re-convergence. When a source node finds that there are mul-

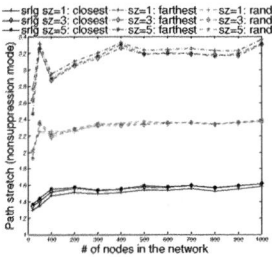
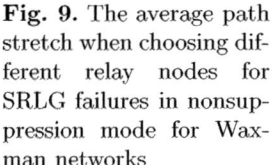
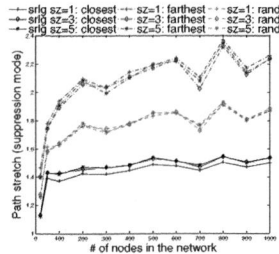
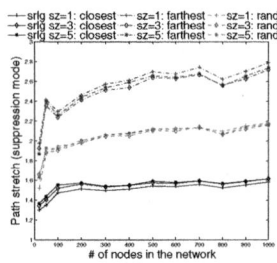

Fig. 9. The average path stretch when choosing different relay nodes for SRLG failures in nonsuppression mode for Waxman networks

Fig. 10. The average path stretch when choosing different relay nodes for SRLG failures in suppression mode for BA networks

Fig. 11. The average path stretch when choosing different relay nodes for SRLG failures in suppression mode for Waxman networks

tiple relay nodes for a reroute link, which relay node should the source choose? We examine three choices in terms of path stretch: the closest relay node to the source, the farthest relay node to the source and a random relay node.

In nonsuppression mode (i.e., suppression is not applied), the average path stretch when choosing difference relay nodes for different size of SRLG failures, in BA and Waxman networks is shown in Figure 8 and 9 respectively. In both networks, no matter what's the size of the SRLG failures, choosing the closest relay node gives the smallest path stretch, less than 1.6 compared to the reconverged shortest path length, while choosing the farthest relay node gives the largest path stretch. A random relay has the stretch in between the above two.

Figure 10 and 11 show the corresponding path stretch under the suppression mode. It is clear that suppression will not affect the path stretch for closest relay nodes. However, since suppression filters some farther relay nodes which tend to have larger stretch, the average path stretch for both farthest relay nodes and the random relay nodes is reduced under suppression mode. So a source can also randomly choose a relay node with stretch lower than or about 2 in both types of networks.

7 Related Work

Recently, IP Fast Reroute (IPFRR) has been proposed to recover from failures as soon as a failure is detected using IP-based schemes [26]. However, existing proposals, except [27] which requires substantial number of additional IP addresses, mainly focus on how to handle a single link failure or dual-link failures [3, 4, 6, 14, 16, 20, 24]. Also, most of existing proposals assume each node has the knowledge of some global topology information [3, 4, 6, 8, 14, 16, 17, 18, 23, 24, 27]. Instead, our work focuses on shared risk link group failures and assumes that each node has neither global connectivity information of the network nor additional global IP addresses information associated with each node. The idea of precomputing backup paths is also explored for BGP [15, 23, 25, 28, 29].

An IPFRR scheme should be able to avoid micro-loops [7, 9, 11]. Francois et al. [10] and Gjoka et al. [13] evaluate the coverage of several IPFRR mechanisms.

Both Loop-free Alternates [4] and U-turn Alternates [3] pre-computes an alternate next hop before a single link failure. Since these two mechanisms find alternates only among next hops, the coverage is not high even for single link failures. Tunnels [6] is more generalized than the above two mechanisms in the sense that it is not limited to only use next hops as tunnel endpoints. But again it can only handle single link failures and is only designed for link state protocols. Also it requires a significant number of computations of shortest paths since it computes a reroute path for each of the neighbors of the sink node.

In Multiple Routing Configurations (MRC) [16], each router pre-computes a number of topology configurations by removing rerouted links. Failure Insensitive Routing (FIR) [24] exploits interface-specific forwarding. Both MRC and FIR focus on single link failures. Failure-Carrying Packets (FCP) [17] uses the packet header to carry the list of failed links and requires potentially expensive dynamic computation to route that packet. Path splicing [23] creates multiple routing trees and allows packets to switch paths by inserting a new packet header. In [20], Li et al. explored the idea of using relay nodes to achieve IP fast reroute around single link failures based only on local information. Kini et. al. [14] proposed an approach to handle two simultaneous link failures by assigning three additional addresses to each node.

Each node in Not-via [27] needs d additional Not-via addresses for all the links for which it is a source node, where d is the degree of that node. These additional IP addresses have to be globally known, even when a link is currently not intended to be a reroute link. This significantly increases the size of the routing table and consequently lower the efficiency of forwarding even when there is no failures. Recent work from Li et al. [18] try to improve the efficiency of Not-via by aggregation, but it requires special allocation schemes of Not-via addresses. Enyedi et al. [8] try to reduce the number of Not-via addresses but they also assume the knowledge of global connectivity information.

8   Concluding Remarks

We have presented an IP fast reroute mechanism for Shared Risk Link Group failures in routing protocols without global topology information. In our mechanism, any node x can advertise that it needs to be able to reroute around a link $x{\rightarrow}y$ when this link fails. Then we leverage a set of relay nodes, computed in advance of any link failures, to tunnel the reroute packets around each failed link right after the detection of a failure. Each node uses a fully distributed algorithm to decide automatically whether it can serve as a relay node for a reroute link or not to avoid all link failures in the same SRLG. Notify messages are sent to the source of a reroute link from relay nodes. We proposed a suppression mode to greatly reduce the number of notify messages. Moreover, our tunneling scheme ensures that loops are never formed even when any number of links fail.

Through simulations on different topologies, we confirmed that our mechanism can achieve close to 100% repair coverage in different types and various size of networks for different SRLG size. The average length of a reroute path is around 1.5 the re-converged optimal paths. As expected, the suppression is quite effective and cut 80% of notify messages in a network of reasonable size (≥ 100).

Our future work includes migrating our IP fast reroute mechanism to interdomain routing protocols. Using our mechanism, each AS can potentially leverage the existing Internet topology to achieve fast reroute around Shared Risk Link Group Failures, without changing the BGP advertising and decision process.

References

1. BRITE: Boston univeristy Representative Internet Topology gEnerator, http://www.cs.bu.edu/BRITE/
2. Enhanced interior gateway routing protocol, http://www.cisco.com/en/US/tech/tk365/ technologies_white_paper09186a0080094cb7.shtml
3. Atlas, A.: U-turn alternates for IP/LDP fast-reroute (February 2006), Internet Draft http://draft-atlas-ip-local-protect-uturn-03.txt
4. Atlas, A., Zinin, A.: Basic specification for IP fast reroute: Loop-free alternates. In: RFC 5286 (September 2008)
5. Boutremans, C., Iannaccone, G., Diot, C.: Impact of link failures on VoIP performance. In: NOSSDAV'02 (May 2002)
6. Bryant, S., Filsfils, C., Previdi, S., Shand, M.: IP fast reroute using tunnels (November 2007), Internet Draft, draft-bryant-ipfrr-tunnels-03
7. Bryant, S., Shand, M.: A framework for loop-free convergence (October 2008), Internet Draft, draft-ietf-rtgwg-lf-conv-frmwk-03
8. Enyedi, G., Szilágyi, P., Rétvári, G., Császár, A.: IP fast reroute: Lightweight not-via without additional addresses. In: IEEE infocom mini-conference (2009)
9. Francois, P., Bonaventure, O.: Avoiding transient loops during IGP convergence in IP networks. In: IEEE Infocom (2005)
10. Francois, P., Bonaventure, O.: An evaluations of IP-based fast reroute techniques. In: CoNext (2005)
11. Francois, P., Bonaventure, O., Shand, M., Bryant, S., Previdi, S.: Loop-free convergence using oFIB (February 2008), Internet draft, draft-ietf-rtgwg-ordered-fib-02
12. Francois, P., Filsfils, C., Evans, J., Bonaventure, O.: Achieving subsecond IGP convergence in large IP networks. In: ACM Sigcomm (2005)
13. Gjoka, M., Ram, V., Yang, X.: Evaluations of IP fast reroute proposals. In: IEEE Comsware (2007)
14. Kini, S., Ramasubramanian, S., Kvalbein, A., Hansen, A.: Fast recovery from dual link failures in IP networks. In: IEEE Infocom (2009)
15. Kushman, N., Kandula, S., Katabi, D., Maggs, B.: R-BGP: Staying connected in a connected world. In: Usenix NSDI (2007)
16. Kvalbein, A., Hansen, A.F., Cicic, T., Gjessing, S., Lysne, O.: Fast IP network recovery using multiple routing configurations. In: IEEE Infocom (2006)
17. Lakshminarayanan, K., Caesar, M., Rangan, M., Anderson, T., Shenker, S., Stoica, I.: Achieving convergence-free routing using failure-carrying packets. In: ACM Sigcomm (2007)
18. Li, A., Francois, P., Yang, X.: On improving the efficiency and manageability of NotVia. In: CoNext (2007)

19. Li, Y., Gouda, M.G.: IP fast reroute in networks with shared risk links. UTCS Technical Report TR-09-38, The University of Texas at Austin (2009)
20. Li, Y., Gouda, M.G.: IP fast reroute without global topology information. UTCS Technical Report TR-09-34, The University of Texas at Austin (2009)
21. Malkin, G.: RFC 2453 - RIP Version 2 (November 1998)
22. Markopoulou, A., Iannaccone, G., Bhattacharyya, S., Chuah, C.-N., Dio, C.: Characterization of failures in an IP backbone. In: IEEE Infocom'04 (2004)
23. Motiwala, M., Elmore, M., Feamster, N., Vempala, S.: Path splicing. In: ACM Sigcomm (2008)
24. Nelakuditi, S., Lee, S., Yu, Y., Zhang, Z.-L., Chuah, C.-N.: Fast local rerouting for handling transient link failures. IEEE/ACM Transaction on Networking 15(2) (April 2007)
25. Olivier Bonaventure, C.F., Francois, P.: Achieving sub-50 milliseconds recovery upon BGP peering link failures. IEEE/ACM Transaction on Networking 15(5) (October 2007)
26. Shand, M., Bryant, S.: IP fast reroute framework (October 2009), Internet Draft, draft-ietf-rtgwg-ipfrr-framework-13.txt
27. Shand, M., Bryant, S., Previdi, S.: IP fast reroute using Not-via addresses (October 2008), draft-ietf-rtgwg-ipfrr-notvia-addresses-03
28. Wang, F., Gao, L.: A backup route aware routing protocol - fast recovery from transient routing failures. In: IEEE Infocom mini-conference (2008)
29. Wang, F., Gao, L.: Path diversity aware interdomain routing. In: IEEE Infocom (2009)

EAU: Efficient Address Updating for Seamless Handover in Multi-homed Mobile Environments[*]

Yuansong Qiao[1,2], Shuaijun Zhang[1], Adrian Matthews[1],
Gregory Hayes[1], and Enda Fallon[1]

[1] Software Research Institute, Athlone Institute of Technology, Ireland
[2] Institute of Software, Chinese Academy of Sciences, China
{ysqiao,szhang,amatthews,ghayes,efallon}@ait.ie

Abstract. Dynamic address configuration is essential when maintaining seamless communication sessions in heterogeneous mobile environments. This paper identifies some significant problems when using the Stream Control Transmission Protocol (SCTP) Dynamic Address Reconfiguration (SCTP-DAR) extension. We illustrate that SCTP-DAR can enter a deadlock state during the handover phase, ultimately resulting in communication failure. Deadlock arises as a result of the basic design rationale of SCTP-DAR, i.e. using a data oriented transmission scheme to transmit address update messages. This paper proposes a new transmission control mechanism for efficiently exchanging up-to-date address information between association endpoints. In particular we introduce an address operation consolidation algorithm which eliminates address operation redundancy. In addition, a priority based transmission re-scheduling algorithm for address updating operations is proposed to detect and remove potential deadlock situations. The above schemes have been verified through experiments.

Keywords: Multi-homing, Heterogeneous Networks, Mobility, Dynamic Address Configuration

1 Introduction

Current mobile communication systems involve a substantial number of multi-homed mobile devices interacting with a pervasive heterogeneous network environment. For example: Wi-Fi, 3G and WiMax are widely deployed in the Internet, and, simultaneously, most smart phones support multiple network connections, such as 3G, Wi-Fi and Bluetooth. Wireless technologies will probably continue to diversify in future. Thus, there has been a significant research and standardization effort focusing on providing general support for multi-homing and mobility, e.g. 4G networks [1], IEEE 802.21 [2], Site Multi-homing by IPv6 Intermediation (SHIM6) [3], Host Identity Protocol (HIP) [4] and Stream Control Transmission Protocol (SCTP) [5].

---

[*] This Research Programme is supported by Enterprise Ireland through its Applied Research Enhancement fund.

M. Crovella et al. (Eds.): NETWORKING 2010, LNCS 6091, pp. 227–238, 2010.
© IFIP International Federation for Information Processing

When a multi-homed mobile node (MN) is roaming across heterogeneous networks, its IP addresses can change frequently. Both mobility and wireless network fluctuations can cause network connections to be disconnected and re-numbered. Even in fixed scenarios, the host addresses may change during network failures if Dynamic Host Configuration Protocol (DHCP) is used to configure IP addresses. The up-to-date IP addresses should be transmitted to the correspondent node (CN) immediately so that communication interruptions can be avoided or reduced. Furthermore, a multi-homed mobile node probably needs to inform the correspondent node of its preferred primary IP address for communication according to its local policies or network conditions.

When the set of addresses in a node is changed, address updating messages can be sent to the correspondent node through the following schemes:

— Transmitting all current addresses of the node to the correspondent node in one message, such as in SHIM6 [3]. This scheme cannot guarantee that Network Address Translation (NAT) middleboxes are traversed correctly. Packets originating from different network addresses of a multi-homed host may pass through different NAT middleboxes on different paths. If a multi-homed host transmits all its addresses in one message, it requires that all NAT middleboxes are synchronized, i.e. one middlebox should translate an address which may pass through other middleboxes in the future. This is difficult to achieve in the current Internet.
— Sending address updating operations (AUOs) to the correspondent node to modify the set of addresses saved in the correspondent node, such as in the SCTP Dynamic Address Reconfiguration (SCTP-DAR) extension [6]. If AUOs are transmitted in separate packets, the protocol can traverse NAT middleboxes correctly [7] [8].

This paper will consider the second address updating scheme as it has broader usage scenarios than the first scheme. The SCTP-DAR standard is an extension for SCTP and therefore it was originally designed for network fault tolerance scenarios. However, researchers immediately found that it provided an ideal way to implement seamless vertical handover between heterogeneous networks. Consequently, there has been much effort to improve the handover performance based on SCTP-DAR, such as in [9] [10] [11].

Most of the current work concentrates on employing some auxiliary functions, such as using link signal strength [11], to enhance SCTP handover performance. SCTP-DAR is used as a fundamental function and it is presumed that it can operate properly.

This paper abandons this assumption and studies the address updating mechanism while using SCTP-DAR in mobile scenarios.

SCTP-DAR defines three AUOs: (1) ADD-IP – Add an address to an association; (2) DELETE-IP – Delete an address from an association; (3) SET-PRIMARY – Request the peer to use the suggested address as the primary address for sending data.

As the received sequence of AUOs can affect the address updating result, the AUOs are reliably transmitted through a First-In-First-Out (FIFO) mechanism which is typically used for data transmission. However AUOs are normally not equally important. Some newly generated operations can override earlier operations. For example, for a specific address, if a DELETE-IP operation is generated after an ADD-IP operation, the final result is that the address is deleted from the SCTP association. If the ADD-IP operation is in the transmission queue, it is not required to be transmitted. Therefore the current SCTP-DAR is not optimized.

This paper will firstly present two scenarios where two SCTP endpoints enter a deadlock state during the handover phase. The association between two SCTP endpoints is finally broken even though there are active IP addresses available for communications. Through analyzing the problems, this paper proposes an efficient address updating scheme (named EAU) which consists of three parts: (1) An ordered & partially reliable transmission scheme for AUOs, which removes the FIFO constraint in SCTP-DAR; (2) An AUO consolidation algorithm, which can merge AUOs to remove redundancy; (3) An AUO re-scheduling algorithm, which can detect and remove potential deadlock situations.

The rest of the paper is organized as follows. Section 2 introduces SCTP-DAR and summarizes related work. Section 3 illustrates the SCTP-DAR performance degradation issues in detail. Section 4 describes the detailed design of the EAU address updating scheme. Section 5 presents test results for the proposed scheme. Conclusions and future work are presented in Section 6.

2 Related Work

SCTP Dynamic Address Reconfiguration (SCTP-DAR) Extension
SCTP [5] is a reliable TCP-friendly transport layer protocol supporting multi-homing and multi-streaming. Two SCTP endpoints can exchange their addresses at the initial stage, but the set of addresses cannot be changed thereafter. Therefore the SCTP-DAR extension [6] was proposed. A new SCTP chunk type called ASCONF (Address Configuration Change Chunk) is defined to convey three AUOs: ADD-IP, DELETE-IP and SET-PRIMARY. In the normal case, only one single ASCONF chunk is transmitted in a packet. During retransmissions, it is allowed to bundle additional ASCONF chunks in a packet. However, only one outstanding packet is allowed. SCTP-DAR defines that the transmission of ASCONF and ASCONF-ACK chunks must be protected by the authentication mechanism defined in [12]. More detailed security considerations are described in [6].

Other Related Work
Various mobile schemes based on SCTP-DAR have been proposed, such as in [9] [10] [11]. SCTP NAT traversing problems are studied in [7] [8]. In [13], two SCTP stall scenarios are presented. The authors identify that the stalls occur as a result of SCTP coupling the logic for data acknowledgment and path monitoring. SCTP Concurrent Multi-path Transfer (CMT-SCTP) is studied in [14].

Apart from SCTP in the transport layer, research has been performed in other layers of the OSI stack in support of multi-homing & mobility. IEEE 802.21 [2] encapsulates the heterogeneity of the underlying networks and provides a unified layer 2 handover support for upper layers. SHIM6 [3] provides multi-homing support in the Network Layer for IPv6. As SHIM6 is designed for IPv6 networks, NAT traversing is not considered in the design. HIP [4] suggests adding a Host Identity Layer into the

OSI stack – consequently, a host is identified by its identity no matter where it is located. In 4G systems, current research [1] is trying to provide integrated support for Multi-homing and Mobility.

3 Two SCTP-DAR Deadlock Scenarios

This section demonstrates two scenarios where the two SCTP endpoints cease to communicate despite there being IP addresses actually available for communications. The scenarios are not deduced from conjectured tests but are observed from real experiments while using Linux to test SCTP mobile functions.

In all the tests of the paper, the address updating policies are set to the same for testing purposes: when the SCTP application receives an interface/address UP event, it sends an ADD-IP and a SET-PRIMARY operation for the new address to its peer; when the application gets an interface/address DOWN event, it sends a DELETE-IP operation to its peer. An unusual phenomenon is observed in the experiments, i.e. an address UP event is always generated when the attached access point is turned off. This is caused by the implementation mechanism in the Linux wireless device drivers. Consequently, an ADD-IP and a SET-PRIMARY operation request are generated by the application. The ADD-IP request is rejected because the address is already in the association. The SET-PRIMARY operation is generated and transmitted to the peer.

3.1 Deadlock Scenario 1: All Addresses Are Changed

When all the IP addresses of an SCTP node are changed and the new addresses are unable to be transmitted to its peer, the association will be broken. An example is where a mobile node has only one network connection and that connection is renumbered (Fig. 1a). In the test, the MN has only one network connection which connects to a Wi-Fi network (AP1). When AP1 is down, the MN connects to AP2 (the network connection is renumbered). The sequence of the events in this experiment is shown in Fig. 1b. A SET-PRIMARY operation for IP1 is generated just after AP1 is turned off. An ADD-IP operation for IP2 is generated after the MN connects to AP2.

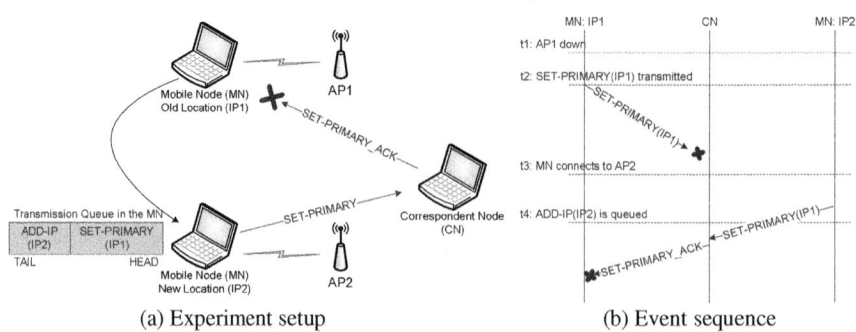

(a) Experiment setup   (b) Event sequence

Fig. 1. Problem 1 – ADD-IP blocked by other ASCONF chunks

In current SCTP-DAR, an ASCONF chunk can be sent to the peer from a new IP address which currently is not in the association. Therefore, the ASCONF chunk with the SET-PRIMARY operation (Fig. 1a) can be sent to the CN through the new IP address (IP2). However, the CN can only send an acknowledgement to an IP address which is already in the association. Consequently the acknowledgement is sent to the old IP address (IP1). The process continues until the SCTP association is broken.

In the experiment, if the ADD-IP (IP2) operation could be sent to the peer, the communication could be recovered. Unfortunately ADD-IP (IP2) is blocked by SET-PRIMARY (IP1) (Fig. 1a). SCTP-DAR defines that only one outstanding ASCONF chunk is allowed in normal situations. Nevertheless, ASCONF chunks can be bundled into one packet during retransmissions according to SCTP-DAR. Apparently, if SET-PRIMARY (IP1) and ADD-IP (IP2) could be bundled in one packet, the communication could be recovered. However, this scheme has two drawbacks: (1) It is not guaranteed that all ASCONF chunks can be allocated to one packet. Path Maximum Transmission Unit (PMTU) varies for different network types. The number of ASCONF chunks is also uncertain; (2) It can cause problems when there are NAT middleboxes between the two SCTP endpoints.

3.2   Deadlock Scenario 2: The Same Address Is Down and Up

This section shows a scenario where an ADD-IP operation cannot be generated when an IP interface goes down and then comes back up. The experiment setup is shown in Fig. 2a. The MN has two IP addresses. The CN has one IP address. The default route in the MN is set to WR1 originally. The link between WR2 and R0 is broken at time t0 in Fig. 2b. Note that the MN cannot detect that Link 2 is broken because SCTP only detects whether peer addresses are active.

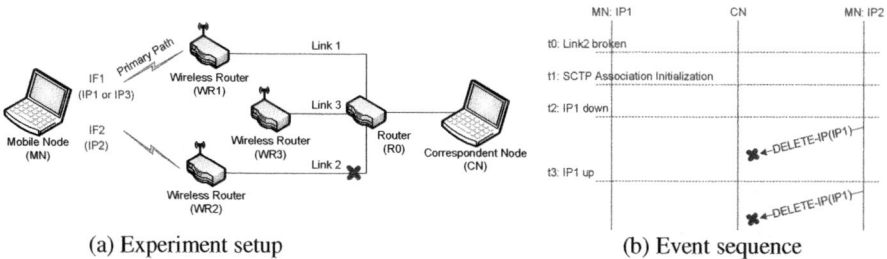

(a) Experiment setup   (b) Event sequence

Fig. 2. Problem 2 – The ADD-IP operation cannot be generated after IP1 is up at time t3 because IP1 is still in the association (under deletion). (Acronyms in the figure: *IP* – IP Address; *IF* – Interface; *WR* – Wireless Router).

The sequence of events in the experiment is shown in Fig. 2b. IP1 goes down at time t2. A DELETE-IP chunk is generated to delete IP1. The MN sends the DELETE-IP chunk through IP2 to the CN. The DELETE-IP chunk or the DELETE-IP_ACK chunk is lost because Path2 is broken.

After a transmission timeout occurs, the DELETE-IP chunk should be retransmitted. According to SCTP-DAR, when an IP address is under deletion, it can be used for the reception of SCTP packets but cannot be used as a source address. Therefore, the DELETE-IP has to be retransmitted via IP2 even though IP1 has become available (time t3). The retransmitted DELETE-IP chunk is consequently lost again. Finally the SCTP association is broken.

The experiment shows three problems in SCTP-DAR: (1) the MN has to transmit an obsolete DELETE-IP operation which cannot reflect the current address status of the MN; (2) the DELETE-IP operation is transmitted on a broken path (Link2). It cannot use the current active address (IP1) because the address is under deletion; (3) the MN cannot create an ADD-IP operation to add IP1 into the association when IP1 is up again (time t2) because IP1 is still in the association (under deletion).

4   Design of EAU

4.1   Design Overview

In the above experiments, the communications are broken because the mobile node fails to send the ADD-IP operations to the peer. In current SCTP-DAR, ASCONF chunks are transmitted in sequence. If the ASCONF chunk at the head of the transmission queue cannot be transmitted successfully, all other ASCONF chunks are blocked. In addition, as SCTP-DAR is designed to transmit the oldest ASCONF chunk first, the internal state in the protocol refuses to generate new AUOs in some circumstances (Problem 2 in Section 3), which can further worsen the situation.

The essential problem in the SCTP-DAR design is that all AUOs are treated equally. In fact, a new operation can obsolete some previous operations. Furthermore, the different types of operations are not of the same priority. The ADD-IP operation is more important than others because it can increase the reachability of the node.

In order to overcome the above problems, the newly generated AUOs, especially ADD-IP operations, should be transmitted as soon as possible. This paper proposes to improve the current SCTP-DAR scheme using the following three steps: (1) Change ordered/reliable transmission to ordered/partially reliable transmission. (Section 4.2); (2) Use a consolidation algorithm to delete obsolete AUOs. (Section 4.3); (3) Use a transmission re-scheduling algorithm to select an AUO with the highest priority for transmission. (Section 4.4)

4.2   ADD-IP/DELETE-IP/SET-PRIMARY Procedure

4.2.1   Ordered/Partially Reliable Transmission Control

As mentioned before, the AUOs have two important characteristics: (1) The transmission sequence can affect the address operation results. Therefore, the AUOs should be transmitted in order; (2) A new AUO can obsolete some old operations. Therefore, these obsolete operations are not required to be transmitted reliably.

Based on these two characteristics, this paper proposes an ordered/partially reliable transmission scheme for transmitting AUOs. The idea is similar to the Partial-Reliability extension for SCTP (PR-SCTP) [15]. As only one ASCONF chunk is outstanding, the transmission control process is much simpler than PR-SCTP.

In the sender side, every ASCONF chunk is assigned a sequence number which is incremented by 1 after being assigned. The sender guarantees that all transmitted ASCONF chunks have consecutive sequence numbers. In normal situations, the sender transmits one ASCONF chunk after the acknowledgement for the previous ASCONF chunk is received (the same as current SCTP-DAR). However, the sender can choose to transmit the next ASCONF chunk without waiting for the acknowledgement when it decides that the outstanding ASCONF chunk should be abandoned (Section 4.3) or should be re-scheduled after another ASCONF chunk (Section 4.4).

On the receiver side, the receiver should be able to receive ASCONF chunks with non-consecutive sequence numbers because the sender may abandon some ASCONF chunks. However, the received sequence number must be in ascending order, i.e. the current receiving sequence number must be greater than the last received sequence number.

4.2.2 Local Address States Definition

In deadlock scenario 2 (Section 3.2), the SCTP node refuses to generate an ADD-IP operation because the address is still in the association. In order to overcome this problem, four new address states are defined for local addresses in the SCTP node (Fig. 3a):

- **INITIAL.** The address is not in the SCTP association.
- **ACTIVE.** The address has been added into the association and is ready for sending and receiving packets.
- **ADDING.** The last operation for the address is ADD-IP. The ADD-IP chunk has been transmitted or queued and has not been acknowledged.
- **DELETING.** The last operation for the address is DELETE-IP. The DELETE-IP chunk has been transmitted or queued and has not been acknowledged.

An IP address can only be in one of these four states. Whenever an IP address goes UP or DOWN, an ADD-IP or DELETE-IP is generated and the address state is set to ADDING or DELETING respectively. Therefore, the address state represents the current status of the address. It ensures an address status change is immediately reflected in the address state.

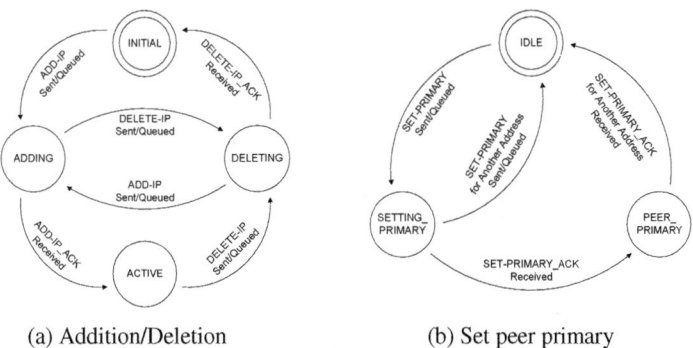

(a) Addition/Deletion   (b) Set peer primary

Fig. 3. State transition diagram for local addresses

Furthermore, three states for setting the peer primary address are defined as follows (Fig. 3b):

- **SETTING_PRIMARY.** The last SET-PRIMARY was sent for this address and has not yet been acknowledged.
- **PEER_PRIMARY.** The acknowledgment of the SET-PRIMARY has been received.
- **IDLE.** The address is in neither of the above two states.

Before changing an address to the SETTING_PRIMARY or PEER_PRIMARY state, the same state for other addresses must be cleared, i.e. only one address can be in the SETTING_PRIMARY or PEER_PRIMARY state.

4.2.3 Address Manipulation Process

Address Addition/Deletion Process
The address adding/deleting process on the sender side is shown in Fig. 3a. When an IP address comes UP, if the address is not in the association or is in the DELETING state, an ADD-IP chunk is generated. The address state is set to ADDING. Conversely, when an IP address goes DOWN, if the address is in the ACTIVE or ADDING state, a DELETE-IP chunk is generated. The address state is set to DELETING.

When the sender receives an acknowledgement for an ADD-IP/DELETE-IP operation, if the address is in the ADDING or DELETING state, an addition or deletion operation is performed respectively. Otherwise the acknowledgement is ignored because it is acknowledging an obsolete operation.

When the receiver receives an ADD-IP/DELETE-IP chunk, it performs the addition/deletion operation if the specified address is not/is in the SCTP association.

Setting Peer Primary Process
The process for setting peer primary in the sender side is shown in Fig. 3b. A SET-PRIMARY operation can be generated only when the IP address is both in the ACTIVE or ADDING state and is not in the SETTING_PRIMARY or PEER_ PRIMARY state. The IP address should be set to the SETTING_PRIMARY state subsequently.

When the sender receives an acknowledgement for a SET-PRIMARY, if the address is in the SETTING_PRIMARY state, the address state is changed to PEER_PRIMARY. Otherwise, neglect the acknowledgement because it is acknowledging an obsolete SET-PRIMARY chunk.

When the receiver receives the SET-PRIMARY chunk, it sets the specified address as peer primary if the address is in the SCTP association.

SCTP Endpoint Synchronization
As the sender of AUOs only saves the latest address updating status, the synchronization between the sender and the receiver should be considered.

An SCTP endpoint saves two sets of addresses, i.e. local address set and peer address set. The states of local addresses are defined in Section 4.2.2. The states of peer addresses are defined as follows:

- **INITIAL.** The address is not in the SCTP association.
- **ACTIVE.** The address is in the SCTP association and has been verified, i.e. a HEARTBEAT chunk has been sent to the address and the HEARTBEAT_ ACK has been received.
- **UNCONFIRMED.** The address has been added into the association but has not been verified.

When a local address is in the ADDING or DELETING state, the address saved in the peer side could be in any state amongst INITIAL, ACTIVE and UNCONFIRMED. Therefore the address should not be used as a source address. However, it should be used for receiving data from the peer.

4.3 Address Operation Consolidation Algorithm

The aim of the algorithm is to delete obsolete AUOs in order to increase the transmission efficiency. The algorithm can be triggered before sending an AUO or just after an AUO is generated.

For each IP address, the last ASCONF chunk contains the final operation for that address. Therefore all previous related ASCONF chunks can possibly be deleted. The general steps to consolidate operations are defined as follows (starting from the tail and working towards the head of the queue): (1) A DELETE-IP operation obsoletes previous ADD-IP, DELETE-IP and SET-PRIMARY operations for that IP address; (2) An ADD-IP operation obsoletes previous ADD-IP and DELETE-IP operations for that IP address; (3) A SET-PRIMARY operation obsoletes any previous SET-PRIMARY operations.

If an outstanding ASCONF chunk is obsolete and should be deleted, the new AS-CONF chunk should be transmitted according to the rules of congestion control, i.e. the new ASCONF chunk should be sent when the acknowledgement of the outstanding ASCONF chunk is received or the transmission timer expires.

The essential idea of the algorithm is to transmit the last operation reliably but to transmit previous operations partially reliably. Irrespective of whether the previous operations have been transmitted or not, the last operation should be kept for transmission to make sure that both SCTP endpoints have the same view of the address set. For example, for a specific address, if a DELETE-IP operation is generated after an ADD-IP operation which is still in the transmission queue, some may think both operations can be deleted. Actually, the DELETE-IP should be kept for transmission. The reason is that there might be a previous DELETE-IP operation which was transmitted unreliably before the ADD-IP operation. Therefore the sender is not sure if the specified address has been deleted in the peer by the unreliably transmitted DE-LETED-IP operation.

4.4 Address Operation Re-scheduling Algorithm

This section describes the algorithm for detecting potential deadlock situations and removing the deadlocks by selecting an appropriate ADD-IP chunk to transmit.

The Re-Scheduling algorithm can be triggered by the following two conditions: (1) All the addresses in the association are not available in the system, e.g. the addresses in the system have been re-numbered; (2) All paths between the two SCTP endpoints are broken, i.e. the ASCONF chunk cannot be sent to the peer successfully. A threshold value called ASCONF_MAX_RTX is defined to detect ASCONF transmission failures. If the number of consecutive transmission timeouts for an ASCONF chunk exceeds the threshold, the Re-Scheduling algorithm is triggered. The threshold is set to 3 in the paper. If it is set to 1, the Re-Scheduling algorithm is triggered for every ASCONF transmission timeout.

In order to maximize the reachability of a mobile node, the priorities of the AUOs are defined as follows:

$$Priority_{ADD-IP} > Priority_{SET-PRIMARY} > Priority_{DELETE-IP}. \qquad (1)$$

When the re-ordering algorithm is triggered, it uses inequality (1) to select an ASCONF chunk to transmit. If there are multiple ADD-IP chunks in the transmission queue, the ADD-IP chunk is selected according to the following descending priorities: (1) The address in the ADD-IP chunk is an active address in the system, i.e. the IP address belongs to a network interface at the SCTP endpoint. (2) The ADD-IP chunk has not been transmitted. (3) The address in the ADD-IP chunk is in SETTING_ PRIMARY state.

5 Implementation and Verification

The proposed scheme has been implemented in Ubuntu 9.04 (with a revised Linux-2.6.27.28 Kernel). Various tests have been executed to cover different network disconnection and interface renumbering scenarios (Six of these tests are listed in Table 1). The main purpose of the tests is to discover whether the proposed scheme can recover the communication when the connections between the MN and the CN experience total disconnection during the handover phase. The experimental network setup is shown in Fig. 2a. The MN and CN keep transmitting data to each other in the tests. All SCTP parameters are set to default. The detailed test setup, physical disconnection time, data interrupt time, and analysis are listed in Table 1. TEST 2 and

Table 1. Experiment setup and results for verifying the function of the proposed scheme. "*PDT*" is the *Physical Disconnection Time* in seconds, which is the duration from the attached *Wireless Router* (*WR*) of an *interface* (*IF*) being turned off to the interface being attached to a new wireless router. "*DIT*" is the *Data Interruption Time* in seconds, which is the duration from the attached wireless router being turned off to data transmission being recovered.

No	IF1 Behaviour	IF2 Behaviour	PDT	DIT
1	IF1 connects to WR1; WR1 is restarted; IF1 re-connects to WR1.	IF2 is always Down.	IF1:53s	53s
2	IF1 connects to WR1; WR1 goes down; IF1 connects to WR3 (IF1 is renumbered).	IF2 is always Down.	IF1:38s	86s
3	IF1 connects to WR1; WR1 is restarted; IF1 re-connects to WR1.	IF2 always connects to WR2. Link2 is broken.	IF1:63s	108s
4	IF1 connects to WR1; WR1 goes down; IF1 connects to WR3 (IF1 is renumbered).	IF2 always connects to WR2. Link2 is broken.	IF1:36s	84s

Table 1. (*continued*)

5	IF1 connects to WR1; WR1 is restarted; : IF1 connects to WR1.	IF2 connects to WR2; : WR2 is restarted; : IF2 connects to WR2.	IF1:57s IF2:136s	85s
6	IF1 connects to WR1; WR1 is restarted; : IF1 connects to WR3 (IF1 is renumbered).	IF2 connects to WR2; : WR2 is restarted; : IF2 connects to WR2.	IF1:38s IF2:132s	84s

Note:
- In **TEST 1**, no ASCONFs are generated because only one IP address is in the association and it cannot be deleted from the association. Data transmission recovers immediately after WR1 goes up.
- In **TEST 2 (The same test as Section 3.1)**, it takes 48s (86-38) for the MN to recover the communication after IF1 connects to WR3. When IF1 connects to WR3, an ADD-IP(IP3) is generated and queued. When the transmission timer of the outstanding SET-PRIMARY(IP1) expires, the re-scheduling algorithm is executed and the ADD-IP(IP3) is arranged to the head of the queue. The queuing time of the ADD-IP(IP3) is 48s.
- In **TEST 3 (The same test as Section 3.2)**, it takes 45s (108-63) for the MN to recover the communication after IF1 re-connects to WR1. When WR1 goes down, a DELETE-IP(IP1) is sent via IF2, which is lost because Link2 is broken. Simultaneously, MN sets IP1 to the DELETING state. When IF1 re-connects to WR1, the default route is set to WR1 and therefore packets are sent through WR1. However, the MN cannot use IP1 as a source address at current stage because IP1 is not in *ACTIVE* state. An ADD-IP(IP1) is generated and queued in the MN. The ADD-IP(IP1) obsoletes the outstanding DELETE-IP(IP1). However, it is not allowed to be sent out until the timeout of the outstanding DELETE-IP(IP1) occurs. The queuing time for the ADD-IP(IP1) is 45s.
- In **TEST 4**, it takes 48s (84-36) for the MN to recover the communication after IF1 connects to WR3. The event procedure is similar to that in TEST 3. The difference is that MN sends ADD-IP(IP3) via WR3 after IF1 connects to WR3.
- In **TEST 5**, it takes 28s (85-57) for the MN to recover the communication after IF1 connects to WR1. The event procedure is similar to that in TEST 3.
- In **TEST 6**, it takes 46s (84-38) for the MN to recover the communication after IF1 connects to WR3. The event procedure is similar to that in TEST 4.

TEST 3 repeat the experiments in Section 3. The results show that the deadlock situations are avoided effectively.

In these tests, after physical connections recover, the MN needs to wait for one transmission timeout of the outstanding ASCONF before sending an ADDIP for the new active address. The waiting time is from 0s to 60s according to the SCTP default configuration. The delay can be reduced by adjusting SCTP parameters.

6 Conclusions and Future Work

This paper studies address updating mechanisms for multi-homed mobile scenarios. It identifies that the design rationale of current SCTP-DAR cannot reflect the characteristics of address updating operations (AUOs). SCTP-DAR uses a data oriented transmission mechanism (First-In-First-Out & Reliable) to transmit AUOs, which significantly degrades handover performance and can cause communication to be broken in certain circumstances.

In order to overcome these problems, this paper proposes a novel address updating mechanism (named EAU) as follows. An ordered/partially reliable transmission scheme is proposed based on the observation that AUOs should be delivered in order but some obsolete operations are not necessary to be transmitted reliably. A set of new address states is defined to reflect the up-to-date address status. A consolidation algorithm is proposed to remove obsolete AUOs. Finally, a re-scheduling algorithm is proposed to detect and remove the deadlock situations presented in the paper.

The proposed address updating mechanism is implemented and verified based on the SCTP module in Linux. However, the mechanism can be used in more general situations because address updating management is a crucial function in many multi-homed mobile applications.

Future work is to test the proposed scheme in more complicated environments, such as in NAT enabled environments.

References

1. Vidales, P., Baliosion, J., Serrat, J., Mapp, G., Stejano, F., Hopper, A.: Autonomic system for mobility support in 4G networks. IEEE Journal on Selected Areas in Communications 23(12), 2288–2304 (2005)
2. IEEE P802.21/D14.0 Media Independent Handover Services (2008)
3. Nordmark, E., Bagnulo, M.: Shim6: Level 3 Multihoming Shim Protocol for IPv6, IETF RFC 5533 (2009)
4. Moskowitz, R., Nikander, P.: Host Identity Protocol (HIP) Architecture, IETF RFC 4423 (2006)
5. Stewart, R. (ed.): Stream Control Transmission Protocol, IETF RFC 4960 (2007)
6. Stewart, R., Xie, Q., Tuexen, M., Maruyama, S., Kozuka, M.: Stream Control Transmission Protocol (SCTP) Dynamic Address Reconfiguration, IETF RFC 5061 (2007)
7. Tüxen, M., Rüngeler, I., Stewart, R., Rathgeb, E.P.: Network Address Translation for the Stream Control Transmission Protocol. IEEE Network 22(5), 26–32 (2008)
8. Hayes, D.A., But, J., Armitage, G.: Issues with Network Address Translation for SCTP. ACM SIGCOMM Computer Communication Review 39(1) (2009)
9. Ma, L., Yu, F.R., Leung, V.C.M.: Performance Improvements of Mobile SCTP in Integrated Heterogeneous Wireless Networks. IEEE Transactions on Wireless Communications 6(10) (2007)
10. Fu, S., Ma, L., Atiquzzaman, M., Lee, Y.: Architecture and Performance of SIGMA: A Seamless Mobility Architecture for Data Networks. In: Proc. ICC '05, Seoul, Korea (2005)
11. Chang, m., Lee, M., Koh, S.: Transport Layer Mobility Support Utilizing Link Signal Strength Information. IEICE Transactions on Communications E87-B(9) (2004)
12. Tuexen, M., Stewart, R., Lei, P., Rescorla, E.: Authenticated Chunks for the Stream Control Transmission Protocol (SCTP), IETF RFC 4895 (2007)
13. Noonan, J., Perry, P., Murphy, S., Murphy, J.: Stall and Path Monitoring Issues in SCTP. In: Proc. of IEEE Infocom, Barcelona (2006)
14. Natarajan, P., Ekiz, N., Amer, P., Iyengar, J., Stewart, R.: Concurrent multipath transfer using SCTP multihoming: Introducing the potentially-failed destination state. In: Proc. IFIP Networking, Singapore (2008)
15. Stewart, R., Ramalho, M., Xie, Q., Tuexen, M., Conrad, P.: Stream Control Transmission Protocol (SCTP) Partial Reliability Extension, IETF RFC 3758 (2004)

Speculative Validation of Web Objects for Further Reducing the User-Perceived Latency

Josep Domenech, Jose A. Gil, Julio Sahuquillo, and Ana Pont

Department of Computing Engineering (DISCA),
Universitat Politecnica de Valencia. Spain
jdomenech@upvnet.upv.es, {jagil,jsahuqui,apont}@disca.upv.es

Abstract. Web caching techniques reduce user-perceived latency by serving the most popular web objects from an intermediate memory. In order to assure that reused objects are not stale, conditional requests are sent to the origin web servers before serving them. Most of the server responses to the conditional requests ratify that the object remains valid and, as a consequence, they do not include the object itself. Therefore, the object transfer time is completely saved when the object is still valid. However, the round-trip time (RTT) of these short responses cannot be saved. This time represents an important fraction of the response time in the current Internet scenario and makes conditional requests save less perceived latency than when they were proposed.

This paper proposes an approach to reduce the amount of conditional requests needed to maintain web cache consistency, thus completely saving both mentioned times (transfer and RTT) taken by such requests. To this end, our system uses a speculative approach similar to the one used in web prefetching which pre-sends freshness labels instead of web objects. The proposed technique has been evaluated using current and representative web traces. Experimental results show that the proposal dramatically reduces up to 55% of both the user-perceived latency and the amount of requests that the server receives.

1 Introduction

The latency perceived by users when they download a web page is affected by different factors like the availability of bandwidth between clients and servers, the server processing time of the request, the round-trip time and the size of the objects composing the page. Since the inception of the Web, many research efforts have concentrated on reducing the perceived latency by improving the infrastructure and by hiding the underlying latencies.

Web caching is based on the fact that users often request objects that were previously requested in the past. Although an object is in the local cache, a conditional request to the origin web server may be required to confirm that the reused object is not stale. In the current web, an important amount of the server responses to such requests certify that the object remains valid. In such a case, responses become much shorter since there is no need to resend the whole object

M. Crovella et al. (Eds.): NETWORKING 2010, LNCS 6091, pp. 239–250, 2010.

again, hence the fact that the object transfer time is removed. However, the user-perceived latency may not be significantly improved since the round-trip time (RTT) cannot be avoided.

Web prefetching is a technique that attempts to reduce the perceived latency by processing a user request before the user actually makes it. In current prefetching systems, the web server keeps track of the user's accesses *to predict* their next requests. Then, the server submits the hint list to the client, which will download the predicted objects in advance.

This paper proposes a new technique that aims to save conditional requests in order to reduce the user-perceived latency by anticipating the ETag of those objects likely to be requested in the near future. To this end, we propose a system similar in the prediction part to a prefetching system, but sending in advance the ETag for those objects included in the hint list (which contains the predictions).

The proposed system extends the prediction engine of the prefetching systems by adding the ETag for each object in the prediction list. Proceeding in this way, server responses allow to *prevalidate* other local objects in cache, although the user has not asked for them. The key performance advantage comes from the fact that subsequent accesses to prevalidated objects will not need to perform a conditional request to the server. Thus, we act on the amount of requests received by the web server and the latency perceived by users. Experimental results show that the prevalidation system dramatically reduces both the amount of requests to the server and the user-perceived latency between 45% and 58%, depending on the workload.

The remainder of this paper is organized as follows. Section 2 introduces some details on caching and prefetching in which our proposal lies. Section 3 presents the particulars of the prevalidation technique. Section 4 describes the experimental environment. Section 5 analyzes performance evaluation results. In Section 6 we discuss some related work in reducing user-perceived latency. Finally, Section 7 presents some concluding remarks.

2   Impact of Caching and Prefetching on User Perceived Latency

This section overviews some caching and prefetching principles. A brief discussion on downloading time issues is also included to provide some insights on performance.

Downloading time issues. A web page is usually considered as the basic unit of a user navigation. For this reason, the time taken to download a page should be considered as the main metric to evaluate the user-perceived latency [1]. Pages are usually composed of an HTML acting as a container of an increasing number of embedded objects like images, javascript code, CSS, etc. [2, 3]. Most common user actions, i.e., clicking on a hyperlink or typing a URI, usually imply the download of an HTML. Embedded objects will only be downloaded when the HTML has been downloaded (or, at least, part of it).

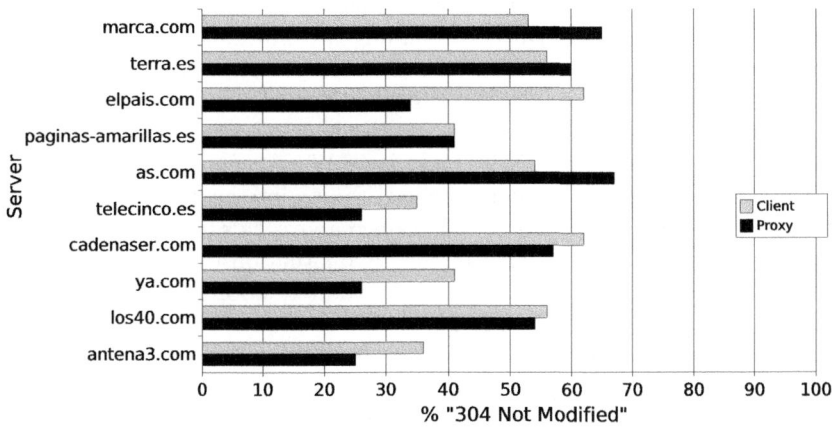

Fig. 1. Percentage of server responses with "304 Not Modified" response code in the top 10 most accessed servers in Spain received by the clients and by the proxy

Due to the fact that web pages are usually composed of several objects, the page downloading time is directly related to the time spent to download each object composing the web page. There are two main components in the downloading time of an object: i) the time taken by the request to reach the server plus the time taken by the response to reach the client (i.e., the round-trip time, RTT), and ii) the object transfer time (which depends on the bandwidth between the client and the server). Advances in technology have reduced both times, but mainly the object transfer time by increasing the available bandwidth. This fact has varied the weight that both components have in the overall downloading time.

Let's use an example in which a client downloads an object of 10KB. In 1996, according to [4], a typical value of bandwidth was 33Kbps and 1,130 ms of RTT. Downloading a 10KB object using the connection of 33Kbps takes 2,483 ms, assuming that there is no overload. So, the latency of downloading that object is 3,613 ms (i.e., 1,130 + 2,483). In this situation, the transfer time is 69% of the total latency.

With current technologies, typical values are 8Mbps of bandwidth and 40 ms of RTT. The transfer time of the same object is 10 ms, so the latency of downloading the object is 50 ms. In this new situation, the transfer time is just 20% of the total latency, which means that the RTT becomes the heaviest part of the current latency when navigating the web. In fact, several research studies show that the downloading time of most web objects does not depend on the file size, but mainly on the network latency [5, 6].

Web caching details. When an object is requested, the server usually provides information about how long its response will remain valid. If the user accesses again to such object before its lifetime expires, the object can be safely used.

However, if the object is stale, the browser must check its freshness before serving it to the user. This is performed by sending a conditional request to the web server: the client requests the object to the server, but it will only be served if the object has changed since the last time the client downloaded it. This request is replied by the server with the object if it has been modified, or with a "304 Not Modified" response otherwise. With these conditional requests, the user saves the transfer time if the object has not changed. Taking into account the components of the downloading time of an object described above, this technique could highly reduce the user-perceived latency in the 1996 situation (69% in the example). However, in the current situation example, the reduction of the user-perceived latency is much more limited (only 20%).

The choice of the expiration date of an object is a critical issue. If a far date is given, it is possible that the browser shows an outdated version of the web page. If the date given is very close, the benefits of web caching are reduced. In practice, web consistency prevails over web caching benefits, so expiration dates are usually closer to the request time or even in the past. As a consequence, the number of conditional requests sent to web servers is higher in current navigations than earlier in the web [3, 7]. This remark can be also probed by analyzing Figure 1, which shows the percentage of server responses with "304 Not Modified" response code for the top 10 most accessed Spanish web servers (according to the EGM Internet audience dated at May 2007 [8]). Data represented in this figure were gathered by the Squid proxy of the Universitat Politècnica de València.

Basic Web prefetching. In order to process user requests in advance, web prefetching is implemented by means of two main elements: the prediction and the prefetching engines. The prediction engine is aimed at guessing the following user accesses. This engine can be located at any part of the web architecture: clients, proxies, and servers, or even in a collaborative way at several elements. To predict future accesses, the most widely used option is to learn from past access patterns. The output of the prediction engine is a hint list, which is composed of a set of URIs likely to be requested by the user in a near future. Due to the fact that in many proposals the hint list is included in the HTTP headers of the server response, the time taken by the predictor to provide this list should be short enough to avoid delaying every user request and, therefore, degrading overall performance.

The prefetching engine is aimed at preprocessing those objects included in the hint list by the prediction engine. By processing the requests in advance, the user-perceived latency when the object is actually demanded is reduced. In the literature, this preprocessing has been mainly concentrated on the transfer of requested objects in advance.

Our proposal takes the prediction engine from web prefetching, However, objects are not speculatively downloaded, so the main cost of prefetching is avoided. This means that prevalidation does not significatively increase neither the network traffic nor the requests to the server.

3   The Prevalidation Technique

We propose the prevalidation technique to further improve caching benefits by acting over the conditional request aimed to ensure the freshness of the cached objects. To this end, the proposal uses a predictive approach similar to the one implemented in web prefetching techniques. Unlike other techniques analyzed below (see Section 6) that focus on the reduction of conditional requests, our proposal addresses specifically the freshness of web resources in the client's cache.

In our system, the web server provides the client with an extended hint list generated by the prediction engine, as shown in Figure 2. It includes the freshness labels of those objects that are likely to be requested in the near future. A freshness label can be any mechanism provided by the HTTP 1.1 for determining whether a resource is stale or not (e.g., last modified time and ETag).

In the implementation evaluated in this paper, the extended hint list is sent to the client piggybacked on the response. Link headers provided by the HTTP 1.1, which are currently being used by prefetching [9], can also be used for piggy-backing the extended hint list. For instance: *Link: <image.gif>; ETag="efsi"; rel=preval.* With the ETag or last modified time included in the link header, the client can validate those cached objects that are still fresh according to the data

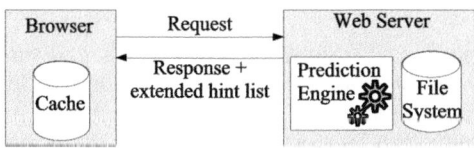

Fig. 2. Architecture of the proposed system

```
 1: Algorithm: Working of a prevalidation-enabled browsing
 2: Input: page: page demanded by the user, cache: current cache contents
 3: Output: cache: updated cache contents
 4: for each object in page do
 5:    if object is fresh in cache then
 6:       display object from cache
 7:    else
 8:       request object to server
 9:       display object from server response
10:       for each predicted_object in predictions attached to the response do
11:          if predicted_object is stale in cache and ETag matches then
12:             extend predicted_object freshness by t seconds
13:          end if
14:       end for
15:    end if
16: end for
```

Fig. 3. Basic working of a prevalidation-enabled browser

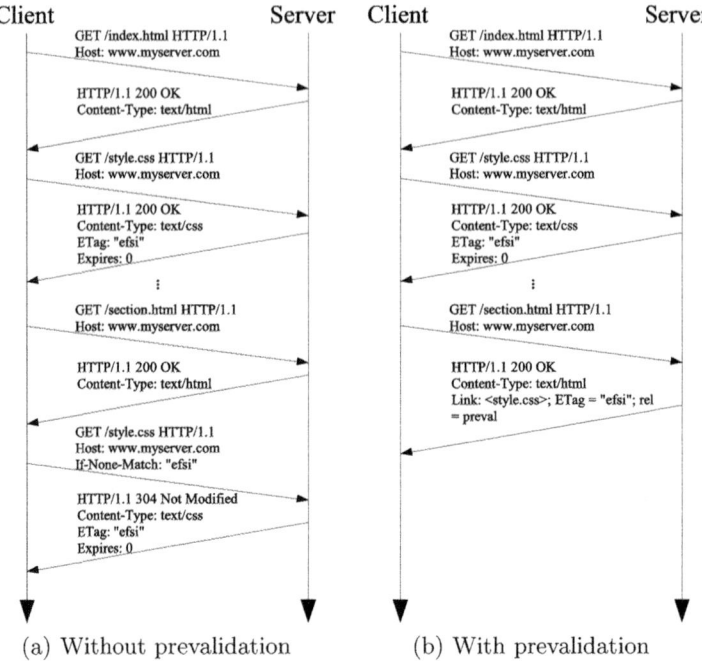

(a) Without prevalidation   (b) With prevalidation

Fig. 4. This figure shows an example of HTTP client - server communication in a system without prevalidation (a) and with prevalidation (b). In both cases, the user demands two web pages (*index.html* and *section.html*) that use the same CSS file (*style.css*). This resource can be stored in the client's cache but must be revalidated due to the *Expires: 0* header, so a conditional request is made in (a). However, in the prevalidation-enabled system, the server provides the client with an extended hint list (i.e., the *Link* header in the last response) that the client uses to prevalidate *style.css*.

provided by the server. To do so, the client compares the last modified time (or ETag) of the object in cache to the last modified time (or ETag) provided in the extended hint list. If the user requests a prevalidated object within t seconds after its validation, it is served from the cache, where t is the validation lifetime. In this way, the client avoids making a conditional GET request to the origin server and waiting for the "304 Not Modified" response. Therefore, the client saves the whole object latency. Notice also that the proposal reduces the amount of requests that the server has to fulfill. Taking into account this implementation, the basic working of a prevalidation-enabled browser is described in Figure 3.

Figure 4 depicts an example of communication between client and server in a system without prevalidation and in a prevalidation-enabled system. This example shows that the server can avoid the last request and, as a consequence, its user-perceived latency by anticipating the freshness label (i.e., the ETag) of the last object.

A side effect is that the prediction engine adds extra computation time for the server to fulfill a request. However, this effect can be mitigated since the prediction can be generated in parallel to the server response in order to keep the computational load of the server unaffected.

Finally, as it is known, many origin servers force the clients to send conditional requests mainly for statistics purposes; for instance, when keeping track of the users that see a banner. Nevertheless, this is not an obstacle for the proposal since, in such cases, those objects could be hosted at a different server (which is the current trend) or the prediction algorithm could have a black list with those objects that are not allowed to be predicted.

To sum up, we claim that the proposed mechanism has a double positive effect. First, it directly benefits the user whose client implements the prevalidation. Second, it indirectly benefits the rest of users because the technique decreases the amount of requests to the server. Even more, as experimental results will show, the proposal does not increase significatively the network traffic between clients and servers.

4   Experimental Environment

Framework. The experimental framework presented in [10] has been extended to implement and evaluate the technique proposed in this paper. The framework, originally developed to test prefetching algorithms, consists of three main parts: the server, the surrogate, and the client. The implementation combines both real and simulated parts in order to provide flexibility and accuracy.

The framework emulates a real surrogate, which is used to access an Apache web server. The surrogate, which provides no caching, acts as a predictor by adding the corresponding HTTP headers to the server response together with the extended hint list, i.e., the URIs provided by the prediction algorithm and their freshness labels. Although the prediction can be made simultaneously to the generation of the response by the server, the surrogate does not start the algorithm for predicting next user accesses until all response headers are received from the server. Since the delivery of the response to the client is delayed by the time taken by the prediction algorithm, this time is also perceived by the client and, therefore, included to compute the user-perceived latency. Notice that our implementation represents the worst-case situation, since the prediction is made after the generation of the response.

The client part represents the behavior of the users surfing the web with a prevalidation enabled browser. To model the set of users that access concurrently to a given server, the simulator is fed by current and real traces. Simulated clients obtain the results of the prediction engine from the response and prevalidate those objects still fresh in cache, as described in Section 3.

The simulator also reproduces the time at which each object was modified. These data were taken from the Squid traces described in Section 4. We assume that an object was modified every time we find a response having a "200 OK" code.

Workload. Two current and real traces, namely *elpais* and *los40*, were used to feed the simulator and to evaluate the prevalidation technique. Since the servers of both traces are included in the top 10 most accessed Internet media according to the EGM Internet audience study [8], therefore they are representative of a Spanish web user. The traces were obtained by filtering the accesses to the servers in the log of the Squid proxy of the Universitat Politècnica de València. Below some characteristics of these traces are summarized.

The trace *elpais* contains accesses to the www.elpais.com web site, which is the 3rd most popular Internet media according to [8]. This trace collects the navigation of 827 users and 535,430 object requests for three days in June 2008. After several tests, we found that the system is, after 300,000 requests, in a steady state, so we used these first 300,000 requests to warm up client caches and to train the prediction algorithm. Some internal ads and banners hosted in this server, whose requests might be required for statistics purposes, were found in this server. In this work, we assume the worst-case, that is, the prediction algorithm never predicts those objects, which barely represent less than 1% of the requested objects.

The trace *los40* contains accesses to the www.los40.com web site, which is 9th most popular Internet media in Spain according to [8]. This trace collects the navigation of 285 users and 242,658 object requests for one week in May 2007. After several tests, we found that the system is in a steady state after 150,000 object requests, so we used these requests to warm up client caches and to train the prediction algorithm. We found no ads or banners hosted in this server that could interfere in the results.

Both websites are rich in dynamic pages since its content is frequently changed because *elpais* and *los40* traces provide up-to-date news regarding general and music information, respectively.

Prediction algorithm. To implement the prediction part of the prevalidation technique, the *dependency graph* (DG) algorithm proposed in [4] has been used. This algorithm has been chosen because of its good performance at predicting embedded objects, as shown in [11]. Embedded objects are good candidates to be prevalidated since they are less often modified than HTML files.

4.1   Performance Metrics

The performance evaluation study is mainly focused on the following metrics, which represent the tradeoff of the prevalidation technique:

Latency reduction. Percentage of reduction in the user-perceived latency achieved by prevalidation compared to the perceived latency when it is not used. We consider as perceived latency the elapsed time between the demand of a page by the user until it is fully displayed, including all its embedded objects [1].

Requests savings. The number of requests, in percentage, that users do not make because the object has been prevalidated (within its validation lifetime). This metric does not include those requests saved due to cache hits.

Traffic ratio. The ratio of bytes transferred through the network when prevalidation is employed divided by the bytes transferred through the network when

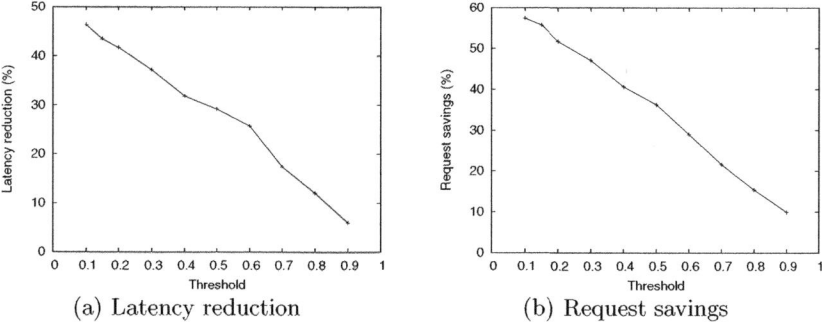

(a) Latency reduction   (b) Request savings

Fig. 5. Benefits of prevalidation as a function of the prediction threshold used under workload *los40*

prevalidation is not used. Prevalidation increases the network traffic with respect to the non-prevalidation case by sending the extended hint list. However, the network traffic is also reduced due to saved requests.

The goal of prevalidation is to achieve high latency reductions and request savings while keeping the traffic ratio close to 1.

5 Performance Evaluation

This section presents the results of the experiments conducted to evaluate the performance of prevalidation. The base configuration for the DG algorithm is a lookahead window size of 8, a threshold of 0.3 and a prevalidation lifetime of 60 seconds. Some of these parameters are modified in the experiments with the aim to explore the performance under different scenarios.

Experimental results when evaluating the proposal are really encouraging, independently of the trace used. Under the trace *los40*, the reduction in the user-perceived latency (see Figure 5(a)) ranges from 5% to 48% depending on the prediction aggressivenes. Results are even better when quantifying the request savings metric, as Figure 5(b) shows. The percentage of requests that prevalidation saves to the server varies between 10% for a 0.9 threshold value and 59% when using a threshold value of 0.1.

The reduction in perceived latencies and number of requests to the server is achieved, under trace *los40*, at the expense of increasing the network traffic slightly. As Figure 6(a) shows, this increase is related to the prediction aggressiveness and falls between 3% and 8%, i.e., traffic ratio is between 1.03 and 1.08. This extra traffic is due to the increase of the response headers size needed to include the prevalidation hint list.

The validation lifetime is a critical parameter because its value is closely related to the consistency of the web when using prevalidation. A high value means that a prevalidated object can be served from cache for a longer time, so it is the upper bound for a user to notice the changes in web objects. However,

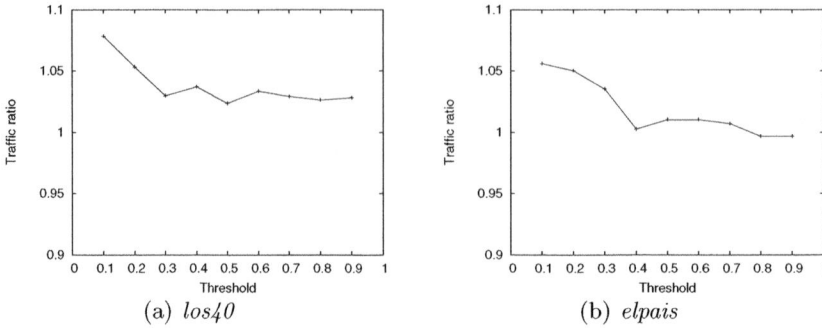

Fig. 6. Traffic ratio of prevalidation depending on the prediction threshold used

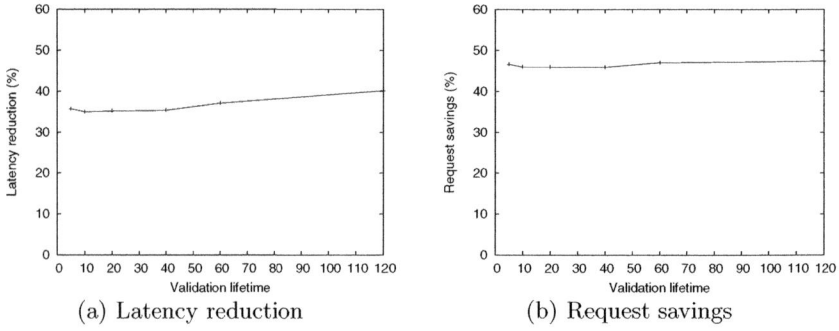

Fig. 7. Effect on performance of changing the validation lifetime under workload *los40*

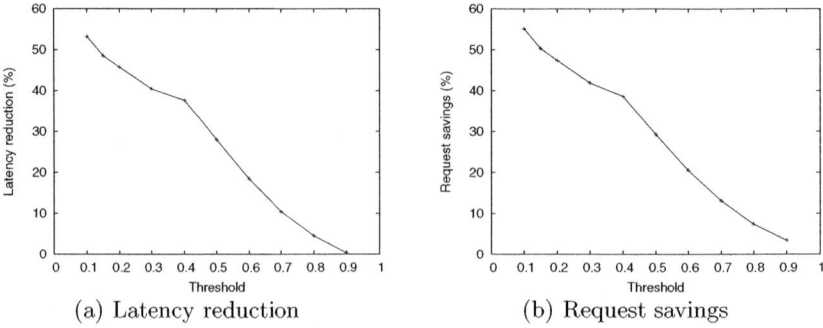

Fig. 8. Benefits of prevalidation as a function of the prediction threshold used under workload *elpais*

the performance of prevalidation is hardly related to this parameter, as Figure 7 shows. This is because most objects are prevalidated few seconds before their use. Common situations are those in which the prevalidation of an embedded object is included in the response of the HTML where it is contained.

We found the best latency reduction of the prevalidation when simulating the workload *elpais*. As shown in Figure 8(a), the latency reduction achieved by the technique ranges from about 54% when using the most aggressive threshold (0.1) to almost no reduction when using the 0.9 threshold. As expected, Figure 8(b) presents similar values for the request savings metric: the reduction of requests to the server ranges from 56% to 4%.

When looking at the traffic costs of the prevalidation, one can observe in Figure 6(b) that, under some configurations (0.8 or 0.9 threshold values), this cost becomes a benefit since the network traffic is below the traffic of the non-prevalidation case (i.e., traffic ratio below 1). In this situation, the amount of bytes not sent due to saved requests (and responses) is higher than the amount of bytes transferred to send the extended hint list. In any case, the increase in network traffic in the other configurations is always below 6%.

6   Related Work

Since the beginning of the web, researchers have been working on reducing the associated latencies using a wide range of techniques. Caching the most popular objects was rapidly adopted by web browsers and extended to other elements of the architecture like proxy servers and surrogates.

Ensuring the freshness and consistency of proxy caches has been a widely researched topic. [12] proposes that servers and proxies keep track of changes in the web by grouping the objects into volumes and using version identifiers. The proxy piggybacks these identifiers in the request. Then, servers return a list of resources that have changed between the supplied and current version, so the proxy can invalidate them. The main drawback of this approach is that servers need to keep track of the evolution of every web resource. Another important issue with this approach is how to group sets of resources into volumes. To deal with volumes, [13] classifies objects according to the frequency and predictability of their changes. To provide strong consistency on caches, the server also provides explicit instructions to caches on how to handle each object, including invalidations for those objects that have changed. Other approach [14] works on reducing the delay of the validation requests for the users. This work proposes different policies for caches to proactively validate objects as they become stale. To do so, each refreshment policy associates a renewal credit with each cached object, so that only those with a positive credit are refreshed. An important disadvantage of this approach is the increase of requests due to the polling done by the proxies to the servers.

Unlike the prevalidation technique, these proposals are addressed to work between servers and proxies and are not appropriate to deal with browser's cache and, therefore, to hide the last-mile latency. However, it is possible to combine them to improve system overall performance.

7   Conclusions

To maintain consistency and to avoid sending stale responses to web users, web caches must confirm the freshness of the cached objects by asking the origin servers before serving them. To do so, conditional requests are used. Unfortunately, because of the current Internet characteristics, the performance benefits due to this conditional requests have been notably reduced. This is because even in the best case –when the object has not been modified– the RTT (which currently represents a high percentage of the perceived latency) cannot be avoided.

To improve web performance and to further reduce the latency perceived by users, we proposed a novel technique that use a predictive approach. The technique permits to prevalidate objects cached without sending conditional requests to the origin server. Results show that the prevalidation technique has a double benefit: it can dramatically decrease the user-perceived latency and reduce the number of requests to the server at the expense of small overhead in network traffic.

References

[1] Domenech, J., Gil, J.A., Sahuquillo, J., Pont, A.: Web prefetching performance metrics: A survey. Performance Evaluation 63 (2006)

[2] Domenech, J., Pont, A., Sahuquillo, J., Gil, J.A.: A user-focused evaluation of web prefetching algorithms. Computer Comm. 30 (2007)

[3] Bent, L., Rabinovich, M., Voelker, G.M., Xiao, Z.: Characterization of a large web site population with implications for content delivery. In: Proc. of the Thirteenth Int. World Wide Web Conf., New York, USA (2004)

[4] Padmanabhan, V.N., Mogul, J.C.: Using predictive prefetching to improve World Wide Web latency. Computer Comm. Review 26 (1996)

[5] Sharma, M., Byers, J.W.: How well does file size predict wide-area transfer time? In: IEEE Global Telecommunications Conf., Taipei, Taiwan (2002)

[6] Murta, C.D., Dutra, G.N.: Modeling http service times. In: IEEE Global Telecommunications Conf. (GLOBECOM), Dallas, USA (2004)

[7] Arlitt, M.F., Williamson, C.L.: Internet web servers: Workload characterization and performance implications. IEEE/ACM Trans. on Networking 5 (1997)

[8] Asociación para la Investigación de Medios de Comunicación (AIMC): Audiencia de Internet EGM, http://www.aimc.es/03internet/31.html

[9] Fisher, D., Saksena, G.: Link prefetching in Mozilla: A server driven approach. In: Proc. of 8th Int. Workshop on Web Content Caching and Distribution (WCW 2003), New York, USA (2003)

[10] Domenech, J., Pont, A., Sahuquillo, J., Gil, J.A.: An experimental framework for testing web prefetching techniques. In: Proc. of 30th EUROMICRO Conf., Rennes, France (2004)

[11] Domenech, J., Gil, J.A., Sahuquillo, J., Pont, A.: DDG: An efficient prefetching algorithm for current web generation. In: Proc. of 1st IEEE Workshop on Hot Topics in Web Systems and Technology, Boston, USA (2006)

[12] Krishnamurthy, B., Wills, C.E.: Piggyback server invalidation for proxy cache coherency. In: Proc. of 7th World Wide Web Conf., Brisbane, Australia (1998)

[13] Mikhailov, M., Wills, C.E.: Evaluating a new approach to strong web cache consistency with snapshots of collected content. In: Proc. of Twelfth Int. World Wide Web Conf., Budapest, Hungary (2003)

[14] Cohen, E., Kaplan, H.: Refreshment policies for web content caches. Computer Networks 38 (2002)

Dynamic Service Placement in Shared Service Hosting Infrastructures

Qi Zhang, Jin Xiao, Eren Gürses, Martin Karsten, and Raouf Boutaba

David. R. Cheriton School of Computer Science,
University of Waterloo, Waterloo, Canada
{q8zhang,j2xiao,egurses,mkarsten,rboutaba}@cs.uwaterloo.ca

Abstract. Large-scale shared service hosting environments, such as content delivery networks and cloud computing, have gained much popularity in recent years. A key challenge faced by service owners in these environments is to determine the locations where service instances (e.g. virtual machine instances) should be placed such that the hosting cost is minimized while key performance requirements (e.g. response time) are assured. Furthermore, the dynamic nature of service hosting environments favors a distributed and adaptive solution to this problem. In this paper, we present an efficient algorithm for this problem. Our algorithm not only provides a worst-case approximation guarantee, but can also adapt to changes in service demand and infrastructure evolution. The effectiveness of our algorithm is evaluated though realistic simulation studies.

1 Introduction

With the abundance of network bandwidth and computing resources, large-scale service hosting infrastructures (e.g. content-delivery networks, service overlays and cloud computing environments) are gaining popularity in recent years. For instance, PlanetLab, as a shared academic overlay network, is capable of hosting global-scale services including content delivery [1] and file storage [2]. More recently, commercial products such as Amazon Elastic Computing Cloud (EC2) [3], Google AppEngine [4] and Microsoft Azure [5] have emerged as attractive platforms for deploying web-based services. Compared to traditional service hosting infrastructures, a shared hosting infrastructure offers numerous benefits, including (1) eliminating redundant infrastructure components, (2) reducing the cost of service deployment, and (3) making large-scale service deployment viable. Furthermore, the ability to allocate and deallocate resource dynamically affords great flexibility, thus (4) making it possible to perform dynamic scaling of service deployment based on changing customer demands.

In general, three types of roles are present in a shared hosting environment: (1) the *infrastructure provider* (InP) who owns the hosting environment; (2) the *service provider* (SP) who rents resources from the infrastructure provider to run its service, and (3) the *client* who is a service customer. In this context, a service provider's objective is to (a) satisfy its performance requirements specified

M. Crovella et al. (Eds.): NETWORKING 2010, LNCS 6091, pp. 251–264, 2010.

in Service Level Agreements (SLAs) such as response time, and (b) minimize its resource rental cost. For example, response time of requests for CNN.com should be less than 2 seconds [6]; Online game services often have stringent response time requirements [7]. Achieving this delay requirement (objective a) is dependent on where the servers are placed, and typically there is a monetary penalty when this requirement is not met. At the same time, achieving resource cost reduction (objective b) is also server location dependent. Thus, we argue there is a strong and consistent incentive for SPs to choose their servers' placement carefully to minimize the total operating cost. This is the case for cloud computing environment as well. Although the number of data centers used by commercial products is relatively small today, this number will grow (as demand for cloud computing grows), and new architectures such as micro [8] and nano [9] data centers will emerge. A distributed solution to this problem is desirable in such a large-scale and dynamic environment, as frequent re-execution of centralized algorithms can incur significant overhead for collecting global information [10,11]. Hence an adaptive and distributed service placement strategy is strongly favored.

In this paper, we present a solution to the dynamic service placement problem (SPP) that minimizes response time violations and resource rental cost at run time. We formulate SPP mathematically and present a distributed local search algorithm that achieves a theoretical worst-case performance guarantee. The effectiveness of our algorithm is experimentally verified using a set of realistic simulation studies.

The remainder of this paper is organized as follows: We present a generic model of a service hosting infrastructure in Section 2. In Section 3, we present the formulation of SPP and introduce the notations. Our distributed algorithm is presented in Section 4. In Section 5, we show that the algorithm achieves an approximation guarantee of 27. Experimental results are presented in Section 6. We discuss related research in Section 7 and conclude our work in Section 8.

2   System Overview

We consider a hosting platform that consists of servers that are diversely situated across geographical locations. A generic architecture of a service hosting platform is depicted in Figure 1. An instance of a service may be installed on one or many servers. To achieve locality awareness and load-balancing, a client request is redirected to a nearby server with enough CPU and bandwidth capacity to handle the request. The component which is responsible for redirecting request is the *request router*. In practice, a request router can be implemented using a variety of techniques, such as DNS-based request redirection, or direct routing of requests [12]. On the other hand, the status of individual servers, including traffic condition, CPU, memory and bandwidth usage are measured at run-time by a *monitoring module* on each server. It should be pointed out that it is entirely possible that monitoring modules are shared among a group of servers, and our model is fully applicable to such a scenario. The *analysis*

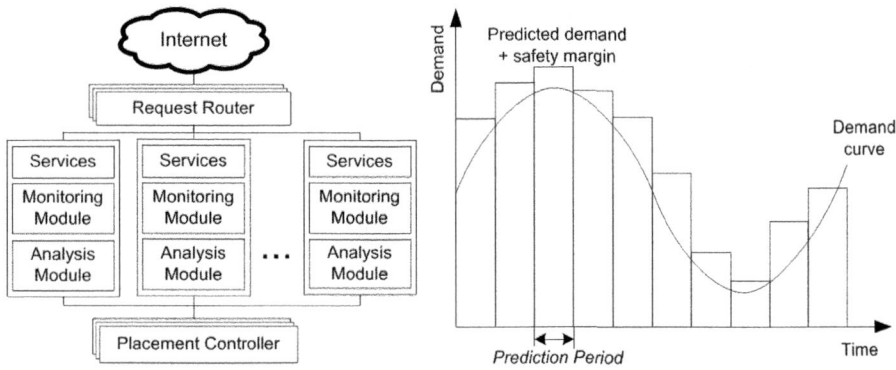

Fig. 1. System Architecture   **Fig. 2.** Dynamic Demand Prediction

modules are responsible for analyzing the current demand and predicting the future demand during the next time interval, as we shall explain below. Finally, the *placement controllers* are responsible for dynamically placing services. This includes copying service software (e.g. virtual machine images) to the server, dynamically starting and stopping service instances. The placement controllers are activated periodically, based on the demand prediction received from the analysis modules. In practice, the placement controllers can be implemented either centralized or distributed. For example, Microsoft Azure uses centralized controllers that are managed by service providers [5]. P2P overlay networks, on the other hand, requires distributed placement controllers that are located on individual peers. In such a scenario, a placement controller can collect relevant information using protocols such as local broadcasting and gossiping [13].

Our main objective is to dynamically adapt placement configurations to changing demands and system conditions. The demand can be modeled as function of time and location. Ideally, we want the placement configuration to be adaptive to individual demand changes. However, this is often impractical since frequent placement reconfiguration can be expensive to implement. Moreover, fine-grained demand fluctuations are usually transient and do not persist for long durations. Therefore, it is necessary to both analyze and periodically predict service fluctuations at coarse-grain levels, as illustrated in figure 2. Furthermore, even when the daily service demand follows some predictable pattern, pre-planned configuration or placement strategies at the infrastructure level are not possible because both the infrastructure and services themselves may change drastically (e.g. unexpected system failures and transient services such as VPN for a sports broadcast). The problem of predicting future service demand has been studied extensively in the past, and many solution techniques (e.g. [14]) have been proposed. It is not our goal to examine better demand prediction methods. Rather, we study how placement configuration should be modified according to a prediction of the future service demand.

3   Problem Definition and Notations

We represent the hosting platform as a bipartite graph $G = (D, F, E)$, where D is the set of clients, F is a set of candidate servers on which service instances can be installed, and E is the set of edges connecting D and F. A service instance may consume two types of resources: (1) *Load-dependent resources*, whose consumption is dependent on the amount of demand served by the service instance. Load-dependent resources typically include CPU, memory and bandwidth. (2) *Load-independent resources*, whose consumption is constant regardless of how much demand the server receives. An example of load-independent resource is storage space of the service software.

We assume every candidate server in F has enough load-independent resources to host the service, since otherwise we can remove it from F. For load-dependent resources, we assume there is a bottleneck capacity cap_s on how many client a service instance $s \in F$ can serve. The bottleneck capacity can be bandwidth, CPU capacity or memory, whichever is more stringent on each server. It should be noted that this bottleneck capacity may change over time. However, as we shall see our local search algorithm can naturally handle this case. In addition, We assume there is a placement cost p_s for renting a service instance on a server s.

Let C_s denote the set of demand assigned to s. For every demand $i \in D$, let s_i represent the service instance who serves i. For every demand $i \in C_{s_i}$, denote by $d(i, s_i)$ the distance between i and s_i that represents the communication latency between i and s_i. Furthermore, let U_{s_i} denote the number of client assigned to a server s_i. Using a simple M/M/1 queuing model [1], We obtain the actual response time $r(i, s_i)$ of a service request i as a function of U_{s_i} and $d(i, s_i)$:

$$r(i, s_i) = d(i, s_i) + \frac{\mu_{s_i}}{1 - U_{s_i}/cap_{s_i}} = d(i, s_i) + \frac{cap_{s_i} \cdot \mu_{s_i}}{cap_{s_i} - U_{s_i}}$$

where μ_s is the service time of a request at the server. The second term essentially models the queuing delay of the request. We construct a cost penalty $cp(i, s_i)$ if the request response time exceeds a threshold value d_{max}. Specifically, we define

$$cp(i, s_i) = a(\frac{r(i, s_i)}{d_{max}})^2$$

where a is a monetary penalty cost. We use a quadratic penalty function here as it reflects the general form of the penalty payout for SLA violation. Higher order polynomial is also possible and does not affect our analysis technique. Our objective is to select a set of servers $S \subseteq F$ to deploy our service as to minimize the sum of hosting cost $c_f(S)$ and service cost $c_s(S)$:

$$c(S) = c_s(S) + c_f(S) = \sum_{i \in D} cp(i, s_i) + \sum_{s \in S} p_s \tag{1}$$

---

[1] More sophisticated model can be easily incorporated by changing the equations for $r(i, s_i)$ and $cp(i, s_i)$.

Algorithm 1. Local Search Algorithm for SPP

1: **while** $\exists$ an add(s), open(s, T) or close(s, T) that reduces total cost **do**
2:     execute this operation

This problem is difficult to solve directly due to the load factor U_s in the equation. We remedy the situation by having a preferred load value l_s for U_s. In this case we define SPP as follows:

Definition 1 (SPP). *Given a demand set D and a server set F , each with a renting cost p_s and a service cost $cp(i, s_i) = a((d(i, s_i) + \frac{\mu_{s_i}}{1-l_{s_i}/cap_{s_i}})/d_{max})^2$, select a set of servers $S \subseteq F$ to host the service and assign the demand to S such that $|C_s| \leq l_s$ for every $s \in S$, minimizing the total cost $c(S) = \sum_{i \in D} cp(i, s_i) + \sum_{s \in S} p_s$.*

It is actually provable that the original problem can be reduced to this formulation (We skip the proof due to space constraints). In fact, setting a preferred load value is more practical for implementation, since typically a service provider may define an expected utilization for each service instance [11].

We observe that SPP is similar to the Capacitated Facility Location Problem (CFLP) [15]. CFLP can be briefly described as follows: Given a graph $G = (V, E)$ and a set of candidate sites $F \subseteq V$ for installing service facilities (each site $s \in F$ has an installation cost f_s and a service capacity U_s), select a subset of sites $S \subseteq F$ to install facilities and assign each client to a facility in S such that the sum of the connection cost $c_s(S) = \sum_{i \in V} d(i, s_i)$ and the installation cost $c_f(S) = \sum_{i \in S} f_s$ is minimized. CFLP is an NP-hard optimization problem [16], for which many centralized approximation algorithms have been proposed. In distributed settings, however, only the uncapacitated case has been studied [17,18].

It can be seen that CFLP is almost identical to SPP except the connection cost is defined as $cp(i, s_i) = d(i, s_i)$. It is easy to show SPP is NP-hard using the same argument for CFLP. Motivated by this observation, we developed an approximation algorithm for SPP based on the local search algorithm for CFLP described in [15]. In our work, we also assume our distance metric satisfies triangular inequality. This is a practical assumption since we can always use the network coordinate services [19] to compute the distance metric. Furthermore, it is also known that Internet triangular inequality is approximately satisfied (i.e. $d(i, j) \leq k(d(i, k) + d(k, j)))$, where k is roughly 3 [20]. It is easy to adjust our prove technique to accommodate this effect.

4   A Local Search Approximation Algorithm for SPP

Local search is a well known meta-heuristic for solving optimization problems. A local search algorithm starts with arbitrary solution and incrementally improves the solution using local operations until the solution can no longer be improved.

Local search algorithms can naturally tolerate dynamic changes in the problem input, as they can start from arbitrary solutions. We believe a local search algorithm for SPP is favorable as it not only tolerates system dynamicity, but also provides a provable worst-case performance guarantee.

Based on the theoretical work by Pal et. al. [15], Algorithm 1 is our proposed local search algorithm for SPP. In our work, we use the term server and facility interchangeably. Two particular terminologies used in the paper are opening and closing a facility t, which means installing a service and uninstalling a service at location t. This algorithm starts from any feasible initial solution, and incrementally improves the solution with one of three types of operations: (1) $add(s)$, which opens a facility s and assign a set of clients to s. (2) $open(s, T)$, which opens a facility s, close a set of facilities T, and assign the clients of facilities in T to s. (3) $close(s, T)$, which closes a facility s, open a set of facilities T and assign the client of s to facilities in T. In all three operations, a facility can be opened multiple times but closed only once.

Due to its simplicity and adaptive nature, this algorithm fits very well with our objective. However, this algorithm cannot be directly implemented in a distribute way, since finding an $open(s, T)$ and a $close(s, T)$ requires solving a knapsack problem and a single node capacitated facility location (SNCFL)[21] problem, respectively. Both problems are NP-hard in general, but can be solved in pseudo-polynomial time using dynamic programming [15]. However, this dynamic programming approach does not apply to distributed settings since it is both expensive and requires global knowledge.

Hence, it is our objective to show that both $open(s, T)$ and $close(s, T)$ can be computed locally and distributedly. In this regard, our algorithm is local in the sense that each operation is computed using neighborhood information. Specifically, Each node s in our algorithm maintains a list of neighborhood servers within a fixed radius. The list of neighbors can be obtained using neighborhood discovery protocols or a gossiping protocol [13]. Since the servers are usually static and do not change often overtime, the list of neighboring servers will not require frequent update.

4.1   Implementing Add(s)

The purpose of the add operation is to reduce the total connection cost. In an $add(s)$ operation, if a facility s is not opened, it will become opened and start serving clients. In this case the cost reduction (CR) of the $add(s)$ operation is:

$$\mathrm{CR}(\mathrm{add}(s)) = \sum_{i \in U} cp(i, s_i) - cp(i, s) - p_s$$

where $U \subseteq V$ is a set of clients. On the other hand, if facility s is already opened, it can invite more clients to join its cluster. In this case we set $f_s = 0$ in the above equation. An add(s) operation can be performed when $\mathrm{CR}(\mathrm{add}(s)) > 0$.

However the straightforward implementation, i.e. Letting s contact the other facilities to find the set of potential clients, is cumbersome to implement. This

Algorithm 2. Local Algorithm for computing an admissible open(s, T) move

1: Sort facilities in decreasing order of their cost efficiencies, $c_s \leftarrow 0$
2: **while** $c_s \leq caps$ and costEff$_{open}(t) \geq 0$ for the next t **do**
3:    $T \leftarrow T \cup t$, $c_s \leftarrow c_s + |C_t|$
4: **return** either T or the next facility $t \notin T$ in the list, whichever is larger

is because s does not know a-priori which facility has clients with high connection cost. Instead, in our implementation, we allow neighborhood facilities to exchange client information and record potential clients. Once s have recorded enough potential clients, s can become open (if not already opened), and start serving these clients. Observe that a client i must satisfy $cp(i, s) \leq cp(i, s_i)$ in order for the reassignment to be beneficial, hence $d(s, s_i) \leq d(i, s_i) + d(i, s) \leq d(i, s_i) + \sqrt{\frac{cp(i,s)}{a}} \cdot d_{max} \leq d(i, s_i) + \sqrt{\frac{cp(i,s_i)}{a}} \cdot d_{max}$. Therefore s_i only needs to exchange client information with servers within radius $d(i, s_i) + \sqrt{\frac{cp(i,s_i)}{a}} \cdot d_{max}$ to guarantee an admissible add(s) operation will be found by a facility s.

4.2 Implementing Open(s, T)

In an open(s, T) operation, a single facility s is opened and a set of facilities T become closed. All the clients of facilities in T get assigned to s. The total reduction of this operation is computed as:

$$\text{CR}(open(s, T)) - \sum_{t \in T}(p_t - |C_t|cp(t, s)) - p_s$$

Again, notice that if s is already opened before the move, then p_s is set to 0. The key challenge here is to determine the suitable set T. The problem of finding the optimal set T that maximizes $\text{CR}(open(s, T))$ can be formulated as a 0-1 knapsack problem: Let s be a knapsack with capacity $caps$, and each open facility t be an item with size$(t) = |C_t|$ and value$(t) = p_t - |C_t|d(s, t)$. The objective is to select a set of items to be packed in the knapsack to maximize the total value. Although it can be solved in pseudo-polynomial time using dynamic programming, it is not practical to implement because it requires global knowledge. Hence, we replace the dynamic programming procedure by a well-known greedy 0.5 approximation algorithm (Algorithm 2), where the cost efficiency of each facility t is defined as $\frac{value(t)}{size(t)}$, which is costEff$_{open}(t) = \frac{p_t}{|C_t|} - cp(s, t)$.

To implement Algorithm 2 distributedly, notice that for any facility t, if $cp(s, t) \geq \frac{p_t}{|C_t|}$, then costEff$(t)$ becomes negative and can be safely ignored in the calculation. Hence each facility t only need to contact neighborhood facilities in radius $\sqrt{\frac{p_t}{a \cdot |C_t|}} \cdot d_{max}$. A neighboring facility s can then compute an admissible open(s, T) once it discovers the entire T.

4.3   Implementing close(s, T)

The objective of the close(s, T) is to close a facility s, open a set of facilities T, and assign all of the clients of s to facilities in T. Again, the main difficulty is to find a suitable set T. The cost reduction of close(s, T) can be computed as:

$$\text{CR}(close(s, T)) = p_s - \sum_{t \in T}(p_t + u_t cp(s, t))$$

Where u_t represents the amount of demand assigned to t after closing s. Similar to open(s, T), if a facility $t \in T$ is already opened, then p_t is set to 0 in the above equation. Computing the optimal T can be formulated as a single-demand capacitated facility location problem (SNCFL)[21]. In this problem, we wish to select a set of facilities T with total capacity at least $|C_s|$, assigning clients of s to facilities in T to minimize

$$cost_{SNCFL} = \sum_{t \in T}(p_t + u_t cp(s, t))$$

This problem is also NP-hard. However, a greedy algorithm is known for this problem [21]. In this algorithm, the facilities are sorted by their cost efficiencies, which in our case, is defined as $\text{costEff}_{close}(t) = \frac{p_t}{cap_t} + cp(s, t)$ This algorithm (Algorithm 3) can be described as follows: the output set T is initially empty. we first sort facilities in increasing order of their cost efficiencies, and then try to add facilities T according to the sorting order. In each step, if adding next facility into T will cause $\sum_{t \in T} cap_t \geq |C_s|$ (i.e. facilities in T have enough capacity to handle demands of s), then T is a candidate solution. Now, if $\text{CR}(close(s, T)) \geq 0$, this is an admissible operation and the algorithm stops. Otherwise, we do not add this facility to T. This process repeats until every facilities in U has been examined. The following lemma is a known result for this algorithm:

Lemma 1. *[21] The greedy algorithm outputs a solution S with $cost_{SNCFL} \leq \sum_{t \in T}(2p_t + u_t cp(s, t))$ for any $T \subseteq F$.*

Clearly, a facility in T can not be outside a radius $\sqrt{\frac{p_s}{a}} \cdot d_{max}$, as in this case the reassignment cost will exceed p_s. Hence close(s, T) operation can be computed locally by s within in this radius.

5   Algorithm Analysis

In this section, we show that our algorithm achieves an approximation factor 27. There are two challenges to this problem. First, our distance function is non-linear, as opposed to the linear function used in CFLP. Second, since we have replaced the dynamic programming procedures with greedy algorithms, we need to show our algorithm can still provide a performance guarantee.

Lemma 2. *If an open operation open(s, T) computed using Algorithm 2 is not admissible, then $2p_s + \sum_{t \in T'}(u_t cp(s, t) - p_t) \geq 0$ for any set of $T' \subseteq F \backslash \{s\}$.*

Algorithm 3. Local Algorithm for computing an admissible close(s, T) move

---

1: $rem \leftarrow |C_s|$, $T \leftarrow \emptyset$, $currentCR \leftarrow 0$, $U \leftarrow$ facilities within radius $\sqrt{\frac{p_s}{a}} \cdot d_{max}$,
2: sort U in increasing order of their cost efficiencies
3: **repeat**
4:     $r = \frac{p_s - currentCR}{rem}$
5:     **while** not every node in U has been examined **do**
6:         $t \leftarrow$ next facility in the sorted list
7:         **if** $rem - cap_t \geq 0$ **then**
8:             $T \leftarrow T \cup t$, $currentCR \leftarrow \sum_{t \in T}(p_t + cap_t d(s,t))$, $rem \leftarrow rem - cap_t$
9:         **else**
10:             **if** $CR(close(s, T \cup t)) \geq 0$ **then**
11:                 **return** $T \cup t$
12: **until** $U == \{\emptyset\}$
13: output there is no admissible close operation

Proof. Since Algorithm 2 is a $\frac{1}{2}$-approximation of the knapsack problem, it is true for any $T' \subseteq F\backslash\{s\}$, $SOL_{ALG2} \geq \frac{1}{2}(\sum_{t \in T'} p_t - |C_t|d(s,t))$. Since it fails to find a admissible operation, $SOL_{ALG2} - p_s \leq 0$ must hold. The claim follows.

Lemma 3. *If a close operation close(s, T) computed using Algorithm 3 is not feasible, then $\sum_{t \in T'}(2p_t + u_t cp(s,t)) - p_s \geq 0$ for any set of $T' \subseteq F\backslash\{s\}$.*

Proof. Since Algorithm 3 is a 2-approximation algorithm for the $SNCFL$, we must have, for any set $T' \subseteq F\backslash\{s\}$, $SOL_{ALG3} \leq \sum_{t \in T'}(2p_t + u_t cp(s,t))$. Since it fails to find an admissible move, we must have $p_s - SOL_{ALG3} \leq 0$. The lemma follows.

Using the above lemmas, we can prove our main result:

Theorem 1. *The distributed algorithm achieves an approximation factor 27.*

Proof. See appendix.

6 Experiments

We have implemented our distributed algorithm in a discrete event simulator and conducted several simulation studies.

We construct topology graphs using the GTITM generator [22], which can generate transit-stub topologies that simulate latencies between hosts on the Internet. We specify the average communication latency at intra-transit, stub-transit and intra-stub domain links to be 20ms, 5ms and 2ms respectively [23]. We randomly pick a subset (10%) of nodes as candidate locations for placing servers. Each candidate location can host up to 5 instances. In our experiments, we set $d_{max} = 400$ms, $\mu_s = 20$ms, $a = 1$ and $p_s = 4$ for 200 node graph, 8 for

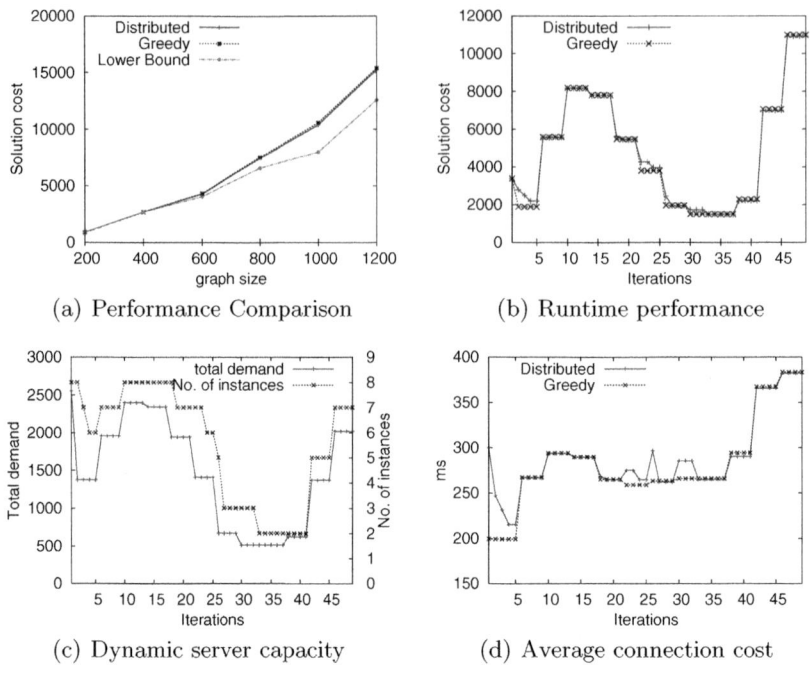

(a) Performance Comparison   (b) Runtime performance

(c) Dynamic server capacity   (d) Average connection cost

Fig. 3. Experimental result

400 node graph and so on. For benchmarking purpose, we have implemented a centralized greedy placement algorithm [24], which has been shown to perform well in practice.

In our first experiment, we evaluate the performance of distributed algorithm in static topologies where there is no demand fluctuation. To estimate the value of optimal solution, we implemented a simple routine that outputs a lower-bound on the optimal solution by ignoring capacity constraints. Our experimental results are shown in Figure 3(a). We observe that although both algorithms perform well compared to the lower-bound values, our algorithm performs better than the greedy algorithm in all cases. This result indicates that, the solution produced by our algorithm is near optimal.

In our next experiment, we evaluate the effectiveness of our algorithm under dynamic conditions, i.e. demand of clients fluctuate overtime and servers may fail randomly. To make experiment realistic, we used a real internet topology graph from the Rocketfuel project [25]. Figure 3(b) shows the cost of the solution at runtime. For comparison purpose, at each time stamp we also show the solution of the greedy heuristic that has access to global topology information. As shown on Figure 3(b), our distributed algorithm is quite adaptive to the dynamic conditions, and in some cases, performs better than the greedy heuristic. Figure 3(c) depicts the change in the number of instances in response to the changes in client population. We observe that the number of instances used, as suggested by our algorithm, follows the service demand very closely. Figure 3(d) illustrates the average distance between clients and their servers at runtime and it is comparable to the centralized greedy solution. These results suggests our algorithm will perform well in practical settings.

7 Related Work

Placing services in large-scale shared service hosting infrastructures has been studied both theoretically and experimentally. Oppenheimer et. al. [26] studied the problem of placing applications on PlanetLab. They discovered that CPU and network usage can be heterogenous and time-varying among the Planet-Lab nodes. Furthermore, a placement decision can become sub-optimal within 30 minutes. They suggested that dynamic service migration can potentially improve the system performance. They also studied a simple load-sensitive service placement strategy and showed it significantly outperforms a random placement strategy. NetFinder [27] is a service discovery system for PlanetLab that employs a greedy heuristic for service selection, based on CPU and bandwidth usage.

Theoretically, Laoutaris et. al. formulated SPP as a uncapacitated facility location problem (UFLP) [10], and presented a local search heuristic for improving the quality of service placement solutions. There are several differences between their work and ours. First, server capacity, such as CPU and bandwidth capacity, are considered in our formulation but not in theirs. Second, their heuristic does not provide a worst-case performance guarantee.

Our problem is also related to the replica placement problem, which has been studied extensively in the literature, primarily in the context of content delivery networks. Most of the early work on this problem focus on centralized cases where the network topology is static [6,24,28,29]. More recent work has also studied dynamic cases [11,30], in which iterative improvement algorithms are proposed. However, none of the above work was able to provide a theoretical worst-case performance guarantee. Our algorithm is as efficient as the above algorithms, and achieves theoretical performance ratio at the same time. A potential research direction is on improving this theoretical performance guarantee.

Another problem related to SPP is the web application placement in data centers. The objective is to achieve load balancing so as to reduce total response time. Several heuristics have been proposed for the problem [31,32], using local algorithms. The main difference between their work and ours is that we consider the distance (which could be latency or hop count) between clients and the servers in our formulation. This is not a concern in their work, since they assume all the servers are located in a single data center.

8 Conclusion

Large-scale shared service hosting infrastructures have emerged as popular platforms for deploying large-scale distributed services. One of the key problems in managing such a service hosting platform is the placement of service instances as to minimize operating costs, especially in the presence of changing network and system conditions. In this paper, we have presented a distributed algorithm for solving this problem. Our algorithm not only provides a constant approximation guarantee, but is also simple to implement. Furthermore, it can automatically adapt to dynamic network and system conditions.

As part of our future work, we would like to extend our work to handle more complex scenarios such as services with interdependencies. We would also like to improve our approximation guarantee. Furthermore, we are also interested in developing a control-theoretic framework for SPP. Most importantly, we would like to conduct a realistic experimentation on a real service hosting infrastructures such as PlanetLab.

References

1. Wang, L., Park, K., Pang, R., Pai, V., Peterson, L.: Reliability and security in codeen content distribution network. In: Proc. USENIX (2004)
2. Park, K., Pai, V.: Scale and performance in the coblitz large-file distribution service. In: Proc. of NSDI (2006)
3. Amazon elastic computing cloud (amazon ec2), http://aws.amazon.com/ec2/
4. Google app engine, http://code.google.com/appengine/
5. Azure services platform, http://www.microsoft.com/azure/default.mspx
6. Tang, X., Jianliang, X.: On replica placement for qos-aware content distribution. In: Proc. of IEEE INFOCOM (2004)
7. Tobias, F., et al.: The effect of latency and network limitations on mmorpgs: a field study of everquest 2. In: Proc. of ACM SIGCOMM Workshop NetGame (2005)
8. Church, K., Greenberg, A., Hamilton, J.: On delivering embarrassingly distributed cloud services. In: ACM HotNets (2008)
9. Valancius, V., Laoutaris, N., Massoulie, L., Diot, C., Rodriguez, P.: Greening the internet with nano data centers. In: ACM CoNext (2009)
10. Laoutaris, N., et al.: Distributed placement of service facilities in large-scale networks. In: Proc. of IEEE INFOCOM (2007)
11. Vicari, C., Petrioli, C., Presti, F.: Dynamic replica placement and traffic redirection in content delivery networks. In: Proc. of MASCOTS (2007)
12. Pathan, A.-M., Buyya, R.: A taxonomy and survey of content delivery networks. Technical Report, University of Melbourne, Australia (2006)
13. Jelasity, M., Montresor, A., Babaoglu, O.: Gossip-based aggregation in large dynamic networks. ACM Trans. on Computer Systems 23
14. Bodik, P., et al.: Statistical machine learning makes automatic control practical for internet datacenters. In: Proc. of USENIX HotCloud (2009)
15. Pal, M., Tardos, T., Wexler, T.: Facility location with nonuniform hard capacities. In: Proceedings of FOCS (2001)
16. Zhang, J., Chen, B., Ye, Y.: Multi-exchange local search algorithm for capacitated facility location problem. In: Math. of Oper. Research (2004)
17. Frank, C., Romer, K.: Distributed facility location algorithms for flexible configuration of wireless sensor networks. In: DCOSS (2007)
18. Moscibroda, T., Wattenhofer, R.: Facility location: Distributed approximation. In: ACM PODC (2005)
19. Dabek, F., Cox, R., Kaashoek, F., Morris, R.: Vivaldi: A decentralized network coordinate system. In: ACM SIGCOMM 2004 (2004)

20. Francis, P., et al.: An architecture for a global internet host distance estimation service. In: IEEE INFOCOM (1999)
21. Gortz, S., Klose, A.: Analysis of some greedy algorithms for the single-sink fixed-charge transportation problem. Journal of Heuristics (2007)
22. Gtitm homepage, http://www.cc.gatech.edu/projects/gtitm/
23. Ratnasamy, S., Handley, M., Karp, R., Scott, S.: Topologically-aware overlay construction and server selection. In: IEEE INFOCOM (2002)
24. Qiu, L., Padmandabhan, V., Geoffrey, V.: On the placement of web server replicas. In: IEEE INFOCOM (2001)
25. Spring, N., Mahajan, R., Wetherall, D., Anderson, T.: Measuring isp topologies with rocketfuel. IEEE/ACM Transactions on Networking, TON (2009)
26. Openheimer, D., Chun, B., Patterson, D., Snoeren, A., Vahdat, A.: Service placement in a shared wide-area platform. In: Proc. of USENIX (2006)
27. Zhu, Y., Mostafa, A.: Overlay network assignment in planetlab with netfinder. Technical Report (2006)
28. Szymaniak, M., Pierre, G., van Steen, M.: Latency-driven replica placement. In: Proc. of Symoposum on Applications and Internet (2005)
29. Karlsson, M., Karamanolis, C.: Choosing replica placement heuristics for wide-area systems. In: International Conference on Distributed Computing Systems, ICDCS (2004)
30. Presti, F., Bartolini, N., Petrioli, C.: Dynamic replica placement and user request redirection in content delivery networks. In: IEEE International Conference on Communications, ICC (2005)
31. Carrera, D., Steinder, M., Torres, J.: Utility-based placement of dynamic web application with fairness goals. In: IEEE/IFIP Network Operations and Management Symposium, NOMS (2008)
32. Tang, S.M., Chunqiang, Spreitzer, M., Pacifici, G.: A scalable application placement controller for enterprise data centers. In: International World Wide Web Conference (2007)

A   Proof of Theorem 1

Let S^* denote the facility set in the optimal solution. We proof our algorithm achieves an approximation ratio of 27.

Lemma 4. $cp(s,o) \leq 3(cp(i,s) + cp(i,o))$ *for any client i and two facilities s,o.*

Proof. By definition, $cp(s,o) = a((d(s,o) + \frac{\mu_o}{1-l_o/cap_o})/d_{max})^2$. Using $d(s,o) \leq d(i,s) + d(i,o)$ to expand this equation, and applying the fact $a^2 + b^2 \geq 2ab$ for any $a, b \in R$, the lemma is proven.

Lemma 4 essentially states that the triangular inequality is 3-satisfied in SPP.

Lemma 5. $c_s(S) \leq c_f(S^*) + 3c_s(S^*)$ *of a local solution S.*

Proof. Similar to lemma 4.1 in [15], except we use the fact the triangular inequality is 3-satisfied.

Now we bound the facility opening cost $c_f(S)$. Similar to [15], we select a set of operations (add, open and close) to convert S to S^*. The key difference between our approach and theirs is that we use greedy algorithms to approximate the optimal open and close operations. Recall that from lemma 2 and 3, $2p_s + \sum_{t \in T'}(u_t cp(s,t) - p_t) \geq 0$ for all feasible open(s,T) operations, and $p_s \leq \sum_{t \in T}(p_t + u_t cp(s,t))$ for all close(s,T) operations. We say an open operation opens s twice, but close each facility in T once. Similarly, an close(s,T) operation close s once and opens each facility in T twice. Hence, using the same set of operations, we can show the following result:

Lemma 6. *With our greedy algorithm 2 and 3, the same set of operations in [15] opens p_t at most 14 times for each $t \in S^*$, and closes p_s exactly once for each $s \in S \backslash S^*$.*

Proof. Follow the same analysis as in Lemma 5.1 of [15], except we use lemma 2 and 3 to bound the cost of open and close operations.

Lemma 7. *The total reassignment cost of all selected operations is at most $2 \cdot 3(c_s(S^*) + c_s(S))$.*

Proof. This proof is identical to Lemma 5.2 in [15], except we use the fact the triangular inequality is 3-satisfied.

Finally, we bound our approximation factor:

Proof (Of Theorem 1). Based on Lemma 6 and 7, the sum of all the inequalities is at most $14c_f(S^*) - c_f(S - S^*) + 2 \cdot 3(c_s(S) + c_s(S^*))$. hence

$$14c_f(S^*\backslash S) - c_f(S\backslash S^*) + 6(c_s(S) + c_s(S^*)) \geq 0$$

Adding $c_s(S) + c_f(S \cap S^*)$ to both sides and using $c_s(S) \leq 3c_s(S^*) + c_f(S^*)$ proved by lemma 5, we obtain $c(S) \leq 23c_f(S^*) + 27c_s(S^*)$.

Evaluating the Impact of a Novel Warning Message Dissemination Scheme for VANETs Using Real City Maps

Francisco J. Martinez[1], Manuel Fogue[1], Manuel Coll[1], Juan-Carlos Cano[2], Carlos T. Calafate[2], and Pietro Manzoni[2]

[1] University of Zaragoza, Spain
{f.martinez,m.fogue,m.coll}@unizar.es
[2] Universidad Politécnica de Valencia, Spain
{jucano,calafate,pmanzoni}@disca.upv.es

Abstract. In traffic safety applications for *Vehicular ad hoc networks* (VANETs), warning messages have to be disseminated in order to increase the number of vehicles receiving the traffic warning information. Hence, redundancy, contention, and packet collisions due to simultaneous forwarding (usually known as the broadcast storm problem) are prone to occur. In the past, several approaches have been proposed to solve the broadcast storm problem in multi-hop wireless networks such as *Mobile ad hoc Networks* (MANETs). Among them we can find counter-based, distance-based, location-based, cluster-based, and probabilistic schemes. In this paper, we present the *enhanced Street Broadcast Reduction* (eSBR), a novel scheme for VANETs designed to mitigate the broadcast storm problem in real urban scenarios. We evaluate the impact that our scheme has on performance when applied to VANET scenarios based on real city maps.

Keywords: VANETs, real city maps, message dissemination.

1 Introduction

Vehicular ad hoc networks (VANETs) are wireless communication networks that do not require any sort of fixed infrastructures, offering a novel networking paradigm to support cooperative driving applications on the road. VANETs are characterized by: (a) constrained but highly variable network topology, (b) specific speed patterns, (c) time and space varying communication conditions (e.g., signal transmissions can be blocked by buildings), (d) road-constrained mobility patterns, and (e) no significant power constraints.

VANETs have many possible applications, ranging from inter-vehicle communication and file sharing, to obtaining real-time traffic information (such as jams and blocked streets), etc. In this work we focus on traffic safety and efficient warning message dissemination, where the objective is to reduce the latency and to increase the accuracy of the information received by nearby vehicles when there is a dangerous situation.

M. Crovella et al. (Eds.): NETWORKING 2010, LNCS 6091, pp. 265–276, 2010.
© IFIP International Federation for Information Processing

In urban vehicular wireless environments, an accident can cause many vehicles to send warning messages, and all vehicles within the transmission range will receive the broadcast transmissions and rebroadcast these messages. Hence, a broadcast storm (serious redundancy, contention and massive packet collisions due to simultaneous forwarding) will occur and must be reduced [1]. In the past, several schemes have been proposed to reduce the broadcast storm problem. However, they have been only validated using simple scenarios such as a highway (several lanes, without junctions) [2,3], or a Manhattan-style grid scenario [4].

In this work, we propose a novel scheme called *enhanced Street Broadcast Reduction* (eSBR), which uses location and street map information to facilitate the dissemination of warning messages in 802.11p based VANETs. We evaluate the performance of our eSBR algorithm in a realistic urban scenario, that is, with a complex set of streets and junctions, and demonstrate how it could improve performance.

This paper is organized as follows: Section 2 reviews the related work on the broadcast storm problem in wireless ad hoc networks. Section 3 presents a brief description of the eSBR scheme and how it works in a real map scenario. Section 4 presents the simulation environment. Simulation results are then discussed in Section 5. Finally, Section 6 concludes this paper.

2   Related Work

In VANETs, intermediate vehicles act as message relays to support end-to-end vehicular communications. For applications such as route planning, traffic congestion control, and traffic safety, the flooding of broadcast messages commonly occurs. However, flooding can result in many redundant rebroadcasts, heavy channel contention, and long-lasting message collisions [1]. Over the years, several schemes have been proposed to address the broadcast storm problem in wireless networks. They are:

1. The *Counter-based scheme* [1]. To mitigate broadcast storms, this scheme uses a threshold C and a counter c to keep track of the number of times the broadcast message is received. Whenever $c \geq C$, rebroadcast is inhibited.
2. The *Distance-based scheme* [1]. In this scheme, authors use the relative distance d between vehicles to decide whether to rebroadcast or not. It is demonstrated that when the distance d between two vehicles is short, the *additional coverage* (AC) of the new rebroadcast is lower, and so rebroadcasting the warning message is not recommended. If d is larger, the additional coverage will also be larger.
3. The *Location-based scheme* presented in [1] is very similar to the distance-based scheme, though requiring more precise locations for the broadcasting vehicles to achieve an accurate geometrical estimation (with convex polygons) of the AC of a warning message. Since vehicles usually have GPS systems on-board, it is possible to estimate the additional coverage more precisely. The main drawback for using this scheme is the high computational cost of

calculating the AC, which is related to calculating many intersection areas among several circles.

4. The *weighted p-persistence*, the *slotted 1-persistence*, and the *slotted p-persistence* techniques presented in [5] are some of the few rebroadcast schemes proposed for VANETs. These three probabilistic and timer-based broadcast suppression techniques are not designed to solve the broadcast storm problem, but they can mitigate the severity of the storm by allowing nodes with higher priority to access the channel as quickly as possible. These schemes are specifically designed for use in highway scenarios.

5. *The Last One* (TLO) scheme, presented in [2], tries to reduce the broadcast storm problem finding the most distant vehicle from the warning message sender, so this vehicle will be the only allowed to retransmit the message. This method uses GPS information from the sender vehicle and the possible receivers to calculate the distance. Although it brings a better performance than simple broadcast, this scheme is only effective in a highway scenario because it does not take into account the effect of obstacles (e.g. buildings) in urban radio signal propagation. Moreover, GPS information must be accurate to achieve good results, and it is not clearly stated how a node knows the position of nearby vehicles at any given time.

6. The TLO approach was extended using a protocol which utilizes adaptive wait-windows and adaptive probability to transmit, named *Adaptive Probability Alert Protocol* (APAL) [3]. This scheme shows even better performance than the TLO scheme, but it is also only validated in highway scenarios.

Note that all these existing schemes alleviate the broadcast storm problem by inhibiting certain vehicles from rebroadcasting, reducing message redundancy, channel contention, and message collisions. In particular, they inhibit vehicles from rebroadcasting when the *additional coverage* (AC) area is very low. In [1], the authors demonstrated that a rebroadcast can only provide up to 61% additional coverage over that area already covered by the previous transmission in the best case (on average, the additional area is of 41%).

3 The Enhanced Street Broadcast Reduction (eSBR) Scheme in Real Maps

In this section, we present the *enhanced Street Broadcast Reduction scheme* (eSBR) - our novel proposal to reduce the broadcast storm problem in real urban scenarios. In urban scenarios, and at the frequency of 5.9 GHz (i.e., the frequency band adopted by the 802.11p standard), radio signals are highly directional and will experience a very low depth of penetration. Hence, in most cases, buildings will absorb radio waves at this frequency, making communication only possible when the vehicles are in line-of-sight.

In our simulations, vehicles operate in two modes: (a) warning, and (b) normal. Warning mode vehicles inform other vehicles about their status by sending warning messages periodically (every T_w seconds). These messages have the

---

Algorithm 1. eSBR_Send()

$P_w = $ AC3;   // set the highest priority
$P_b = $ AC1;   // set default priority
$ID = 0;$ // initialize sequence number of messages
while *(1)* **do**
    if *(vehicle$_i$ is in warning mode)* **then**
        create message m;
        set m.priority $= P_w$;
        set m.seq_num $= ID++$;
        broadcast warning message (m);
        sleep (T_w);
    else
        create message m;
        set m.priority $= P_b$;
        broadcast beacon (m);
        sleep (T_b);

---

highest priority at the MAC layer. Normal mode vehicles enable the diffusion of these warning packets and, periodically (every T_b seconds), they also send *beacons* with information such as their positions, speed, etc. These periodic messages have lower priority than warning messages and are not propagated by other vehicles. With respect to warning messages, each vehicle is only allowed to propagate them once for each sequence number, i.e., older messages are dropped.

Algorithms 1 and 2 describe our eSBR scheme, where *vehicle$_i$* indicates each vehicle in the scenario; m indicates each message sent or received by each vehicle; *warning* represents a warning message generated by a warning mode vehicle; *beacon* represents a normal message generated by an normal vehicle; T_w is the interval between two consecutive warning messages; T_b is the interval between two consecutive normal messages; P_w indicates the priority of the warning messages and P_b indicates the priority of the normal messages.

When *vehicle$_i$* starts the broadcast of a message, it sends m to all its neighbors. When another vehicle receives m for the first time, it rebroadcasts it by further relaying m to its neighbors. Depending on their characteristics, every vehicle repeats *send(warning)* or *send(beacon)* operations periodically with different periods (T_w and T_b, respectively). When a new message m is received, the vehicle tests whether m has already been received. To achieve this, each vehicle maintains a list of message IDs. An incoming warning message ID is inserted in the list if m is received for the first time (i.e. its ID has not been previously stored in the list), and if so it is rebroadcasted to the surrounding vehicles only when the distance d between sender and receiver is higher than a distance threshold D, or the receiver is in a different street than the sender. We consider that two vehicles are in a different street when: (i) both are indeed in different roads (this information is obtained by on-board GPS systems with integrated street maps), or (ii) the receiver, in spite of being in the same street, is near to an intersection. Hence, warnings can be rebroadcasted to vehicles which are traveling on other

---

Algorithm 2. eSBR_OnRecv()

---

for *(every received message)* **do**

    if *(m* is *is a warning* **and** *m.seq_num received for the first time)* **then**

        if *(distance between sender and receiver > D* **or** *both vehicles are in different streets)* **then**

            | rebroadcast(*m*);

        else

            discard(*m*);

            /* warnings are only rebroadcasted when additional coverage area is high or they can be propagated to different streets */

    else

        discard(*m*);

        // duplicated warnings and beacons are not rebroadcasted

---

streets, overcoming the radio signal interference due to the presence of buildings. If the message is a *beacon*, it is simply discarded since we are not interested in the dissemination of beacons.

Figure 1 shows an example in a real map scenario where shaded polygons represent buildings. When vehicle A broadcasts a warning message, it is only received by neighboring vehicles B, C, and D because buildings interfere with the radio signal propagation. In this situation, if we use distance or location-based

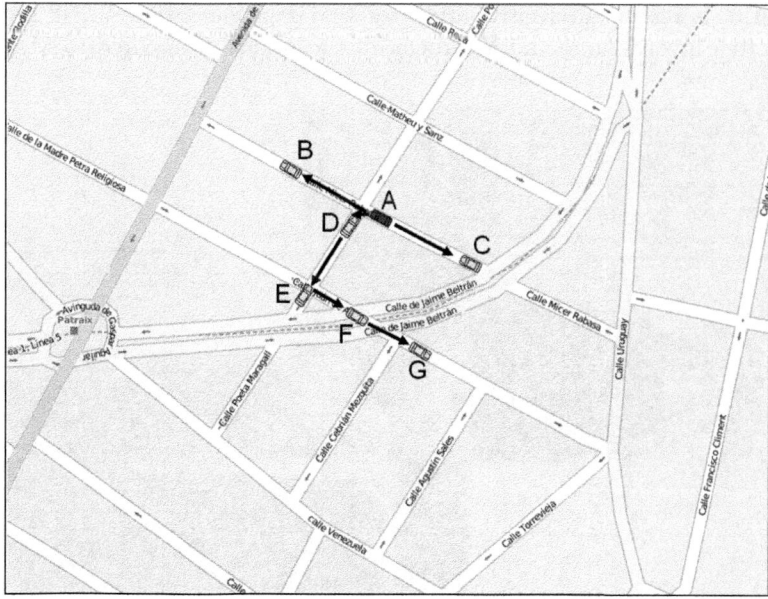

Fig. 1. The enhanced Street Broadcast Reduction scheme: example scenario taken from the city of Valencia in Spain

schemes, vehicles B, C, and D will rebroadcast the message only if distances $d1$, $d2$ and $d3$, respectively, are large enough (i.e., the distance is larger than a distance threshold D), or its additional coverage areas are wide enough (i.e., the AC is larger than a coverage threshold A). So, supposing that only vehicle C meets this condition, the warning message could still not be propagated to the rest of vehicles (i.e., E, F, and G).

Our eSBR scheme solves this problem as follows. In eSBR, vehicle D will rebroadcast the warning message since vehicle D is in a different street than vehicle A. In this way, the warning message will arrive to all the vehicles represented in only four hops. In modern *Intelligent Transportation Systems* (ITS), vehicles are equipped with on-board GPS systems containing integrated street maps. Hence, location and street information can be readily used by eSBR to facilitate dissemination of warning messages. When the additional coverage area is wide enough, vehicles will rebroadcast the received warning message. However, when the additional coverage area is very low, vehicles will rebroadcast warning messages only if they are in a different road.

Note that distance and location schemes can be very restrictive, especially when buildings interfere with radio signal propagation. Without eSBR, warning messages will not arrive at vehicles E, F and G due to the presence of buildings.

4   Simulation Environment

Deploying and testing VANETs involves high cost and intensive labor. Hence, simulation is a useful alternative prior to actual implementation. VANET simulations often involve large and heterogeneous scenarios. Compared to MANETs,

Fig. 2. Images of the experiment to determine the radio signal attenuation due to distance between vehicles

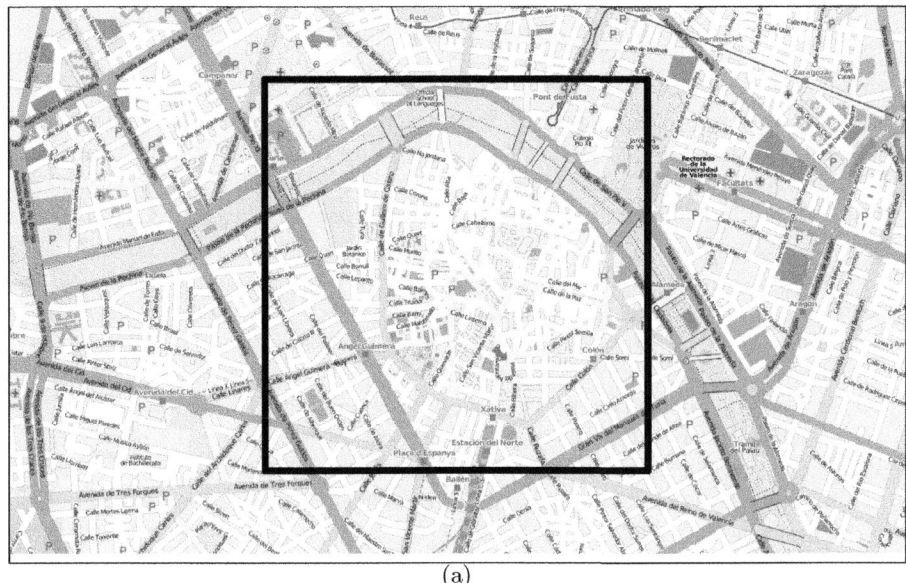

(a)

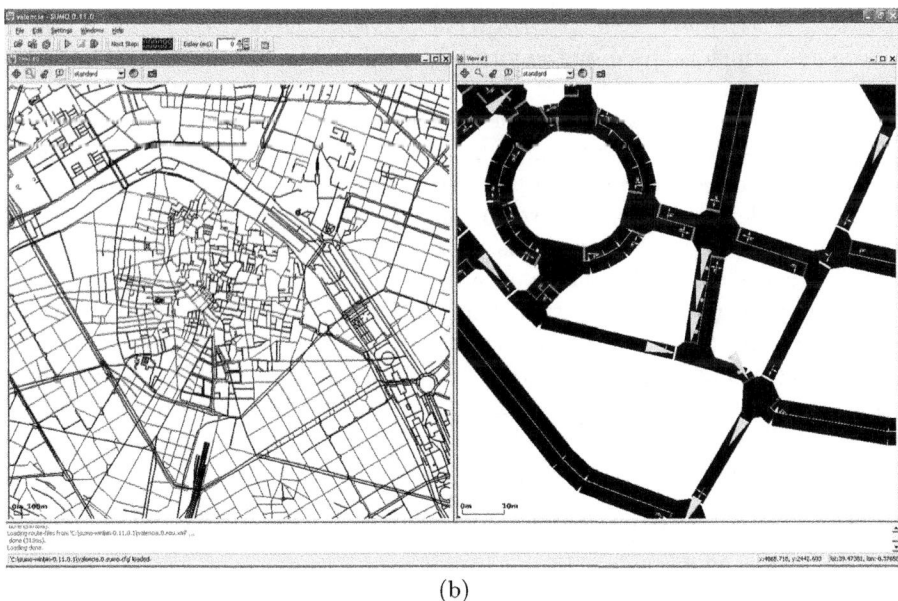

(b)

Fig. 3. Simulated scenario of Valencia city: (a) OpenStreetMap layout, and (b) the SUMO converted version. The box shows our simulated area.

VANET simulations must account for some extra characteristics that are specific to vehicular environments [6].

In this section, we present our VANET simulation setup. Simulation results presented in this paper were obtained using the ns-2 simulator. We modified the simulator to follow the upcoming *Wireless Access in Vehicular Environments* (WAVE) standard closely. Achieving this requires extending the ns-2 simulator to implement IEEE 802.11p, which is a draft amendment to the IEEE 802.11 standard that defines enhancements to support *Intelligent Transportation Systems* (ITS) applications.

In terms of the physical layer, the data rate used for packet broadcasting was fixed at 6 Mbit/s, i.e., the maximum rate for broadcasting in 802.11p when assuming a 20 MHz channel. The MAC layer is based on the IEEE 802.11e *Enhanced Distributed Channel Access* (EDCA) *Quality of Service* (QoS) extensions. Therefore, application messages are categorized into different *Access Categories* (ACs), where AC0 has the lowest, and AC3 the highest priority. The contention parameters used for the *Control Channel* (CCH) are shown in [7]. Thus, in our warning message dissemination mechanism, warning messages have the highest priority (AC3) at the MAC layer, while *beacons* have lower priority (AC1). Moreover, since we are simulating real city maps with buildings, we have modified the ns-2 simulator to model the impact of distance and obstacles in signal propagation. The Radio propagation model used was the *Real Building and Distance Attenuation Model* (RBDAM) model, a new model we have implemented inside ns-2 which is based on the formerly proposed BDAM model [8]. RBDAM considers the signal attenuation due to the distance between vehicles based on real data obtained from experiments in the 5.9 GHz frequency band using the IEEE 802.11a standard (see Figure 2), with a maximum transmission range of 400 meters, and also accounts for the presence of buildings (i.e., communication among vehicles is only possible when they are within line-of-sight).

To perform realistic simulations, it is specially important that the chosen mobility generator will be able to obtain a detailed microscopic traffic simulation and to import network topologies from real maps. Our mobility simulations are performed with SUMO [9], an open source traffic simulation package which has interesting microscopic traffic capabilities such as: collision free vehicle movement, multi-lane streets with lane changing, junction-based right-of-way rules, traffic lights, etc. (see Figure 3b). SUMO can also import maps directly from map databases such as OpenStreetMap [10] and TIGER [11].

The simulated topology represents the downtown area of Valencia city in Spain, and it was obtained from OpenStreetMap. Figure 3 shows the layout used in our simulations.

5 Simulation Results

In this section, we perform a detailed analysis to evaluate the impact of the proposed eSBR scheme on the overall system performance. We compare the impact of our scheme in two different scenarios: a Manhattan grid scenario, and a realistic city map, i. e., the city of Valencia in Spain. Since performance results are highly related to the specific scenarios used, and due to the random nature of the mobility model, we performed fifteen simulations to obtain reasonable confidence intervals. All the results shown here have a 90% confidence interval. Each simulation lasted for 450 seconds. In order to achieve a stable state before gathering data traffic, we only started to collect data after the first 60 seconds.

We evaluated the following performance metrics: (a) percentage of blind vehicles, (b) warning notification time, and (c) number of packets received per vehicle. The percentage of blind vehicles is the percentage of vehicles that do not receive the warning messages sent by accidented vehicles. These vehicles remain blind because of their positions, due to packet collisions, or due to signal propagation limitations. The warning notification time is the time required by normal vehicles to receive a warning message sent by a "warning mode" vehicle (a vehicle that broadcasts warning messages). Table 1 shows the simulation parameters used.

In this paper, we evaluated the performance of our eSBR proposed scheme with respect to: (i) a distance-based scheme, and (ii) a location-based scheme, using two different scenarios: (i) a Manhattan grid-style, and (ii) a real map of Valencia, a city from Spain, using an average density of 75 vehicles/km². The impact of other parameters affecting VANET warning message dissemination, such as the density of vehicles, the priority and periodicity of messages, etc., was previously studied in [12].

Table 1. Parameter values for the simulations

Parameter	Value
number of vehicles	300
map area size	$2000m \times 2000m$
maximum speed	$50\ km/h$
distance between streets (in Manhattan)	$100m$
number of warning mode vehicles	3
warning packet size	$256bytes$
normal packet size	$512bytes$
packets sent by vehicles	$1\ per\ second$
warning message priority	$AC3$
normal message priority	$AC1$
MAC/PHY	802.11p
Radio Propagation Model	RBDAM
maximum transmission range	$400m$
eSBR distance threshold (D)	$300m$

Figure 4 shows the warning notification times obtained. When the simulation scenario is a Manhattan grid (see Figure 4a), the system needs only one second to inform at least 75% of the vehicles for all the schemes, and the message dissemination process ends in less than 2 seconds. The percentage of vehicles receiving the warning message is an 8% higher when using eSBR compared to both the distance-based and the location-based schemes. Concerning the

Manhattan scenario, warning messages reach longer distances because of the lack of direction changes in the straight streets that characterize this map.

When simulating a real map scenario (see Figure 4b), some noticeable differences appear. First of all, the percentage of vehicles receiving warning messages decreases from about the 90-95% in a Manhattan scenario to less than 50%, since the complexity of the layout and the arrangement of the buildings of real maps significantly interfere with the signal propagation process. Thus, the highly directional radio signal reduces drastically the probability of reaching longer distances. As for the warning notification time, we obtained better results using the eSBR scheme. The warning messages reach 40% of the vehicles in 0.45 seconds. Also notice that the message dissemination process ends before 1 second for all schemes. This time is 50% lower than the obtained in the Manhattan scenario because, in a real scenario, the information is only spread to nearby vehicles.

As for the percentage of blind nodes, Figure 5a shows how it will largely depend on the chosen simulation scenario. There are few blind vehicles when simulating a Manhattan scenario for all schemes (less than 15%), and the percentage of blind nodes is reduced by half when using eSBR. Nevertheless, less than 50% of the vehicles are aware of warning messages when simulating a real city map. If we use this more realistic scenario we find that, when the eSBR scheme is adopted, the percentage of vehicles which receive warning messages increases by 8% when compared to the distance and location-based techniques.

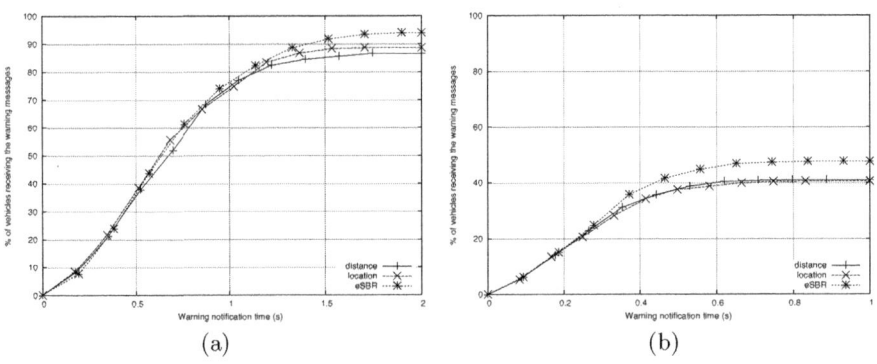

Fig. 4. Average propagation delay when varying the simulation scenario: (a) using a Manhattan layout, and (b) using a real map

Finally, Figure 5b shows the total number of packets received per vehicle, which is a measure of the degree of contention in the channel. In the Manhattan scenario the signal propagates easily due to the streets' position, and so the vehicles receive many duplicated messages. However, in a real map scenario, the more complex layout makes difficult signal propagation, and so the number of messages received decreases. These messages are received only by vehicles which are likely to face the dangerous situation, i.e. they are in the same street or in

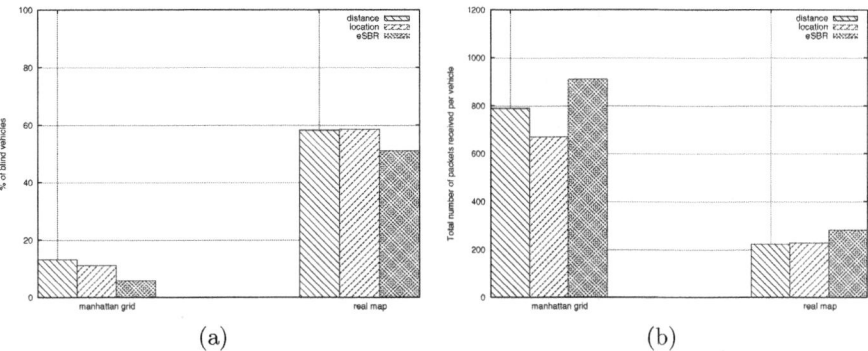

Fig. 5. (a) Percentage of blind vehicles vs. scenario layout, and (b) Total number of packets received per vehicle vs. scenario layout, both accounting for a Manhattan and a real map scenario

nearby ones. When simulating the real city map, the number of packets received per vehicle is almost the same for the distance and location-based schemes; when using eSBR the number of packets received increases slightly, but the percentage of vehicles which receive warning messages increases to a greater extent.

Authors in [1] demonstrated that the location-based scheme was more efficient than the distance-based scheme, since it reduces redundancy without compromising the number of vehicles receiving the warning message. Nevertheless, the main drawback for using this scheme is the high computational cost of calculating the additional coverage. Our simulation results demonstrate that eSBR outperforms both these schemes without introducing much complexity and calculations.

6   Conclusion

Achieving efficient dissemination of messages is of utmost importance in vehicular networks to warn drivers of critical road conditions. However, broadcasting of warning messages in VANETs can result in increased channel contention and packet collisions due to simultaneous message transmissions. In this paper, we introduce the *enhanced Street Broadcast Reduction* (eSBR) scheme to reduce broadcast storm in real map urban scenarios and to improve the performance of warning message dissemination. Simulation results show that eSBR outperforms other schemes in high density urban scenarios, yielding a lower percentage of blind vehicles while drastically alleviating the broadcast storm problem, being thus suitable for real scenarios. Our experiments also highlight that the message propagation behavior in realistic scenarios based on maps of actual cities differs greatly from more traditional Manhattan-style scenarios. Thus, we consider that the results obtained using unrealistic scenarios should be revised, and we recommend the adoption of real maps whenever possible.

Acknowledgments

This work was partially supported by the *Ministerio de Educación y Ciencia*, Spain, under Grant TIN2008-06441-C02-01, and by the *Fundación Antonio Gargallo*, under Grant 2009/B001.

References

1. Tseng, Y.-C., Ni, S.-Y., Chen, Y.-S., Sheu, J.-P.: The broadcast storm problem in a mobile ad hoc network. Wireless Networks 8, 153–167 (2002)
2. Suriyapaibonwattana, K., Pomavalai, C.: An effective safety alert broadcast algorithm for VANET. In: International Symposium on Communications and Information Technologies (ISCIT), October 2008, pp. 247–250 (2008)
3. Suriyapaiboonwattana, K., Pornavalai, C., Chakraborty, G.: An adaptive alert message dissemination protocol for VANET to improve road safety. In: IEEE Intl. Conf. on Fuzzy Systems (FUZZ-IEEE), August 2009, pp. 1639–1644 (2009)
4. Korkmaz, G., Ekici, E., Ozguner, F., Ozguner, U.: Urban multi-hop broadcast protocols for inter-vehicle communication systems. In: Proceedings of First ACM Workshop on Vehicular Ad Hoc Networks (VANET 2004) (October 2004)
5. Wisitpongphan, N., Tonguz, O.K., Parikh, J.S., Mudalige, P., Bai, F., Sadekar, V.: Broadcast storm mitigation techniques in vehicular ad hoc networks. IEEE Wireless Communications 14, 84–94 (2007)
6. Martinez, F.J., Toh, C.-K., Cano, J.-C., Calafate, C.T., Manzoni, P.: A survey and comparative study of simulators for vehicular ad hoc networks (VANETs). Wireless Communications and Mobile Computing (October 2009)
7. Eichler, S.: Performance evaluation of the IEEE 802.11p WAVE communication standard. In: Proceedings of the Vehicular Technology Conference (VTC 2007 Fall), Baltimore, MD, USA (September 2007)
8. Martinez, F.J., Toh, C.-K., Cano, J.-C., Calafate, C.T., Manzoni, P.: Realistic Radio Propagation Models (RPMs) for VANET Simulations. In: IEEE Wireless Communications and Networking Conference (WCNC), Budapest, Hungary (2009)
9. SUMO, Simulation of Urban MObility (2009), http://sumo.sourceforge.net
10. OpenStreetMap, collaborative project to create a free editable map of the world (2009), http://www.openstreetmap.org
11. TIGER, Topologically Integrated Geographic Encoding and Referencing (2009), http://www.census.gov/geo/www/tiger
12. Martinez, F.J., Cano, J.-C., Calafate, C.T., Manzoni, P.: A Performance Evaluation of Warning Message Dissemination in 802.11p based VANETs. In: IEEE Local Computer Networks Conference (LCN), Switzerland (October 2009)

Resource Optimization Algorithm for Sparse Time-Driven Sensor Networks

María Luisa Santamaría, Sebastià Galmés, and Ramon Puigjaner

Dept. of Mathematics and Computer Science, Universitat de les Illes Balears
Cra. de Valldemossa, km. 7.5, 07122 Palma, Spain
{maria-luisa.santamaria,sebastia.galmes,putxi}@uib.es

Abstract. Time-driven sensor networks are devoted to the continuous reporting of data to the user. Typically, their topology is that of a data-gathering tree rooted at the sink, whose vertexes correspond to nodes located at sampling locations that have been selected according to user or application requirements. Thus, generally these locations are not close to each other and the resulting node deployment is rather sparse. In a previous paper, we developed a heuristic algorithm based on simulated annealing capable of finding an optimal or suboptimal data-gathering tree in terms of lifetime expectancy. However, despite the enhanced lifetime, the overall link distance is not optimized, fact that increases the need for additional resources (relay nodes). Therefore, in this paper we propose the Link Distance Reduction algorithm, with the goal of reducing link distances as long as network lifetime is preserved. The benefits of this new algorithm are evaluated in detail.

Keywords: Time-driven sensor networks, TDMA, spanning tree, minimum spanning tree, network planning.

1 Introduction

In *time-driven* or *continuous monitoring* sensor networks, communication is triggered by nodes, which regularly deliver sensed data to the user [1]. Thus, the traffic generated by these networks usually takes the form of a continuous and predictable flow, and therefore the use of a TDMA-based protocol becomes especially appropriate [2]. Furthermore, it is common that time-driven sensor networks are deployed in a structured manner [1][2], either by selecting strategic locations or by adopting some regular sampling pattern, and that data are routed to the sink through multi-hop predetermined paths [1]. These paths form a *reverse multicast* or *convergecast* structure called the *data-gathering tree* [2].

Because the strategic locations can well be far to each other or because the monitored variable can exhibit low spatial variability (as it is usually the case), it is not surprising that the resulting application-driven deployment is rather sparse. Sparse manually-deployed sensor networks have received less attention in literature (in contrast to dense and randomly-deployed networks), in spite of putting forward some interesting challenges. In essence, these come from the fact that large inter-node distances can be impractical or unattainable for sensor nodes. Inevitably, given the

M. Crovella et al. (Eds.): NETWORKING 2010, LNCS 6091, pp. 277–290, 2010.

limited communication range of nodes, any solution demands the introduction of additional resources, basically in the form of relay nodes (nodes with their sensing capability deactivated). This could be done by randomly scattering them over the whole sensor field, but this would yield to prohibitively large amounts of nodes precisely in sparse areas. Thus, we suggest that the problem can be better addressed from a network planning perspective, which tries to control the number of additional resources as long as the functionality and lifetime of the network are preserved.

As part of this strategy, in [3] we developed a heuristic method based on *simulated annealing* (SA) that finds an optimal (or suboptimal) data-gathering tree in terms of lifetime expectancy. However, in general the overall link distance of this tree can be substantially improved (decreased), fact that would contribute to reduce the amount of relay nodes required to make the network operational. Thus, in this paper we propose the *Link Distance Reduction* (LDR) algorithm, which starts from the data-gathering tree delivered by the SA-based algorithm, and then progressively reduces its overall link distance as long as the lifetime obtained from SA is not decreased.

Accordingly, the rest of the paper is organized as follows. In Section 2, we describe our network planning approach to the problem of topology construction for sparse deployments, along with related work. In Section 3, we motivate the need for an algorithm that reduces the sum of link distances of the SA graph. In Section 4, we describe the new algorithm (LDR) in detail. Then, in Section 5, we evaluate its performance via simulation. This also includes the issue of computational complexity. Finally, in Section 6, we draw the main conclusions and suggestions for further research.

2   Network Planning Approach and Related Work

The problem of constructing a data-gathering tree can be faced up from a network planning perspective or as part of a protocol-oriented design. In the latter case, a real-time distributed algorithm is designed, in order to be embedded into an adaptive routing protocol capable of supporting frequent topology updates. This strategy has been typically used in dense, randomly-deployed networks. In contrast, in the former case, the goal is to find an optimized static routing tree, and thus the algorithm to be designed is not subject to real-time requirements (although computation time is still of concern) and can be executed centrally. This approach becomes especially appropriate for sparsely-deployed sensor networks or for sensor networks where maintaining and updating a routing table at each node is computationally too expensive.

Despite its importance, the network planning approach has received less attention than the approach centered on adaptive routing. Yet, some works deserve a description. For instance, the work presented in [4] considers the problem of designing a data-gathering tree with the goal of reducing the total amount of data transmitted. Hence, in contrast to our premises (see Section 3), its main focus is on a packet aggregation scheme. Analogously, the emphasis of [5] is also on packet aggregation. A common feature to these works, as stated by [6], is that their ultimate goal is to minimize the total energy consumption, thus ignoring the fact that some nodes may deplete their energy faster than others and cause loss of coverage. Thus, the algorithm proposed in [6] focuses on maximizing the lifetime of the network, defined as the time until first node death (optimization with strict coverage requirements). However, this work adopts several assumptions that significantly simplify the analysis of the problem. One of them is that nodes cannot adjust their transmission power, which is not always the case. This assumption eliminates the dependence on distance and thus it reduces the complexity of the proposed algorithm. Another simplifying assumption is the adoption of a packet aggregation scheme that reduces the number of packets to be forwarded by a node to the number of direct descendents it has. Compared to this work, our particular approach adopts the same definition of lifetime, but it uses a more complete energy consumption model that does not rely on the abovementioned assumptions.

Another simplification present in [6] is the assumption that the initial network is fully connected. Since this is not necessarily true, especially in sparse deployments, the network planning perspective has also considered the possibility to introduce additional resources in the network, in the form of relay nodes. In this context, several works assume that nodes have a maximum transmission range, and then formulate the problem of introducing the fewest number of relay nodes so that a certain degree of network connectivity is achieved. Examples of these works are [7] and [8], which consider flat topologies, and [9] and [10], which assume a two-tiered architecture where relay nodes always belong to the backbone network. Essentially, the theoretical formulation of these topological problems is that of a *Steiner tree with minimum number of Steiner points* (ST-MSP), a problem that is known to be NP-hard. However, as stated above, the focus of these works is on network connectivity rather than on lifetime (at most, lifetime is partially addressed by considering transmission ranges and the so-called *node degrees*). By contrast, our network planning approach addresses both network lifetime and minimization of the number of Steiner points (resource optimization) by means of the following two stages:

- *Elementary tree construction.* The construction of an elementary data-gathering tree consists of determining a spanning tree that connects all regular or duty nodes (nodes at locations of interest) to the sink, with a view to maximizing the lifetime of the sensor network. In this process, we assume that these nodes have full power control capability with the only limit of their own energy resources (battery). In other words, the likely existence of a maximum transmission range is ignored at this stage. Under this assumption, the work presented in [3] showed that the problem of finding a data-gathering tree with optimal lifetime, by including in the analysis the workload of every node to a maximum extent, is NP-hard. Then, a heuristic method based on simulated annealing was proposed, which it was shown to yield, in general, a spanning tree with near-optimal lifetime in linear time.
- *Placement of additional nodes.* In the case of sparse deployments, the data-gathering tree that results from the previous stage usually contains link distances that exceed some maximum transmission range. In this case, our strategy consists of regularly inserting the necessary number of relay nodes into the uplink of every regular node. This is the technique of *steinerized edges*, which was also considered in [7], [8], [10] and [11]. In the latter, the number of inserted nodes was related to the corresponding lifetime enhancement.

3   Problem Motivation

The work presented in this paper is halfway between the two stages of the topology construction process described in the previous section. The reason is that although the

SA-based algorithm generates a spanning tree with optimal or suboptimal lifetime, in general the sum of the resulting link distances can be substantially decreased without penalizing such lifetime. This, in turn, may contribute to reduce the amount of additional resources required in the second stage. In this section, we illustrate these ideas by means of a simple test scenario; however, for the sake of completeness, we first set up a lifetime model for a time-driven sensor network with N nodes.

3.1 Lifetime Model

A lifetime model requires an energy consumption model. The energy consumption model used in the present paper was first developed in [3] (on the basis of a radio model described in [12]), under the following premises:

- Only the energy wasted by the node transceivers is considered.
- No data aggregation, since the spatial correlation among samples (readings) in sparse scenarios tends to be low. In addition, by ignoring data aggregation a worst-case lifetime analysis is performed.
- Some low duty-cycle MAC protocol, such as TDMA, governs the access of nodes to the wireless medium.
- Time is divided into rounds or reporting periods, whose duration is specified at the application or user levels and is typically much larger than the duration of packets and TDMA frames. In addition, this allows for neglecting the delays incurred by TDMA, whatever slot assignment scheme is used and despite the absence of a packet aggregation scheme.

Then, in order to characterize the traffic workload of each node $i = 1...N$, two magnitudes were introduced:

- $g(i)$: number of packets to be generated by node i per communication round (*generating parameter*). This specification takes into account heterogeneous scenarios where the sampling process needs to be more intensive in some locations than in others.
- $\sigma(i)$: number of packets to be forwarded by node i during every communication round (*forwarding parameter*).

Note that these magnitudes become time-independent when dealing with static data-gathering trees, as it is the case of the present paper. Note also that if $CH(i)$ denotes the set of child nodes of node i, the following equality holds:

$$\sigma(i) = \sum_{j \in CH(i)} g(j) + \sum_{j \in CH(i)} \sigma(j), \forall i . \tag{1}$$

Now, let $E_G(i)$ ($E_F(i)$) be the energy consumed by node i to generate and transmit (receive and forward) a packet. Then, the total energy consumed by this node during a communication round, namely $E(i)$, can be expressed as follows:

$$E(i) = g(i) \cdot E_G(i) + \sigma(i) \cdot E_F(i) , \forall i . \tag{2}$$

Furthermore, if E_R is the energy consumed by any node to receive a packet and $E_T(i)$ is the energy consumed by node i to transmit a packet at distance $d(i)$, we can set up the following equations:

$$E_G(i) = E_T(i) , \forall i . \tag{3a}$$

$$E_F(i) = E_R + E_T(i) , \forall i . \tag{3b}$$

The energy consumed to receive a packet is the energy dissipated by the node transceiver circuitry to perform such operation. Because all nodes are assumed to be provided by the same vendor, this energy can be considered constant throughout the network. On the other hand, the energy wasted to transmit a packet includes a distance-dependent component in addition to the energy dissipation, and thus it generally varies from node to node (in spite of being provided by the same vendor). These statements are made explicit via the so-called radio model [12]:

$$E_R = E_{elec} \cdot m . \tag{4a}$$

$$E_T(i) = E_{elec} \cdot m + E_w \cdot m \cdot d^f(i) , \forall i . \tag{4b}$$

Here, E_{elec} stands for the energy dissipated by the transceiver circuitry to transmit or receive a single bit, $E_w \cdot d^f(i)$ is the energy radiated to the wireless medium to transmit a single bit over a link of distance $d(i)$, f is the path-loss exponent and m is the packet size in bits. In turn, the distance-dependent component depends on the distance itself:

$$d \leq d_0 \Rightarrow E_w = E_{fs} , f = 2 . \tag{5a}$$

$$d > d_0 \Rightarrow E_w = E_{mp} , f > 2 . \tag{5b}$$

Equation (5a) models free space propagation, which is valid below the reference distance (d_0) (although beyond the so-called far-field region of the transmitting antenna [12]), whereas equation (5b) reflects multi-path fading effects. In this latter case, a typical value for the path-loss exponent is 4. By combining equations (2), (3a)-(3b) and (4a)-(4b), the total energy wasted by a node in every communication round can be rewritten in the following way:

$$E(i) = (g(i) + 2\sigma(i)) \cdot E_{elec} \cdot m + (g(i) + \sigma(i)) \cdot E_w \cdot m \cdot d^f(i) , \forall i . \tag{6}$$

Note that in this analysis the energy wasted to wake up a node has been neglected. This is an acceptable and usual approximation, which in addition makes the analysis independent of the particular slot assignment scheme in the case of TDMA. It can also be noticed that expression (6) reveals that the energy consumed by a node depends on three major factors: workload, represented by the traffic parameters $g(i)$ and $\sigma(i)$, target distance and packet size.

The energy consumption model given by (6) can now be embedded into a lifetime model. By assuming that lifetime is defined as the time until first node death, we can

---

express it in rounds as follows (B is the energy initially available at all nodes, that is, the battery):

$$L = \frac{B}{\max\{E(i), i = 1 \dots N\}} = \frac{B}{E(i)|_{max}}. \tag{7}$$

3.2 Illustrative Example

We consider the scenario detailed in Fig. 1, where 8 regular nodes are supposed to send 1 packet to the sink per round of communication, that is, $g(i) = 1, i = 1 \dots N$ (for simplicity, with no loss of generality). For the radio model, we take the data from a realistic radio module previously introduced in [13]: $E_{elec} = 50$ nJ/bit, $E_w = 10$ pJ/bit/m^2 and f = 2 if the transmission distance is below 75 m (the so-called reference distance), $E_w =$ 0.0013 pJ/bit/m^4 and $f = 4$ if the transmission distance is above 75 m, and $B = 15$ kJ. For the packet size, we assume a slightly large value of 125B, although the choice for this parameter is not relevant to our analysis. Under all these assumptions, Fig. 1 also shows the graph generated by the SA algorithm developed in [3]. This graph yields a lifetime of 722168 rounds, determined by node 8 – see expression (7). Besides, this graph reveals one of the main features of the SA algorithm: it focuses on maximizing the lifetime of the network as defined by (7), but not on enhancing the lifetimes or transmission distances of particular nodes unless this could have a positive impact on the final result. For instance, it is very probable that node 4 could have been connected to node 2 instead of node 5 without perturbing the lifetime of the network.

In this context, a reference algorithm is the Minimum Spanning Tree, which finds the spanning tree that minimizes the overall link distance (if link cost is defined as the Euclidean distance). In particular, when applied to the node deployment of Fig. 1, the result is the graph shown in Fig. 2 in solid lines. The average link distance has decreased from 207.970 m (with SA) to 148.302 m, which is the minimum achievable value. This represents a reduction of approximately 30%. However, the lifetime of the network has also decreased to 514451 rounds (this time determined by node 7), which is also about 30% smaller. The reason is that, by focusing exclusively on the total link distance, the MST algorithm ignores the workload supported by nodes and thus it does not optimize the lifetime of the network.

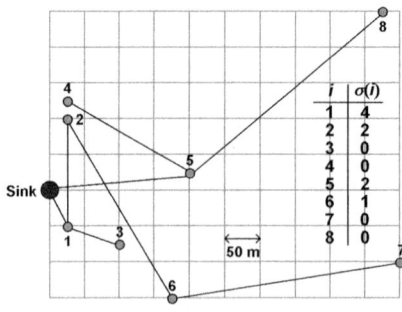

Fig. 1. Node deployment and graph generated by the SA algorithm. Forwarding parameters are indicated.

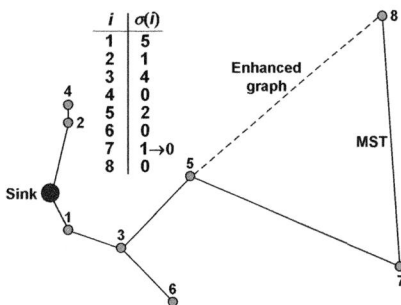

i	$\sigma(i)$
1	5
2	1
3	4
4	0
5	2
6	0
7	$1\to0$
8	0

Fig. 2. MST and enhanced graphs. Forwarding parameters are indicated, including the change of $\sigma(7)$ from one graph to the other.

Obviously, a question that arises is whether an intermediate solution is feasible. In this respect, the graph shown in Fig. 2 – with the dashed line instead of line 8-7, constitutes a positive answer. This graph, which results from applying the combined SA+LDR algorithm (LDR is described in the next section), exhibits the same lifetime as the SA graph (722168 rounds, again determined by node 8), but its average link distance is now 148.855 m, very close to the minimum achieved by the MST algorithm. In fact, it can be noticed that the two graphs in Fig. 2 are very similar, although their small dissimilarity has an important effect on network lifetime.

4   Link Distance Reduction Algorithm

Starting from the spanning tree generated by the SA algorithm, the proposed Link Distance Reduction algorithm is a heuristic iterative method that is intended to transform this tree into another one with less average link distance, as long as lifetime is not degraded. In essence, the strategy is to check the connections of all nodes to their parents, trying to find closer parents while preserving the tree structure (no cycles are formed) and maintaining or even increasing the lifetime of the network (recall that, in general, the SA algorithm yields a near-optimal solution).

Before proceeding to the specification of the algorithm, let us introduce some useful definitions and naming conventions:

- Let $t = (V, E)$ denote a particular spanning tree, where V is the set of $N+1$ vertices representing the N (regular) nodes and the sink, and E is a collection of two-element subsets of V that represent connections between pairs of nodes. Let T denote the set of all trees spanning all vertices in V.
- Given that t is a spanning tree, every edge in E can be expressed as a pair $(i, p(i)), i = 1 \ldots N$, where $p(i)$ denotes the parent of node i (that is, the node to which node i sends packets on the way to the sink). Thus, $p(i) = 0$ if node i is directly connected to the sink (it is assumed that node 0 is the sink).
- As stated above, the algorithm explores potential parents for each node. Thus, we introduce $p^*(i)$ to denote the parent that is tried for node i at a particular stage of the execution process.

- In our formal specification of the algorithm, a state corresponds to a particular spanning tree. Accordingly, three variables are managed. The first one is `Initial State`, which denotes the spanning tree obtained from the SA algorithm. The LDR algorithm starts from this state. Then, `Current State` and `New State` denote the spanning trees at the beginning and end of each iteration, respectively. Therefore, `New State` is different from `Current State` as long as at least one parent of any node is changed during the corresponding iteration.
- Let $d(i, j)$ be the Euclidean distance between any two vertices $i, j = 0 \ldots N$. Accordingly, let **A** be an NxN matrix such that $\mathbf{A}(u, v) = i$ and $\mathbf{A}(u, v') = j$ with $v < v'$ means that $d(u, i) \leq d(u, j)$, with $u, v, v' = 1 \ldots N$, $i, j = 0 \ldots N$ and $i, j \neq u$. This matrix depends exclusively on the deployment of the network and thus it needs to be calculated only once. Also, let **B** be a row vector with N components such that $\mathbf{B}(v) = i$ and $\mathbf{B}(v') = j$ with $v < v'$ means that $d(i, p(i)) \geq d(j, p(j))$, with $i, j, v, v' = 1 \ldots N$. Essentially, for a given spanning tree, **B** is the list of nodes in decreasing order of their uplink distances. Thus, it depends on the particular spanning tree. In the algorithm, **B** is re-calculated at the beginning of each iteration.
- Finally, `No Cycle` and `Lifetime Not Degraded` are two self-explanatory logical variables that represent the two conditions that must be fulfilled in order to accept a potential parent. The former implies determining that the connection of a node to a potential parent does not form any cycle. Under this condition, the latter implies the re-evaluation of network lifetime according to (6) and (7) (note that the two conditions should be checked in this order).

Based on these definitions, the pseudo-code for the algorithm can be written as follows:

```
Link Distance Reduction Algorithm
1  Set A;
2  New State ← Initial State; {Initial State = Spanning tree
   obtained from SA}
3  repeat
4     Current State ← New State;
5     Set B;
6     for i = 1 to N do
7        j ← 1;
8        p*(B(i)) ← A(B(i), 1);
9        while p*(B(i)) <> p(B(i)) do
10           if (No Cycle == True) and (Lifetime Not Degraded
              == True) then
11              p(B(i)) ← p*(B(i))
12           else
13              j ← j+1;
14              p*(B(i)) ← A(B(i), j);
15           end if
16        end while
17     end for
18 until New State == Current State
```

In this algorithm, lines 3-18 correspond to a single iteration. During an iteration, the connections of all nodes to their parents are checked in decreasing order of their link distances (lines 6-17), and, for each node, potential parents are examined in increasing order of their distance to such node, that is, starting with the closest one (lines 9-16). Thus, for any given node, in the worst case this process yields the same parent node as the previous iteration. Also note that, as stated before, the algorithm executes a new iteration as long as at least one change is registered in the previous one. In effect, any change requires revisiting the situation of all nodes in a subsequent iteration; for instance, if node i was checked before node j in iteration k, and the connection of node j was changed during such iteration, maybe a potential parent for node i that was rejected in iteration k can now be accepted in iteration $k+1$.

5 Simulation Results

In order to validate the proposed algorithm and evaluate its computational complexity, we performed several simulation experiments using the *Mathematica* software. The scenario for these experiments was a 1000m x 1000m field, with coordinate $x \in [0,1000]$ and coordinate $y \in [-500,500]$. We placed the sink at $(0,0)$ and for the radio model we used the values of subsection 3.2. Each simulation experiment corresponded to a given number of nodes, which was varied between 10 and 100 in steps of 10. The number of runs in all simulation experiments was adjusted so as to produce 95% confidence intervals below 10%. In turn, each run consisted of generating a random deployment with the corresponding number of nodes, and then executing, on the one hand, the MST algorithm, and, on the other hand, the SA followed by the LDR algorithms. The result of each run consisted of the graphs generated by the MST, SA and SA+LDR algorithms, as well as several performance metrics evaluated on these graphs. Specifically, the most significant performance metrics were the following ones:

- *Average link distance.* For any data-gathering tree, this is the sum of link distances divided by the number of nodes (there are as many links as nodes). In order to reduce the amount of additional resources that could be required, the goal is to reduce the average link distance as much as possible.
- *Lifetime.* This is the network lifetime as given by (7).
- *Number of relay points.* This refers to the total number of relay points (nodes) that should be marked on the graph of the network in order to fulfill some maximum transmission range requirement.

Fig. 3 shows the results obtained for the first two metrics. In particular, Fig. 3(a) shows the downward trend of the average link distance (in meters) with the number of nodes, irrespective of the algorithm. This behavior was completely predictable, because the greater the amount of nodes deployed in a fixed-size region, the closer they are to each other. Moreover, this figure shows that the lowest average link distance corresponds to the MST algorithm, which is obvious since the only goal of MST is to minimize the sum of link distances. On the other hand, since the SA algorithm only takes link distances into account as long as they contribute to increase the lifetime of

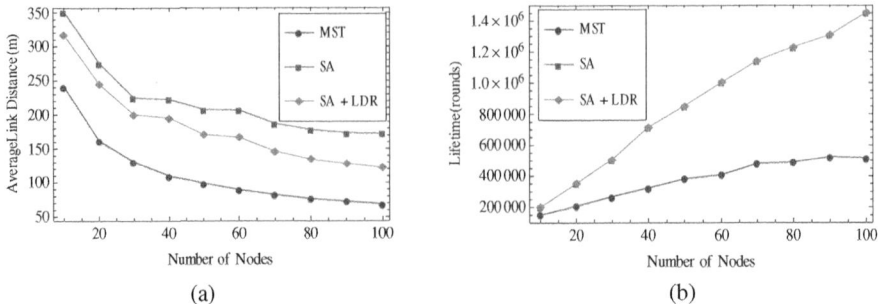

Fig. 3. Average link distance (a) and lifetime (b) versus number of nodes. The SA and SA+LDR curves in (b) are in full agreement.

the network, it generates an average link distance substantially greater than that of MST. Between these two extreme cases, the combined SA+LDR algorithm represents the best tradeoff, as it tries to reduce the overall link distance as long as the lifetime obtained from SA is not degraded. Specifically, it can be noticed that the gap between the SA and SA+LDR curves is significant, fact that validates the role of the LDR algorithm. Finally, another observation is that the MST curve is smoother than the others. This is due to the inherent randomness of the simulated annealing technique on which the SA and SA+LDR algorithms are based, which generally leads to a larger solution space than with the MST algorithm (in this case the solution space is typically a single graph or at most a very small set of graphs).

The downward trend of curves in Fig. 3(a) is in line with the upward trend of the lifetime curves shown in Fig. 3(b). This is due to the fact that transmission distance plays a dominant role in energy consumption, and its progressive decrease with the number of nodes has more impact than the corresponding increase in traffic load (represented by the forwarding parameters) – see expression (6). As it can also be noticed, the curves corresponding to SA and SA+LDR are completely superposed, which is not surprising as the LDR algorithm does not degrade the lifetime of the SA graph, and usually does not improve it either. Finally, another observation about Fig. 3(b) is the growing value of the gap between the two curves. This is explained by the fact that only the SA and SA+LDR algorithms focus on network lifetime.

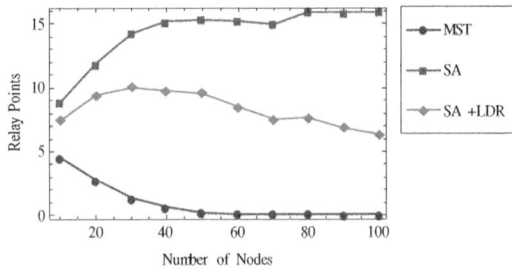

Fig. 4. Number of relay points versus number of nodes

Fig. 4 plots the evolution of the number of relay points as the number of nodes increases, for a maximum transmission range of 250m. In this case, the three algorithms experience different trends. These can be explained by viewing the average link distance and the number of nodes as separate factors: on the one hand, reducing the average link distance (obviously as a consequence of increasing the number of nodes) contributes to reducing the number of relay points required for network connectivity; on the other hand, since there can be one or more relay points per (regular) node, there is some positive correlation between these two magnitudes. In the case of MST, which focuses exclusively on reducing the overall link distance, the former factor dominates. However, the opposite happens with SA, since the whole set of link distances play a secondary role in this algorithm. Again, the SA+LDR curve is in an intermediate position, and thus it exhibits a local maximum at a relatively low number of nodes. In particular, the gap between the SA and SA+LDR curves as well as the decreasing trend of the latter beyond its local maximum validate the significance of LDR as a resource optimization algorithm.

In addition to guaranteeing network connectivity, the introduction of relay nodes enhances the lifetime of the network beyond the values shown in Fig. 3(b) (according to the energy consumption model considered so far). Certainly, this is applicable to both the SA+LDR and MST algorithms, but the difference is that the starting point is generally better for the former than for the latter. Thus, the enhanced lifetime is expected to be greater with SA+LDR than with MST, at least in a statistical sense.

In addition to the above metrics, we introduced two complementary measures:

- *Average energy consumption.* This is the average energy consumed by a node per round of communication. It is the result of dividing the total energy consumed by the network during a round of communication by the number of nodes. Although the definition of lifetime used in this paper relegates this measure to a secondary role, it could achieve more importance from alternative viewpoints. For instance, if electrical power supply for nodes was feasible (as it is the case of some scenarios), lifetime would not be crucial anymore, but preserving energy consumption would still be a relevant issue (for environmental protection).

- *Number of cross-points.* This is the number of crossing or intersection points among the edges of the graph that represents the sensor network. It can be viewed as a rough estimate of the entropy of such graph. Although this measure may also appear to be secondary, it could become more meaningful in some scenarios. For instance, in case directive antennas were used for inter-node communication, the number of cross-points would have an impact on the amount of interferences.

Fig. 5 plots the evolution of these additional metrics as the number of nodes increases, for the three algorithms considered so far. In particular, the results shown in Fig. 5(a) (expressed in nJ) are completely correlated with those in Fig. 3(a). Again, this is due to the fact that transmission distance plays a dominant role in energy consumption. So, we can conclude that MST outperforms SA and SA+LDR when the emphasis is put on the overall energy consumption, in contrast to what is shown in Fig. 3(b), which focuses on network lifetime. Regarding Fig. 5(b), we should recall that the SA algorithm focuses on maximizing the lifetime of the network as defined by (7), but not on enhancing the connections of the nodes that do not determine this lifetime.

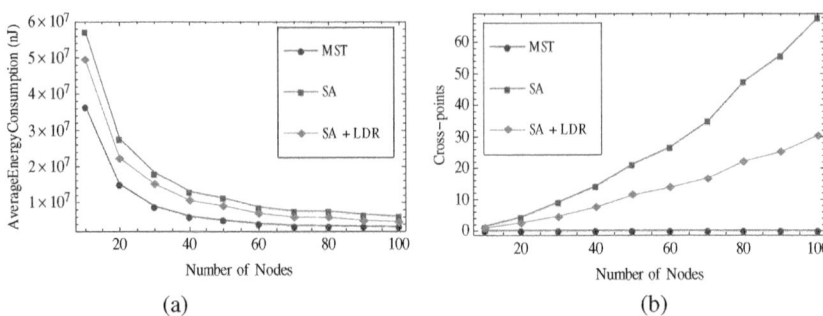

Fig. 5. Average energy consumption (a) and number of cross-points (b) versus number of nodes

This, in addition to the inherent randomness of this algorithm, makes the resulting graph rather chaotic, with a large amount of cross-points that obviously increases with the number of nodes. From this point of view, the MST algorithm exhibits an opposite behavior: it produces a graph with no cross-points, as shown in Fig. 5(b). Between these two extremes, the SA+LDR curve also shows an upward trend, but smoother than in the SA case.

Finally, we also evaluated the computational complexity of the proposed algorithm. Although from a network planning perspective the topology design algorithms are not subject to strong real time requirements, the issue of computational complexity is still of concern, since these algorithms may have to be executed from time to time for network reconfiguration. In this sense, the results obtained for LDR are very satisfactory, since the trend of the regression curve shown in Fig. 6 is practically linear. Specifically, this curve represents the evolution of the average number of iterations (see Section 4) as the number of nodes increases. In addition to its linear trend, this average takes on relatively moderate values, as it varies from approximately 30 for 10 nodes to around 1800 for 100 nodes.

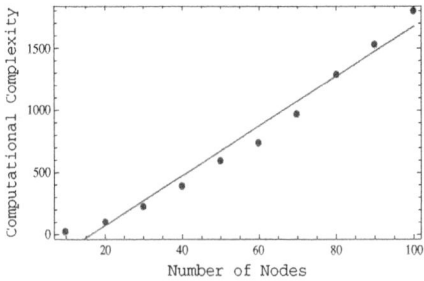

Fig. 6. Computational complexity of LDR as a function of the number of nodes

6   Conclusions and Further Research

In this paper, we have developed the Link Distance Reduction algorithm, a heuristic method that minimizes the overall link distance of a time-driven sensor network graph while preserving or even improving its lifetime. This algorithm uses hill-climbing without backtracking, and thus it could stop at a local optimum. However, in contrast, this limitation has favored its simplicity and computational efficiency, and at the same time it has not precluded the algorithm from yielding significant reductions in average link distance and number of relay points.

The work presented in this paper can be extended to other energy-consumption models, as well as to scenarios where the deployment density is kept constant or the maximum transmission range of nodes is varied. A more advanced research would consist of developing a modified version of the original SA algorithm that included the average link distance minimization, in order to enable the achievement of the global optimum (although at the expense of increasing the computational complexity). Another issue would be the distributed implementation of the proposed algorithms. Also, the issues of connectivity, lifetime enhancement and fault-tolerance could be jointly treated by replacing single nodes with clusters of cooperative transmitting nodes (see [14] for a survey on MIMO/MISO techniques). Finally, our research results are progressively being integrated in *NetLife*, a software tool for planning time-driven sensor networks.

Acknowledgments. This work has been supported in part by the Spanish Ministry of Science and Technology under contract TIN2009-11711.

References

1. Akkaya, K., Younis, M.: A Survey on Routing Protocols for Wireless Sensor Networks. Ad Hoc Networks 3, 325–349 (2005)
2. Krishnamachari, B.: Networking Wireless Sensors. Cambridge University Press, Cambridge (2005)
3. Santamaría, M.L., Galmés, S., Puigjaner, R.: Simulated Annealing Approach to Optimizing the Lifetime of Sparse Time-Driven Sensor Networks. In: 2009 IEEE International Symposium on Modeling, Analysis & Simulation of Computer and Telecommunication Systems (MASCOTS), pp. 193–202. IEEE, Inc., Los Alamitos (2009)
4. Goel, A., Estrin, D.: Simultaneous Optimization for Concave Costs: Single Sink Aggregation or Single Source Buy-at-Bulk. In: ACM Symposium on Discrete Algorithms (2003)
5. Enachescu, M., Goel, A., Govindam, R., Motwani, R.: Scale Free Aggregation in Sensor Networks. In: First International Workshop on Algorithmic Aspects of Wireless Sensor Networks (2004)
6. Wu, Y., Fahmy, S., Shroff, N.B.: On the Construction of a Maximum-Lifetime Data Gathering Tree in Sensor Networks: NP-Completeness and Approximation Algorithm. In: 27th Conference on Computer Communications, INFOCOM (2008)
7. Cheng, X., Du, D., Wang, L., Xu, B.: Relay Sensor Placement in Wireless Sensor Networks. Wireless Networks 14(3), 347–355 (2008)
8. Chen, D., Du, D., Hu, X., Lin, G., Wang, L., Xue, G.: Approximations for Steiner Trees with Minimum Number of Steiner Points. Journal of Global Optimization 18(1), 17–33 (2000)
9. Tang, J., Hao, B., Sen, A.: Relay Node Placement in Large Scale Wireless Sensor Networks. Computer Communications 29(4), 490–501 (2006)

10. Kashyap, A., Khuller, S., Shayman, M.: Relay Placement for High Order Connectivity in Wireless Sensor Networks. In: 25th Conference on Computer Communications, INFO-COM (2006)
11. Galmés, S.: Lifetime Planning for TDMA-Based Proactive WSN with Structured Deployment. In: 2007 IFIP/ACM Latin American Networking Conference, LANC (2007)
12. Rappaport, T.S.: Wireless Communications: Principles and Practice. Prentice-Hall, Englewood Cliffs (2002)
13. Heinzelman, W.B., Chandrakasan, A.P., Balakrishnan, H.: An Application-Specific Protocol Architecture for Wireless Microsensor Networks. IEEE Trans. on Wireless Communications 1(4), 660–670 (2002)
14. Ahmad, M.R., Dutkiewicz, E., Huang, X.: Performance Evaluation of MAC Protocols for Cooperative MIMO Transmissions in Sensor Networks. In: 5th ACM Symposium on Performance Evaluation of Wireless Ad Hoc, Sensor, and Ubiquitous Networks (2008)

Routing Protocol for Anycast Communications in a Wireless Sensor Network

Nancy El Rachkidy, Alexandre Guitton, and Michel Misson

LIMOS-CNRS, Clermont University
Complexe scientifique des Cézeaux,
63177 Aubière cedex, France
{nancy,guitton,misson}@sancy.univ-bpclermont.fr

Abstract. In wireless sensor networks, there is usually a sink which gathers data from the battery-powered sensor nodes. As sensor nodes around the sink consume their energy faster than the other nodes, several sinks have to be deployed to increase the network lifetime. In this paper, we motivate the need of anycast communications in wireless networks, where all the sinks are identical and can gather data from any source. To reduce interference and congestion areas on the wireless medium, the path from a source to a sink has to be distant from the path connecting another source to another sink. We show that determining distant paths from sources to sinks is an NP-hard problem, and we propose a linear formulation in order to obtain optimal solutions. Then, we propose a sink selection and routing protocol called S4 and based on realistic assumptions and we evaluate it through simulations. Finally, we conclude that anycast routing protocols in wireless sensor networks should not compute paths independently for each source, but rather consider all the sources simultaneously.

Keywords: Sink selection and routing protocol, anycast communications, wireless sensor networks.

1 Introduction

In the past few years, wireless sensor networks have been used in several monitoring applications. Battery-powered nodes collect data with their sensors, and send it in a wireless manner to a data-gathering station, called the sink. The sink usually has a large memory capacity, and sometimes does not even have energy limitations, contrarily to the sensor nodes. The sink has several roles: it can store historical data, analyze the data to detect discrepancies or emergency situations, or act as a gateway providing connectivity with a wired network.

As the traffic of the network converges to the sink, nodes close to the sink consume their energy faster than farther nodes. When all the nodes around the sink have depleted their energy, the sink is not able to receive any data from the sensors, and gets disconnected from the network. When this situation happens, the whole network is considered to be down. A solution to this problem is to

M. Crovella et al. (Eds.): NETWORKING 2010, LNCS 6091, pp. 291–302, 2010.

deploy several sinks. If traffic is balanced among the sinks, the network lifetime can be significantly increased since the energy consumption will be almost equal for all the nodes in the network.

The paradigm of anycast communications, also termed one-to-any communications, becomes very important in a network with multiple sinks: when a sensor node produces data, it has to send it to any sink available. A sink selection strategy is to choose for each source a sink arbitrarily. An alternative strategy is to route data to the closest sink. Assuming that the sources and the sinks are uniformly distributed in the network, this simple strategy is assumed to balance the energy consumption.

In this paper, we show that in order to minimize interference and to reduce the congestion areas between the different paths, all the sources have to be considered simultaneously. Figure 1 shows a topology with two sources s_1 and s_2 and two sinks d_1 and d_2, with three sink selection strategies, in order to give an insight of why considering sources simultaneously is important. On part (a) of the figure, each source is connected to the closest sink, which generates contention around sink d_2. On part (b), each source is connected to a different sink in order to balance the sink load. However, the path (s_1, d_2) intersects with the path (s_2, d_1). Contention on the wireless medium is generated around the area where the paths meet. On part (c), paths (s_1, d_1) and (s_2, d_2) are distant from each other. The wireless traffic generated on one path has little impact on the traffic generated on the other path (provided that those paths are distant enough). This third sink selection strategy can only be achieved by considering all the sources and sinks simultaneously. Also note that the sink selection and the routing have to be performed at the same time.

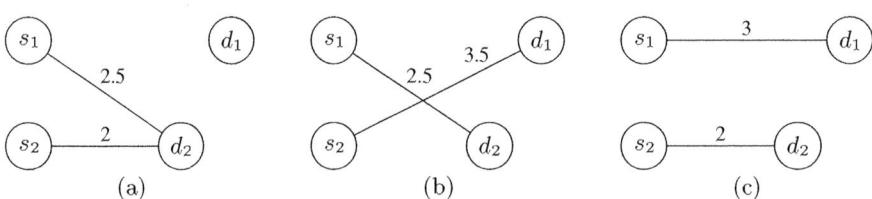

Fig. 1. To minimize interference and reduce congestion on the paths from each source to a sink, all the sources and sinks have to be considered simultaneously, as in (c)

The paper is organized as follows. Section 2 describes the related works. Section 3 formally describes the problem of finding distant paths for a set of anycast communications. We prove that the problem is NP-hard, and we propose an integer linear formulation to obtain optimal solutions. Section 4 describes our strategy, based on pivot routing in order to ensure that paths are distant. Section 5 describes our simulation environment and settings, and provides the simulation results we obtained. Finally, we conclude our work in Sect. 6.

2   Related Work

In this section, we focus on three main topics: the deployment of multiple sinks, the anycast paradigm which allows sources to send data to any sink, and the use of multipath in wireless routing protocols.

2.1   Multi-sink Deployment

Multi-sink deployment refers to a wireless network architecture where several sinks are deployed, each of them having identical functional capabilities. Recently, interest is emerging towards scenarios with multiple sinks in order to improve the network lifetime and to ensure a fair delivery of data among sinks [1,2]. Another advantage of deploying multiple sinks is to improve the data gathering by reducing the communication delay from sensors to sinks [3,4] or the total communication cost [5].

2.2   Anycast Communications

Anycast is a one-to-any communication paradigm where a source communicates with a single sink, chosen among a set of possible sinks. It has been shown that the lifetime of a wireless sensor network can be increased by deploying several sinks, and accessing them using an anycast protocol [6,7]. In [8], the authors also showed that anycast forwarding schemes can significantly reduce the expected packet delivery delays.

2.3   Multipath Routing

Multipath routing is a feature that enables a source to send packets to a destination through multiple different paths at the same time. The main advantage of this feature is to improve the reliability of the packet delivery: even if one path becomes blocked due to a node failure for instance, the destination is still able to receive packets as long as at least one path is active. Using multipath also helps balancing the energy consumption among the network and therefore extends network lifetime. Most of the multipath routing protocols are based on classic on-demand single path routing methods [9,10].

Disjoint multipath routing methods try to determine disjoint paths, *i.e.*, paths that do not have nodes or edges in common. As stated in [11], using disjoint multipaths does not remove the potential for collisions, resulting in large packet loss rates and reduced data transmission performance. The main reason is that wireless transmissions might interfere communications between distant nodes. In [12], the authors aim to find zone-disjoint multipaths using directional antennas. A promising approach is described in [13], where authors proposed an energy efficient and collision aware disjoint multipath routing algorithm. The flooding required to determine such paths is limited to nodes close to the main discovery route.

In this paper, we do not use multipath to route traffic from a source to a sink. We rather aim to determine a collection of paths (one per source-sink pair) that are distant from each other. However, as shown in the next section, the problem of finding disjoint multipaths between a source and a sink, and between two source-sinks pairs are closely related.

In [14], the authors proposed a probabilistic proactive routing protocol called PiRAT and based on pivots. When an emergency situation occurs, several geographically close sensors might produce alarm messages that have to be forwarded to a sink. The paths followed by all the alarms becomes congested and several alarm packets might be dropped. By selecting randomly distant pivot nodes for each source, PiRAT is able to reduce the congestion areas and to improve the network performance in terms of delay and packet loss. Indeed, it allows diversity in routing and avoids congested areas of the network, which contributes to balance the traffic load and the energy consumption between nodes. In Sect. 4, we use a similar pivot approach in order to obtain distant paths.

3   Problem Statement and Modelling

In this section, we study the problem of determining a sink for each source. We also study the related problem of finding a path from each source to its assigned sink, so that the paths from all the sources are distant from each other. We focus on providing a formal description of the sink selection and routing strategy for anycast communications.

Let us consider a set of sensors V forming a wireless sensor network $G = (V, E)$. E is defined in the following way: if x and y can communicate with each other, we have $(x, y) \in E$ and $(y, x) \in E$. Let $S \subset V$ denote a set of sources, and $D \subset V$ denote a set of sinks. Finally, let us denote by $h(x, y)$ the hop count between two nodes x and y. The minimum distance between two paths p_1 and p_2, denoted by $h(p_1, p_2)$, can be defined as:

$$h(p_1, p_2) = \min_{x \in p_1, y \in p_2} h(x, y).$$

Definition 1. *The sink selection and routing problem for anycast wireless communications (SSRPAW) consists in finding a set of paths $\{p_i\}$ that connects each source $s_i \in S$ to a sink $d_{f(i)} \in D$, such that $h(p_{i_1}, p_{i_2}) \geq \delta$ for any $i_1 \neq i_2$ and for a given $\delta > 0$.*

The rationale behind the SSRPAW problem is to find a sink selection strategy (characterized by the function f) and a routing strategy (characterized by the choice of paths $\{p_i\}$) that ensures that paths are distant enough from each other to avoid contention in the medium. The number of hops between two different paths is at least δ. δ depends on the propagation conditions. In a dense network, interference is often negligible after two hops, and thus δ is often 2.

In the remainder of this section, we show that SSRPAW is NP-hard. Then, we propose an integer linear program that allows to compute optimal solutions. Finally, we show by simulations that only small instances have optimal solutions. This motivates the need of a heuristic that can work with limited computational capabilities and realistic assumptions.

3.1   Proof of the NP-Completeness of SSRPAW

In order to prove the NP-completeness of SSRPAW, we first have to define a similar problem.

Definition 2. *The set-to-set disjoint path problem takes as input a graph $G = (V, E)$, a set of k sources S and a set of k destinations D. It consists in determining if there are k mutually node-disjoint paths $\{p_i\}$, such that p_i is a path from s_i to d_{j_i}, for $1 \leq i \leq k$ and j a permutation of $\{1, \ldots, k\}$. Two mutually node-disjoint paths have no node in common, except for the source and the destination.*

Theorem 1. *The set-to-set disjoint path problem is NP-complete [15]. It is similar to the node-to-node disjoint path problem.*

Theorem 2. *SSRPAW is NP-complete for any δ.*

Proof. The proof of the NP-completeness of SSRPAW for any $\delta > 0$ is by reduction to the set-to-set disjoint path problem. We show in the following that if one is able to solve SSRPAW on a specific graph $\bar{G}$ in polynomial time, one has solved the set-to-set disjoint path problem in a general graph G in polynomial time (which is unlikely, unless $P = NP$).

Let $G = (V, E)$ be an arbitrary general graph. The construction of $\bar{G} = (\bar{V}, \bar{E})$ is the following. Each node $n \in V$ is also a node of $\bar{V}$. Each edge $e = (x, y) \in E$ becomes a path of δ edges in $\bar{E}$, connecting $x \in \bar{V}$ to $y \in \bar{V}$.

Let us now assume that SSRPAW can be solved in polynomial time in a graph $\bar{G}$. This means that there are $|S| = k$ paths $\{\bar{p}_i\}$ in $\bar{G}$, for $1 \leq i \leq k$, such that each path $\bar{p}_i$ connects a source $s_i \in S$ to a sink $d_{j_i} \in D$. Moreover, for any $i_1 \neq i_2$, $h_{\bar{G}}(\bar{p}_{i_1}, p_{i_2}) \geq \delta$, by definition of SSRPAW. By construction of $\bar{G}$, each path $\bar{p}$ in $\bar{G}$ can be translated into a path p in G. Thus, we have k paths $\{p_i\}$ in G such that each path p_i connects a source $s_i \in S$ to a sink $d_{j_i} \in D$. These paths are such that $h_G(p_{i_1}, p_{i_2}) \geq \delta/\delta = 1$, which means that they are node disjoint. Thus, we have solved the set-to-set disjoint path problem between S and D in polynomial time, which completes the proof.

3.2   Integer Linear Formulation

The goal of this subsection is to define optimal solutions by an integer linear program. This program takes as input a set of nodes V, a set of sources $S \subset V$, a set of sinks $D \subset V$ and a set of binary variables $e_{x,y}$ representing the edges. The objective of the integer linear program is to find a set of paths $\{p_s\}$, one per source $s \in S$ and to any sink $d \in D$, such that paths are distant from each other and the total number of edges used is minimized. Each path is defined as a set of binary variables $p_s(x, y)$, such that $p_s(x, y)$ is 1 if path p_s uses edge (x, y), and 0 otherwise. The resulting objective function is given on Table 1.

The problem constraints are given on Table 1. Inequality (1) states that it is forbidden to use an edge (x, y) in a path if x and y are not neighbors (or

Table 1. Integer linear constraints

minimize $\sum_{s \in S} \sum_{x \in V} \sum_{y \in V} p_s(x, y)$

such that $\forall s \in S, x \in V, y \in V$	$p_s(x, y) \leq e_{x,y}$	(1)
$\forall s \in S$	$\sum_{y \in V} p_s(s, y) \geq 1$	(2)
$\forall s \in S$	$\sum_{x \in V} \sum_{d \in D} p_s(x, d) \geq 1$	(3)
$\forall s \in S, x \in V\backslash D, y \in V\backslash D$	$p_s(x, y) \leq \sum_{z \in V\backslash\{x\}} p_s(y, z)$	(4)
$\forall x \in V$	$\sum_{s \in S} \sum_{y \in V} p_s(x, y) \leq 1$	(5)

$\forall s_1 \in S, s_2 \in S\backslash\{s_1\}, x \in V, x' \in V$

$$\sum_{y \in V} p_{s_1}(x, y) + \sum_{y \in V} p_{s_2}(x, y) \leq \#S - (\#S - 1)e_{x,x'} \qquad (6)$$

same as previous with $p_{s_1}(x, y)$ and $p_{s_2}(x', y)$ (7)
same as previous with $p_{s_1}(x', y)$ and $p_{s_2}(x, y)$ (8)
same as previous with $p_{s_1}(x', y)$ and $p_{s_2}(x', y)$ (9)

equivalently, if there is no edge (x, y) in the graph, that is if $e_{x,y} = 0$). Constraint (2) indicates that for any source s, there is at least one edge that leaves s in p_s. Constraint (3), symmetrically, indicates that each path p_s should terminate in a node of D. This constraint corresponds to anycast communications. Constraint (4) is a connectivity constraint: it states that for each edge (x, y) of a path p_s (except for sinks), there is at least one edge (y, z) (with $z \neq x$). In other words, it says that each edge (x, y) on a path p_s is followed by an edge (y, z) on the same path. Constraint (5) states that each edge (x, y) is used by at most one path. Constraints (6), (7), (8) and (9) indicate that the distance between paths should be of at least two or more. Thus, the program only[1] applies to $\delta = 2$. More specifically, it says that:

(i) If an edge (x, x') exists in the graph, there is at most one path that uses node x or x', since x and x' are neighbors. Indeed, if the edge (x, x') exists, $e_{x,x'}$ is equal to 1, and the right part of the equation is equal to $\#S - (\#S - 1) = 1$ (where $\#S$ represents the cardinal of S).
(ii) If the edge (x, x') does not exist in the graph, the number of paths that uses nodes x and x' is not limited. In this case, $e_{x,x'} = 0$ and the right part of the equation is equal to the maximum number of paths $\#S$. As $\#S$ is a natural limit to the number of paths, the inequality does not bring restriction in this case.

These four equations cannot be merged into one, because it is possible for a single path p_{s_1} to use the edge (x, x'). In this case, the summation would count two edges for this path (one is (x, x') and the other is (x', y)), and the result would not anymore be smaller than 1. Note however that Constraint (5)

---

[1] A similar approach can be applied for larger values of δ, but drastically increases the number of constraints.

(corresponding to $\delta = 1$) is not required when $\delta > 1$, as it is included in Constraints (6) to (9).

3.3 Computation of Optimal Solutions

In this subsection, we study the optimal solutions found by our integer linear program, using the GLPK (GNU Linear Programming Kit) solver. We generated a grid topology of 7×7 nodes, such that each node can communicate to its four direct neighbors only. We tried to find paths from each source to any sink, with a minimum distance of $\delta = 2$ between paths. Sources and sinks are chosen randomly such that all the sources are at a distance of δ of each other, and all the sinks are at a distance of δ of each other. We varied the number of sources and the number of sinks. Results are averaged over 50 simulations.

Figure 2 shows that the average optimal path cost increases with the number of sources and sinks. When the number of sources and sinks becomes larger, the paths are longer in order to ensure disjoint paths. Figure 3 shows the percentage of optimal paths found as a function of the number of sources and sinks. When the number of sources is smaller than four, optimal solutions are always found. When the number of sources is five or six, there are no optimal solutions. This means that it is not possible to find distant paths with $\delta = 2$. The given integer linear formulation is too restrictive, and is not applicable in a realistic setup.

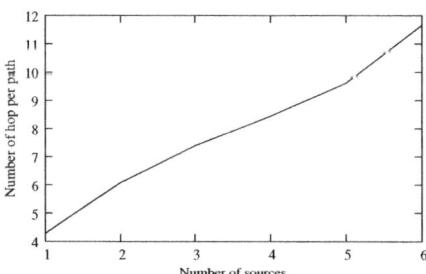

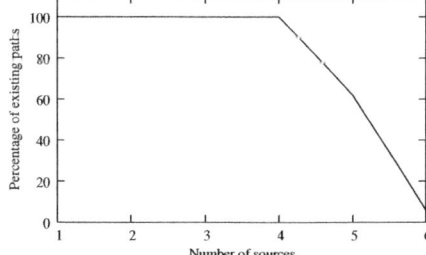

Fig. 2. Average optimal path cost per source-sink pair as a function of the number of sources, when the number of sinks is equal to the number of sources

Fig. 3. Percentage of topologies having an optimal solution as a function of the number of sources, with the number of sinks equal to the number of sources

Simulations are run on a standard personal computer and they took about two hours and 35 minutes. The results show that even without limited computational capabilities, optimal solutions cannot always be found, even on small instances. In a wireless sensor network, where nodes have very limited computational capabilities, there is a strong need for a simple heuristic that is able to provide approximated solutions using realistic assumptions.

4   Heuristic Sink Selection Strategies

As we have shown in the previous section, it is not realistic to find optimal disjoint paths from sources to sinks. Moreover, the optimal formulation makes assumptions about the network that are not realistic. This is why we propose in this section an heuristical approach called simultaneous sink selection strategy, combined with pivot routing. Before describing our strategy, we present two commonly used strategies.

Random sink selection (RSS): In RSS, sinks are randomly chosen. Packets are routed using the shortest path from the source to the selected sink. We expect this strategy to perform badly in terms of packet loss, as it might incur congestion in several areas of the networks. Congestion also increases the delay due to retransmission attempts.

Closest sink selection strategy (CSSS): In CSSS, each source is connected to the closest sink. The distance from the source to each sink can be computed according to the geographical coordinates or using hop count. CSSS does not take into account the fact that the areas around sinks can be congested. It uses the shortest path between the source and the selected sink.

Simultaneous sink selection strategy (S4): S4 uses a greedy algorithm to select sinks and to compute paths: each source is considered sequentially. For each source, a pivot node is selected in order to make paths as disjoint as possible. The pivot selection works as follows. When considering a source s, S4 considers all the nodes as potential pivots and all the sinks as potential destinations. For each potential pivot x and potential destination d, S4 determines the hop count $h(s, x)$ between s and x, and the hop count $h(x, d)$ between x and d. S4 also determines the number of nodes in common between the path $p(s, x, d)$ from s to d via x and $\mathcal{P}$, the union of all the paths already chosen by S4 for the previous sources. For a given source s, S4 has several candidate paths. S4 chooses one of the paths that minimizes the number of nodes in common with $\mathcal{P}$. If there are still several candidate paths, S4 chooses one of the paths of minimum length among the candidates. Notice that S4 combines a sink selection strategy with a routing mechanism. Thus, S4 reduces the energy consumption since it balances the traffic in the whole network.

The hop count between two nodes can be computed using a simple signalling protocol, or using properties of hierarchical addresses (such as those used in IEEE 802.15.4 for example). This process is out of scope of this paper. The computation of the number of nodes in common between two paths can be computed by having s sending a first message to x, and a second message via x to d. Each node on both path can send a notification back to s, which is then able to count the number of common nodes. Another possibility is again to use properties of hierarchical addresses, which allows a node s to determine the path between two nodes a and b, provided that the addresses of a and b are known. This process is also out of scope of this paper. We assume here that s has a centralized knowledge of the topology.

In order to operate S4, we consider that there is a centralized entity which knows the whole topology (as mentioned previously) and computes paths for all the sources. For each source s, this entity has to compute the shortest paths from s to any node, and the shortest paths from any sink to any node. The first computation requires $\mathcal{O}(n + m)$ operations, where n is the number of nodes and m is the number of edges, and the second requires $\mathcal{O}(n + m)$ too (note that a single computation is performed for all the sinks). Then, all the nodes are considered as pivots and all the sinks are considered as destinations. Thus, the overall complexity is $\mathcal{O}(|S|(n+m)|D|)$. As $|D|$ and $|S|$ are supposed to be rather small, the burden of the centralized entity in S4 is reasonable.

5   Evaluation

In this section, we describe the simulations we ran in order to compare S4 with RSS and CSSS. We used the NS-2 simulator, version 2.31. We used the IEEE 802.15.4 physical and MAC layers with the non-beacon enabled mode. The propagation model used was the two ray ground model, with default parameters. The transmission power was set to a realistic value of -25 dBm, and the radio range was set to 25 m. The size of the nodes queue was set to 50 packets.

In our simulations, we considered for simplicity reasons a set of 49 nodes uniformly distributed on a grid of 70×70 square meters. Each node is located at a distance of 10 m of its neighbors. All sensors were full function devices with routing capabilities. The PAN coordinator was located at the center of the area. We waited for the network to be fully associated before injecting data packets. Data packets of 77 bytes (at the physical layer) were generated during 50 seconds, at a fixed rate of 2 packets per second. The routing protocol we used for the three strategies was AODV[2] [16]. Notice that before AODV can send packets to an unknown destination, it has to establish a path through reply and request messages, which introduces delay[3]. In order to have stable results for RSS, results are averaged over 500 repetitions.

5.1   Performance Metrics

Our algorithm is evaluated and compared to RSS and CSSS (detailed in Sect.4), according to two performance metrics:

(i) Packet loss: the packet loss is defined as the number of packets received by the sink nodes over the number of packets generated by the source nodes. Thus, the packet loss metric takes into account the losses due to collisions and the losses due to a large number of retransmission attempts.
(ii) End-to-end delay: the end-to-end delay is the time interval between the transmission of a packet by the source and the reception of the same packet

---

[2] AODV was slightly modified in S4 in order to allow pivot nodes.
[3] For RSS and CSSS, AODV has to establish paths from each source to its assigned destination. For S4, AODV has to establish paths from each source to its pivot, and from this pivot to the sink.

by the sink, at the application layer. The time required by AODV to establish routes, as well as the delay introduced by retransmissions, are taken into account into the end-to-end delay. However, the end-to-end delay only takes into account the packets that are correctly received by the sink.

5.2   Packet Loss

Figure 4 and Fig. 5 show the average packet loss as a function of the number of sources and sinks respectively. We notice that the packet loss for the three strategies increases consistently with the number of sources, and decreases with the number of sinks. When the number of sources in the network is large, the traffic load is large too and the medium is overloaded by the generated packets. S4 is able to significantly reduce the packet loss compared to RSS and CSSS. Indeed, S4 aims to build distant paths by selecting pivots for each source-sink pair, which contributes to balance the traffic between nodes and to reduce congestion on the medium. S4 reduces by approximately 36% the packet loss probability of RSS and CSSS, for five sources and five sinks. S4 outperforms the other two strategies when the number of sinks is one: S4 reduces by approximately 41% the packet loss probability of RSS and CSSS for five sources and one sink.

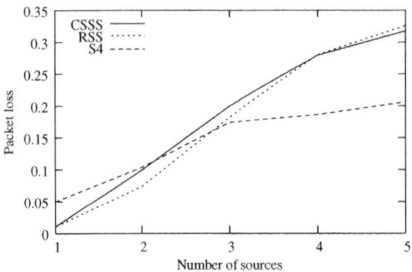
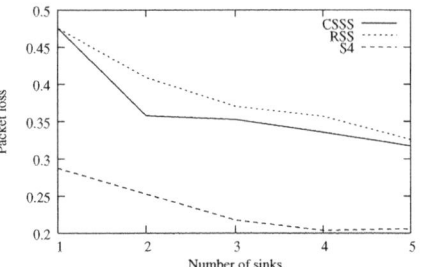

Fig. 4. Packet loss probability as a function of the number of sources, with as many sinks as sources

Fig. 5. Packet loss probability with five sources, as a function of the number of sinks

5.3   End-to-End Delay

Figure 6 shows the average end-to-end delay for the three strategies as a function of the number of sources. For RSS, the delay increases with the number of sources and sinks and becomes stable when there are more than four sources in the network. The delay is positively affected by the number of sinks (as the average distance between a source and a sink decreases) and negatively affected by the traffic load (as the medium becomes congested). It can be noticed that CSSS induces larger delays than RSS. This is explained by the fact that CSSS tends to select the same sink for several close sources, which yields to congested areas around the sinks. RSS balances the sink usage by choosing sinks randomly.

While CSSS and RSS have almost the same packet loss, the impact on delay is significant: packets with large delay are more likely to be dropped in RSS, which reduces the average packet loss for this strategy. S4 has the best behavior of the three strategies. This proves that it is important to consider all sources simultaneously for the sink selection, and to build distant paths during the routing process. S4 reduces the average end-to-end delay over CSSS by 70% and over RSS by 50%, for five sources and five sinks.

Figure 7 shows the average end-to-end delay for the three strategies, for five sources, as a function of the number of sinks. For the three strategies, we notice that the delay decreases with the number of sinks. With a large number of sinks, there are less congested areas in the network, and thus the number of packet retransmissions decreases (because the packet loss decreases too, see Fig. 5). S4 outperforms the two other strategies, even with one sink: in this case, it reduces the end-to-end delay of CSSS and RSS by 47%.

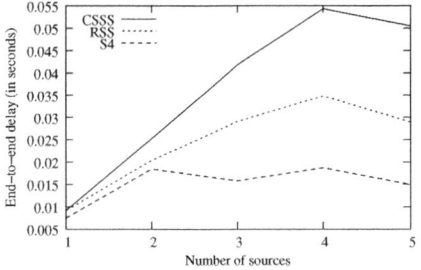

Fig. 6. End to end delay as a function of the number of sources, with as many sinks as sources

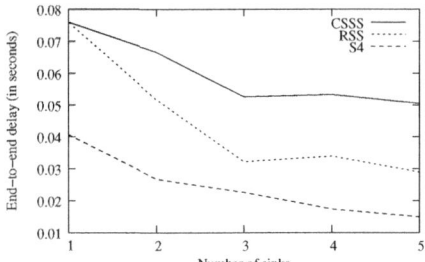

Fig. 7. End to end delay with five sources, as a function of the number of sinks

6   Conclusion

In this paper, we showed that determining distant paths from sources to sinks is an NP-hard problem. Then, we proposed an integer linear program that computes the optimal solutions. We proposed an heuristic called S4 based on realistic assumptions. S4 is a centralized approach that selects sinks and pivots in order to provide distant paths between source-sink pairs and to reduce congestion in the network. Simulation results showed that S4 outperforms the existing strategies in terms of delay (which is reduced by up to 50% in our scenarios) and packet loss (which is reduced by up to 41% in our scenarios). The perspectives of this work include the enhancement of S4 (including the order in which sources are selected). We aim to have a distributed strategy that is able to provide distant paths without requiring the knowledge of the whole topology. Moreover, we plan to simulate S4 for other representative topologies.

Acknowledgment. This work has been partially supported by a research grant from the Lebanese National Council for Scientific Research (LNCSR).

References

1. Kim, H., Seok, Y., Choi, N., Kwon, T.: Optimal multi-sink positioning and energy-efficient routing in wireless sensor networks. Information Networking (2005)
2. Oyman, E.I., Ersoy, C.: Multiple sink network design problem in largescale wireless sensor networks. In: IEEE ICC (2004)
3. Chang, J., Tassiulas, L.: Maximum lifetime routing in wireless sensor networks. IEEE/ACM Transactions on Networking 12 (2007)
4. Buratti, C., Orris, J., Verdone, R.: On the design of tree-based topologies for mutli-sink wireless sensor nerworks (September 2006)
5. Kalantari, M., Shayman, M.: Design optimization of multi-sink sensor networks by analogy to electrostatic theory. In: IEEE WCNC (2006)
6. Hu, W., Bulusu, N., Jha, S.: A communication paradigm for hybrid sensor/actuator networks. Springer International Jounal of Wireless Information Networks 14(3) (2005)
7. Thepvilojanapong, N., Tobe, Y., Sezaki, K.: Har: Hierarchy-based anycast routing protocol for wireless sensor networks. In: SAINT (2005)
8. Kim, J., Lin, X., Shroff, N.B.: Minimizing Delay and Maximizing Lifetime for Wireless Sensor Networks With Anycast. In: IEEE INFOCOM (2008)
9. Marina, M.K., Das, S.R.: On-demand multipath distance vector routing in ad hoc networks. In: ICNP (2001)
10. Lee, S.J.: gerla, M.: Split multipath routing with maximally disjoint paths in ad hoc networks. In: IEEE ICC (2001)
11. Pearlman, M.R., Haas, Z.J., Sholander, P., Tabrizi, S.S.: On the impact of alternate path routing for load balancing in mobile ad hoc networks. In: ACM Mobile Ad Hoc Networking and Computing (2000)
12. Saha, D., Toy, S., Bondyopadhyay, S., Ueda, T., Tanaka, S.: An adaptive framework for multipath routing via maximally zone-disjoint shortest paths in ad hoc wireless networks with directional antenna. In: Proc. Global Telecommunications (2003)
13. Wang, Z., Bulut, E., Szymanski, B.K.: Energy Efficient Collision Aware Multipath Routing for Wirelss Sensor Networks. In: IEEE ICC (2009)
14. El Rachkidy, N., Guitton, A., Misson, M.: Pirat: Pivot Routing for Alarm Transmission in Wireless Sensor Networks. IEEE Local Computer Networks (2009)
15. Qian-Ping, G., Satoshi, O., Shietung, P.: Efficient algorithms for node disjoint path problems. In: Proceedings of Electronics, Information and Communication Engineers Conference, vol. 2 (1994)
16. Perkins, C., Belding-Royer, E., Das, S.: Ad hoc on-demand distance vector (AODV) routing. Request For Comments 3561, IETF (2003)

Fault-Tolerant Power-Aware Topology Control for Ad-Hoc Wireless Networks

Harichandan Roy, Shuvo Kumar De, Md. Maniruzzaman, and Ashikur Rahman

Department of Computer Science and Engineering
Bangladesh University of Engineering and Technology, Dhaka, Bangladesh
{hari252bd,shuvo.buet,monir085}@gmail.com, ashikur@cse.buet.ac.bd

Abstract. Minimizing energy consumption and ensuring fault tolerance are two important issues in ad-hoc wireless networks. In this paper, we describe a distributed topology control algorithm which minimizes the amount of power needed to maintain bi-connectivity. The algorithm selects optimum power level at each node based on local information only. The resultant topology has two properties: (1) it preserves the minimum energy path between any pair of nodes and (2) it ensures fault tolerance by maintaining bi-connectivity. By presenting experimental results, we show the effectiveness of our proposed algorithm.

Keywords: Ad-hoc Network, Topology control, Minimum-energy, Fault tolerance.

1   Introduction

Ad-hoc wireless networks are getting widespread with the recent development of wireless communication system. Since the basic components of multi-hop wireless networks are mostly battery-operated devices, power conservation is one of the key issues of such networks. It is not energy efficient to use the communication networks where each node transmits with its maximum power. So, power control is needed which deals with the problem of choosing the minimum power level by each node to minimize the energy consumption for the whole network. To ensure minimum power level, the topology has to preserve minimum-energy paths between the nodes. Besides preserving minimum-energy path, the topology control automatically maintains some properties such as reduced average node degree, smaller average transmission power etc. The network built in this way has profound effect on the performance of routing layer. Power control also results in extending battery life of the nodes.

On the other hand, by reducing the number of links in the network, topology control algorithms actually decrease the degree of routing redundancy. As a result, the topology thus derived is more susceptible to node failures/departures. Besides this, failure of nodes is a common phenomena in ad-hoc wireless network. This problem can be mitigated if an adequate level of fault tolerance can be properly ported into topology control. Fault tolerance increases the robustness

M. Crovella et al. (Eds.): NETWORKING 2010, LNCS 6091, pp. 303–314, 2010.

of the network maintaining connectivity in case of any breakdown or an increase in load at any vicinity of the network.

In this paper, we propose a *distributed* topology construction algorithm based on *local information* only. By the term *local information* we mean that a node only has information about the position of one hop or two hops neighbors. Requiring more than two hops neighbor information a much overhead is incurred which will subdue the benefit of the topology control. Finally, the proposed algorithm preserves all minimum-energy paths between every pair of nodes and ensures fault tolerance by maintaining global bi-connectivity.

1.1 The Problem Statement

We use the same model as [7] which considers, a n-node, multi-hop, ad-hoc wireless network deployed on a two-dimensional plane. Suppose that each node is capable of adjusting its transmission power up to a maximum denoted by P_{max}. Such a network can be modeled as a graph $G = (V, E)$, with the vertex set V representing the nodes, and the edge set E defined as follows:

$$E = \{(x, y) | (x, y) \in V \times V \land d(x, y) \leq R_{max}\} \tag{1}$$

where $d(x, y)$ is the distance between nodes x and y and R_{max} is the maximum distance reachable by a transmission at the maximum power P_{max}. The graph G defined this way is called the maximum powered network. Note that the graph constructed this way is a visual representation of the inherent topology of the network. That is why we use the term *topology* and *graph* interchangeably throughout this paper.

Formally, the aim of this paper is to construct a graph $G' \subseteq G$ in a distributed fashion based on local information, where for any node pair u and v the minimum-energy path between u and v in G is also preserved in G' and moreover it provides fault tolerance as G' has at least two vertex-disjoint paths between any two nodes. Controlling topology in this way has a benefit to maintain connectivity through another backup path and hence make the topology more resilient to any node failures or departures.

2 Related Work

A significant amount of research has been directed at power control algorithms for wireless mobile networks but a very few consider the problem of minimizing energy consumption and providing fault tolerance simultaneously.

Ramanathan et al. [8] considered the problem of adjusting the transmission powers of nodes and presented two *centralized algorithms* CONNECT and BICONN-AUGMENT. They introduced two heuristics to deal with the dynamics of the mobile environment. But neither heuristic absolutely preserves connectivity, even if it is achievable in principle. Cone-Based Topology Control (CBTC), proposed by Li et al. [5], generates a graph structure. A serious drawback of the algorithm is the need to decide on the suitable initial power level and the increment at each step. Bahramgiri et al. [1] augmented the CBTC algorithm [5] to provide fault tolerance. However there is no guarantee that the proposed modification preserves minimum-energy paths.

Rodoplu and Meng [9] addressed a work targeting significant reductions in energy consumption. They introduced the notion of relay region based on a specific power model. Their work provides a distributed position-based network protocol optimized for minimum-energy consumption in mobile wireless networks. However the algorithm does not consider fault tolerance. Li [4] modified the algorithm of Rodoplu and Meng [9]. The sub-network constructed by their algorithm is provably smaller than that constructed by Rodoplu and Meng [9]. But the algorithm has the problem of partially-enclosed nodes. If a node is partially-enclosed, it has to use its maximum transmission power, which will soon drain out its battery power. Also this algorithm does not consider fault tolerance.

Shen et al. [10] proposed a distributed topology control algorithm, which preserves minimum-energy property and bi-connectivity. But the algorithm uses another algorithm presented in [3] to identify cut vertices which can not be done using *local information*. Acquiring global topology information is expensive and wastes more energy. Hajiaghayi et al. [2] addressed minimizing power while maintaining k-fault tolerance. However their distributed algorithm uses MST for which no locally computable algorithm is available. So, their distributed algorithm is not a local algorithm. Ning Li and Jennifer C. Hou [6] considered k-connectivity of wireless network to meet the requirement of fault tolerance. They presented a centralized greedy algorithm, called Fault-tolerant Global Spanning Subgraph ($FGSS_k$) and based on $FGSS_k$, proposed a localized algorithm, called Fault-tolerant Local Spanning Subgraph ($FLSS_k$). But their work, not necessarily, contains minimum-energy paths between any two node.

3   Definitions

3.1   Power Model

We use the same power model as [7]. Here, we assume the well known, generic, two-way, channel path loss model where the minimum transmission power is a function of distance. To send a packet from node x to node y, separated by distance $d(x, y)$, the minimum necessary transmission power is approximated by

$$P_{trans}(x, y) = t \times d^\alpha(x, y) \qquad (2)$$

where $\alpha \geq 2$ is the path loss factor and t is a constant. Signal reception is assumed to cost a fixed amount of power denoted by r. Thus, the total power required for one-hop transmission between x and y becomes

$$P_{trans}(x, y) = t \times d^\alpha(x, y) + r \qquad (3)$$

The model assumes that each node is aware of its own position with a reasonable accuracy, e.g., via a GPS device.

3.2   Smallest Minimum-Energy Path-Preserving Subgraph of G

We say that a graph $G' \subseteq G$ is a minimum-energy path-preserving graph or, alternatively, that it has the minimum-energy property, if for any pair of nodes (u, v) that are connected in G, at least one of the (possibly multiple) minimum-energy paths between u and v in G also belongs to G'. Minimum-energy path-preserving graphs were first defined in [4]. Typically, many minimum-energy path-preserving graphs can be formed from the original graph G. The smallest of such subgraphs of G is $G_{min} = (V, E_{min})$, where $(u, v) \in E_{min}$ iff there is no path of length greater than 1 from u to v that costs less energy than the energy required for a direct transmission between u and v. Let $G_i = (V, E_i)$ be a subgraph of $G = (V, E)$ such that $(u, v) \in E_i$ iff $(u, v) \in E$ and there is no path of length i that requires less energy than the direct one-hop transmission between u and v. Then G_{min} can be formally defined as follows:

$$G_{min} = G_2 \bigcap G_3 \bigcap G_4 \bigcap ..... \bigcap G_{n-1} \tag{4}$$

Any subgraph G' of G has the minimum-energy property iff $G' \supseteq G_{min}$. Thereby, each of $G_i \supseteq G_{min}$, for any $i = 2, 3, ..., n-1$ is a minimum-energy path preserving graph.

3.3   Cover Region and Cover Set

Consider a pair of nodes (s, f), such that f lies within the transmission range of s, i.e., is reachable by s at P_{max}. Consider the set of all points that can possibly act as relays between s and f, such that it would be more power efficient for s to use an intermediate node located at one of those points instead of sending directly to f. We will call it the cover region of s and f and denote by $C(s, f)$. The collection of all nodes falling into the cover region of s and f will be called the cover set of s and f. Formally the cover region and cover set, are described by the following definition. The Cover Region $C(s, f)$ (Figure 1) of a pair of nodes (s, f) in G, where f is reachable from s is defined as:

$$C(s, f) = \{< x, y > | td^\alpha(s, < x, y >) + td^\alpha(< x, y >, f) + r \leq td^\alpha(s, f)\} \tag{5}$$

where $a \geq 2$. In the above equation, $d(s, < x, y >)$ denotes the distance between node s and a hypothetical node located at $< x, y >$ and r is the fixed receiving power. The Cover Set of the same pair (s, f) in G is defined as:

$$\xi_G(s, f) = \{v | v \in V \land Loc(v) \in C(s, f)\} \tag{6}$$

where $Loc(v)$ is the location of the node positioned at $< x, y >$ in the network.

3.4   Articulation Point

A vertex v in a connected graph G is an articulation point iff the deletion of vertex v together with all edges incident to v disconnects the graph into two or more nonempty components. A graph G is bi-connected iff it contains no articulation points.

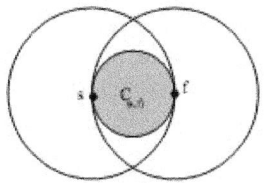

Fig. 1. The cover region between s and f

3.5 Minimum-Energy Path-Preserving Bi-connected Graph

A graph G' is a minimum-energy path-preserving bi-connected graph iff it contains no articulation points and it keeps pairwise minimum-energy paths. A graph $G' = (V, E')$ is a minimum-energy path-preserving bi-connected subgraph of a bi-connected graph $G = (V, E)$ iff $E' \in E$ and G' is bi-connected and for each pair of $(u, v) \in V$, minimum-energy paths between u and v in G are also in G'.

A variety of minimum-energy path-preserving graphs can be created if one or two hops' neighbor information is available. For example, let us consider a graph $G_2^1 = (V, E_2^1)$ which is a subgraph of $G = (V, E)$ such that $(u, v) \in E_2^1$ iff $(u, v) \in E$ and there is no path of length two that requires less energy than the direct path between u and v. Shortly, $(u, v) \in E_2^1$ iff $\xi_G(u, v)$ is empty. Note that such graph can be locally constructed if only one hop neighbors' position information is available to a node because a path of length two between u and v can only be created by using a neighbor z of u which is also a neighbor of v.

Another minimum-energy path-preserving graph is denoted as $G_2^2 = (V, E_2^2)$, which is another subgraph of $G = (V, E)$ such that $(u, v) \in E_2^2$ iff $(u, v) \in E$ and there are no two or more vertex-disjoint paths of length two requiring less energy than the direct path between u and v. Shortly, $(u, v) \in E_2^2$ iff $|\xi_G(u, v)| < 2$ where $|\xi_G(u, v)|$ represents number of elements in $\xi_G(u, v)$. Such a graph can be constructed if a node knows the position of all single hop neighbors.

Now let us define another graph $G_3^2 = (V, E_3^2)$, a subgraph of $G = (V, E)$ such that $(u, v) \in E_3^2$ iff $(u, v) \in E$ and there are no two or more vertex-disjoint paths of length three built with the neighbors of either u or v that require less energy than the direct path between u and v. However, to construct such a graph a node must know all its one hop and two hop neighbors exact location.

Mixing G_2^2 and G_3^2 we define $G_{2|3}^2 = (V, E_{2|3}^2)$ as the subgraph of $G = (V, E)$ such that $(u, v) \in E_{2|3}^2$ iff $(u, v) \in E$ and there are no two or more vertex-disjoint paths of length *two or three* requiring less energy than the direct path between u and v.

In this paper, we propose distributed algorithms to construct G_2^1, G_2^2, and $G_{2|3}^2$. We also prove that G_2^2 and $G_{2|3}^2$ are *minimum-energy path-preserving bi-connected* graphs.

4   Algorithms

In this section we describe distributed algorithms for constructing G_2^1, G_2^2, and $G_{2|3}^2$ topologies for a given topology G. The notations used in this section indicate the same meaning as in definition section. As we are considering distributed algorithm, the following algorithms run at each node as per necessity.

Consider a node s is constructing either of G_2^1, G_2^2, and $G_{2|3}^2$. At first s broadcasts a single *neighbor discovery message* (NDM) at the maximum power P_{max}. All nodes receiving the NDM from s send back a reply. While s collects the replies of its neighbors, it learns their identities and locations. It also constructs the cover sets with those neighbors. Initially, all those sets are empty. The set $N_G(s)$, which also starts with empty set, keeps track of all the nodes discovered in the neighborhood of s in G. Whenever s receives a reply to its NDM from a node v, it executes the algorithm *updateCoverRegion(s, v)* described in Algorithm 1. After running *updateCoverRegion(s, v)*, node s executes one of the following algorithms presented in 4.1, 4.2 and 4.3 depending on the subgraph it likes to construct.

Algorithm 1. *updateCoverRegion(s, v)*

1: **for each** $w \in N_G(s)$ **do**
2:    **if** $Loc(v) \in C(s, w)$ **then**
3:        $\xi_G(s, w) = \xi_G(s, w) \cup \{v\}$
4:    **else if** $Loc(w) \in C(s, v)$ **then**
5:        $\xi_G(s, v) = \xi_G(s, v) \cup \{w\}$
6:    **end if**
7:    $N_G(s) = N_G(s) \cup \{v\}$
8: **end for**

4.1 G_2^1 Topology Construction

Assume the algorithm (Algorithm 2) is running on node s. For each neighbor v of s in G this algorithm checks whether the cover set between s and v is empty or not. If it is empty then v is included into the neighbor set of s in G_2^1 since it indicates that there is no node that can be used as relay to transmit message using lower energy than the direct path between s and v. Otherwise, v is not included into the neighbor set of s in G_2^1.

Algorithm 2. G_2^1 TOPOLOGY CONSTRUCTION

1: **for each** $v \in N_G(s)$ **do**
2:    **if** $\xi_G(s, v)$ is empty **then**
3:        $N_{G_2^1}(s) = N_{G_2^1}(s) \cup \{v\}$
4:    **end if**
5: **end for**

4.2 G_2^2 Topology Construction

Assume the algorithm (Algorithm 3) is running on node s. For each neighbor v of s in G this algorithm checks whether number of nodes in the cover set is less than 2 or not. If the number of nodes in the cover set is less than 2 then v is included into the neighbor set of s in G_2^2 since it indicates that there are no two or more nodes that can be used as relay to transmit message using lower energy than the direct path between s and v. Otherwise, v is not included into the neighbor set of s in G_2^2.

Algorithm 3. G_2^2 TOPOLOGY CONSTRUCTION

1: **for** each $v \in N_G(s)$ **do**
2:     **if** $|\xi_G(s,v)| < 2$ **then**
3:         $N_{G_2^2}(s) = N_{G_2^2}(s) \cup v$
4:     **end if**
5: **end for**

4.3 $G_{2|3}^2$ Topology Construction

Assume the algorithm (Algorithm 4) is running on node s. Initially it is assumed that all the links in G are also in $G_{2|3}^2$. Then for each neighbor v of s in G this algorithm checks whether number of nodes in the cover set is greater than or equal to 2 or not.

If the number of nodes in the cover set $\xi_G(s,v)$ is greater than or equal to 2 then the vertex v is removed from $N_{G_{2|3}^2}(s)$ as it indicates that there are two or more lower energy paths than the direct path between s and v. If the cover set consists of only one node z, then the algorithm checks the presence of any path of length 3 built with neighbors of s excluding z. This path of length 3 is denoted by P_{-z} in the algorithm where $-z$ indicates that z is excluded and P_{-z} is boolean. If P_{-z} is true then there exists such path of length 3 otherwise not. So, if P_{-z} is true then v is removed from $N_{G_{2|3}^2}(s)$.

Algorithm 4. $G_{2|3}^2$ TOPOLOGY CONSTRUCTION

1: $N_{G_{2|3}^2}(s) = N_G(s)$
2: **for** each $v \in N_G(s)$ **do**
3:     **if** $|\xi_G(s,v)| \geq 2$ **then**
4:         $N_{G_{2|3}^2}(s) = N_{G_{2|3}^2}(s) - \{v\}$
5:     **else if** $\xi_G(s,v)$ is $\{z\}$ and P_{-z} is true **then**
6:         $N_{G_{2|3}^2}(s) = N_{G_{2|3}^2}(s) - \{v\}$
7:     **else if** $\xi_G(s,v)$ is empty and at least two vertex-disjoint paths of length 3 consisting of the neighbors of s exist **then**
8:         $N_{G_{2|3}^2}(s) = N_{G_{2|3}^2}(s) - \{v\}$
9:     **end if**
10: **end for**

If the cover set is empty, v is removed from $N_{G^2_{2|3}}(s)$ if there exists at least two vertex-disjoint paths of length 3. These two vertex-disjoint paths must be built with neighbors of s.

5   Theorems

Lemma 1. For a given topology $G = (V, E)$ each pair of $(u, v) \in V$, $\xi_G(u, v)$ is empty iff $(u, v) \in E$ is a minimum-energy path.

Proof. Since $(u, v) \in E$ is a minimum-energy path it indicates that there is no node in their cover region $C(u, v)$ which can be used as relay to transmit information between u and v with less energy than the direct transmission between u and v. So, $\xi_G(u, v)$ is empty.

On the other hand, if $\xi_G(u, v)$ is empty then there is no node in their cover region which implies no node exists that can be used as relay to transfer information between u and v with less energy than the direct transfer between u and v. So, $(u, v) \in E$ is a minimum-energy path.

Lemma 2. G^2_2 keeps all edges of G^1_2.

Proof. Each edge $(u, v) \in E(G)$ is preserved in G^1_2, iff the cover set $\xi_G(u, v)$ is empty. In case of G^2_2 two cases arise:

Case 1: $(u, v) \in E(G)$ is preserved when the cover set $\xi_G(u, v)$ is empty.
Case 2: $(u, v) \in E(G)$ is also preserved when the cover set $\xi_G(u, v)$ contains only one node. So, G^2_2 keeps all edges of G^1_2 by case 1.

Lemma 3. $G^2_{2|3}$ keeps all edges of G^1_2.

Proof. Each edge $(u, v) \in E(G)$ is preserved in G^1_2, iff the cover set $\xi_G(u, v)$ is empty. In $G^2_{2|3}$, three types of cases arise:

Case 1: The edge $(u, v) \in E(G)$ is preserved in $G^2_{2|3}$ when the cover set $\xi_G(u, v)$ is empty.
Case 2: The edge $(u, v) \in E(G)$ is preserved in $G^2_{2|3}$ iff there are no less costly two paths, one of length 2 and another of length 3.
Case 3: The edge $(u, v) \in E(G)$ is preserved in $G^2_{2|3}$ iff there are no less costly two paths of length 3. So, $G^2_{2|3}$ preserves all edges of G^1_2 by case 1.

Theorem 1. G^1_2 topology preserves minimum-energy paths.

Proof by Induction. Base Case. Let $(u, v) \in E(G)$ and is a minimum-energy path between u and v. Applying Lemma 1 $\xi_G(u, v)$ is empty. So, $(u, v) \in E(G^1_2)$.

Induction Hypothesis. Let $u \sim w$, a minimum-energy path in G, is also in G^1_2 where $u \in V$ and $w \in V$ where V indicates vertex set of G and G^1_2.

Induction Step. Let $x \in V$. If $u \sim x$ is a minimum-energy path in G consisting of $u \sim w$ and (w, x), then (w, x) must be a minimum-energy path. Since

$(w, x) \in E(G)$ and is a minimum-energy path, from base case we can say that $(w, x) \in E(G_2^1)$. And from Induction Hypothesis, $u \sim w$, a minimum-energy path in G, is also in G_2^1. Merging these conditions, $u \sim x$ is a minimum-energy path and it is contained in G_2^1. So, G_2^1 topology preserves minimum-energy path.

Theorem 2. G_2^2 and $G_{2|3}^2$ topologies preserve minimum-energy path.

Proof. It is clear that $E(G_2^1) \subseteq E(G_2^2)$ by applying Lemma 2. And $E(G_2^1) \subseteq E(G_{2|3}^2)$ by applying Lemma 3. Since G_2^1 preserves minimum-energy path from Theorem 1, G_2^2 and $G_{2|3}^2$ topologies must preserve minimum-energy path.

Theorem 3. G_2^1 topology may not ensure bi-connectivity.

Proof by Case. Let, $G = (V, E)$ is a bi-connected graph where $V = \{1, 2, 3\}$ and $E = \{a, b, c\}$ (Figure 2(a)). And two distinct paths between 1 and 3 are $< 1, 2, 3 >$ and $(1, 3)$.

If $\xi_G(1, 3) = \{2\}$ then according to G_2^1 topology construction algorithm $(1, 3) \notin E(G_2^1)$. Then now, G_2^1 topology has a path $< 1, 2, 3 >$ where 2 is an articulation point (Figure 2(b)). So, Bi-connectivity may not ensure.

(a) (b)

Fig. 2. Proof by case: (a) Bi-connected (b) Not bi-connected

Theorem 4. G_2^2 topology ensures bi-connectivity.

Proof by contradiction. Let $G = (V, E)$ is a bi-connected graph. Consider any vertex u and v is a neighbor of u. There exists less costly two paths $< u, x, v >$ and $< u, y, v >$ where $\{x, y\} \subseteq \xi_G(u, v)$. So, according to G_2^2 topology construction algorithm $(u, v) \notin E(G_2^2)$. Now, we have to prove that $G - (u, v)$ is still bi-connected. Let $G' \equiv G - (u, v)$ is not bi-connected and there must be at least one articulation point w. So, $G' - \{w\}$ is not connected. Three types of cases arise in choosing w:

Case 1. If $w \notin \{x, y\}$ and $w \notin \{u, v\}$ then $G' - \{w\}$ has $< u, x, v >$ and $< u, y, v >$ paths.
Case 2. If $w \in \{x, y\}$ then $G' - \{w\}$ has at least one path between u and v. Such as if $w = x$ then $< u, y, v >$ still exists and if $w = y$ then $< u, x, v >$.
Case 3. If $w \in \{u, v\}$ then all vertices in $G' - \{w\}$ are in same component.

So, u and v are always connected. This implies that there is no possibility to find out such w as an articulation point which contradicts our assumption that

w is an articulation point. So, $G - (u, v)$ is bi-connected and G_2^2 topology ensures bi-connectivity.

Theorem 5. $G_{2|3}^2$ topology ensures bi-connectivity.

Proof by contradiction. Let $G = (V, E)$ is a bi-connected graph. Consider any vertex u and v is a neighbor of u. There exists less costly two paths $< u, X, v >$ and $< u, Y, v >$ where X and Y are the set of relay nodes. So, according to $G_{2|3}^2$ topology construction algorithm $(u, v) \notin E(G_{2|3}^2)$. Now, we have to prove that $G - (u, v)$ is still bi-connected. Let $G' \equiv G - (u, v)$ is not bi-connected and there must be at least one articulation point w. So, $G' - \{w\}$ is not connected. Three types of cases arise in choosing w:

Case 1. If $w \notin \{X \cup Y\}$ and $w \notin \{u, v\}$ then $G' - \{w\}$ has $< u, X, v >$ and $< u, Y, v >$ paths.
Case 2. If $w \in \{X \cup Y\}$ then $G' - \{w\}$ has at least one path between u and v. Such as if $w \in X$ then $< u, Y, v >$ still exists and if $w \in Y$ then $< u, X, v >$.
Case 3. If $w \in \{u, v\}$ then all vertices in $G' - \{w\}$ are in same component.

So, u and v are always connected. This implies that there is no possibility to find out such w as an articulation point which contradicts our assumption that w is an articulation point. So, $G - (u, v)$ is bi-connected and $G_{2|3}^2$ topology ensures bi-connectivity.

6   Simulation Output

To evaluate performance of our algorithm we have created some sample ad hoc networks where nodes are deployed under uniform distribution. Initially we

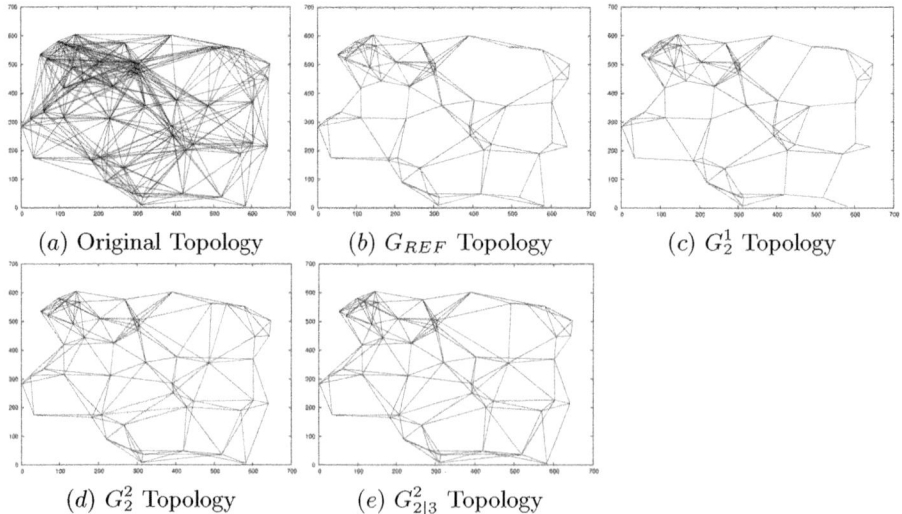

(a) Original Topology   (b) G_{REF} Topology   (c) G_2^1 Topology

(d) G_2^2 Topology   (e) $G_{2|3}^2$ Topology

Fig. 3. Topologies for 50 nodes

Table 1. Average number of edges in different topologies

Number of nodes	Original G_{REF}	G_2^1	G_2^2		$G_{2\|3}^2$
50	391	146	140	187	184
75	435	184	173	238	237
100	556	245	234	315	314

Table 2. Percentage of node pairs having vertex-disjoint backup path besides minimum-energy path

Number of nodes	Original G_{REF}	$G_2^1(\%)$	$G_2^2(\%)$	$G_{2\|3}^2(\%)$
50	100.0	100.0	93.75	99.30	98.30
75	94.68	94.68	87.75	93.92	93.92
100	100.0	100.0	92.79	97.99	97.98

deployed $50 - 100$ nodes over a flat square area of $650m \times 650m$. Figure 3(a) shows a typical deployment of 50 nodes, each having a maximum communication range of $250m$. This is the starting graph G for our algorithm. The later one in Figure 3(b) is our reference graph G_{REF} which is created for comparison purpose with G_2^1, G_2^2 and $G_{2|3}^2$. G_{REF} is constructed in the following way. **First**, we find the minimum-energy path between a pair of vertices such as u, v by Dijkstra's algorithm. **Then** we find the second minimum-energy path between u, v by Dijkstra's algorithm which is exclusive of the vertices of the former one. This process continues for all pairs of the vertices. **Finally** we remove all the edges not included in the paths chosen by this process. The graphs found by G_2^1, G_2^2, and $G_{2|3}^2$ topology construction algorithms from graph G are in Figure 3(c), Figure 3(d) and Figure 3(e) respectively.

Similarly, we have run simulations for 75 nodes and 100 nodes to compare the outputs with G_{REF}. The result is presented in Table-1, 2. From Table-1 we see that the edge reduction in G_2^1 is the most. But it may not ensure bi-connectivity. Among two bi-connected graphs $G_{2|3}^2$, and G_2^2, the former one reduces more edges than the later one. We now compare the issue of fault tolerance among various topology construction algorithms. Fault tolerance means maintaining network connectivity in case of any breakdown of the network. In case of bi-connected graphs, there are two disjoint paths between any two nodes. When one path gets destroyed, another path can still sustain the connectivity. So if a topology is bi-connected then it can ascertain fault tolerance. Since G_2^1 topology may not ensure bi-connectivity, so it will not always ensure fault tolerance. But G_2^2 and $G_{2|3}^2$ are both bi-connected. So, they ensure fault tolerance.

Although G_2^2 and $G_{2|3}^2$ are bi-connected and they ensure two vertex-disjoint paths in any pair of vertices and hence fault-tolerant but none of them can be minimum-energy path. In Table-2 we show what percentage of node pairs have a vertex-disjoint backup path besides the minimum-energy path.

7 Conclusion

The algorithm we have presented to build minimum-energy path-preserving bi-connected graph is a distributed algorithm that maintains minimum-energy paths as well as provides fault tolerance in ad-hoc wireless networks. We have proposed G_2^1, G_2^2 and $G_{2|3}^2$ topology construction algorithms of which G_2^1 seems less robust but G_2^2 and $G_{2|3}^2$ have higher fault tolerance and robustness. Simulation results show that $G_{2|3}^2$ has less number of edges than G_2^2 on the average. However G_2^2 preserves vertex-disjoint backup paths in addition to minimum-energy paths among more node pairs.

References

1. Bahramgiri, M., Hajiaghayi, M., Mirrokni, V.S.: Fault-tolerant and 3-dimensional distributed topology control algorithms in wireless multi-hop networks. In: Proc. Eleventh International Conference on Computer Communications and Networks (ICCCN), October 2002, pp. 392–397 (2002)
2. Hajiaghayi, M., Immorlica, N., Mirrokni, V.S.: Power optimization in fault-tolerant topology control algorithms for wireless multi-hop networks. IEEE/ACM Trans. Netw. 15(6), 1345–1358 (2007)
3. Kazmierczak, A., Radhakrishnan, S.: An optimal distributed ear decomposition algorithm with applications of biconnectivity and outerplanarity testing. IEEE Trans. Parallel Distrib. Syst. 11(2), 110–118 (2000)
4. Li, L., HalpernRahman, J.: Minimum energy mobile wireless networks revisited. In: Proc. IEEE International Conference on Communications (ICC) (June 2001)
5. Li, L., Hlapern, J.Y., Bahl, P., Wang, Y., Watenhofer, R.: Analysis of a con-based topology control algorithm for wireless multi-hop networks. In: ACM Symposium on Principle of Distributed Computing, PODC (2001)
6. Li, N., Hou, J.C.: Flss: A fault tolerant topology control algorithm for wireless networks. In: Proc. of the 10th annual international conference on Mobile computing and networking, pp. 275–286 (2004)
7. Rahman, A., Gburzynski, P.V.: Mac-assisted topology control for ad-hoc wireless network. International Journal of communication Systems 19(9), 976–995 (2006)
8. Ramanathan, R., Rosales-Hain, R.: Topology control of multihop wireless networks using transmit power adjustment. In: Proc. of IEEE INFOCOM 2000, Tel Aviv, Israel, March 2000, pp. 404–413 (2000)
9. Rodoplu, V., Meng, T.: Minimum energy mobile wireless networks. IEEE Journal of Selected Areas in Communications 17, 1333–1344 (1999)
10. Shen, Z., Chang, Y., Cui, C., Zhang, X.: A fault-tolerant and minimum-energy path-preserving topology control algorithm for wireless multi-hop networks. In: Proc. of the International Conference on Computational Intelligence and Security-I (CIS), pp. 864–869 (2005)

Server Guaranteed Cap:
An Incentive Mechanism for Maximizing
Streaming Quality in Heterogeneous Overlays

Ilias Chatzidrossos, György Dán, and Viktória Fodor

School of Electrical Engineering,
KTH, Royal Institute of Technology,
Osquldas väg 10, 100-44, Stockholm, Sweden
{iliasc,gyuri,vfodor}@kth.se

Abstract. We address the problem of maximizing the social welfare in a peer-to-peer streaming overlay given a fixed amount of server upload capacity. We show that peers' selfish behavior leads to an equilibrium that is suboptimal in terms of social welfare, because selfish peers are interested in forming clusters and exchanging data among themselves. In order to increase the social welfare we propose a novel incentive mechanism, Server Guaranteed Cap (*SGC*), that uses the server capacity as an incentive for high contributing peers to upload to low contributing ones. We prove that *SGC* is individually rational and incentive compatible. We also show that under very general conditions, there exists exactly one server capacity allocation that maximizes the social welfare under *SGC*, hence simple gradient based method can be used to find the optimal allocation.

Keywords: p2p streaming, incentive mechanisms, social welfare.

1 Introduction

The goal of peer-to-peer (p2p) streaming systems is to achieve the maximum possible streaming quality using the upload capacities of the peers and the available server upload capacity. In general, the achievable streaming quality depends heavily on the aggregate upload capacity of the peers [1]. Hence, a key problem of p2p streaming systems is how to give incentives to selfish peers to contribute with all their upload capacity. Numerous schemes were proposed to solve this problem (e.g., [2,3]). These schemes relate peers' contribution with the streaming quality they receive: the more a peer contributes, the better streaming quality it can potentially receive. The correlation of peer contribution to the quality it receives is based on the assumption that all peers are capable of contributing but refrain from doing so.

Nevertheless, peers might be unable to have a substantial contribution with respect to the stream rate because of their last-mile connection technology. Most DSL and cable Internet connections are asymmetric, hence peers may have sufficient capacity to, e.g., download high definition video but insufficient for forwarding it. Similarly, in high-speed mobile technologies, such as High Speed Downlink

M. Crovella et al. (Eds.): NETWORKING 2010, LNCS 6091, pp. 315–326, 2010.

Packet Access (HSDPA), the download rates are an order of magnitude higher than the upload rates [4]. Peers using assymetric access technologies would receive poor quality under incentive schemes that offer a quality proportional to the level of peer contribution.

Furthermore, using such incentive schemes, high contributing peers maximize their streaming quality if they prioritize other high contributing peers when uploading data. As a consequence, peers with similar contribution levels form clusters and exchange data primarily among themselves. While high contributing peers can achieve excellent streaming quality this way, the quality experienced by low contributing peers is low, and the average streaming quality in the p2p system is suboptimal.

In order to increase the average streaming quality in the system, we propose a mechanism that gives incentives to high contributing peers to upload to low contributing ones. The mechanism relies on reserving a portion of the server capacity and providing it as a safety resource for high contributing peers who meet certain criteria. We show that high contributing peers gain by following the rules set by the incentive mechanism, and they fare best when they follow the rules truthfully. We also show that due to some basic properties of p2p streaming systems our mechanism can easily be used to maximize the streaming quality.

The rest of the paper is organized as follows. In Section 2, we motivate our work by studying the effect of selfish peer behavior in a push-based p2p streaming overlay. In Section 3, we describe our incentive mechanism and provide analytical results. We show performance results in Section 4. In Section 5 we discuss previous works on incentives in peer-to-peer streaming systems. Finally, Section 6 concludes our paper.

2 Motivation

We consider mesh-based p2p streaming systems to evaluate the effect of selfish peer-behavior. Due to their flexibility and easy maintenance, mesh-based systems received significant attention in the research community [5,6,7,8], and are underlying the majority of commercial streaming systems (e.g., [9], [10]).

2.1 Case Study: A Mesh-Based p2p Streaming System

The streaming system we use as an example was proposed in [7] and was subsequently analyzed in [8,11]. The system consists of a server and N peers. The upload capacity of the server is m_t times the stream rate. For simplicity we consider two types of peers: peers with high upload capacity, called contributors, and peers without upload capacity, called non-contributors. The upload capacity of the contributors is c times the stream rate, while that of non-contributors is zero. We denote by α the ratio of non-contributors in the overlay.

Each peer is connected to d other peers, called its neighbors. The server is neighbor to all peers. Every peer maintains a playout buffer of recent packets, and exchanges information about its playout buffer contents with its neighbors periodically, via so called buffer maps. The server sends a copy of every packet to m_t randomly chosen peers. The peers then distribute the packets among each other according to a forwarding algorithm. The algorithm takes as input the information about packet availability in neighboring peers (known from the buffer maps) and produces a forwarding decision, consisting of a neighbor and a packet sequence number. In this work we consider the Random Peer - Random Packet ($RpRp$) forwarding algorithm. This algorithm has been shown to have a good playback continuity - playback delay tradeoff ([7,11]). According to this algorithm, a sending peer first chooses randomly a neighbor that is missing at least one of the packets the sending peer possesses, then it selects at random the missing packets to send. Peers play out data B time after they were generated by the server, and we refer to this as the playback delay.

To study the impact of peer cooperation in the overlay, we introduce the notion of the generosity factor, which we denote by β. This parameter shows how generous a peer is towards its non-contributing neighbors, and can be expressed as the ratio of the probability of uploading to a non-contributor over the ratio of a peer's non-contributing neighbors. The generosity factor takes values in the interval $[0, 1]$. When $\beta = 1$, the peers are completely generous and they upload to their neighbors regardless of whether they, on their turn, are uploading or not. When $\beta = 0$, peers are not generous at all, or equivalently completely selfish, and will only upload to peers that upload as well.

The generosity level affects the playout probabilities of the contributing and non-contributing peers. At $\beta = 1$ the two playout probabilities are equal. As β decreases, capacity is subtracted from the non-contributors and added to the contributors, and consequently the playout probability of the contributors increases, while that of non-contributors decreases.

2.2   Playout Probability, Individual Utility and Social Welfare

The performance of p2p streaming systems is usually measured in terms of the playout probabilities of the peers, i.e., the probability p_i that peer i receives packets before their playout deadlines [7,11,8]. The impact of the playout probability p_i on the peers' satisfaction is, however, typically influenced by the loss resilience of the audiovisual encoding. To allow for a wide range of encodings, we use utility functions to map the playout probability to user satisfaction. Formally, the utility function is a mapping $u : [0, 1] \rightarrow [0, 1]$. We consider three kinds of utility functions.

Linear function: Utility function of the form $y = a \cdot p_i + b$. An improvement in the playout probability yields the same increase in utility regardless of the already achieved playout probability.

Concave function: Utility is a concave function of the playout probability, that is, the marginal utility is a non-increasing function of the playout probability.

Step function: There is an instantaneous transition from a zero utility to a utility of a unit upon reaching a threshold p_i^*. The peer is only satisfied above the threshold playout probability.

We measure the aggregate utility of the peers, called the social welfare, using the utilitarian welfare model. In the utilitarian welfare model the social welfare is the sum of the utilities, which is equivalent to the average peer utility

$$SWF = (1 - \alpha) \cdot u(p_c) + \alpha \cdot u(p_{nc}), \qquad (1)$$

where p_c and p_{nc} denote the playout probability of contributors and non-contributors respectively.

2.3 The Effects of Selfish Behavior

In the following we show the effects of selfish behavior on the social welfare. The numerical results we show, were obtained using an analytical model and via simulations. The analytical model is an extension of our previous work [11], where we developed a model of the playout probability in a push-based system with homogeneous peer upload capacities. We extended the model to incorporate two types of peers, contributors and non-contributors. The extended model can be found in [12]. The simulation results were obtained using the packet-level event-driven simulator used in [11]. In the simulations, nodes join the overlay at the beginning of the simulation, and are organized into a random d-regular graph. After the overlay with N peers is built, the data distribution starts according to the forwarding algorithm described in Section 2.1. The algorithm is executed in time slots in a way that contributors with capacity c make c forwarding decisions per slot. All results presented in the following refer to an overlay of one server with upload capacity $m_t = 11$ times the stream rate and $N = 500$ peers, where each peer is connected to $d = 30$ neighbors and contributors have an upload capacity of $c = 2$ times the stream bitrate.

Fig. 1a and 1b show the effect of the generosity factor on the playout probability of the contributors and the non-contributors obtained using the model and simulations. The figures also show the average playout probability in the overlay (dotted lines). The ratio of non-contributors is $\alpha = 0.5$.

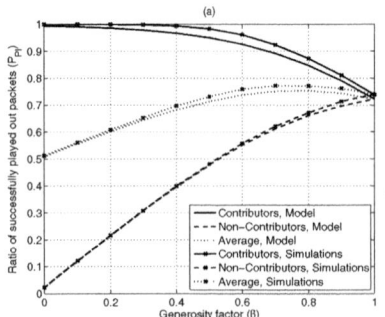 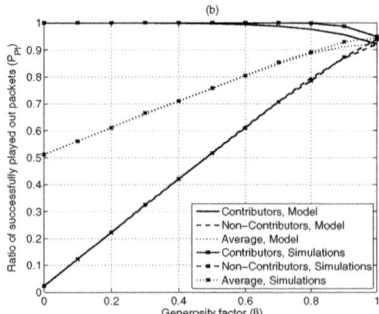

Fig. 1. Ratio of successfully played out packets vs generosity factor β for playback delay of (a) $B = 10$ and (b) $B = 40$. Analytical and simulation results.

Fig. 1a shows a system where the playback delay is small. Clearly, contributors maximize their playout probabilities for $\beta = 0$, but the average playout probability is suboptimal in this case. The average playout probability is suboptimal for $\beta = 1$ as well. For a larger playback delay (Fig. 1b) the average playout probability is maximized for $\beta = 1$.

From the figures we can conclude that the optimal generosity factor depends on the playback delay. At low playback delays high capacity is needed for on time delivery, and therefore contributors can receive and efficiently forward packets only at low values of β. At high playback delays though, increased capacity at contributors leads to marginal gains only, and therefore $\beta = 1$ is optimal. This inefficiency of the forwarding algorithm is due to the lack of coordination between the peers [13]. Under push-based algorithms, like the one consid-

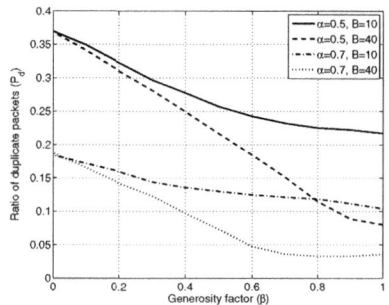

Fig. 2. Ratio of duplicate transmissions vs generosity factor (β) for $B = 10$ and $B = 40$. Simulations.

ered here, the lack of coordination leads to duplicate packet receptions at peers, i.e., a peer receives the same data from more than one of its neighbors. Fig. 2 shows how the probability of receiving duplicate packets increases together with the capacity allocated for forwarding among contributors. This in turn leads to playout probabilities below 1 even when contributors are completely selfish ($\beta = 0$). Similarly, in the case of pull-based systems the lack of coordination may lead to request collision at the peers, with the consequence that some of the requests can not be served and the packet miss ratio can become substantial [14]. Based on these findings, we argue that performance degradation due to peer clustering is intrinsic to uncoordinated p2p dissemination algorithms.

Next, we proceed with our utility based analysis of the overlay. For linear utility function we use $u(p_i) = p_i$, so the utility curve coincides with the curve presented in Fig. 1a and 1b. For concave utility we use a logarithmic function, $u(p_i) = log_{10}(1 + 9p_i)$. For the step function we set the threshold to $p_i^* = 0.95$. Our conclusions do not depend on the the particular values of the parameters and the choice of the logarithmic function.

Fig. 3a and 3b show the social welfare versus the generosity factor for the three kinds of utility functions and for playback delays of $B = 10$ and $B = 40$ respectively. In the case of small playback delay ($B = 10$) the social welfare for the linear and the concave utility functions attains its maximum for $\beta < 1$. For the step function the social welfare equals 0 for high values of β, when contributors are not able to receive at least with probability $p_i^* = 0.95$. As β decreases, there is a transition in utility, but the contributors do not gain anything by becoming more selfish after the transition, and the social welfare remains constant. In the case of large playback delay ($B = 40$), we see that the

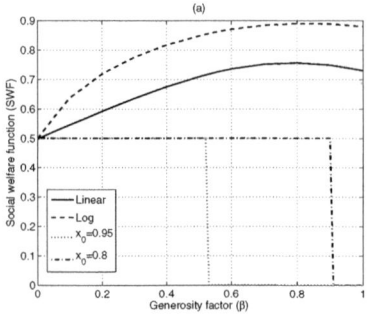 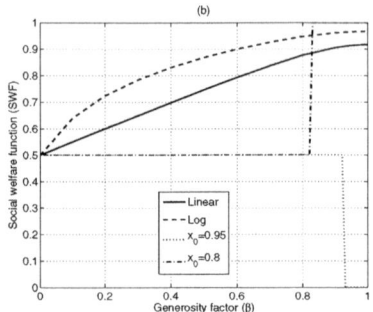

Fig. 3. Social welfare vs. generosity factor β for playback delays of (a) $B = 10$ and (b) $B = 40$. Overlay with $\alpha = 0.5$. Analytical results.

social welfare for linear and concave utility functions attains its maximum for $\beta = 1$. For the step function we observe a similar transition of the social welfare as for $B = 10$, but at a higher value of the generosity factor β. The transition occurs where the contributors achieve a playout probability of $p_i^* = 0.95$. To understand the importance of the threshold value, let us consider $p_i^* = 0.8$. We see in Fig. 1b that in this case, the social welfare becomes maximal for $\beta \geq 0.8$, as both contributors and non-contributors achieve playout probabilities above the threshold. To summarize, we draw two conclusions from these figures. First, for the linear and concave utility functions the value of β that maximizes the social welfare is a function of the playback delay, but in general $\beta = 0$ is far from optimal. Second, for the step function the threshold value p_i^* plays an important role in whether $\beta = 0$ is optimal.

3   The SGC Incentive Mechanism

Our work is motivated by the observation that the peers' selfish behavior leads to a loss of social welfare. In our solution, we exploit the inability of contributors to achieve the maximum playout probability by being selfish and offer them seamless streaming if they increase their generosity, that is if they serve non-contributors as well. In the following, we describe our incentive mechanism, called Server Guaranteed Cap (SGC).

Under SGC, there are two types of data delivery: *p2p dissemination* and *direct delivery* from the server. Fresh data is distributed in the overlay using *p2p dissemination*: the peers forward the data according to some forwarding algorithm. Contributors can also request data *directly from the server* if they do not exceed a threshold playout probability of T_p via p2p dissemination. In our scheme the server ensures that by combining p2p dissemination and direct delivery the contributors possess all data with probability 1. In order to be able to serve the requests for direct delivery, the server reserves m_r of its total upload capacity m_t for direct delivery. m_r has to be large enough to cap the gap between the threshold probability T_p and 1. Given the number of contributors in the overlay and the reserved capacity m_r, the server can calculate the threshold value of the playout probability below which it would not be able to cap all contributors

$$T_p = 1 - \frac{m_r}{(1 - \alpha) \cdot N}. \tag{2}$$

The server advertises the threshold value T_p as the maximum playout probability that contributors should attain through p2p dissemination. In turn the peers report their playout probabilities p_i achieved via p2p dissemination to the server. Based on these reports, the server knows which are the contributors with $p_i \leq T_p$, that is, which contributors are entitled for direct delivery.

3.1 Properties of SGC

In the following we show two important properties of the proposed mechanism: ex-post individual rationality and incentive compatibility [15].

Ex-post individual rationality means that a contributing peer does *not* achieve *inferior* performance by following the rules of the mechanism irrespective of the playout probability it would achieve without following the mechanism.

Proposition 1. *The SGC mechanism is ex-post individually rational.*

Proof. Consider that the server advertises a threshold probability of T_p. All contributors that receive up to $p_i \leq T_p$ via p2p dissemination are entitled to pull the remaining $1 - T_p$ directly from the server. Hence a peer with $p_i = T_p$ receives data with probability $P_i = p_i + (1 - T_p) = 1$, which is at least as much as it would achieve by not following the rules of the mechanism.   □

Since *SGC* relies on peers reporting their playout probabilities to the server, it is important that peers do not have an incentive to mis-report their playout probabilities. In the following we show that *SGC* satifies this criterion, i.e., it is incentive compatible.

Proposition 2. *The SGC mechanism is incentive compatible.*

Proof. Let us denote the playout probability of peer i by p_i and the probability it reports by $\bar{p}_i$. As before T_p is the threshold probability that the contributors must not exceed in order to be directly served by the server, and m_r is the corresponding reserved capacity at the server. Contributors can receive $m_r/(1 - \alpha)N = 1 - T_p$ share of the stream from the server directly if they report $\bar{p}_i \leq T_p$. Consequently, if peer i achieves $p_i \leq T_p$ and reports it truthfully ($\bar{p}_i = p_i$), it receives data with probability $P_i = min(1, p_i + (1 - T_p))$. If $p_i > T_p$ and peer i reports truthfully, it receives with probability $P_i = p_i$.

Clearly, peer i can not benefit from over-reporting its playout probability, so we only have to show that it has no incentive for under-reporting it either. In order to show this we distinguish between three cases.

- $\bar{p}_i < p_i \leq T_p$: the playout probability that the peer will finally receive will be $P_i = min(1, p_i + 1 - T_p) \leq 1$, which is the same that it would receive if it were telling the truth.

- $\overline{p_i} \leq T_p < p_i$: the playout probability that the peer will finally receive will be $P_i = min(1, p_i + 1 - T_p) = 1$. The peer could achieve the same by having $p_i = T_p$ and reporting $\overline{p} = p$.
- $T_p < \overline{p_i} < p_i$: the peer is not entitled to direct delivery, so $P_i = p_i$.   □

3.2 Optimal Server Capacity Allocation

A key question for the implementation of the mechanism is how to determine the advertised probability threshold T_p, that is, how to find the reserved capacity m_r, that maximizes the social welfare. Since the server capacity is fixed, the choice of m_r affects the server capacity available for the p2p dissemination, and hence the efficiency of the data delivery through p2p dissemination.

In the following we show that for a wide class of p2p streaming systems there is a unique value of m_r that maximizes the social welfare, and this class is characterized by the fact that the marginal gain of increasing the upload capacity in the system is non-increasing.

Let us express the playout probability achieved through p2p dissemination as a function of the overlay size N and the p2p upload capacity. We denote the p2p upload capacity by C, and it is the sum of $m_t - m_r$ and the aggregate upload capacity of the contributors. We define the mapping $f : (\mathbb{N}, \mathbb{R}) \to [0, 1]$ of number of peers in an overlay and the p2p upload capacity, to the average playout probability of the peers. Clearly, f depends on the implemented forwarding algorithm.

Definition 1. *A p2p streaming system is called efficient if the playout probability of the peers is a concave function of the p2p upload capacity C.*

We only consider *linearly scalable* systems, where the efficiency of the forwarding algorithm does not depend on the overlay size for a given ratio of peers over p2p upload capacity, that is, $f(k \cdot N, k \cdot C) = f(N, C), \forall k \in \mathbb{R}$. Given that, we formulate the following proposition.

Proposition 3. *The construction of an efficient p2p streaming system is always possible regardless of the characteristics of the forwarding algorithm used.*

Proof. Suppose that f is strictly convex in an interval in its domain. Formally, there exist p2p upload capacity values C_1, C_2, with $C_1 < C_2$, for which it holds

$$\lambda f(N, C_1) + (1 - \lambda)f(N, C_2) > f(N, \lambda C_1 + (1 - \lambda)C_2), \forall \lambda \in (0, 1). \quad (3)$$

Let us consider a system with N peers and p2p upload capacity C, $C_1 < C < C_2$. We split the overlay into two partitions, one with size λN and server capacity λC_1 and the other with $(1 - \lambda)N$ peers and $(1 - \lambda)C_2$ server capacity, such that $C = \lambda C_1 + (1 - \lambda)C_2$. For the two overlays we have that

$$f(\lambda N, \lambda C_1) = f(N, C_1) \quad (4)$$
$$f((1 - \lambda)N, (1 - \lambda)C_2) = f(N, C_2). \quad (5)$$

Consequently, for the original overlay we have $f(N,C) = f(N, \lambda C_1 + (1-\lambda)C_2) = \lambda f(N, C_1) + (1-\lambda)f(N, C_2)$, which contradicts (3). That is, by splitting the overlay and applying the same forwarding algorithm in the two parts independently, we can create an efficient p2p streaming system.   □

For a given server capacity, m_t, SGC requires that the server caps the gap between the playout probability p_i achieved via p2p dissemination and 1. Therefore, the value of m_r should be such that contributors can achieve T_p for some $\beta \in [0,1]$.

Definition 2. *The feasible range for the implementation of SGC is then defined as* $\mathbb{M}_r = \{m_r \in (0, m_t) : m_r \geq (1 - f((1-\alpha) \cdot N, C)) \cdot (1-\alpha) \cdot N\}$, *where C is the p2p upload capacity.*

Proposition 4. *For an efficient system, a feasible range of server upload capacities* $\mathbb{M}_r$ *and a concave utility function, the social welfare is a concave function of the reserved server capacity* m_r.

Proof. The social welfare of the system is given as

$$SWF = (1-\alpha) \cdot u\left(f((1-\alpha) \cdot N, C_c) + \frac{m_r}{(1-\alpha) \cdot N}\right) + \alpha \cdot u\left(f(\alpha \cdot N, C_{nc})\right), \quad (6)$$

where C_c and C_{nc} are the upload capacities allocated to contributors and to non-contributors respectively (as a function of β), and $C = C_c + C_{nc}$. Under the SGC mechanism, the contributors receive the stream with probability 1, so the above equation becomes

$$SWF = (1-\alpha) \cdot u(1) + \alpha \cdot u\left(f(\alpha \cdot N, C_{nc})\right). \quad (7)$$

The first part of the sum is constant in m_r, so we have to show that the second part of the sum is a concave function of m_r. First we evaluate C_{nc}. Since our system is efficient the playout probability p_i is concave with respect to C_c, or equivalently C_c is convex in $p_i = T_p$. Since $C_{nc} = C_t - C_c$, C_{nc} is concave in $1 - T_p$, which in turn is linear in m_r. Therefore, C_{nc} is a concave function with respect to m_r. Consequently the composite function $f(C_{nc})$ is concave as well with respect to m_r, as a composition of non-decreasing concave functions [16]. For the same reason $u \circ f$ is concave with respect to m_r, which proves the proposition.   □

A consequence of *Proposition 4* is that the social welfare function SWF has exactly one, global, maximum on $\mathbb{M}_r$. Hence, the server can discover the optimal amount of reserved capacity m_r by using a gradient based method starting from any $m_r \in \mathbb{M}_r$.

4   Numerical Results

In the following we present numerical results that quantify the gain of the proposed incentive mechanism. The social welfare with respect to the reserved capacity by the server is shown in Fig. 4. The total capacity of the server is $m_t = 11$.

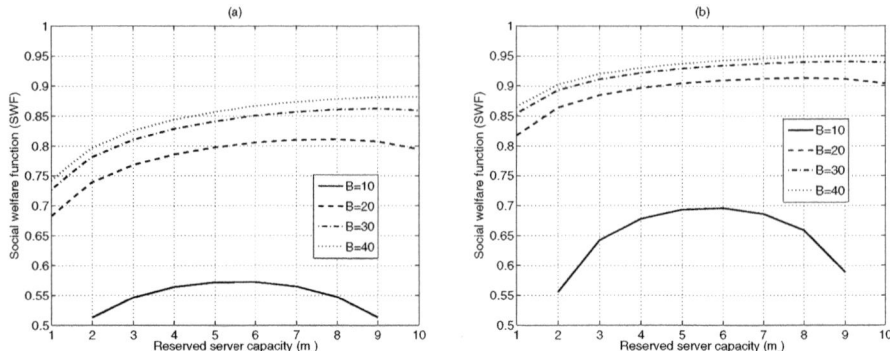

Fig. 4. Social welfare versus reserved server capacity for different playback delays. Linear (a) and logarithmic (b) utility functions. Overlay with $N = 500$, $d = 30$ and $m = 11$.

We can see that the feasible region of SGC depends on the playback delay for the system. For B=10, it holds that $m_r \in [2, 9]$, while for larger playback delays $m_r \in [1, 10]$. The increase of m_r triggers two contradicting effects. On one side, it increases the playout probability of contributors through the direct delivery. On the other side, it decreases the efficiency of the p2p dissemination phase, since the amount of server capacity dedicated to that type of dissemination is decreased. The social optimum is at the allocation where the rate of decrease of the efficiency of p2p dissemination becomes equal to that of the increase achieved through the direct delivery.

Finally, we note that even using SGC there is a *loss of social welfare* compared to the *hypothetical case* when β is optimized by generous peers to maximize the social welfare. We can observe this loss by comparing the maximum social welfare obtained in Figs. 4 to that in Fig. 3 (for $B = 10$ and $B = 40$). This loss of social welfare is the *social cost of the selfishness of peers*.

5   Related Work

A large number of incentive mechanisms was proposed in recent years to solve the problem of free-riding in p2p streaming systems. These mechanisms are either based on pairwise incentives or on global incentives.

Pairwise incentive schemes were inspired by the tit-for-tat mechanism used in the BitTorrent protocol [3,17]. However, tit-for-tat, as used in BitTorrent, was shown not to work well in live streaming with random neighbor selection [3,17]. The authors in [17] proposed an incentive mechanism for neighbor selection based on playback lag and latency among peers, achieving thus better pairwise performance. In [18], the video was encoded in layers and supplier peers favored neighbors that uploaded back to them, achieving thus service differentiation as well as robustness against free-riders.

Global incentive schemes take into account the total contribution of a peer to its neighbors. In [2], a rank-based tournament was proposed, where peers are ranked according to their total upload contribution and each peer can choose as neighbor any peer that is below itself in the ranked list. Thus, peers that have high contribution have also higher flexibility in selecting their neighbors. In [19], the authors proposed a payment-based incentive mechanism, where peers earn points by uploading to other peers. The supplier peer selection is performed through first price auctions, that is, the supplier chooses to serve the peer that offers her the most points.

All the aforementioned incentive mechanisms assume that peers are always capable of contributing but, due to selfishness, refrain from doing so. We, on the contrary, consider peers that are unable to contribute because of their access technologies. Associating streaming quality with contribution unnecessarily punishes these weak peers. Therefore our goal is to maximize the social welfare in the system, by convincing high contributing peers to upload to low contributing peers as well. In this aspect, our work is closely related to [20], where a taxation scheme was proposed, based on which high contributing peers subsidize low contributing ones so that the social welfare is maximized. However, in contrast to [20], where it is assumed that peers voluntarily obey to the taxation scheme and they can only react to it by tuning their contribution level, we prove that our mechanism is individually rational and incentive compatible. To the best of our knowledge our incentive scheme is unique in these two important aspects.

6   Conclusion

In this paper we addressed the issue of maximizing the social welfare in a p2p streaming system through an incentive mechanism. We considered a system consisting of contributing and non-contributing peers and studied the playout probability for the two groups of peers. We showed that when contributing peers are selfish the system operates in a state that is suboptimal in terms of social welfare. We proposed an incentive mechanism to maximize the social welfare, which uses the server's capacity as an incentive for contributors to upload to non-contributing peers as well. We proved that our mechanism is both individually rational and incentive compatible. We introduced the notion of efficient p2p systems and proved that for any efficient system there exists exactly one server resource allocation that maximizes the social welfare. An extension of our scheme to several classes of contribution levels is subject of our future work.

References

1. Kumar, R., Liu, Y., Ross, K.: Stochastic fluid theory for p2p streaming systems. In: IEEE INFOCOM (2007)
2. Habib, A., Chuang, J.: Service differentiated peer selection: An incentive mechanism for peer-to-peer media streaming. IEEE Transactions on Multimedia 8, 610–621 (2009)

3. Silverston, T., Fourmaux, O., Crowcroft, J.: Towards an incentive mechanism for peer-to-peer multimedia live streaming systems. In: IEEE International Conference on Peer-to-Peer Computing (2008)
4. Chahed, T., Altman, E., Elayoubi, S.E.: Joint uplink and downlink admission control to both streaming and elastic flows in CDMA/HSDPA systems. Performance Evaluation 65, 869–882 (2008)
5. Rejaie, R., Magharei, N.: Prime: Peer-to-peer receiver-driven mesh-based streaming. In: Proc. of IEEE INFOCOM (2007)
6. Tang, Y., Zhang, M., Zhao, L., Luo, J.-G., Yang, S.-Q.: Large-scale live media streaming over peer-to-peer networks through the global internet. In: Proc. ACM Workshop on Advances in peer-to-peer multimedia streaming, P2PMMS (2005)
7. Bonald, T., Massoulié, L., Mathieu, F., Perino, D., Twigg, A.: Epidemic live streaming: Optimal performance trade-offs. In: Proc. of ACM SIGMETRICS (2008)
8. Liang, C., Guo, Y., Liu, Y.: Investigating the scheduling sensitivity of p2p video streaming: An experimental study. IEEE Transactions on Multimedia, 11, 348–360 (2009)
9. PPLive, http://www.pplive.com/en/about.html
10. UUSee, http://www.uusee.com
11. Chatzidrossos, I., Dán, G., Fodor, V.: Delay and playout probability tradeoff in mesh-based peer-to-peer streaming with delayed buffer map updates. In: P2P Networking and Applications (May 2009)
12. Chatzidrossos, I., Dán, G., Fodor, V.: Server Guaranteed Cap: An incentive mechanism for maximizing streaming quality in heterogeneous overlays. Technical Report TRITA-EE 2009:059, KTH, Royal Institue of Technology (December 2009)
13. Picconi, F., Massoulié, L.: Is there a future for mesh-based live video streaming? In: IEEE International Conference on Peer-to-Peer Computing (2008)
14. Liang, C., Guo, Y., Liu, Y.: Is random scheduling sufficient in p2p video streaming? In: Proc. of the 28th International Conference on Distributed Computing Systems, ICDCS (2008)
15. Fudenberg, D., Tirole, J.: Game Theory. MIT Press, Cambridge (1991)
16. Boyd, S., Vandenberghe, L.: Convex Optimization. Cambridge University Press, Cambridge (2004)
17. Pianese, F., Perino, D.: Resource and locality awareness in an incentive-based P2P live streaming system. In: Proc. of the Workshop on Peer-to-peer Streaming and IP-TV (2007)
18. Liu, Z., Shen, Y., Panwar, S.S., Ross, K.W., Wang, Y.: Using layered video to provide incentives in p2p live streaming. In: Proc. of Workshop on Peer-to-Peer Streaming and IP-TV (2007)
19. Tan, G., Jarvis, S.A.: A payment-based incentive and service differentiation mechanism for peer-to-peer streaming broadcast. In: Proc. of 14th International Workshop on Quality of Service, IWQoS (2006)
20. Chu, Y., Chuang, J., Zhang, H.: A case for taxation in peer-to-peer streaming broadcast. In: Proc. of the ACM SIGCOMM workshop on Practice and theory of incentives in networked systems (2004)

End-to-End Throughput with Cooperative Communication in Multi-channel Wireless Networks

Zheng Huang[1], Xin Wang[1,*], and Baochun Li[2]

[1] School of Computer Science
Shanghai Key Laboratory of Intelligent Information Processing
Fudan University, Shanghai, 200433, P.R. China
`xinw@fudan.edu.cn`
[2] Department of Electrical and Computer Engineering
University of Toronto

Abstract. Although cooperative communication has been proposed at the physical layer to address multi-path fading effects, how physical layer gains with cooperative communication can translate to tangible performance benefits in end-to-end flows remains to be an open problem. This paper represents one step forward towards a deeper understanding of the interplay between end-to-end throughput and physical layer cooperative communication, in the general context of multi-hop multi-channel wireless networks. Based on a decode-and-forward physical layer design with rateless codes, we reformulate the problem of routing and channel assignment to account for physical layer cooperation. We design a distributed protocol to solve the new problem. Our simulation results have validated the effectiveness of our protocol to offer a substantial gain with respect to stabilizing the offered aggregate throughput in the network.

Keywords: cooperative communication, channel assignment, multi-hop multi-channel wireless networks.

1 Introduction

Cooperative communication has been proposed as a powerful physical layer technique to combat fading [1] or to increase the physical layer capacity [2] in wireless relay networks. A basic model studied in the research of cooperative strategies is a "triangle" network [3], which consists of a source S, a destination D, and a relay node R. In the transmission from S to D, the relay R can cooperate with S to jointly forward the packets to D. Cooperation can improve the channel capacity when the channel (S, R) and (R, D) have a higher quality than the channel (S, D).

While there is a large body of literature focusing on various cooperative communication strategies, most of these results are limited to the triangle network

---

* Corresponding author.

M. Crovella et al. (Eds.): NETWORKING 2010, LNCS 6091, pp. 327–338, 2010.

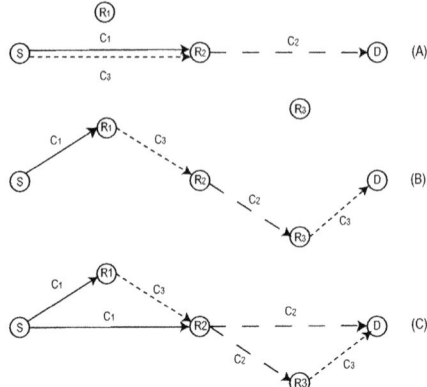

Fig. 1. An example showing the benefits of cooperative communication in multi-hop wireless networks. (A) (B) Conventional channel assignment with no cooperation in two fading cases. (C) Channel assignment with cooperation.

or its generalizations. Such triangle networks have a two-hop topology, which is difficult for these cooperative communication strategies to be extended directly to its multi-hop counterpart. On the other hand, as pointed out in [4], while there exists a large volume of literature on cooperative communication strategies in the physical layer, there are very few higher layer protocols that can take advantage of physical layer cooperation in a multi-hop network setting.

In this paper, our objective is to make use of cooperative communication strategies in *multi-hop* and *multi-channel* wireless networks. In multi-channel networks, each node is equipped with several wireless interfaces, each of which can be tuned to a channel from an orthogonal set of channels. We seek to improve the aggregate end-to-end network throughput by allowing nodes to cooperate in the physical layer, which is a challenge not fully explored in the literature.

As a preamble of our work, a motivating example that involves a multi-hop and multi-channel network can be described as follows. Consider a wireless network with five nodes, labeled as S, R_1, R_2, R_3 and D in Fig. 1. We assume that there is a unicast traffic from S to D. We have three orthogonal channels C_1, C_2 and C_3 to assign on nodes' interfaces, with an objective of maximizing the unicast throughput. Channel capacity is dependent on which link the channel is placed, *i.e.*, channel diversity is considered. Nodes can transmit simultaneously via orthogonal channels without interference. By conventional channel assignment with no cooperation, we may assign C_1, C_3 on (S, R_2) and C_2 on (R_2, D), as illustrated in Fig. 1(A). The throughput is thus bounded by the minimum of the channel capacity of C_2 and the sum of that of C_1 and C_3 on respective links. However, it can occur that the channel (S, R_2) or (R_2, D) has a very poor quality. In this case, a better alternative may exist to assign channels on (S, R_1), (R_2, R_3), (R_1, R_2) and $(R_3 - D)$ as in Fig. 1(B).

Cooperative communication in the physical layer, on the other hand, provides a new insight to this example scenario. As in Fig. 1(C), S, R_1 and R_2 can form a local "triangle" for cooperation, while R_2, R_3 and D forming another. In particular, S (R_2) can broadcast the packets to R_1 (R_3) and R_2 (D) by a common channel C_1 (C_2) at first, and then R_1 (R_3) helps to forward the packets to R_2 (D) by another channel. Because the capacity from S to R_2 and R_2 to D can both increase via cooperation, the unicast throughput from S to D is thus improved.

From this example, it is clear that, although physical layer cooperation improves end-to-end throughput in multi-hop and multi-channel networks, it is non-trivial to design distributed protocols to realize such a performance gain. There is a tradeoff between throughput improvement and temporary increase of network congestion.

In this paper, we seek to make use of physical layer cooperative communication strategies to improve the aggregate end-to-end throughput in multi-hop and multi-channel networks. To our knowledge, this has not been fully explored in prior work. Towards this objective, we reformulate the routing and channel assignment problem to account for our physical layer model for cooperation. In this context, we propose a new concept, *cooperative link*, as the component of a cooperative routing path. We have designed a decentralized protocol to maximize aggregate end-to-end throughput.

2   Related Work

Our work builds upon cooperative communication strategies that have been studied thoroughly at the physical layer, such as amplify-and-forward [5], decode-and-forward [6], compress-and-forward [7] and compute-and-forward [8]. Most of these studies are from an information-theoretic perspective. In contrast, the objective of this paper is to translate the physical-layer capacity improvement via cooperation to network-layer throughput benefits. In this sense, our work is not directly related to recent advances in cooperative diversity (e.g., [1]).

From the perspective of the network layer, we select decode-and-forward rather than the other three strategies as the underlying strategy for cooperative communication, because with decode-and-forward it is flexible enough to incorporate cooperation into the multi-hop and multi-channel network model. Specifically, we use a coding scheme based on rateless codes in [9] to formulate the underlying cooperative model.

We note that there has been existing work on translating physical layer gains of cooperative communication to the high layer performance benefits, all of which uses the decode-and-forward strategy in the system model. In [4], a cross-layer approach has been proposed to exploit cooperative diversity in single-channel ad hoc networks. It provides few insights on how cooperative communication could be used to improve the network performance. In contrast, we identify a clear underlying cooperative model that could be analyzed quantitatively from the perspective of a higher layer. In addition, we consider multi-channel rather than a single-channel networks.

One common feature in most existing works in the area of multi-radio or multi-channel networks is that packets are forwarded along a chain of point-to-point links. We believe that considering cooperation in multi-channel networks is inherently attractive. More recently, there has been work that considered network coding in multi-channel networks [10]. Compared with its system model, we used a different underlying cooperative communication strategy, and assume that a channel can have different capacities on different links due to fading effects at different locations. In addition, our use of dynamic channel assignment is more flexible than static channel assignment in [10].

3   System Model

3.1   Network Model

We consider a wireless network $N = (V, E)$ with a set of stationary wireless nodes. There are a total of K orthogonal channels denoted by $\mathcal{C} = \{c_1, c_2, ..., c_K\}$ in the network. Each wireless node $v \in V$ is equipped with κ_v channel interfaces. A *channel assignment* $\mathcal{A}$ assigns a collection A_v of κ_v channels to node v, with each interface on v tuned to a channel from A_v, $A_v \subset \mathcal{C}$. A half-duplex model is assumed on each channel.

We assume each wireless node uses a fixed transmission power. It follows that there is a fixed *transmission range* R_T and *interference range* $R_I > R_T$ associated with each node. As each channel may have a different capacity at different locations in the network, we denote R_{ij}^c as the capacity of channel c on the link (i, j), provided that there is no interference. We use the disk model [11] to account for the interference relationships (Fig. 2).

For traffic flows, we assume that there is a collection $\mathcal{S}$ of elastic unicast flows in which each session $m \in \mathcal{S}$ runs concurrently between a pair of wireless nodes m_s and m_t. We denote the achieved throughput of a unicast flow m by σ_m.

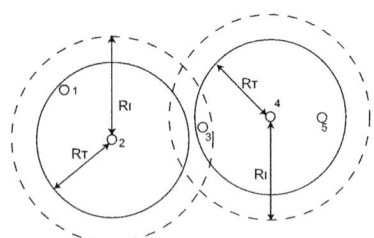

Fig. 2. An example exhibiting the interference relationship. The solid and dash circles indicate the transmission range and interference range, respectively. Link $(3, 4)$ interferes with link $(1, 2)$. However, link $(4, 5)$ can operate simultaneously with link $(1, 2)$ via a common channel since there is no interference.

3.2   Cooperative Communication Model

A basic cooperative communication opportunity involves the operation of three nodes, namely, s, r and t as in Fig. 3. Let node s communicate with node t.

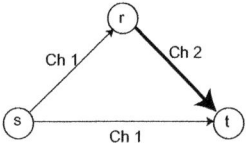

Fig. 3. The underlying cooperative communication model. Node s broadcasts the packets via a common channel 1. Node r, once decoding the message, helps to re-encode and forward to node t via another channel 2. Rateless codes are used as the coding scheme.

If no cooperation is present, node s uses a block coding scheme to encode the information, and forwards the coded packets on channel 1. At the other side of channel 1, node t decodes the packet and recovers the original information. The achievable rate of traffic is denoted as R_{st}^1. Alternatively, when node s forwards a packet to t, node r can overhear this packet since channel 1 is a "broadcast" channel. If we assume $R_{st}^1 < R_{sr}^1$, it follows that node r manages to decode the packet before node t. In the remaining time, node r can help node s to re-encode and forward the packet to node t via channel 2.

The achievable rate with this cooperative communication strategy has been proved in [12] and later extended in [9]: Let $f := \begin{cases} \frac{R_{st}^1 + R_{rt}^2}{R_{sr}^1 + R_{rt}^2} & \text{if } R_{st}^1 < R_{sr}^1, \\ 1 & \text{otherwise.} \end{cases}$ and let $R := fR_{sr}^1 = fR_{st}^1 + (1-f)(R_{st}^1 + R_{rt}^2) = R_{st}^1 + (1-f)R_{rt}^2$. Then for any $\delta > 0$, there exists a block coding scheme at rate $R - \delta$ such that with increasing block length, the decoding error probability is driven arbitrarily close to 0.

Rateless code has been argued in [9] to facilitate node s to choose such a rate, without channel state information. *Fountain codes*, for example Raptor [13] and LT [14] codes, are typical forms of rateless codes.

4   Routing and Channel Assignment with Cooperative Communication

4.1   Cooperative Routing

We propose the notion of *cooperative links* to describe the cooperative routing path. A *cooperative link* consists of three parts: there is a single source s, a single receiver t, and a potential set B of relay nodes. We denote the set of cooperative links as $E_c = \{(i, B, j), i \in V, j \in V, B \subset V\}$. For the sake of simplicity, in the following analysis, we restrict to the case when $|B| = 1$.

For each unicast session $m \in S$, a *cooperative path* between m_s and m_t can be expressed as a chain of links from $E \cup E_c$. For example, the cooperative path from node 1 to 7 as in Fig. 4 can be written as $\{(1,2), (2, \{3\}, 4), (4,5), (5, \{6\}, 7)\}$.

We assume that the entire system operates according to a "universal" clock, which divides time into slots of unit length. Each link in $E \cup E_c$ is associated with a queue. Let $\{q_l(t) : l \in E \cup E_c\}$ denote the number of packets queued at link l

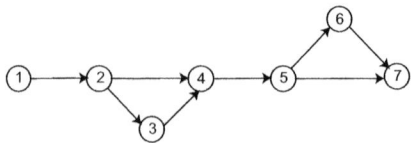

Fig. 4. A routing path from node 1 to 7 can be expressed as $\{(1,2),(2,\{3\},4),(4,5),(5,\{6\},7)\}$. Here $(2,\{3\},4)$ and $(5,\{6\},7)$ are cooperative links.

waiting to be served before time t. Suppose we allow for multi-path routing for each unicast session m. Let $K(m)$ denote the set of routing paths between m_s and m_t, and P_{mk} denote the fraction of traffic of m that is routed on the path $k \in K(m)$. Here, $\sum_{k \in K(m)} P_{mk} = 1$ for all m. We denote the routing matrix as G. If the kth route of the unicast session m passes through link l, then $G^l_{mk} = 1$. Otherwise, $G^l_{mk} = 0$.

The cooperative routing problem considered in multi-channel and multi-hop networks is to determine the cooperative path of each unicast flow and determine the routing matrix G^l_{mk} and the fraction P_{mk}, if multi-path routing is considered.

4.2 Cooperative Channel Assignment

We adapt a dynamic channel assignment model as in [15,16,17]. That is, the interface on nodes can switch to a different channel dynamically from one time slot to the next. We use $\mathcal{A}(t)$ to denote the channel assignment in time slot t. It is required that under each $\mathcal{A}(t)$, there is no interference in the network.

5   A Distributed Cooperative Protocol

Our distributed protocol consists of two parts. The first part is cooperative routing, which facilitates wireless nodes to discover local cooperative communication opportunities in a distributed manner. The output of this part is a chain of direct and cooperative links that could be used as a routing path. Also, for multi-path routing, the traffic fraction P_{mk} and routing matrix G^l_{mk} are determined. The second part is to assign channels on direct and cooperative links that have been determined in the first part.

5.1   Cooperative Routing Protocol

Our routing protocol involves two stages. The first stage is a direct link routing discovery that operates similar to a traditional single-path routing scheme. Specifically, for a given unicast session m, we use the hop-count as our routing metric, and assume that each wireless node can measure its distance to the destination m_t. The source node m_s broadcasts a probing packet which all of the neighbors of the source in the transmission range could overhear. A neighbor chooses to forward (in a broadcast manner) the probing packet if its distance to the destination is less than that of the last-hop predecessor. This process continues until the destination node is reached. It follows that a single path of a chain of direct links can thus be established between the source m_s and destination m_t.

The second stage is to find the potential cooperative opportunities, which is the key feature of the cooperating routing discovery. In doing so, each node along the path (formed in the first stage) picks up the node from its neighbors with the shortest distance from the current node, and selects this neighbor to form a cooperative link. We describe the detailed steps of these two stages in Algorithm 1.[1]

Algorithm 1. Cooperative Routing Discovery

1: Input: A unicast session m with source node m_s and destination node m_t.
2: Output: A chain P of direct or cooperative links as a single routing path.
3: Stage 1: Generate the direct link routing path.
4: Stage 2: Discover local cooperative communication opportunities.
5: $current = m_s$
6: **loop**
7:    **if** $current == m_t$ **then**
8:       **return**
9:    **end if**
10:   search in the neighborhood of $current$ node $N(current)$ for a neighbor i with the smallest distance $d(i, m_t)$ to the destination m_t.
11:   $next\ hop = i$
12:   Add $(current, next\text{-}hop)$ to P.
13:   search in $N(current)$ for a neighbor j with the smallest distance $d(current, j)$ to the $current$ node
14:   **if** j exists and $j \neq next\text{-}hop$ **then**
15:      Replace $(current, i)$ with the cooperative link $(current, \{j\}, i)$ in P
16:   **end if**
17:   $current = next\text{-}hop$
18: **end loop**.

5.2   Cooperative Channel Assignment Protocol

The difficulty in generalizing the distributed algorithm of channel assignment under the direct link scenario to the cooperative link scenario arises from the inherent feature of cooperative links. In the direct link scenario, there is one channel assigned on one direct link. In contrast, in order to let one cooperative link, say link (i, B, j), be active, one needs to assign 2 channels on this link if $|B| = 1$, as indicated in Section 3. This 2 (channel)-to-1 (link) mapping relationship makes it almost impossible to directly generalize the existing algorithm.

---

[1] Although we use the hop-count as the metric here, the algorithm is easily extended to other metrics such as the ETX model.

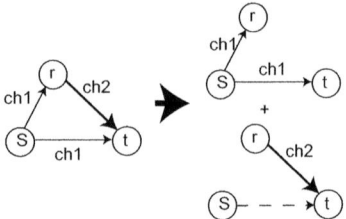

Fig. 5. A cooperative link $(s, \{r\}, t)$ can be decomposed into two "virtual" links. One is a virtual broadcast link $(s, \{\{r\}, t\})$, and the other is a virtual multiple-access link $(\{s, \{r\}\}, t)$. Only one channel is needed for each virtual link.

Our key idea to address this difficulty is to decompose a cooperative link (i, B, j) into two "virtual" links: one is a virtual broadcast link $(i, \{B, j\})$ and the other is a virtual multiple access link $(\{i, B\}, j)$. As in Fig. 5, initially a cooperative link $(s, \{j\}, k)$ needs two channels ($ch1$ and $ch2$) to perform the cooperation. Now we decompose it into two virtual channels: a virtual broadcast link $(s, \{\{r\}, t\})$ and a virtual multiple access link $(\{s, \{r\}\}, t)$. Each link now needs only one channel. Note that on the virtual link $(\{s, \{r\}\}, t)$, we do not need to assign an extra channel on the link from s to t (indicated by the dash line) since once we have assigned the channel on the virtual broadcast counterpart, we can use the same channel on s to t for the virtual multiple access link.

By the decomposition of cooperative links in E_c, we now have three kinds of links E, E_{cb} and E_{cm} in the network. E is the direct link set, E_{cb} is the virtual broadcast link set decomposed from E_c, and E_{cm} is the virtual multiple access link set from E_c. The interference relationship among these links can be generalized directly from that under the case of conventional direct links.

We define that a link l in $E \cup E_{cb} \cup E_{cm}$ interferes with a link $(i, \{B, j\})$ in E_{cb} or $(\{i, B\}, j)$ in E_{cm}, if at least one of the end points of l (there are two end points if $l \in E$ and three end points for $l \in E_{cb} \cup E_{cm}$) is located in the interference disk formed by $D_i \cup D_j \cup (\cup_{k \in B} D_k)$. We denote $\mathcal{I}$ as the *interference degree* of the network [15]. The interference degree $\mathcal{I}(i, j)$ defined on link (i, j) is the largest number of links in the interference range I_{ij} that do not interfere with each other. The interference degree $\mathcal{I}$ is then the largest $\mathcal{I}(i, j)$ over all links in the network.

Through the definition of the interference relationship, we can derive an *cooperative interference degree* (denoted by $\mathcal{I}_c$) of the network with cooperative links. Compared with $\mathcal{I}$, which is the interference degree of the same network but without cooperative links, we have $\mathcal{I}_c \leq \mathcal{I}$, since cooperative links replace direct links in the routing path, and one cooperative link itself is made up of several direct links in E.

Our algorithm is a generalization of the algorithm in [15] with the concept of virtual broadcast and multiple access links. We let each link $l \in E \cup E_{cb} \cup E_{cm}$ maintain $K + 1$ queues. There is a queue q_l and a total of K channel queues u_l^c, one for each channel $c \in C$ at link l. The channel assignment algorithm involves two steps.

We focus on the case in which there is only one single cooperative routing path for each session $m \in \mathcal{S}$. Since for multi-path cooperative routing, the set of routing paths has been generated in the cooperating routing phase, what we need to do is just to run the channel assignment in single path case iteratively on each of the cooperative link. Thus the focus on single-path case is sufficient.

Cooperative Channel Assignment Protocol with Single-path Routing

- Step 1: Let $x_l^c(t)$ be the maximum number of packets that could be assigned on link l through channel c in time slot t. For each link $l \in E \cup E_{cb} \cup E_{cm}$,

$$x_l^c(t) = \begin{cases} R_l^c & \text{if } \frac{q_l}{\alpha_l} \geq \frac{1}{R_l^c}(CA_l^c(\frac{u}{R}) + CE_l(\frac{u}{R})), \\ 0 & \text{otherwise.} \end{cases} \qquad (1)$$

Here, α_l is a positive constant chosen for link l, R_l^c is the capacity of link l when channel c is assigned on it, $CA_l^c(\frac{u}{R})$ is the *congestion cost* [15] at link l to use channel c, and $CE_l(\frac{u}{R})$ is the *interface cost* [15] at link l. For link $l = (i, j) \in E$, $CA_{ij}^c(\frac{u}{R}) = \sum_{e \in I_{ij}} \frac{u_e^c}{R_e^c}$. Note that link e, which interferes with link (i, j) may come from $E_{cb} \cup E_{cm}$. $CA_l^c(\frac{u}{R})$ for $l \in E_{cb} \cup E_{cm}$ can be defined in a similar manner.

The calculation of R_l^c for link $l \in E_{cb} \cup E_{cm}$ is a little tricky. For $l = (i, \{\{j\}, k\}) \in E_{cb}(B = \{j\})$, we have $R_{i,\{\{j\},k\}}^c = R_{ij}^c$, since during the first phase of broadcast in the underlying cooperative communication strategy, only node j finishes decoding. For $l = (\{i, \{j\}\}, k) \in E_{cm}(B = \{j\})$, we have $R_{\{i,\{j\}\},k}^c = R_{jk}^c$ since there is actually one channel c assigned between j and k in the virtual multiple access link. The interface cost $CE_l(\frac{u}{R})$ for link $(i, j) \in E$ is defined as $\frac{1}{\kappa_i} \sum_{e \in E(i)} \sum_{d=1}^{K} \frac{u_e^d}{R_e^d} + \frac{1}{\kappa_j} \sum_{e \in E(j)} \sum_{d=1}^{K} \frac{u_e^d}{R_e^d}$. Here, the set $E(i)$ represents the set of links in $E \cup E_{cb} \cup E_{cm}$ that are adjacent with node i. For link $l = (i, \{\{j\}, k\}) \in E_{cb}$ or $l = (\{i, \{j\}\}, k) \in E_{cm}$, $CE_l(\frac{u}{R})$ is defined as $\frac{1}{\kappa_i} \sum_{e \in E(i)} \sum_{d \in \mathcal{C}} \frac{u_e^d}{R_e^d} + \frac{1}{\kappa_j} \sum_{e \in E(j)} \sum_{d \in \mathcal{C}} \frac{u_e^d}{R_e^d} + \frac{1}{\kappa_k} \sum_{e \in E(k)} \sum_{d \in \mathcal{C}} \frac{u_e^d}{R_e^d}$.

In each time slot t, link l assigns $y_l^c(t) \in [0, x_l^c(t)]$ to each link channel u_l^c. The queueing dynamics of $q_l(l \in E \cup E_{cb} \cup E_{cm})$ is thus expressed as:

$$q_l(t+1) = q_l(t) + \sum_{m \in \mathcal{S}} G_m^l \sigma_m - \sum_{c \in \mathcal{C}} y_l^c(t). \qquad (2)$$

where $\sum_{c \in \mathcal{C}} y_l^c(t) = \min\{q_l(t), \sum_{c \in \mathcal{C}} x_l^c(t)\}$.
- Step 2: Based on the queue length at u_l^c, a maximal schedule $\mathcal{A}^c(t)$ is calculated by the distributed algorithm in [18]. Then at the end of time slot t, the queueing dynamics of $u_l^c(t)$ is expressed as:

$$u_l^c(t+1) = u_l^c(t) + y_l^c(t) - R_l^c 1_{\{l \in \mathcal{A}^c(t)\}}. \qquad (3)$$

The following main result demonstrates the efficiency ratio of the above two-step algorithm.

Proposition 1: Assume each session $m \in \mathcal{S}$ uses a single cooperative routing path, and the routing matrix is given by G_m^l. The efficiency ratio of the proposed algorithm is $\frac{1}{\mathcal{I}_c+2}$, where $\mathcal{I}_c$ is the cooperative interference degree of the network.

The proof could be generated as a direct extension of that in [15]. We omit it here due to space constraints.

6   Simulations

To evaluate the performance of our protocol, we compare it with the best non-cooperative distributed algorithm in the literature [15]. To make a fair comparison, our simulation is based on a similar grid topology as shown in Fig. 6. There are 25 nodes, represented by the circles, and 60 direct links, represented by the dash lines. Compared with the topology in [15], we add nodes 17 to 25 to facilitate cooperative communication. As for interference relationship, we use the node-interference model as in [15,19].

There are 8 orthogonal channels ($K = 8$) in the system. Each node is equipped with 8 interfaces ($\kappa_i = 8, i \in V$). The parameters α_l in Equation (1) are set as 10 and 100, for the cooperative and non-cooperative protocol, respectively. The capacity R_l^c is randomly chosen from 1 to 5, according to a uniform distribution. For traffic patterns, the unicast sessions are represented by the arrows in Fig. 6. Each session is assumed to have a uniform throughput of σ. The chain of arrows in Fig. 6 could be regarded as the conventional routing path generated in the first stage of our cooperative routing discovery protocol.

We plot the comparison of per-link mean queue backlog in Fig. 7. In our data statistics, we do not include those links with no traffic passing through, *e.g.*, link $(1,5)$. From Fig. 7, two protocols behave at the same level when the offered throughput σ ranges from 3 to 5.6. However, from $\sigma = 6$, the backlog of the non-cooperative protocol in [15] increases to infinity quickly. As indicated in [15], this throughput σ could be viewed as the boundary of the capacity region. In contrast, as the offered throughput increases, the queue backlog with our cooperative protocol increases in a much slower manner. At $\sigma = 6$, the backlog is 5162 (vs. 1866 in the non-cooperative protocol); at $\sigma = 6.4$, the backlog is

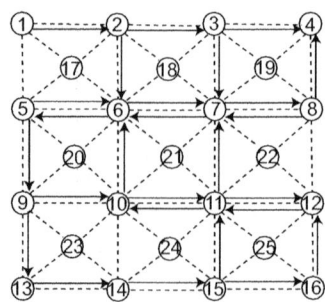

Fig. 6. The grid topology

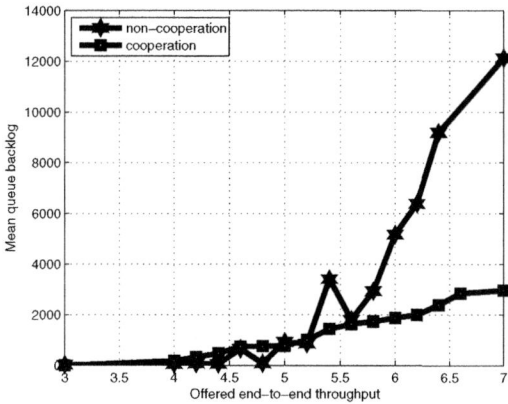

Fig. 7. Comparison of our cooperative protocol with the non-cooperative protocol

9197 (vs. 2386); and at $\sigma = 7$, the backlog is 12126 (vs. 2954). The improvement thus ranges from 100% to 300%.

7 Conclusion

This paper explores how physical layer gains using cooperative communication strategies can translate to tangible performance for end-to-end flows at higher layers, in a general context of multi-channel networks. Based on a specific physical layer cooperative communication model with rateless codes, we reformulate the conventional routing and channel assignment problem. Moreover, we provide an efficient distributed protocol to solve these problems. Our simulation results have shown that, by using physical layer cooperation, our protocol performs 100% to 300% better, with respect to stabilizing the offered aggregate throughput, as compared to the best non-cooperative distributed protocol in the literature.

Acknowledgement

This work was supported in part by NSFC under Grant No. 60702054, Shanghai Municipal R&D Foundation under Grant No. 09511501200, the Shanghai Rising-Star Program under Grant No. 08QA14009.

References

1. Laneman, J.N., Tse, D.N.C., Wornell, G.W.: Cooperative diversity in wireless networks: Efficient protocols and outage behavior. IEEE Trans. Inf. Theory 50, 3062–3080 (2004)
2. Kramer, G., Gastpar, M., Gupta, P.: Cooperative strategies and capacity theorems for relay networks. IEEE Transactions on Information Theory 51(9), 3037–3063 (2005)

3. der Meulen, E.C.V.: Three-terminal communication channels. Adv. Appl. Probab. 3, 120–154 (1971)
4. Jakllari, G., Krishnamurthy, S.V., Faloutsos, M.: A framework for distributed spatio-temporal communications in mobile ad hoc networks. In: Proc. IEEE INFOCOM (2006)
5. Nosratinia, A., Hunter, T.E., Hedayat, A.: Cooperative communication in wireless networks. IEEE Communications Magazine 42(10), 74–80 (2004)
6. Laneman, J.N., Wornell, G.W., Tse, D.: An efficient protocol for realizing cooperative diversity in wireless networks. In: Proc. IEEE Inter. Symp. Inform. Theory (June 2001)
7. Cover, T., Gamal, A.E.: Capacity theorems for the relay channel. IEEE Transactions on Information Theory 25(5), 572–584 (1979)
8. Nazer, B., Gastpar, M.: Compute-and-forward: Harnessing interference with structured codes. In: Proceedings of the IEEE International Symposium on Information Theory (ISIT 2008), Toronto, Canada (July 2008)
9. Castura, J., Mao, Y.: Rateless coding over fading channels. IEEE Comm. Lett. 10 (January 2006)
10. Zhang, X., Li, B.: On the benefits of network coding in multi-channel wireless networks. In: Proc. IEEE SECON (2008)
11. Tang, J., Xue, G., Zhang, W.: Interference-aware topology control and qos routing in multi-channel wireless mesh networks. In: Proc. ACM MobiHoc (2005)
12. Mitran, P., Ochiai, H., Tarokh, V.: Space-time diversity enhancements using collaborative commnications. IEEE Trans. Inf. Theory 51(6), 2041–2057 (2005)
13. Shokrollahi, A.: Raptor codes. In: Proc. IEEE Inter. Symp. Inform. Theory, p. 36 (2004)
14. Luby, M.: LT codes. In: Proc. 43rd Annual IEEE Symp. Fundations Computer Science (FOCS), pp. 271–282 (2002)
15. Lin, X., Rasool, S.: A distributed joint channel-assignment, scheduling and routing algorithm for multi-channel ad hoc wireless networks. In: Proc. IEEE INFOCOM (2007)
16. Kodialam, M., Nandagopal, T.: Characterizing the capacity region in multi-radio multi-channel wireless mesh networks. In: MobiCom '05: Proceedings of the 11th annual international conference on Mobile computing and networking, pp. 73–87. ACM Press, New York (2005)
17. Bahl, P., Chandra, R., Dunagan, J.: SSCH: slotted seeded channel hopping for capacity improvement. In: IEEE 802.11 Ad-Hoc Wireless Networks, in ACM Mobicom, pp. 216–230 (2004)
18. Hańćkowiak, M., Karoński, M., Panconesi, A.: On the distributed complexity of computing maximal matchings. In: SODA '98: Proceedings of the ninth annual ACM-SIAM symposium on Discrete algorithms, Philadelphia, PA, USA, pp. 219–225. Society for Industrial and Applied Mathematics (1998)
19. Hajek, B., Sasaki, G.: Link scheduling in polynomial time. IEEE Trans. Inf. Theory 34(5), 910–917 (1998)

Cost Bounds
of Multicast Light-Trees in WDM Networks

Fen Zhou[1], Miklós Molnár[1], Bernard Cousin[2], and Chunming Qiao[3]

[1] IRISA / INSA, Campus de Beaulieu, Rennes 35042, France
{fen.zhou,molnar}@irisa.fr
[2] IRISA / University of Rennes 1, Campus de Beaulieu, Rennes 35042, France
bernard.cousin@irisa.fr
[3] Computer Science Department, State University of New York at Buffalo, USA
qiao@computer.org

Abstract. The construction of light-trees is one principal subproblem for multicast routing in sparse splitting Wavelength Division Multiplexing (WDM) networks. Due to the light splitting constraint and the absence of wavelength converters, several light-trees may be required to establish a multicast session. However, the computation of optimal multicast light-trees is NP-hard. In this paper, we study the wavelength channel cost (i.e., total cost) of the light-trees built for a multicast session. An equal cost of 1 *unit hop-count cost* is assumed over all the fiber links in the network. We prove that the total cost of a multicast session is tightly lower limited to K and upper bounded to (1) $K(N-K)$ when $K < \frac{N}{2}$; (2) $\frac{N^2-1}{4}$ or $\frac{N^2}{4}$ respectively when $K \geq \frac{N}{2}$ and N is odd or even, where K is the number of destinations in the multicast session and N is the number of nodes in the network. Classical sparse splitting multicast routing algorithms such as Reroute-to-Source and Member-Only [3] also follow these bounds. And particularly in WDM rings, the optimal multicast light-tree has a cost inferior to $N - \lceil \frac{N}{K+1} \rceil$.

Keywords: WDM Networks, All-Optical Multicast Routing (AOMR), Light-trees, Sparse Splitting, Cost Bounds.

1 Introduction

Multicast routing in WDM networks is to determine a set of lightpaths on a physical topology from a source to a set of destinations involved in a multicast communication. The light-tree concept is introduced in [1] to minimize the number of wavelength channels and transceivers used for a multicast session. Branching nodes in a light-tree should be equipped with light splitters. However, in sparse splitting [2] WDM networks, there are two kinds of nodes: multicast capable nodes (MC [2], i.e. nodes equipped with light splitters) and multicast incapable nodes (MI [2], i.e. nodes without light splitters). An MC node is capable of replicating the data packets in the optical domain via light splitting and sends them to several ports, while an MI node cannot split but generally has

M. Crovella et al. (Eds.): NETWORKING 2010, LNCS 6091, pp. 339–350, 2010.

the *Tap-and-Continue* (TaC [4]) capacity which permits to tap a small amount of optical power from the incoming light beam and forward it to only one outgoing port. Although one tree is sufficient to span all the multicast destinations in a network without constraint, minimizing the cost of the multicast tree is already a Steiner-Problem which is proven to be NP-complete. In face of sparse splitting, lack of wavelength converters, as well as continuous wavelength and distinct wavelength constraints [5], one light-tree may not be enough to cover the entire multicast group members but several ones may be required, i.e., a light-forest [3]. As a result, it is even harder to optimize the total number of wavelength channels used for a multicast session.

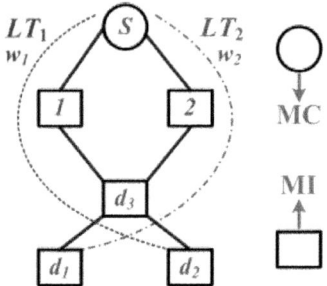

Fig. 1. An example sparse splitting WDM network

Although many light-tree computation heuristics have been proposed in recent works [3,9,10,11], no literature has addressed the cost bound of multicast light-trees in sparse splitting WDM networks, let alone the approximation of the light-trees built by heuristics towards the optimal solution. Since the wavelength channel cost is a very important metric for the selection of the multicast light-trees, it is very interesting and imperative to know at least the cost bound of the light-trees, which could be referenced when designing a WDM network. In [11], a heuristic is proposed to construct multicast light-trees with QoS guarantee and the cost upper bound of the light-trees is given. However, in [11] it is supposed that all the network nodes are equipped with costly light splitters, while it is not realistic in large WDM mesh networks due to the high price and complex architecture of light splitters. Literature [12] also gives a cost upper bound of $\frac{N^2}{4}$ for the multicast light-trees in sparse splitting WDM networks, where N denotes the number of nodes in the network. However, the cost bound in [12] have the following two shortcomings. First it is derived on the hypothesis that the set of multicast light-trees computed for a multicast session still retain a tree structure in the IP layer (i.e., when all these light-trees are merged together). In fact, this hypothesis is not always correct as demonstrated in the following example. A multicast session with source s and destinations d_1, d_2 and d_3 is required in a sparse splitting optical network shown in Fig. 1 with solid line. Since node d_3 is an MI node, two light-trees (i.e., LT_1 (dotted line) and LT_2 (dashed line)) on two different wavelengths may be computed. As we can see the IP layer of the merged LT_1 and LT_2 are drawn in Fig. 1 with solid line, which is the same as the network topology. Obviously, it is not a tree but it exists a cycle. Second, the bound $\frac{N^2}{4}$ in [12] seems too large for small size multicast sessions, e.g., a multicast session with a source and only two destinations.

For the reasons above, in this paper we give a much better bound for wavelength channel cost of multicast light-trees. It is valid for most of the multicast routing algorithms under sparse splitting constraint, even if the IP layer of the set of multicast light-trees does not retain the tree structure (e.g, the iterative multicast routing algorithms as Member-Only [3]). Costly and complex wavelength converters are supposed to be unavailable, and an equal cost of 1 *unit hop-count cost* is assumed over all the fiber links in the network. We prove that the total cost of a multicast session is upper bounded to (1) $K(N - K)$, when $K < \frac{N}{2}$; (2) $\frac{N^2-1}{4}$, when $K \geq \frac{N}{2}$ and N is odd; (3) $\frac{N^2}{4}$ when $K \geq \frac{N}{2}$ and N is even, where K is the number of destinations in the multicast session and N is the number of nodes in the network. Besides, the total cost is lower limited to K. Moreover, in WDM rings the optimal multicast light-tree has a total cost inferior to $N - \lceil \frac{N}{K+1} \rceil$.

The rest of this paper is organized as follows. System model is given and the multicast routing problem is formulated in Section 2. Then the cost bound of multicast light-trees in WDM mesh network is discussed in Section 3. After that, the cost bound of multicast light-trees in WDM rings is investigated in Section 4. Numerical results are obtained through simulation in Section 5. Finally, we conclude the paper in Section 6.

2   Multicast Routing with Sparse Splitting

2.1   Multicast Routing Problem

Multicast routing involves a source and a set of destinations. In sparse splitting WDM networks, a set of light-trees is employed to distribute messages from the source to all the group members simultaneously. The objective of studying multicast routing in WDM networks is to minimize the wavelengths channel cost while fulfilling a multicast session. The computation of light-trees for a multicast session generally has the following principles.

1. Due to sparse splitting and absence of wavelength conversion, the degree of an MI node in the light-tree cannot exceed two. In consequence not all destinations could be included in the same light-tree. Thus, several light-trees on different wavelengths may be required for one multicast session.
2. Among the light-trees built for a multicast session, one destination may be spanned (used to forward the incoming light beam to other destination nodes) by several light-trees, but it should be served (used to receive messages from the source) by only one light-tree. (e.g., d_3 in Fig. 1 is spanned by both LT_1 and LT_2 to forward the incoming light beam to d_2 and d_1 respectively. But it must tap the light beam only once for recovering multicast messages either in LT_1 or in LT_2).

3. Since the number of wavelengths supported per fiber link is limited, the maximum number of wavelengths required and the traffic congestion in a fiber link should be taken into account during the selection of multicast light-trees. Thus, if a set of destinations D have been spanned by a light-tree LT_1, $D \subseteq LT_1$, it is entirely useless to construct another light-tree LT_2 to serve and only serve the destinations in subset D_i, with $D_i \subseteq D$. The reason is that destinations in D_i could be served directly in LT_1.

2.2 System Model

A sparse splitting WDM network can be modeled by an undirected graph $G(V, E, c)$. V represents the vertex-set of G, $|V| = N$. Each node $v \in V$ is either an MI or an MC node.

$$V = \{v| \ v = MI \ or \ v = MC\} \tag{1}$$

E represents the edge-set of G, which corresponds to the fiber links between the nodes in the network. Each edge $e \in E$ is associated with a cost function $c(e)$. Function c is additive over the links of a lightpath $LP(u, v)$ between two nodes u and v, i.e.,

$$c\big(LP(u,v)\big) = \sum_{e \in LP(u,v)} c(e) \tag{2}$$

We consider a multicast session $ms(s, D)$, which requests for setting up a light distribution structure (i.e., light-tree) under optical constraint (i.e., wavelength continuity, distinct wavelength, sparse splitting and lack of wavelength conversion constraints) from the source s to a group of destinations D. Let K be the number of destinations, $K = |D|$. Without loss of generality, it is assumed that k light-trees $LT_i(s, D_i)$ are required to serve all the destinations involved in multicast session $ms(s, D)$, where $i \in [1, k]$. It holds true that

$$1 \leq k \leq K \leq N - 1 \tag{3}$$

Although the ith light-tree $LT_i(s, D_i)$ may span some destinations already spanned in the previous light-trees, D_i is used to denote exclusively the set of newly served destinations in $LT_i(s, D_i)$. Since all the destinations in D are served by k light-trees, we obtain

$$D = \bigcup_{i=1}^{k} D_i \tag{4}$$

These k sets of destinations D_i are disjoint, i.e.,

$$\forall i, j \in [1, k] \ and \ i \neq j, \ D_i \cap D_j = \emptyset \tag{5}$$

Let a positive integer $K_i = |D_i|$ denote the size of the subset D_i, then we have

$$\sum_{i=1}^{k} K_i = |D| = K \tag{6}$$

The total cost of multicast session $ms(s, D)$ is defined as the wavelength channel cost of the light-trees built to serve all the destinations in set D. It can be calculated by

$$c\big(ms(s,D)\big) = \sum_{i=1}^{k} c\big[LT_i(s, D_i)\big]$$

$$= \sum_{i=1}^{k} \sum_{e \in LT_i(s, D_i)} c(e) \tag{7}$$

In this paper, we only investigate the cost bounds in link equally-weighted WDM networks. It is assumed that all links have the same cost function

$$c(e) = 1 \; unit \; \text{hop-count-cost} \tag{8}$$

Thus,

$$c\big(ms(s,D)\big) = \sum_{i=1}^{k} \sum_{e \in LT_i(s, D_i)} 1 \tag{9}$$

3   Cost Bounds of Multicast Light-Trees in WDM Mesh Networks

In this section, we will study the cost bounds of light-trees in WDM networks with two different light splitting configurations: full light splitting and sparse splitting. Let $SR = N_{MC}/N$ be the ratio of MC nodes in the network. For the full light splitting case $SR = 1$, and for the sparse splitting case $0 \leq SR < 1$. Next, we first invest the cost bounds in the full splitting WDM networks.

3.1   Full Light Splitting WDM Networks

In the case that all network nodes are equipped with light splitters, each node could act as a branching node in a light-tree. Hence, one light-tree is sufficient to span all the multicast members. It is a Steiner-problem which tries to find a minimum partial spanning tree covering the source and all the multicast members. In a light-tree, there are at most N nodes when all the networks nodes are spanned ($\{n|n \in LT\} = V$), and at least $K + 1$ nodes if and only if the light-tree just contains the source and the multicast members ($\{n|n \in LT\} = \{s\} \cup D$). So, the cost of the multicast light-tree is bounded to

$$K \leq c\big(ms(s,D)\big) \leq N - 1 \tag{10}$$

To minimize the total cost in full light splitting case, the Minimum Path heuristic [6] and the Distance Network heuristic [8] can be good choices, since they are guaranteed to get a light-tree with a total wavelength channel cost no more than $2\big(1 - \frac{1}{K+1}\big)$[7,8] times that of the optimal Steiner tree. i.e.,

$$c\big(ms(s,D)\big) \leq 2\big(1 - \frac{1}{K+1}\big) \times C_{Opt} \tag{11}$$

where C_{Opt} denotes the wavelength channel cost of the Steiner tree.

344   F. Zhou et al.

3.2   Sparse Splitting WDM Networks

In this subsection, the cost bounds in sparse splitting WDM networks are studied. In the case of sparse splitting, only a subset of nodes can act as branching nodes in a light-trees. One light-tree may not be sufficient to accommodate all the group members simultaneously. Generally, several light-trees should be employed.

Lemma 1. *Given* $i, j \in [1, k]$ *and* $i \neq j$, *at least* $\exists$ *a destination* $d \in D_i$ *such that* $d \notin LT_j(s, D_j)$.

Proof. The aim of constructing the ith light-tree $LT_i(s, D_i)$ is to serve the destinations in the subset D_i, and the jth light-tree $LT_j(s, D_j)$ is used for serving the destinations in subset D_j. Let us suppose proof by contradiction that all the destinations in D_i are also included in $LT_j(s, D_j)$, i.e., $D_i \subseteq LT_i(s, D_i)$ and $D_i \subseteq LT_j(s, D_j)$. Then, all the destinations in set $D_i \cup D_j$ can be served by only one light-tree, i.e., $LT_j(s, D_j)$. According to the third principle of multicast light-tree computation, it is entirely useless to employ an additional light-tree to re-serve the destinations in D_i. As a result, $LT_i(s, D_i)$ can be eliminated and only $k - 1$ light-trees are required for multicast session $ms(s, D)$, which contradicts with the assumption. Hence, *Lemma 1* is proved.

Lemma 2. $\forall j \in [1, k]$, *the cost of the* j^{th} *light-tree holds*

$$K_j = |D_j| \leq c\big(LT_j(s, D_j)\big) \leq N - k \qquad (12)$$

Proof. According to equation (5), all the k subsets of destinations D_i, $i \in [1, k]$, are disjoint. Based on *Lemma 1*, at leat $k - 1$ destinations are not spanned in a light-tree. The number of nodes in a light-tree is consequently no more than $N - (k - 1)$. Furthermore, if no other nodes are included in the jth light-tree except the source s and the destinations in D_j (i.e. $\{n | n \in LT_j(s, D_j)\} = \{s\} \cup D_j$), then the number of nodes in the jth light-tree is minimal and equals $K_j + 1$. Hence, the cost bounds of a light-tree can be obtained as

$$K_j \leq c\big(LT_j(s, D_j)\big) \leq N - k \qquad (13)$$

Theorem 1. *In sparse splitting WDM networks, the total cost of the light-trees built for the multicast session* $ms(s, D)$ *satisfies*

$$K \leq c\big(ms(s, D)\big) \leq \begin{cases} K(N - K), & K < \frac{N}{2} \\ \frac{N^2}{4}, & K \geq \frac{N}{2} \text{ and } N \text{ is even} \\ \frac{N^2 - 1}{4}, & K \geq \frac{N}{2} \text{ and } N \text{ is odd} \end{cases} \qquad (14)$$

Proof. According to *Lemma 2* and equation (7), the total cost of the light-trees built for a multicast session $ms(s, D)$ holds

$$c\big(ms(s, D)\big) \leq \sum_{i=1}^{k} (N - k)$$
$$\leq k(N - k) \qquad (15)$$
$$\leq -(k - \frac{N}{2})^2 + \frac{N^2}{4}$$

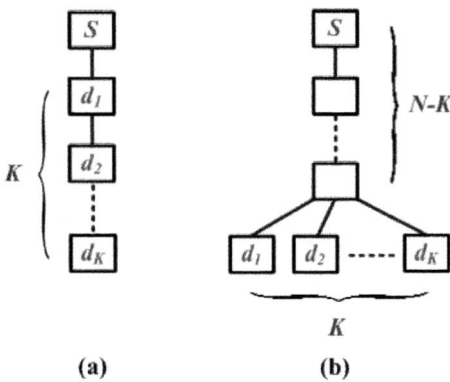

Fig. 2. (a)The best case; (b)The worst case

Regarding k is an integer and $1 \leq k \leq K$, we obtain

$$c\big(ms(s, D)\big) \leq \begin{cases} K(N - K), & K < \frac{N}{2} \\ \frac{N^2}{4}, & K \geq \frac{N}{2} \text{ and } N \text{ is even} \\ \frac{N^2 - 1}{4}, & K \geq \frac{N}{2} \text{ and } N \text{ is odd} \end{cases} \quad (16)$$

Moreover, according to *Lemma* 2, it is also true that

$$c\big(ms(s, D)\big) \geq \sum_{i=1}^{k} K_i = K \quad (17)$$

In fact the cost bounds given in *Theorem* 1 are tight. In the following we give two examples to show their accuracy. It is not difficult to imagine that the case with the minimal cost appears when all and only all the destinations are involved in the light-tree computed for multicast session $ms(s, D)$, as shown in Fig. 2(a). That is to say $\{n|n \in LT\} = \{s\} \cup D$. It is obvious that the lower bound K is tight.

And the worst case may happen when the network topology is like that in Fig. 2(b), where K lightpaths on different wavelengths are needed to serve K destinations to the source. Here, it is observed that the cost of the optimal light-trees equals $K(N - K)$. If the multicast session group size is equal to $|D| = K = \frac{N}{2}$ (K is even), an exact wavelength channel cost of $\frac{N^2}{4}$ should be consumed to establish the multicast session $ms(s, D)$. This example verifies that the exact accuracy of the upper bound given in *Theorem* 1.

4 Cost Bound of Multicast Light-Trees in WDM Rings

4.1 Multicast Light-Tree in WDM Rings

In WDM rings, all the nodes are mandatorily equipped with TaC [4] capability, one light-tree is able to span all the multicast members. The multicast light-tree in a WDM ring consists of either a single lightpath or two edge disjoint lightpaths originated from the same source. In a N nodes WDM ring, the cost of the multicast light-tree for multicast session $ms(s, D)$ complies

$$K \leq c\big(ms(s, D)\big) \leq N - 1 \tag{18}$$

4.2   Optimal Multicast Light-Tree in WDM Rings

Different from WDM mesh networks, minimizing the cost of multicast light-tree in a WDM ring is very simple. The minimum spanning tree for the multicast members is the optimal solution. Here, we use the concept gap introduced in [13,14]. A gap is a path between two adjacent multicast members in $\{s\} \cup D$ such that no other members are involved in this path nor is the source node. The optimal multicast light-tree can be obtained by removing the biggest gap from the ring [13].

Theorem 2. *In a WDM ring, the cost of the optimal light-tree for multicast session $ms(s, D)$ complies*

$$K \leq c\big(ms(s, D)\big) \leq N - \lceil \frac{N}{K+1} \rceil \tag{19}$$

Proof. Beginning from the source node s, we index the destination nodes from d_1 to d_K in the clockwise manner. Let g_1 denote the length of the gap between source s and d_1, g_i be the length of the ith gap, i.e., the gap between d_{i-1} and d_i, and g_{K+1} be the gap between source s and d_K as shown in Fig. 3. In a WDM ring of N nodes, we obtain

$$\sum_{i=1}^{K+1} g_i = N \tag{20}$$

And the cost of the optimal multicast light-tree for multicast session $ms(s, D)$ can be determined by

$$c\big(ms(s, D)\big) = N - \max_{1 \leq i \leq K+1} g_i \tag{21}$$

In order to obtain the cost bound of the light-tree, we have to determine the value range of $\max_{1 \leq i \leq K+1} g_i$. Note that all g_i are positive integers and satisfy equation (20). We obtain the following inequality

$$\max_{1 \leq i \leq K+1} g_i \geq \lceil \frac{N}{K+1} \rceil \tag{22}$$

This result corresponds to the case that multicast members are evenly distributed in the WDM ring. Thus we obtain

$$c\big(ms(s, D)\big) \leq N - \lceil \frac{N}{K+1} \rceil \tag{23}$$

Besides, if all the multicast group members stick together one by one, the optimal light-tree thus only consists of the source and the destinations. Then, we can obtain the lower bound

$$c\big(ms(s, D)\big) \geq K \tag{24}$$

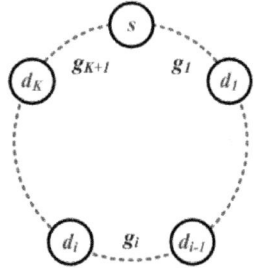

Fig. 3. The gaps in a WDM ring

5 Simulation and Numerical Results

As proven in Section 3.2, the cost bound of $Theorem$ 1 derived in sparse slitting WDM networks is valid for a serial of algorithms which respect the sparse splitting constraint plus the 3 principles mentioned in section 2. In this section, we try to obtain some numerical results to compare the proposed cost bounds and the cost of multicast light-trees computed by some classical heuristic algorithms, such as Reroute-to-Source (R2S) [3] and Member-Only (MO) [3]. Althought the cost bounds in $Theorem$ 1 is proven to be strictly tight, the cost of multicast light-trees depends a lot on the network topologies and the distribution of the multicast session members. Hence, here we do not mean to verify the accuracy of the proposed cost bound, but just to demonstrate the difference between the cost bound and the real cost of heuristic algorithms over a simple mesh topology.

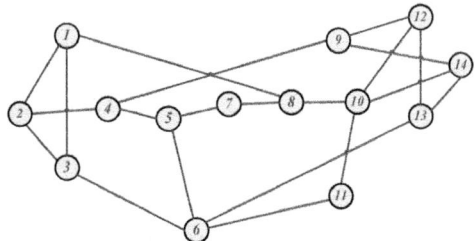

Fig. 4. The simulation topology: NSF Network

In the simulation, Reroute-to-Source and Member-Only algorithms are implemented by using C++ with LEDA package [15] in the 14 nodes NSF network (refer to Fig. 4). The network is configured without light splitters. The source and multicast members are assumed to be distributed uniformly over the topology.

The cost bounds of the multicast light-trees computed by MO and R2S heuristics are demonstrated in Fig. 5 when the multicast group size $K+1$ varies from 2 (Unicast) to 14 (Broadcast). 5000 multicast sessions are randomly generated for

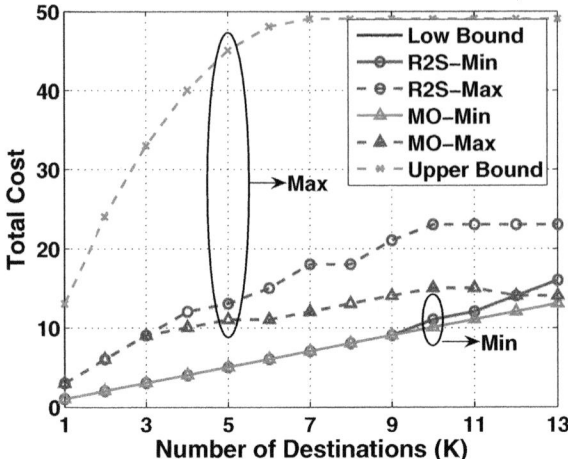

Fig. 5. The Cost Bound of multicast light-trees when the number of destinations K varies in NSF Network

a given multicast group size, meanwhile Member-Only and Reroute-to-Source algorithms are employed to compute the multicast light-forest for each session. Among 5000 light-forests, the biggest cost of the light-forests (denoted by R2S-Max and MO-Max) and smallest cost of the light-forests (denoted by R2S-Min and MO-Min) are figured out and plotted in Fig. 5. The low bound and the upper bound provided in $Theorem$ 1 are compared with the simulation result. According to the figure, it is observed that the proposed low bound is covered by MO-Min since they are almost the same. The low bound is also very near to R2S-Min. But we can also find that the upper bound is much bigger than the biggest costs obtained (MO-Max and R2S-Max) by the simulation.

Table 1. Cost Bound Comparison in NSF Network

$\|D\| = K$	LB	ILP	MO	R2S	UB
2	2	3.2	3.2	3.6	24
3	3	4.5	4.6	5.2	33
4	4	5.7	5.7	6.7	40
5	5	6.7	6.9	8.2	45
6	6	8.2	8.5	9.1	48
7	7	8.3	8.5	10.9	49
8	8	8.7	9.3	11.7	49
9	9	9.6	10.1	12.3	49
10	10	10.8	11.1	15	49
11	11	11.3	11.7	17.3	48
12	12	12	12	17.3	49
13	13	13	13.1	18.9	49

The cost bound and the average cost of 20 multicast sessions is shown in Table 1, for each $K \in [2, 13]$. In the table, **LB** denotes the lower bound, **UB** denotes the upper bound, and **ILP** denotes the average cost of the cost optimal multicast light-trees for 20 multicast sessions. The optimal multicast light-trees are computed by the integer linear programming (ILP) method by using Cplex [16]. We can see as the group size grows, the cost of any solutions (heuristic solutions and optimal ILP solutions) increases also. While the cost bound augments at the beginning, but keeps constant after $K \geq= \frac{N}{2} = 7$.

Does the numerical results obtained above mean that the proposed upper bound is too big? No! This is because the simulation results depend on the simulation topology. The proposed upper bound is valid for all the algorithms which complies the three rules mentioned in section 2. As discussed in subsection 3.2, given the network topology in Figs. 2(a) and (b), both the low bound and the upper bound are always tight.

6   Conclusion

Multicast routing in all-optical WDM mesh networks is an important but challenging problem. It is NP-complete to minimize the wavelength channel cost consumed per multicast session under the sparse splitting constraint. Although many papers have focused on the algorithms of multicast light-trees computation, no one addresses the cost bound of light-trees nor the approximation ratios of heuristic algorithms. Hence, in this paper we investigate the bound of wavelength channel cost consumed by a multicast session. An equal cost of 1 unit hop-count cost is assumed over all the fiber links in the network. We find that it is tightly lower limited to the number of destinations K, and strictly upper bounded to (1) $K(N - K)$ when $K < \frac{N}{2}$; (2) $\frac{N^2-1}{4}$ or $\frac{N^2}{4}$ respectively when $K \geq \frac{N}{2}$ and N is odd or even, where K is the number of destinations in the multicast session and N is the number of nodes in the network. Source-oriented multicast light-trees computation heuristic algorithms like Reroute-to-Source [3] and Member-Only [3] follow this cost bound, as they respect the three principles for light-trees computation mentioned in Section 2. In a particular situation, where the network topology is a WDM ring, the optimal multicast light-tree can be determined by removing the biggest gap from the ring. We found its cost is inferior to $N - \lceil \frac{N}{K+1} \rceil$. In the future, the approximation ratios of some classical heuristic algorithms of all-optical multicast routing will be investigated.

References

1. Sahasrabuddhe, L.H., Mukherjee, B.: Light-trees: optical multicasting for improved performance in wavelength-routed networks. IEEE Communications Magazine 37(2), 67–73 (1999)
2. Malli, R., Zhang, X., Qiao, C.: Benefit of multicasting in all-optical networks. In: SPIE Proceeding on All-Optical Networking, vol. 2531, pp. 209–220 (1998)

3. Zhang, X., Wei, J., Qiao, C.: Constrained multicast routing in WDM networks with sparse splitting. IEEE/OSA Journal of Lightware Technology 18(12), 1917–1927 (2000)
4. Ali, M., Deogun, J.S.: Cost-effective Implementation of Multicasting in Wavelength-Routed Networks. IEEE/OSA Journal of Lightwave Technology 18(12), 1628–1638 (2000)
5. Mukherjee, B.: WDM optical communication networks: progress and challenges. IEEE Journal on Selected Areas in Communications 18(10), 1810–1824 (2000)
6. Takahashi, H., Matsuyama, A.: An approximate solution for the Steiner problem in graphs. Math. Japonica 24(6), 573–577 (1980)
7. Winter, P.: Steiner Problem in Networks: A Survey. Networks 17, 129–167 (1987)
8. Kou, L., Markowsky, G., Berman, L.: A fast algorithm for Steiner trees. Acta Informatica 15, 141–145 (1981)
9. Zhou, F., Molnár, M., Cousin, B.: Avoidance of Multicast Incapable Branching Nodes in WDM Netwoks. Photonic Network Communications 18(3), 378–392 (2009)
10. Zhou, F., Molnár, M., Cousin, B.: Is light-tree structure optimal for multicast routing in sparse light splitting WDM networks. In: The 18th International Conference on Computer Communications and Networks, San Francisco, USA, August 2, pp. 1–7 (2009)
11. Jia, X., Du, D., Hu, X., Lee, M., Gu, J.: Optimization of wavelength assignment for QoS multicast in WDM networks. IEEE Transaction on Communications 49(2), 341–350 (2001)
12. Lin, H., Wang, S.: Splitter placement in all-optical WDM networks. In: Proceeding of IEEE GLOBECOM, pp. 306–310 (2005)
13. Scheutzow, M., Seeling, P., Maier, M., Reisslein, M.: Multicasting in a WDM-upgraded resilient packet ring. Journal of Optical Networking 6(5), 415–421 (2006)
14. Scheutzow, M., Seeling, P., Maier, M., Reisslein, M.: Shortest path routing in optical WDM ring networks under multicast traffic. IEEE Communications Letters 10(7), 564–566 (2006)
15. http://www.algorithmic-solutions.com/leda/index.htm
16. http://www.ilog.com/products/cplex/

Bidirectional Range Extension for TCAM-Based Packet Classification

Yan Sun and Min Sik Kim

School of Electrical Engineering and Computer Science,
Washington State University,
Pullman, Washington 99164-2752, U.S.A.
{ysun,msk}@eecs.wsu.edu

Abstract. Packet classification is a fundamental task for network devices such as edge routers, firewalls, and intrusion detection systems. Currently, most vendors use Ternary Content Addressable Memories (TCAMs) to achieve high-performance packet classification. TCAMs use parallel hardware to check all rules simultaneously. Despite their high speed, TCAMs have a problem in dealing with ranges efficiently. Many packet classification rules contain range specifications, each of which needs to be translated into multiple prefixes to store in a TCAM. Such translation may result in an exponential increase in the number of required TCAM entries. In this paper, we propose a bidirectional range extension algorithm to solve this problem. The proposed algorithm uses at most two TCAM entries to represent a range, and can be pipelined to deal with multiple range fields in a packet header. Since this algorithm assumes a non-redundant rule set, i.e., no range overlap between different rules, which can be obtained by applying our previous work on redundancy removal in TCAM using a tree representation of rules. Our experiments show a more than 75% reduction in the number of TCAM entries by applying the bidirectional range extension algorithm to real-world rule sets.

Keywords: packet classification, TCAM, bidirectional range extension.

1 Introduction

There are a number of network services that require packet classification, such as policy-based routing, firewalls, provision of differentiated qualities of service, and traffic billing. In each case, it is necessary to determine which flow an arriving packet belongs to, for example, where to forward it, whether to forward or filter it, what class of service it should receive, or how much should be charged for transporting it. As packet classification has been widely deployed on the Internet, demand for efficient packet classification grows. The function of a packet classification system is to map each packet to a decision according to a sequence of rules, which is called a packet classifier. The rules specified in a packet classifier may or may not be mutually exclusive; two rules may overlap. When it happens with no explicit priorities specified, we follow the convention that a

M. Crovella et al. (Eds.): NETWORKING 2010, LNCS 6091, pp. 351–361, 2010.

Table 1. A Simple Header Rule Set

Rule	Type	Source IP	Destination IP	Source Port	Destination Port	Decision
r_1	TCP	*	192.168.0.0/16	<1024	*	accept
r_2	TCP	*	192.168.14.1	*	139	discard
r_3	UDP	192.168.0.0/16	*	*	700:900	accept
r_4	TCP	*	*	*	*	discard

rule closer to the top of the list takes priority. Table 1 shows a simple packet classifier of four rules.

Perhaps the most popular method for high-speed packet classification in practice is to use a Ternary Content Addressable Memory (TCAM) [1]. A TCAM is a memory chip where each entry can store a packet classification rule in ternary form. It stores data patterns in the form of (value, bit mask) pairs. A query key is simultaneously compared against all the patterns stored in a TCAM. A key q is said to match a stored pattern (v, m) if $q\&m = v\&m$, where "&" is the bit-wise logical AND operator. Given a packet, the TCAM hardware can compare the packet with all stored rules in parallel and then return the decision of the first rule that the packet matches through a priority encoder. Thus, it takes $O(1)$ time to find the decision for any given packet. Because of their high speed, TCAMs have become the industrial standard for high speed packet classification [2]. The architecture of a TCAM used in packet classification is shown in Fig. 1.

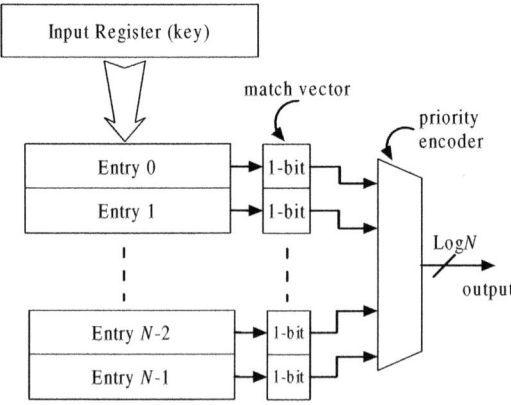

Fig. 1. TCAMs used in the packet classifications

A key (Protocol, Source IP address, Destination IP address, Source Port, and Destination Port) is stored in the input register and the rules are stored in the TCAM entries. The key compares with all the entries in parallel and the results are stored in the match vector, where 1's represent that the corresponding entries match the key, and the priority encoder chooses the match with the highest priority. At last, the output signal is used to find the corresponding action.

Despite their high speed, TCAMs have two major drawbacks when used in packet classifiers. First, they consume a large amount of power and have high hardware cost. Thus, their capacity in packet classifiers is often limited. Second, they are inefficient when applied to packet classifiers with port number ranges, because TCAMs can only store rules in ternary form, which means that port numbers need to be converted to one or more prefixes before being stored in TCAMs. This may lead to a significant increase in the number of TCAM entries needed to encode a rule. For example, 30 prefixes are needed to represent a single source port range [1, 65534], and 20 prefixes are needed to represent a destination port range [1, 2046]. Thus, $30 \times 20 = 600$ TCAM entries are required to represent a single rule with these two ranges. We observe that packet classifiers typically have at most one port range in each rule, and rules specifying two port ranges are rare. However, a small number of such rules can consume most of the TCAM entries and the number of such rules is increasing. Therefore, minimizing the number of required TCAM entries is crucial.

In this paper, we first introduce our previous work on redundancy removal in TCAM using a tree representation of rules [3], which removes all redundancy in the original rule set to make sure each packet matches and only matches a single new rule. Then we propose a bidirectional range extension algorithm to solve the range explosion problem based on the non-redundant rule set. The proposed algorithm uses at most two TCAM entries to represent a range, and can be pipelined to deal with multiple range fields in a packet header. In our experiments, we achieve a total reduction of 84.27% in the number of TCAM entries consumption.

The remainder of the paper is organized as follows. Section 2 presents previous work related to this paper. Section 3 discusses our previous work on removing redundancies in packet classification rules, and Section 4 proposes our bidirectional range extension algorithm to reduce the number of TCAM entries. Then the experimental results are presented in Section 5. Finally, we conclude in Section 6.

2   Related Work

Previous work exploring solutions to deal with the range expansion problem falls into two major categories: hardware-based solutions, which require changing TCAM hardware circuits, and software-based solutions, which do not require such changes. Below, we review previous work in these two categories.

Hardware-based solutions: The basic idea of hardware-based solutions is to modify TCAM circuits and architecture [2,4,5,6,7]. For example, van Lunteren et al. proposed a method of adding comparators at each entry to better accommodate range matching in packet classifiers [6]. While this allows to use TCAMs more efficiently, any solution from this research line has some drawbacks such as the cost of hardware modification.

Software-based solutions: Software-based solutions are more likely to be adopted by networking vendors and ISPs because they do not require changing TCAM hardware or existing packet classification systems. Many software-based solutions have been proposed [8,9,10,11,12,13,14,15,3] to reduce the TCAM entry consumption. Their basic idea is to preprocess ranges that appear in a packet classifier or convert a given packet classifier to another semantically-equivalent packet classifier that requires fewer TCAM entries, and then store the new rule set in a TCAM. Hence, the TCAM circuits need not be modified to implement range storage. Although these methods can typically achieve a 40–60% reduction in the number of TCAM entries by reducing redundancy in the rule set or combining multiple TCAM entries into a single TCAM entry, more reduction is desirable because of the low capacity and hist cost of TCAMs. Our work first reduces the redundancy in the rule set, and then extends the range in both directions, upward and downward, storing a single range into at most two TCAM entries. To the best of our knowledge, this article is the first attempt at using a range extension method to reduce the TCAM entries consumption.

3   Redundancy Removal Tree

Packet classifiers usually check the following five fields in each packet header: protocol type, source port number, destination port number, source IP address, and destination IP address. When a TCAM is used to implement a packet classifier, all rules stored in TCAM entries must be represented as exact binary values or binary values with wildcard bits. However, in a typical packet classification rule, some fields such as source and destination port numbers are represented as integer ranges rather than exact values or binary prefix values. Thus, we need to convert a rule with fields represented as integer ranges into one or more binary prefixes, which may lead to range expansion. During the process of range expansion, each field of a rule should be expanded separately. For example, if a 3-bit field of a rule is $[1, 6]$, the corresponding minimum set of prefixes covering the range includes 001, 01*, 10*, and 110. In the worst-case, range expansion of a w-bit integer range yields $2w - 2$ prefixes [16]. The next step to the range expansion is to compute the cross-product of obtained prefix sets, resulting in an exponential increase of the number of prefixes needed to replace a single rule.

The problem can be mitigated by removing redundancy in the original rule set. Two rules in a packet classifier may overlap, which means that one packet may match two or more rules. Besides, two rules in a packet classifier may conflict with each other. In other words, two overlapping rules may have different decisions. Many packet classifiers resolve conflicts by choosing the first match, which has a higher priority. In firewalls, typical decisions include "accept," "discard," "accept with logging," and "discard with logging."

Our goal is to reduce redundancy in a given packet classifier by removing redundant rules and overlapping parts. To achieve this goal, we build a Minimum Range Tree T for a packet classifier C: $(r_1, r_2, \ldots, r_n)$ over fields $F_1, \ldots, F_d$. The tree T must satisfy the followings:

- The height of the tree is equal to the number of fields in the packet classifier.
- Edges of each depth of the tree store the ranges of the corresponding field. All edges in the same depth cover the whole range of the field, and there is no overlap between any pair of them.
- A directed path from a leaf node to the root is called a *decision path*. For a given packet, the tree has exactly one matched decision path.
- Each leaf node is labeled with the decision associated with the corresponding decision path.

Fig. 3(a) shows a range tree for the simple packet classifier in Fig. 2. In this example, we assume every packet has only two fields, F_1 and F_2, and the domain of each field is $[0, 9]$.

$$r_1 : F_1 \in [0, 4] \land F_2 \in [0, 9] \rightarrow \text{accept}$$
$$r_2 : F_1 \in [0, 4] \land F_2 \in [4, 9] \rightarrow \text{accept}$$
$$r_3 : F_1 \in [5, 9] \land F_2 \in [7, 9] \rightarrow \text{accept}$$
$$r_4 : F_1 \in [5, 9] \land F_2 \in [0, 2] \rightarrow \text{discard}$$
$$r_5 : F_1 \in [0, 9] \land F_2 \in [0, 9] \rightarrow \text{discard}$$

Fig. 2. A simple packet classifier

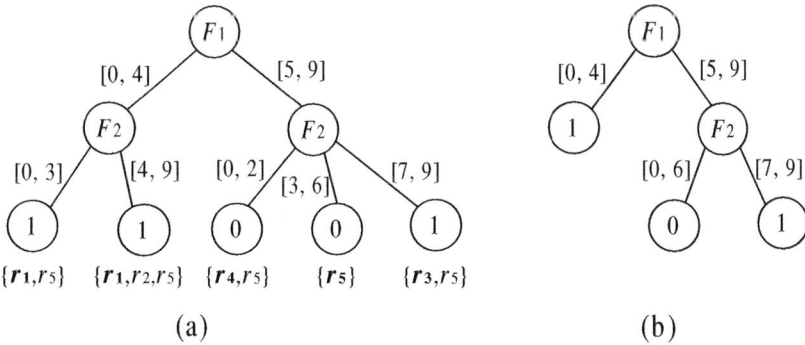

Fig. 3. Constructing a Minimum-Range-Tree for the Packet Classifier in Fig. 1

In Fig. 3, each edge represents a range in the corresponding field, and we use number 1 as a shorthand for "accept" and number 0 as a shorthand for "discard" in labeling leaf nodes. We first build a tree as in Fig. 3(a) according to the packet classifier in Fig. 2, then we combine two neighboring leaf nodes if they have the same decision and share the same parent node. The Minimum-Range-Tree is shown in Fig. 3(b). Now we get three new non-redundant rules based on the Minimum-Range-Tree. Please see previous work [3] for details.

4   Proposed Algorithm

After removing redundancies with the minimum range tree, there remains no overlap between different rules, and thus we can sort all the new rules by their ranges in a field. However, this does not solve the range explosion problem; it only mitigates it. Still, a single range may need multiple TCAM entries to represent itself. In the worst case, $2m - 2$ entries are needed to store a single m-bit range. For a 16-bit port range, it may need 30 TCAM entries. In order to further reduce the number of TCAM entries, we propose a bidirectional range extension, which only needs at most two TCAM entries to represent a single range regardless of the length of the range.

4.1   Bidirectional Range Extension

Given a list of ranges, we assume that every value belongs to exactly one range, that a range appearing earlier in the list has a higher priority than a range appearing later, and that the output of range matching for each range is binary, either 1 or 0. Note that applying the minimum range tree algorithm described in Sec. 3 and sorting resulting ranges by the boundary values yield a list of ranges satisfying these assumptions. Assuming these, we take the following steps to reduce the number of required TCAM entries.

We divide each range into two parts, each of which is represented as a single prefix. One part is called an upward extension and the other a downward extension. It is demonstrated in Fig. 4. Given a range, e.g., [100001, 110101], find a number N that belongs to the range and has the most consecutive 0's starting at the least significant bit (LSB). In our example, N will be 110000. Convert all the rightmost consecutive 0's of N into "don't care" bits. The result will be the upward extension (master entry). If N is equal to the lower bound of the original range, this range does not need a downward extension. Otherwise, the numeric prefix (excluding "don't care" bits) of the upward extension minus 1 (10 in our example), followed by the same number of "don't care" bits will be the downward extension (slave entry) as shown in Fig. 4(a). We use the set of all upward extensions as the master set and the set of downward extensions as the slave set. Note that the master set has the same number of entries as the number of ranges in the original list. On the other hand, the slave set only contains entries for those ranges where the original range is not a subset of the upward extension.

Below, We prove that the original range is a subset of the master and slave entries combined.

Proof: Assume that the range is $[N_{\mathrm{L}}, N_{\mathrm{H}}]$. Suppose $N_{\mathrm{L}} = \sum_{i=0}^{n-1} b_i^{\mathrm{L}} 2^i$ and $N_{\mathrm{H}} = \sum_{i=0}^{n-1} b_i^{\mathrm{H}} 2^i$, where $b_i^{\mathrm{L}}, b_i^{\mathrm{H}} \in \{0, 1\}$. Let N be the number in $[N_{\mathrm{L}}, N_{\mathrm{H}}]$ that has the longest consecutive zeros (n_0 bits) starting at LSB. Thus, we have $N = 2^{n_0} + \sum_{i=n_0+1}^{n-1} b_i 2^i$, where $b_i \in \{0, 1\}$. Then the master entry is a range $[N, N_{\mathrm{master}}]$, where $N_{\mathrm{master}} = N + \sum_{i=0}^{n_0-1} 2^i$ with at least consecutive $n_0 + 1$ 1's starting at LSB. Therefore, $N_{\mathrm{master}} + 1 = 2^{n_0+1} + \sum_{i=n_0+1}^{n-1} b_i 2^i$, which has at

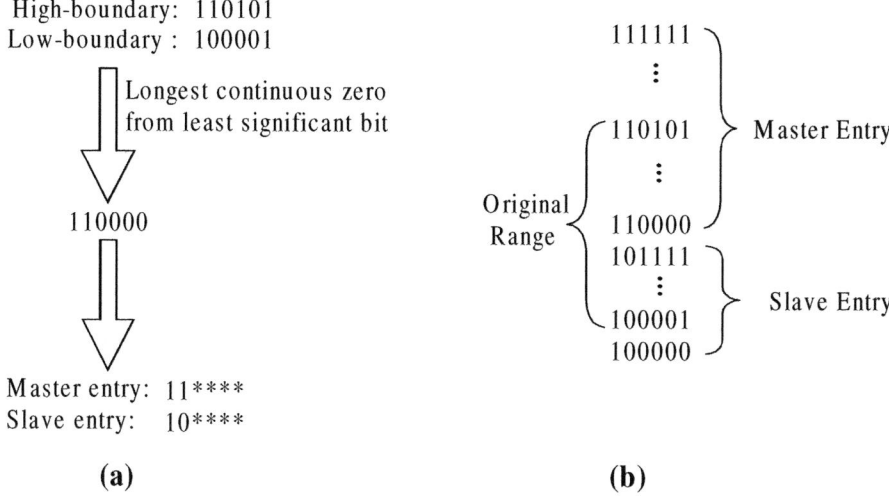

Fig. 4. An example of how to divide a range and generate master/slave entries

least $n_0 + 1$ consecutive 0's starting at LSB, or more 0's than N, and thus should not be in $[N_L, N_H]$. Therefore,

$$N_H \leq N_{\text{master}}. \tag{1}$$

Similarly, the slave entry, if we have one, is a range $[N_{\text{slave}}, N-1]$, where $N_{\text{slave}} = N - 2^{n_0} = \sum_{i=n_0+1}^{n-1} b_i 2^i$. Note that N_{slave} has at least $n_0 + 1$ consecutive 0's starting at LSB, or more 0's than N, and thus should not be in $[N_L, N_H]$. Therefore,

$$N_L > N_{\text{slave}}. \tag{2}$$

By Eq. 1 and Eq. 2, $[N_L, N_H]$ is a subset of $[N_{\text{slave}}, N_{\text{master}}]$.

Note that each of the master and slave entries can be represented by a single prefix, and thus by a single TCAM entry.

4.2 Matching Using Extended Ranges

A field of an incoming packet is compared against both the master and slave sets; each set is stored in a separate TCAM. After knowing the decision from the master set, we need to find out whether the matched entry has a corresponding entry in the slave set. If there is no such entry, the decision in the master set becomes the final decision; otherwise, we need to consider the corresponding slave entry. In the slave set, we obtain the decision and the low-boundary (N_L) of the corresponding entry. If both decisions, one from the master set and the other from the slave set, are identical, that becomes the final decision; otherwise we compare the low-boundary from the slave set with the input key. If the low-boundary is smaller than the key, we choose the decision from the slave set; otherwise we choose the decision from the master set.

For the worst case, the single interval [1, 65534] requires only 2 TCAM entries in our algorithm instead of 30 entries, which reduces about 93.3%.

5 Simulation Results

5.1 Experimental Results on Minimum Range Tree

For experiments, we collected rules from actual packet classifiers. Because rule sets vary across different applications, we gathered as many rules as we could and then randomly selected one thousand rules from them. For the Minimum Range Tree, we compared the number of TCAM entries used by the original rules and the number of TCAM entries used after removing redundancies using the Minimum Range Tree. The simulation result showed that our algorithm reduces 66.4% of TCAM entries in total [3].

5.2 Analysis of Bidirectional Range Extension

In this subsection we analyze the performance of the proposed bidirectional range extension algorithm on a single range, such as the destination port range. Given the number of bits for the field, we generate every possible range whose length is greater than 1 with the same probability. Thus, the generated rule set may include ranges [0, 1], [0, 2], ..., [0, $2^n - 1$], [1, 2], [1, 3],...,[$2^n - 2$, $2^n - 1$], where n is the number of bits of this field. And the simulation results are shown in Table 2.

Table 2. Analysis of a Single Range

Range length (bits)	# of TCAM entries with range extension	# of TCAM entries w/o range extension	Saving (%)
1	1	1	0.00
2	9	10	10.00
3	49	65	24.62
4	225	355	36.62
5	961	1831	47.52
6	3969	9103	56.40
7	16129	43935	63.29
8	65025	206911	68.57
9	261121	955007	72.66
10	1046529	4335871	75.86
11	4190209	19421695	78.43
12	16769025	86033407	80.51
13	67092481	377595903	82.23
14	268402689	1644400639	83.68
15	1073676289	7114039295	84.91
16	4294836225	30602706943	86.00

The percentages of entries saved with different field widths are shown in Fig. 5. We can see that the percentage of entries saved by our algorithm increases as the field width increases, and we can save 86.00% of TCAM entries of the 16-bit port number field. For the real-world rule sets we collected[1], our algorithm can save 77.47% and 74.81% entries for the destination port field and source port field, respectively, excluding non-range rules. For a single 16-bit port number, our approach reduces the average number of entries from 14.25 to 2.

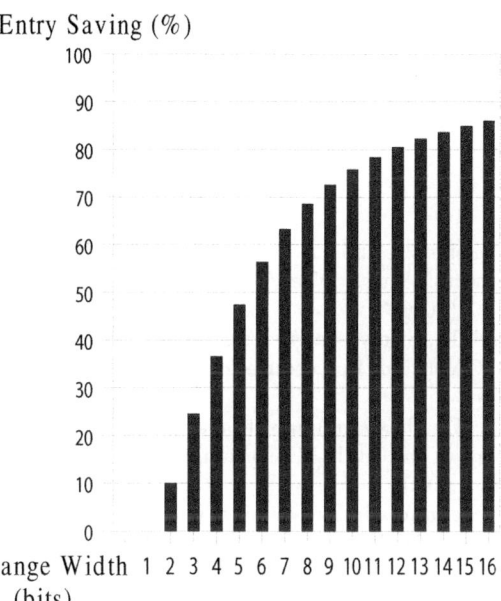

Entry Saving (%)

Range Width 1 2 3 4 5 6 7 8 9 10 11 12 13 14 15 16
(bits)

Fig. 5. The Entries saved with Different Field Width

For a rule with both destination port and source port ranges, our algorithm can further save TCAM entries because of the multiplication effect. For the scenarios shown in Table 2, our algorithm saves 98.04% of TCAM entries for randomly generated rules with both destination port and source port fields, and saves 82.78% for the ranges included in the real-world rule sets. The difference between them is mainly caused by the fact that some popular ranges, such as [0, 1023], can be represented by a single prefix in real-world rule sets. If we include rules without ranges in the simulation, our algorithm can save 53.18% TCAM entries for the real-world rule sets.

From both redundancy removal using the Minimum Range Tree and the bidirectional range extension, we can reduce 84.27% of entries for the real-world rule sets with additional circuits such as two adders and a little latency.

---

[1] http://www.routeviews.org/

6   Conclusion

In this paper, we proposed a bidirectional range extension algorithm to solve the range explosion problem in TCAM. The proposed algorithm assumes a non-redundant rule set, which can be achieved by previous work. Our algorithm first divides a range into two ranges, and then extends the ranges upward and downward to make each extended range consumes at most one TCAM entry. The result is that each range consumes at most two TCAM entries. Our algorithm significantly reduces the number of TCAM entries needed by a packet classifier. In our experiments, after removing redundancies, we observed a reduction of 86.00% on average in the number of TCAM entries, and an overall reduction of 84.27% for real-world rule.

References

1. Pagiamtzis, K., Sheikholeslami, A.: Content-addressable memory (CAM) circuits and architectures: A tutorial and survey. IEEE Journal of Solid-State Circuits 41(3), 712–727 (2006)
2. Karthik, L., Anand, R., Srinivasan, V.: Algorithms for advanced packet classification with ternary CAMs. In: Proceedings of ACM SIGCOMM '05, August 2005, pp. 193–204 (2005)
3. Sun, Y., Kim, M.S.: Tree-based minimization of TCAM entries for packet classification. In: Proceedings of the 7th IEEE Consumer Communications and Networking Conference (January 2010)
4. Huan, L.: Efficient mapping of range classifier into ternary-CAM. In: Proceedings of the 10th Symposium on High Performance Interconnects, August 2002, pp. 95–100 (2002)
5. Lunteren, J., Engbersen, T.: Fast and scalable packet classification. IEEE Journal on Selected Areas in Communications 21(4), 560–571 (2003)
6. Ed, S., David, T., Jonathan, T.: Packet classification using extended TCAMs. In: Proceedings of the 11th IEEE International Conference on Network Protocols, November 2003, pp. 120–131 (2003)
7. Faezipour, M., Nourani, M.: CAM01-1: A customized TCAM architecture for multi-match packet classification. In: IEEE Global Telecommunications Conference, November 2006, pp. 1–5 (2006)
8. Sumeet, S., Florin, B., George, V., Jia, W.: Packet classification using multidimensional cutting. In: Proceedings of ACM SIGCOMM '03, August 2003, pp. 213–224 (2003)
9. Liu, A.X., Meiners, C.R., Zhou, Y.: All-match based complete redundancy removal for packet classifiers in TCAMs. In: Proceedings of the 27th IEEE INFOCOM, April 2008, pp. 111–115 (2008)
10. Applegate, D.A., Calinescu, G., Johnson, D.S., Karloff, H., Ligett, K., Wang, J.: Compressing rectilinear pictures and minimizing access control lists. In: Proceedings of the 18th Annual ACM-SIAM Symposium on Discrete Algorithms, January 2007, pp. 1066–1075 (2007)
11. Qunfeng, D., Suman, B., Jia, W., Dheeraj, A., Ashutosh, S.: Packet classifiers in ternary CAMs can be smaller. In: Proceedings of ACM SIGMETRICS/Performance 2006, June 2006, pp. 311–322 (2006)

12. Meiners, C.R., Liu, A.X., Torng, E.: Topological transformation approaches to optimizing TCAM-based packet classification systems. In: Proceedings of the eleventh international joint conference on Measurement and modeling of computer systems, pp. 73–84 (2009)
13. Bremler-Barr, A., Hendler, D.: Space-efficient TCAM-based classification using gray coding. In: Proceedings of the 26th IEEE International Conference on Computer Communications, May 2007, pp. 6–12 (2007)
14. Che, H., Wang, Z., Zheng, K., Liu, B.: DRES: Dynamic range encoding scheme for TCAM coprocessors. IEEE Transactions on Computers 57(7), 902–915 (2008)
15. Pao, D., Li, Y.K., Zhou, P.: Efficient packet classification using tcams. Computer Networks: The International Journal of Computer and Telecommunications Networking 50(18), 3523–3535 (2006)
16. Gupta, P., Mckeown, N.: Algorithms for packet classification. IEEE Network 15(2), 24–32 (2001)

Estimating the Economic Value of Flexibility in Access Network Unbundling

Koen Casier, Mathieu Tahon, Mohsan Ahmed Bilal, Sofie Verbrugge, Didier Colle, Mario Pickavet, and Piet Demeester

Ghent University – IBBT – IBCN, Gaston Crommenlaan 8, 9050 Gent, Belgium
{koen.casier,mathieu.tahon,sofie.verbrugge,
didier.colle,mario.pickavet,piet.demeester}@intec.ugent.be

Abstract. By means of local loop unbundling (LLU), the incumbent operator opens the access network connecting the customer to the central office to other operators. Other licensed operators (OLO) will use this possibility to provide customers their services. The price of the LLU-offer is set by a regulator, aiming to allow fair competition in a monopolistic market. This is typically fixed at the lowest possible price covering all of the incumbent's costs in this network. Now, the OLO always has the choice between the regulated LLU-offer, own installation or not providing access to the customer. In order to remove any unfair competition advantages, the regulator should incorporate a flexibility bonus in its LLU-pricing strategy. In this paper we show how the economic technique of real option valuation can be used for estimating a numeric value for this flexibility advantage.

Keywords: Local Loop Unbundling, Techno-Economic, Real Options, Telecom Regulator.

1 Origin, Context and Impact of Local Loop Unbundling

In Europe a lot of the telecom networks have evolved out of community or country-wide installed copper networks. Considering the importance of those telephone networks, this company was placed under the control of the government, which gave additional stimuli to grow this network country-wide, connecting every inhabitant. While the networks were originally installed for providing telephony signal transmission, later on they also allowed for data transmission. With the advent of ADSL, the importance of digital data transmission over this network grew. Soon data took over as the driving factor for network upgrades. Clearly at the other side, the network connected to a very diverse set of services. At this point in time, the publicly owned telecommunication network keeps off all competition and innovation in this field.

A large privatization wave decoupled most of the so-called incumbent network operators in Europe from direct control of the government. Still the existing network infrastructure, which had been fully installed and paid for in the public era, poses a large advantage to the use of these incumbent operators. Without clearing this unfair

M. Crovella et al. (Eds.): NETWORKING 2010, LNCS 6091, pp. 362–372, 2010.

historic advantage, new entrants will never have a chance to compete in the former monopolistic market. To counter this negative effect on competition, the national telecom regulators have imposed the incumbent network operators the obligation to open up their existing infrastructure to the new entrants often called other licensed operators (OLO). When considering the access network, this obligation is called the local loop unbundling (LLU) and the regulator will fix the price based on an estimation of the maintenance costs of this network for the incumbent. In extension the operator might also be forced to open up at the data link layer which is often called bitstream access. Opening the access network at any possible place to the OLOs is referred to as LLU in the remainder of this paper.

In this study we considered the case of a copper based network operator who is faced with a regulatory fixed price for giving the OLOs access to his network. As mentioned this price is based on an estimation of the costs for maintenance (and other costs) of this selected part of the network for the incumbent. It is considered a fair price covering the costs the incumbent indulges. We get a better picture of the full interaction and economic impact when looking from the point of view of the OLO. The OLO can choose, at each point in time and at each location, to connect customers. In this choice he will invest a small amount on the equipment to be collocated in the central office (CO). For each customer connected he will pay the fixed price directly covering the costs of the incumbent, and as such the OLO will have a competitive case against the incumbent for each customer. On top of that, the OLO has a higher flexibility as he can choose where and when to start connecting and how many customers to connect. He can also pull out from certain regions when they are no longer profitable and shift equipment and attention to other locations. When used optimally this higher flexibility allows the OLO to maximize revenues, minimize costs and risks, and have a more profitable overall business case than the incumbent.

The economic value of such flexibility is less straightforward to estimate in the business case and is as such often neglected. In order to judge on the impact of this flexibility, one should use more advanced calculations. In this paper we show how we used the theory of real options to calculate the economic value of the flexibility, in the remainder of the paper referred to as options, of the OLO. This economic evaluation technique has been used abundantly in the past to estimate the impact of uncertainty and managerial flexibility on the outcome of a business case [1], [2], etc. In [3] and [4] the same issue of estimating the value of LLU for an OLO through the use of real options is addressed. In our paper we go one step further, providing a quantitative full-fledged study combined with realistic figures for the calculation of the OLOs advantage.

In section 2 we build a full quantitative model for the calculation of the costs for a copper based OLO. We also indicate what unbundling offers this operator can use in the construction of his own business case. In order to reflect the economic impact of the choices the OLO has at hand, in section 3 this business case is enhanced with real options. We also indicate which uncertain factors will have the highest impact on the case and on the choices to be taken. The outcome is an extended business case for the OLO in which he will always choose the optimal installation path given the uncertain input. Confronting this model with the original business case gives an idea of the value of the options in LLU. The second part of this section shows several realistic results. Finally section 4 aggregates all information available in the paper and distils some general and more detailed conclusions from this.

2   Forecasting the Costs for an OLO

Access to existing network infrastructure can be opened up at different levels. Theoretically the network can be opened at each level of functionality, for instance as defined in the OSI layered model. In the context of a copper based network, the network is typically unbundled in the following ways, numbered according to the OSI layer at which they are opened:

0.   Full Unbundling (FU): The incumbent patches the copper pair running to a customer directly to the equipment of the OLO. This requires the equipment of the OLO to be placed in collocation with the incumbent's installation in the CO. In this approach, the OLO has full access to the physical medium at OSI layer 0.

1.   Shared Line (SL): The incoming copper pair is led through a splitter, where voice is separated from data. Both signals are brought to either incumbent or OLO. At this point, the physical (OSI) layer 1 is fully defined for both voice (PSTN) and data (xDSL).

2.   Bitstream access (BA): The incumbent splits voice from data in the CO. The data is sent further into the metro network and is transported to a fewer number of aggregation switches. At these aggregation switches (called point of presence or POP), the OLO can collocate his equipment in order to reach the customer. The connection from this point up to the customer is handled on OSI layer 2 or 3.

Figure 1 gives an overview of these three possibilities for opening up the network.

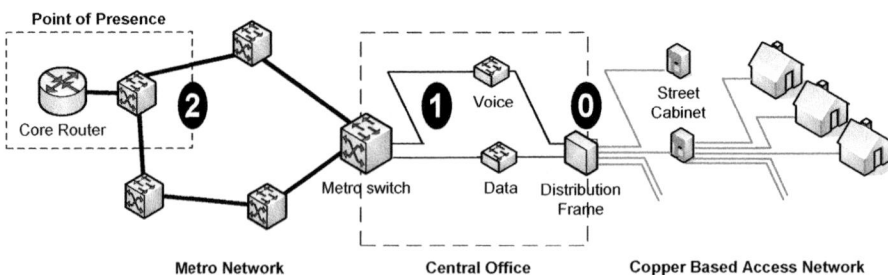

Fig. 1. Overview of the network from access (right) to core (left) with an indication of the unbundling locations

2.1   Forecasting the Market

In general the OLO takes his share of the total telecom market. Still the incumbent operator often holds a major part of the market and leaves only a small margin for the different OLOs to capture. In the context of this research we considered the Belgian market, in essence a duopoly in which the cable and copper operator are in fierce competition. We focus on the copper network segment, in which Belgacom is the incumbent and is regulated nationally by the Belgian regulator BIPT. The OLO market in Belgium represents a share somewhere below 15% or less than 500k residential connections [5]. The same source also contains an overview of the occurrences of the different types of LLU used by the OLOs. For estimating the market for the OLO, we started from the growth of broadband in Belgium, modeled using a Gompertz curve (sales indicated by S(t)) (1) fitted to the values found in [5] (resulting in $[m,a,b]$ = [3.6M, 5.25, 0.32]). Within this overall customer potential, we modeled a market segment for all OLOs in the market, initially set at 15%. In following years, this market segment has an uncertainty of growth based on the previous year. We limited the growth to between -5% and 5% of the customer base in the market segment.

$$S(t) = m \cdot e^{-e^{-b \cdot (t-a)}} \quad \text{with} \quad \begin{array}{ll} m & \text{market potential} \\ a & \text{infliction point (37\% adoption)} \\ b & \text{slope of the adoption} \end{array} \tag{1}$$

2.2 Installation of Equipment

In the given situation the market should be split in the smallest individual regions in which the OLO can compete with the incumbent. As mentioned before the CO location is the first place at which the OLO can connect customers (FU or SL). In each of those individual markets, the OLO can select its strategy and build a smaller economic model. In order to connect customers here, the OLO will install copper connected data and/or voice equipment in the CO of the incumbent. In addition, the OLO will also have to pay a LLU rental fee per actual customer to the incumbent. Additionally the OLO will also have a growth in its overall operations as a result of the new installation in this region. On the positive side, the OLO will also gain income from the customers connected in the region.

In case the OLO would rather prefer to connect part of the customers using BA, he will place data equipment at a higher aggregation point in the POP. Again he will have to pay an LLU rental fee to the incumbent on a per customer base.

2.3 The Overall Cost Model

The overall cost model is obtained by combining the different building blocks and is shown in Figure 2. The background of how to build the techno-economic model and how to calculate process and infrastructure costs can be found in [6]. A very important parameter in this context is the cost of goods sold (COGS) which is actually the tariff the OLO will have to pay to the incumbent for using the LLU offer for connecting the customers. We used this model on an installation set of 500 COs (FU and SL) and one aggregation POP (BA) at which the OLO can connect. In [5] and [7] the proportion of FU, SL and BA are known and we spread this proportion randomly over the 500 COs. Additionally we have randomized the customer base for each CO according to a logarithmic cumulative function in which 35 COs connect 50% of the customers and 100 COs connect 70% of the customers. The combination of the costs and revenues in all COs with collocated equipment (FU or SL) and the results for the different POPs and associated customers gives the final cashflow of the business case for each year. Summing these cashflows, while taking into account the time value of money, gives the net present value (NPV) of the OLOs business case. We considered a time span of 5 years for the evaluation of the business case. Of course the selection whether to use BA or FU/SL can be optimized for each region and in each year. This flexibility will be dealt with in the following chapter.

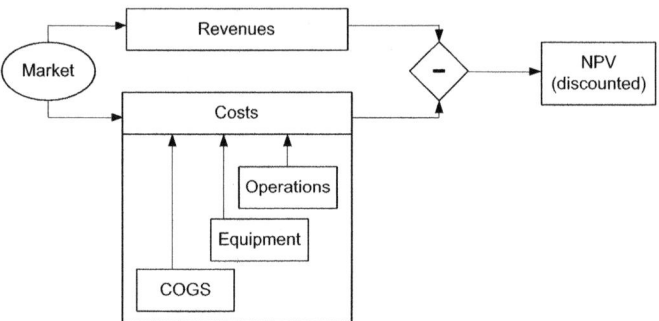

Fig. 2. Overall cost model used for calculating the outcome for the OLO

Although this case has been constructed in the context of the Belgian situation, it did not use all real input values and can as such not be translated to reality without considerable changes in the input values. The conclusions expressed in this paper are not statements explicitly valid for Belgium but only valid for our theoretical model with its specific hypothesis and cost elements.

3 Introducing Flexibility and Estimating Its Value

The NPV calculation presented above cannot capture flexibility. Once a project is set out, the project path is fixed. However, most real business cases have built-in flexibility. The management can at every point of time decide to follow a different path, e.g. abandoning the project after one year when the payoff does not meet the requirements. The economic theory proposed an extension of the NPV model to implement flexibility, the real option approach. The real option valuation methodology tries to capture (and include) the value of managerial flexibility present in a business case, much in the same way the flexibility presented in financial options (over stocks) is valued. A financial option gives the right to buy or sell over a limited period the underlying asset for a predetermined exercise price. As it is a right (and not obligation) the value of an option will always be positive. The technique of Real Options was defined in 1977 [8] and applies option pricing theory to the valuation of investments in real assets. It proved especially useful in investment decisions consisting of different (optional) phases. As it adds flexibility to the business case, it alleviates (partly) the estimation of the risk by means of the discount factor as in the calculation of the net present value.

The value of a project should now be extended by the value of the options, and is defined as the summation of the original NPV of the project, as calculated in Section 2, with the value of each of the options. Several option valuation techniques are in use in economic literature. Black and Scholes and Binomial tree valuation (and extensions) are used in theoretical real option studies. These valuation techniques are generally not feasible for larger studies in which several options are modeled and several uncertainties are taken into account. In this case a Monte Carlo simulation is used on the model taking into account all options and uncertainties. The model is simulated 10000 times, every time generating a NPV. The result of such a simulation is a distribution of all results. The outcome mean NPV of this model will already reflect the summation of original NPV and option values. [9] provides a more extensive introduction to Real Options theory, with a lot of practical examples.

3.1 Integrating Real Options with the Techno Economic Model

In the setup of the cost model, we already indicated some of the most important uncertainties for an OLO. We incorporated some additional uncertainties into the real option evaluation of the business case and indicate below their average value and standard deviation. We also indicate additional information on the expected probability distribution (assumptions) where necessary (in general we assumed a normal distribution).

- *Average Return per User (ARPU)*
  Average ARPU of €30 per month per user with a standard deviation of €1. A higher ARPU will of course lead to a more positive business for the OLO.
- *Churn rate*
  Average Churn of 15% with a standard deviation of 1.5%.
- *Regulatory context*
  Binary chance (50/50) for a change in the regulation (LLU pricing) in the coming years. Consecutive changes have been inversely correlated in order to reduce instability with year on year changes in regulations.
- *Equipment pricing*
  We estimated a price erosion on the equipment, diminishing the costs yearly. To reflect the uncertainty on equipment pricing, the yearly calculated cost is changed additionally using a multiplier. This multiplier is modeled by a normal distribution on average 100% and with a standard deviation of 2%. (cheaper or more costly)
- *OLO Market*
  Both the full OLO market and the position for one OLO within this market are uncertain. The full OLO market is expected to follow the 15% of the total broadband adoption. On top of this we expect both the full market and the position for one OLO in this market to change using a multiplier. This multiplier is modeled by a normal distribution on average 100% and with a standard deviation of 2%.

In real options we try to estimate the effect of managerial flexibility to counter the negative effects and gain from the positive, considering the given uncertainties. This flexibility is incorporated in the model by means of options, choices that can be made and will alter the outcome of the business case. Here the OLO always has the option to select the type of LLU for each separate area - CO in case of FU or SL, or all CO's connected to the same POP in case of BA – separately. In order to keep the calculations feasible, we restricted the options in two ways. First we did not expect an OLO to switch from CO collocation to BA for a given area. This is a reasonable assumption as switching back would involve a lot of upfront investments (CO-equipment) to go up in smoke. Once the CO-equipment is installed, FU or SL will also almost always be the cheaper option. Secondly we also restricted the different options to be taken at the end of each year instead of continuously executable options. Figure 3 gives a visual representation of the remaining options.

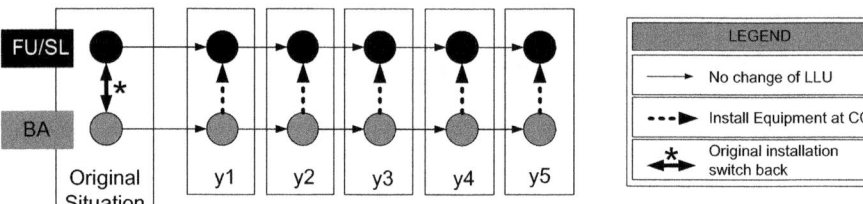

Fig. 3. A visual overview of the connection options for the OLO at the end of each year

Considering the complex nature of the cost estimation model and the large amount of uncertain factors influencing the outcome of this case, we choose to calculate the value of the different options in the business case using a Monte Carlo simulation based approach. Of course we calculated the outcome in the case of a full optimal managerial decision in each case. Additionally, we calculated the outcome in case the OLO would (FU) only be able to provide access over FU or SL, (BA) only be able to provide BA, (No) be stuck with its existing installation, (Thresh.) start from the existing situation and switch using a threshold of 480 users per CO. To be complete we refer to the full optimal use of the options as (Opt.). In this latter case, the operator can also make use of a free correction in the first year (as if he is a new OLO).

3.2 Results of Real Options

One of the main parameters in the business case is the position of the considered OLO within the full OLO market. Figure 4 gives a view on the outcome of the business case for an OLO with a market share of 10% up to 100%.

The first thing to notice is how an OLO with less than 30% of our expected (15%) OLO market, meaning 4%-5% of the total Broadband Market will have real difficulties in reaching a positive result. Once the OLO has more than 30% of the OLO market, he has a good chance of reaching a positive business case and its margins are increasing for an increasing market share.

Next the results also clearly show how the full optimal strategy has a good advantage over the non-flexible situation (either FU, BA or No) and Figure 55 shows a value of the options in a full optimal situation. We find an option value of €9 per customer for an OLO with a positive business case (40% market share) when comparing to FU. This value is rapidly decreasing to between €2 and €1 for an

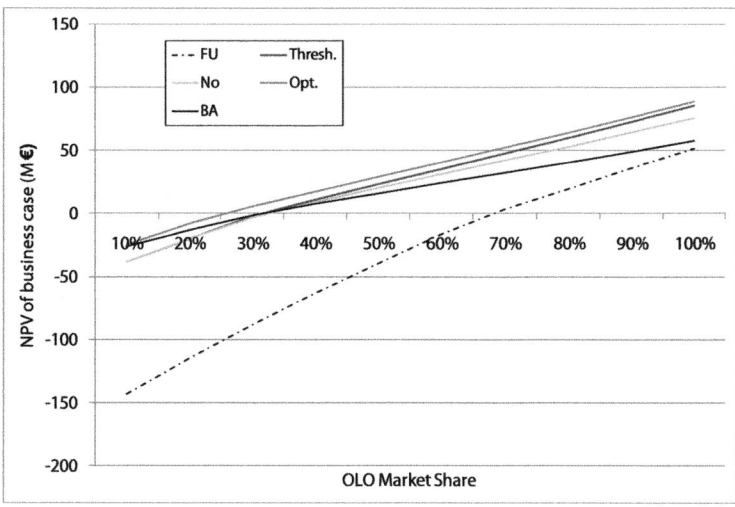

Fig. 4. NPV of the business case for an OLO using the different options

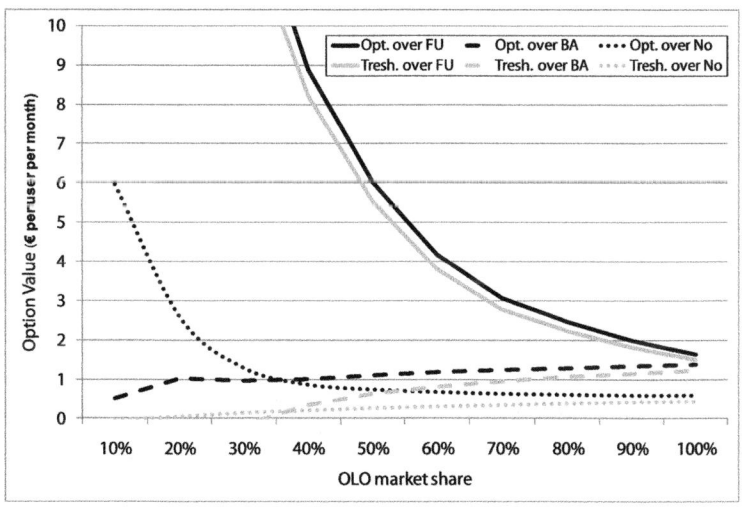

Fig. 5. The value of the different options at the OLO's disposal

increasing market share. When comparing to BA we find an almost flat option value of €1 between 40% and full market share.

Basing the option of switching on a predefined threshold is only slightly worse and will still give a good option value over a CO-based installation. However for a market share of less than 30% the value of the option becomes negative. As the estimation of the managerial response is based on some upfront decision, rather than the selection of the real optimum, this calculation is strictly speaking not a real option valuation. Still this simulation of a likely managerial response in a threshold based rule gives valuable information. It requires less knowledge and outlook to make the decision and still incorporates some intelligence towards the decision (at least above a market share of 40%). In this context it is important to note that the threshold based selection will start from the existing situation and can as such not switch back in the first year (as is allowed for Opt. and is assumed in BA. Allowing this based on the predefined threshold, for which the results are shown in Figure 66, makes the rule more generally applicable and gives a positive (average) option value in all considered cases.

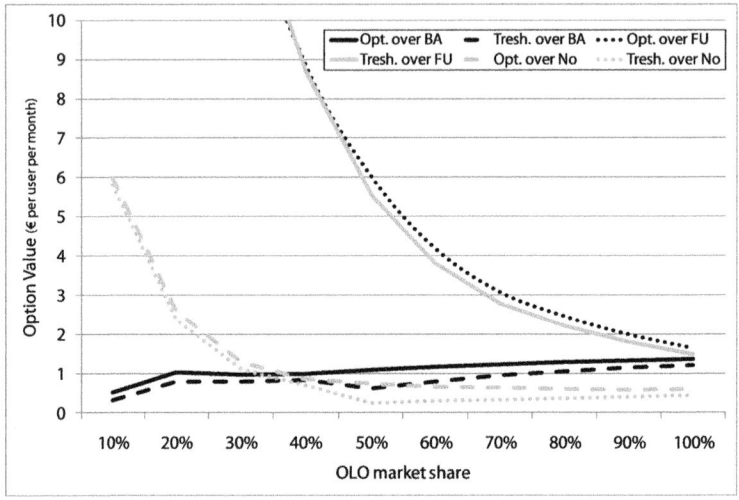

Fig. 6. The value of the different options with improved threshold based option

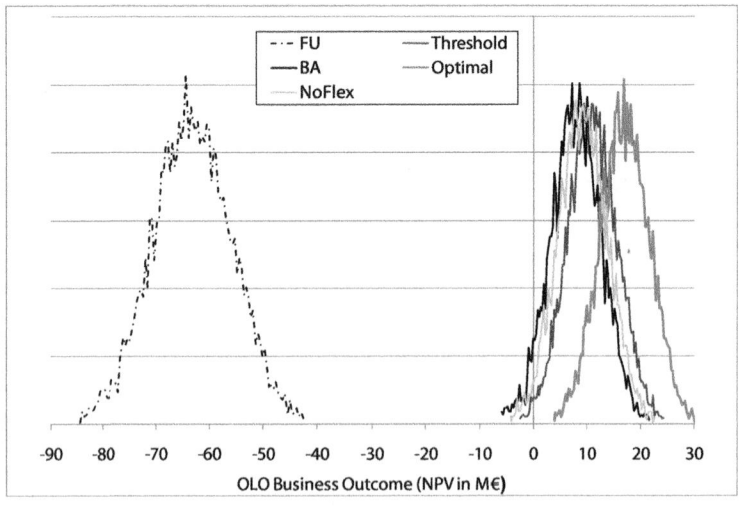

Fig. 7. NPV distribution for an OLO with 40% market share

As mentioned before, the business case becomes (on average) positive for an OLO on condition he has a market share at least 40% of the OLO market (6% of the total market) in most strategies. The outcome NPV in this case is shown in Figure 7 for all different strategies for a variation in input as used in the Monte Carlo simulation. Clearly chances of a negative business case are very low in case at least BA is provided to the OLO. In case both LLU offers are provided, a much higher result is attainable. The highest result is logically found by selecting the highest result in each case as shown in Opt..The sensitivity analysis performed in this case also gives important information on the most important uncertainty parameters for the simulations. Market share is without any question the most important.

4   Conclusions

In many European countries, incumbent telephony operators had a monopoly on the telecom market. Historically they were rolled out under monopoly on a country-wide scale. The copper infrastructure has already been fully installed and paid for the in the past and this network provides the incumbent with a intangible advantage on the market. By means of local loop unbundling (LLU), the regulator forces the incumbent operator to allow other licensed operators (OLO) to connect customers over the existing network. The OLO will have to pay the incumbent on a per customer base for this usage, reflecting the incumbents costs over this network. In such a regulated market, the OLO has a high flexibility of choosing where to connect which customers. Traditional investment analysis techniques cannot take this flexibility into account and are not useable for measuring the value of this flexibility for the OLO. Still this value is very important. In the first place, this additional flexibility undermines the fair market competition for the incumbent player and when no counter measures are taken it gives no incentives to the OLOs to rollout new own equipment. A solution to this problem is the use of real options to incorporate the value of flexibility and managerial options into the business case.

We have constructed a techno-economic model for estimating the costs of an OLO for providing connections to its customers on top of an existing copper based network. We used the real option valuation approach in the paper and were able to calculate the option value for the OLO. For the hypothetical case we estimated an option value up to €9 per line per month, a value which can clearly not be neglected. The business case of an OLO already becomes profitable once he reaches 40% or more of the OLO market. As such, up to two OLOs can perfectly co-exist with the existing incumbent. In case the market has a considerable size, a third OLO focusing on this opportunity could still make a positive business case.

These results clearly show that the value of this flexibility cannot be neglected and should be taken into account by a regulator when fixing the prices and options for LLU. Setting the tariff too high will of course scare the OLOs from the market. Setting the tariff too low on the other hand, might break down future competition as it gives no incentive to follow the ladder of investment for an OLO and might squeeze the margins of the incumbent and kill all incentives there for upgrading his network or rolling out his own infrastructure.

The current approach considered a copper based network, but the question of LLU stays as important for new network installations. A valuable extension to this study would be the combination of both to estimate the flexibility of LLU in a new fiber network. Also mobile networks for which we already constructed some economic models [10], [11] could be extended with this line of reasoning. Even beyond the field of telecommunications, other infrastructures such as electricity, gas and water which were recently privatized in Europe, exhibit similar possibilities for opening access to the transport network. Finally the case considered in this paper only mentions one incumbent player, whereas in a duopoly installation, future networks or international connections, more than one player could be offering LLU over the same (or overlapping) region. In this case OLOs could gain even more from the additional flexibility they have for switching operators.

References

1. Camacho, F.T., Menezes, F.M.: Access Pricing and Investment: A Real Options Approach. Competition and Regulation in Network Industries CRNI, JEL L51 (2008)
2. Sadowski, B.M., Verheijen, M., Nucciarelli, A.: From Experimentation to Citywide Rollout: Real Options for a Municipal WiMax Network in the Netherlands, Communications & strategies, No. 70, 2nd Q (2008)
3. Pindyck, R.: Mandatory Unbundling and Irreversible Investment in Telecom Networks. NBER Working Paper No. 10287, JEL No. L0 (2004),
   http://web.mit.edu/rpindyck/www/Papers/
   Mandatory_Unbundling.pdf
4. Haussman, J.A.: Regulated Costs and Prices in Telecommunications. In: The handbook of Telecommunications Economics, Edward Elgar, vol. II (2000)
5. Organisation for Economic Co-operation and Development, Broadband statistics,
   http://www.oecd.org/
6. Verbrugge, S., Casier, K., Van Ooteghem, J., Lannoo, B.: Practical steps in techno-economic evaluation of network deployment planning. In: IEEE Global Communications Conference, Exhibition and Industry Forum (Globecom 2009), Honolulu, Hawaii, USA, November 30-December 4 (2009)
7. Nationaal Instituut voor de Statistiek, http://www.statbel.fgov.be
8. Myers, S.: Determinants of Corporate Borrowing. Journal of Financial Economics 5(2), 147–175 (1997)
9. De Maeseneire, W.: The real options approach to strategic capital budgeting and company valuation, Financiële Cahiers, Larcier (2006) (ISBN: 2-8044-2318-2)
10. Van Ooteghem, J., Verbrugge, S., Casier, K., Pickavet, M., Demeester, P.: Future proof strategies towards fiber to the home. In: ICT-Public & Private Interplay Next Generation Communication conference in Sevilla, December 10-12 (2008)
11. Van Ooteghem, J., Verbrugge, S., Casier, K., Lannoo, B., Pickavet, M., Demeester, P.: Competition and Interplay Models for Rollout and Operation of Fiber to the Home Networks. Working paper (2009)

Intercarrier Compensation between Providers of Different Layers: Advantages of Transmission Initiator Determination

Ruzana Davoyan and Wolfgang Effelsberg

Department of Mathematics and Computer Science,
University of Mannheim, Germany
{ruzana.davoyan,wolfgang.effelsberg}@informatik.uni-mannheim.de

Abstract. This paper addresses the important issue of providing balanced allocation of the interconnection costs between networks. We analyze how beneficial is the determination of the original initiator of a transmission to the providers of different layers. The introduced model, where intercarrier compensation is based on the differentiated traffic flows, was compared with the existing solution, which performs cost compensation based on the traffic flows. For our analysis we considered both unilateral and bilateral settlement arrangements.

Keywords: Interconnection arrangement, intercarrier compensation, Internet economics.

1 Introduction

The Internet is a system of interconnected networks, which are connected either through a direct link or an intermediate point to exchange traffic. Currently, the Internet provides two basic types of interconnections such as peering and transit, and their variations. Peering is the arrangement of traffic exchange on the free-settlement basis, so that Internet service providers (ISPs) do not pay each other and derive revenues from their own customers. In transit, a customer ISP pays a transit ISP to deliver the traffic between the customers. Emergence of new types of ISPs (with large number of customers and great amount of content) led to appearance of new types of models, such as paid peering and partial transit.

Traditionally, before interconnecting, provider calculates whether the interconnection benefits would outweigh the costs [1]. Simple economic principle suggests sharing the costs between all parties. In the case of telephony, the study [2] argued that both calling and called parties benefit from the call, and consequently, should share the interconnection costs. In the Internet, under symmetry of traffic flows, the termination costs are set to zero, since it is assumed that the termination fees are roughly the same, and a peering arrangement is used. However, because no termination cost is charged, settlement-free model is considered inefficient in terms of cost compensation [3]. Generally, if providers are asymmetric in terms of size, peering is not appropriate, since providers incur different

M. Crovella et al. (Eds.): NETWORKING 2010, LNCS 6091, pp. 373–384, 2010.

costs and benefit differently. In such a case, an interconnection arrangement is governed by the financial compensation in a bilaterally or unilaterally negotiated basis. In the bilateral settlement arrangements, the payments are based on the net traffic flow. In the unilateral settlement arrangements, a customer provider pays for sent and received traffic, even though traffic flows in both directions. This causes the existence of imbalance in allocation of the interconnection costs. In particular, smaller providers in high cost areas admit higher subscription fees. There exists a large body of literature that discusses interconnection challenges [1, 4-6]. Various pricing schemes have attempted to provide sustainable conditions for smaller ISPs [7-8]. These models make different trade-offs between the two objectives of interconnection pricing, viz., competition development and profitability. Hence, no single model has a clear advantage over the others. As cited in [9], it was recommended to establish bilateral settlement arrangements and to compensate each provider for the costs that it incurs in carrying traffic generated by the other network. However, it was argued that traffic flows are not a reasonable indicator to share the costs, since it is not clear who originally initiated a transmission and, therefore, who should pay for the costs. In other words, compensation between providers cannot be solely done based on the traffic flows, which provide a poor basis for cost sharing [9]. Recently provided analytical studies in [10-11], investigated the impact of the original initiator of a transmission at the wholesale and retail levels in the case of private peering arrangements. Further, we extended studies by examining the benefits of the customer providers only, which purchased transit services [12]. However, the remaining literature on the economics of interconnection considers the intercarrier compensation based only on the flows of traffic [2, 13-16].

The main objective of this paper is to explore the role of the determination of a transmission initiator on ISPs of different layers. The paper differs from the prior reported studies in that it considers how beneficial is the traffic differentiation to *all providers*, such as transit and customer, and it examines customer ISPs, which *operate in different cost areas*. Our studies involve the earlier introduced model, called Differentiated Traffic-based Interconnection Agreement (DTIA) that distinguishes traffic into two types to determine the transmission initiator in the IP networks and to compensate the costs. In contrast to the existing solutions [17], in which the payments are based on traffic flows (TF), we compensate differently for a particular traffic type. Unlike telephony, where the transmission initiator covers the entire costs, imposing uniform retail pricing, the proposed model distributes the joint costs between all parties and supports the diversity of existing retail pricing schemes in the Internet. Comparative studies were provided for the agreements based on the traffic flows and differentiated traffic flows compensations. We considered both unilateral and bilateral settlement models. The rest of the paper is organized as follows. Section 2 discusses existing financial settlements. Section 3 describes the motivation for traffic differentiation. Section 4 provides analytical studies. Section 5 concludes this paper.

2   Financial Settlements

Generally, providers arrange financial settlements in order to determine the distribution of the interconnection costs [17]. Before examining financial settlements within the Internet, we consider the telephone system. As an example, assume a scenario, where Alice makes a call to Bob. Accepting the call, Bob incurs termination costs to its provider that should be covered either directly by billing Bob or indirectly by billing the calling party's carrier. As cited in [3], "existing access charge rules and the majority of existing reciprocal compensation agreements require the calling party's carrier, [...], to compensate the called party's carrier for terminating the call". Thus, the initiator of the call, i.e., Alice pays to a subscribed provider for the entire call since Alice asked to reserve the circuit. In contrast to the telephony example, establishing a connection in the Internet does not require any reservation of a circuit. Usually packets between Alice and Bob are routed independently, sometimes even via different paths. Therefore, as cited in [18], "it is very important to distinguish between the *initiator* and the *sender*, and likewise between the *destination* and the *receiver*". The initiator is the party that initiates a call or a session, and the destination is the party that receives a call. In contrast, the sender (the originator) is the party that sends traffic, and the receiver (the terminator) is the party that receives traffic. In telephony, the initiator is considered to be the originator and is charged based on the transaction unit, namely a "call minute" for using the terminating network. Even though it may be argued that a TCP session can be considered as a call, where the initiator of a session pays for the entire traffic flow, such a model deals with technical issues, considerable costs, and implies uniform retail pricing. Currently, the Internet uses the packet-based accounting model, under which the volume of the exchange traffic in both directions is measured, and adopts a small set of interconnection arrangements. Specifically, in the *service-provider* (unilateral) settlement, namely transit and paid peering business relationships, a customer ISP pays to a transit ISP for sent and received traffic. In the *settlement-free* agreement, namely peering relationships, providers do not pay each other. In some cases ISPs adopt the *negotiated-financial* (bilateral) settlement where the payments are based on the net flow of traffic. For detailed discussion see [17-19].

3   Motivation for Traffic Differentiation

The principle that we follow is that both parties derive benefits from the exchange of traffic and, therefore, should share the interconnection costs. Considering a system without externalities [20], the costs should be shared based on the benefits obtained by each party. However, in the real world, it is impossible to measure the benefits of parties and so to share the costs. If content is not equally distributed between providers, traffic imbalance occurs, and hence, costs and revenues are not shared evenly. As cited in [21], traffic flow is dominant towards a customer requested the content and generates 85% of the Internet traffic. This implies that inbound traffic is much more compared to outbound traffic of content request. In telephony for example, it is acceptable that more than 50% of rural network's revenue could come from the incoming calls. In contrast, in the Internet, customer networks pay for the entire traffic flows. It was recommended to compensate each provider for the costs that it incurs in carrying traffic based on the traffic flows. However, traffic flows are not a good measure for costs sharing, since "it is impossible to determine who originally initiated any given transmission on the Internet" and therefore, provide a poor basis for cost sharing [9]. On the other hand, providers are unwilling to inspect the IP header of a packet, since "the cost of carrying an individual packet is extremely small, and the cost of accounting for each packet may well be greater than the cost of carrying the packet across the providers" [19].

The DTIA model presented in [22] manages inter-provider cost compensation considering the original initiator of a transmission. In order to determine a party that originally initiated a transmission, we differentiated traffic into two types, referred to as *native*, which is originally initiated by the provider's own customers, and *stranger* that is originally initiated by the customers of any other network. Indeed, outgoing traffic of ISP_i that is the same as adjacent provider's incoming traffic may be i) either a part of a transmission initiated by a customer of ISP_i, ii) or a part of a transmission initiated by a customer of any other network. Furthermore, we suggest that providers compensate differently for a particular type of traffic, where stranger traffic is charged at a lower rate than native traffic. More specifically, each provider settled DTIA compensates the cost of carrying traffic according to the differentiated traffic flows. For detailed description of the DTIA model and its traffic management mechanism see [22].

4   The Model of Interconnection

In our analysis we follow an assumption done in [13] to capture traffic imbalance and therefore, consider two types of the customers, namely websites (*who host information and content*) and consumers (*who use information and content provided on websites*). Actually, traffic is exchanged between consumers, between websites, from websites to consumers, and from consumers to websites. According to the proposed approach, a node (customer) in a P2P network is considered as *a consumer as well as a website simultaneously*, since it can act as a client and as a server. Thus, traffic generated from websites to consumers and vice versa along with Web, FTP and streaming media traffic captures P2P traffic. Traffic between consumers captures VoIP traffic that tends to be symmetric, and email exchange that is much smaller than traffic generated from websites to consumers. To focus on explicit monetary transfers between providers and traffic asymmetry in its simplest way, we consider traffic exchange i) from consumers to websites, and ii) from websites to consumers. To simplify analytical studies the following assumptions were made throughout the paper:

Assumption 1. *Let $\alpha_i \in (0,1)$ network i's market share for consumers and $\beta_i \in (0,1)$ its market share for websites. The market consists of only one transit and two customer ISPs, i and j, where $i \neq j = 1, 2$ and $\alpha_i + \alpha_j = \alpha$, $\beta_i + \beta_j = \beta$.*

Assumption 2. *Balanced calling pattern, where each consumer requests any website in any network with the same probability is considered[1]. Each consumer originates one unit of traffic per each request of website. The number of consumers and websites in the market is given by N and M respectively.*

We examine a scenario, in which ISP_i and ISP_j exchange traffic through the transit ISP_k. The amount of the differentiated traffic originated from ISP_i with destination to ISP_j and vice versa is given by

$$t_{ik}^{nat} = \alpha_i \beta_j NM \qquad t_{jk}^{nat} = \alpha_j \beta_i NM \tag{1}$$

$$t_{ik}^{str} = \alpha_j \beta_i NMx \qquad t_{jk}^{str} = \alpha_i \beta_j NMx \tag{2}$$

where t_{ik}^{nat} (t_{jk}^{nat}) denotes the amount of outgoing native traffic (exchanged from consumers to websites) and t_{ik}^{str} (t_{jk}^{str}) is the amount of stranger traffic (exchanged from websites to consumers) with respect to ISP_i (ISP_j). The variable x denotes the average amount of traffic caused by requesting a website. It is known that P2P traffic asymmetry is typically caused by less capacity provisioned in the upstream direction. Thus, upstream/downstream P2P traffic flows can be asymmetric, which implies that x is different for the customers subscribed to different ISPs. However, this does not affect the results of our studies. The total amount of traffic originated by ISP_i and ISP_j are $t_{ik} = t_{ik}^{nat} + t_{ik}^{str}$ and $t_{jk} = t_{jk}^{nat} + t_{jk}^{str}$.

4.1 Unilateral Settlement Arrangements

We start by considering a unilateral settlement arrangement, where transit ISP charges the customer providers for every unit of traffic sent and received. Let c_i^k and c_j^k are the marginal costs of the connectivity of ISP_i and ISP_j correspondingly. These providers operate in different cost areas so that $c_i^k < c_j^k$, where marginal costs exhibit increasing returns to scale (i.e., $ISP_i > ISP_j$). ISP_k charges the customer ISPs a_k and b_k for every unit of native and stranger traffic respectively, where $a_k > b_k$ (ISPs pay less for stranger traffic). In particular, in DTIA we consider that a customer ISP i) compensates fully the imbalance in the connectivity costs between endpoints, if the exchanged traffic is native, and ii) does not compensate this difference, if the originated traffic is stranger. The difference in the costs of the exchanged traffic between the points is defined by

$$\Delta = c_j^k - c_i^k \tag{3}$$

Proposition 1. *The access charge for stranger traffic is set to the lowest cost of the connectivity, i.e., $b_k = c_i^k$.*

Proof. Interconnection costs between the customer providers are covered by the access charges. Since native traffic for ISP_i is stranger for ISP_j, the sum of fees for native and stranger traffic are equal to the whole costs, that is

$$c_i^k + c_j^k = a_k + b_k \tag{4}$$

---

[1] Due to the lack of mathematical models on how traffic between ISPs is distributed, many works make a statistical assumption, such as balanced calling pattern.

In the DTIA model, a provider compensates the imbalance in the costs expressed by (3) fully only for native traffic. This cost difference is not compensated for the stranger traffic. Consequently, it can be written that

$$a_k = b_k + \Delta \tag{5}$$

By substituting (3) and (5) in (4), it can be obtained that access rate for stranger traffic is set to the lowest cost of the connectivity, that is $b_k = c_i^k$. Obviously, that the access charge for native traffic is defined by $a_k = c_i^k + \Delta$. □

We investigate the payments of the customer providers in the classical and DTIA models. The payments of ISP_i and ISP_j to transit ISP in DTIA are given by

$$f_{ik} = a_k(t_{ik}^{nat} + t_{jk}^{str}) + b_k(t_{ik}^{str} + t_{jk}^{nat}) \tag{6}$$
$$f_{jk} = a_k(t_{jk}^{nat} + t_{ik}^{str}) + b_k(t_{jk}^{str} + t_{ik}^{nat}) \tag{7}$$

The sum of these payments presents the incremental revenue of the transit ISP $\pi_k = f_{ik} + f_{jk}$. The net payments of the customer ISPs according to the traffic flow based compensation are denoted by $\check{f}_{ik}$ and $\check{f}_{jk}$ and calculated as follows

$$\check{f}_{ik} = c_i^k(t_{ik} + t_{jk}) \tag{8}$$
$$\check{f}_{jk} = c_j^k(t_{ik} + t_{jk}) \tag{9}$$

Proposition 2. *The payments of larger (smaller) providers are higher (less) in DTIA than these in the classical model.*

Proof. Considering the net payments of the larger ISP_i, from (6) and (8) follows that $\check{f}_{ik} - f_{ik} = (b_k - a_k)(t_{ik}^{nat} + t_{jk}^{str}) < 0$, i.e., $f_{ik} > \check{f}_{ik}$. Similarly, comparing the payments of ISP_j in the DTIA and classical models given by (7) and (9) we obtain that $\check{f}_{jk} - f_{jk} = (a_k - b_k)(t_{ik}^{nat} + t_{jk}^{str}) > 0$. This gives that $f_{jk} < \check{f}_{jk}$. □

4.2   Bilateral Settlement Arrangements

This subsection examines bilateral settlement models, where each provider (including customer ISP) is compensated for the costs of carrying traffic.

Reciprocal Access Charges. In the following lines we explore the case when the customer providers charge the transit provider reciprocal access charges and vice versa. Let b be the access payment that ISP_k subsidizes ISP_i and ISP_j for every unit of sent traffic, where $b < c_j^k$. The marginal connectivity costs of the customer providers charged by ISP_k can be written as follows

$$c_i^k + c_j^k = c_k + \sigma \tag{10}$$

where c_k is the marginal transportation cost of ISP_k; σ is an arbitrary constant.

Proposition 3. *The access charge for stranger traffic set by ISP_k is equal to $b_k = c_k + b$ (i.e., the total costs of ISP_k).*

Proof. The costs of ISP_k are comprised of the marginal transmission cost and the payment to access customer provider's infrastructure, i.e., $c_k + b$. The bilateral settlement model is attractive to ISP_k only if its own costs are covered. These costs correspond to the minimum level of access charge set by ISP_k, that is

$$c_k + b = \min\{a_k, b_k\}$$

In DTIA provider compensates less the costs of carrying stranger traffic, thus

$$b_k = c_k + b \tag{11}$$

Obviously, that the access charge for native traffic set by the transit provider is increased by the arbitrary constant and calculated as follows $a_k = b_k + \sigma$.   □

The net interconnection payments from ISP_i and ISP_j to ISP_k are defined by

$$f_{ik} = a_k t_{ik}^{nat} + b_k t_{ik}^{str} \qquad f_{jk} = a_k t_{jk}^{nat} + b_k t_{jk}^{str} \tag{12}$$

Analogously, the net transfers of ISP_k to the customer providers are given by

$$f_{ki} = b(t_{jk}^{nat} + t_{jk}^{str}) \qquad f_{kj} = b(t_{ik}^{nat} + t_{ik}^{str}) \tag{13}$$

It can be noticed that the transit ISP is charged based on the rate for stranger traffic, because we consider that it does not have any customers of its own.

Before examining the payments of the customer ISPs in the DTIA and classical models with bilateral settlements, we consider access charges and net payments in the classical solution. Let $\check{b}$ be the payment paid by ISP_k to the customer providers for sending traffic. In the model with bilateral settlements, the access charge set by the transit provider is defined by $\check{a}_k = c_i^k + c_j^k + \check{b}$. Assume that ISP_k has users, thus b (in DTIA) is the rate charged by the customer ISPs for unit of stranger traffic only, while $\check{b}$ (in the classical model) is payment for unit of traffic. As a result, we obtain that $\check{b} \geq b$. The payments of ISP_i and ISP_j are

$$\check{f}_{ik} = \check{a}_k t_{ik} \qquad \check{f}_{jk} = \check{a}_k t_{jk} \tag{14}$$

Proposition 4. *The net payments of the customer providers in the DTIA model are less than these in the classical model.*

Proof. Considering the payments of ISP_i and ISP_j given by (12) and (14) follows

$$\check{f}_{ik} - f_{ik} = t_{ik}^{nat}(\check{a}_k - a_k) + t_{ik}^{str}(\check{a}_k - b_k) > 0 \tag{15}$$
$$\check{f}_{jk} - f_{jk} = t_{jk}^{nat}(\check{a}_k - a_k) + t_{jk}^{str}(\check{a}_k - b_k) > 0 \tag{16}$$

□

Non-reciprocal Access Charges. We continue with examination of the bilateral settlement model with asymmetric access fees. Let b_i and b_j ($b_i < b_j$) are the access rates for every unit of traffic received by ISP_i and ISP_j correspondingly.

Following the results of Proposition 3, fees that the transit provider charges the customer ISPs for native and stranger traffic can be rewritten as

$$a_{ik} = c_i^k + c_j^k + b_j \qquad a_{jk} = c_i^k + c_j^k + b_i$$
$$b_{ik} = c_k + b_j \qquad\qquad b_{jk} = c_k + b_i$$

The net payments from ISP$_i$ and ISP$_j$ to the transit ISP$_k$ and vice versa are

$$f_{ik} = a_{ik}t_{ik}^{nat} + b_{ik}t_{ik}^{str} \qquad f_{jk} = a_{jk}t_{jk}^{nat} + b_{jk}t_{jk}^{str} \qquad (17)$$
$$f_{ki} = b_i\left(t_{jk}^{nat} + t_{jk}^{str}\right) \qquad f_{kj} = b_j\left(t_{ik}^{nat} + t_{ik}^{str}\right) \qquad (18)$$

The following lines explore the payments of customer ISPs in the classical and DTIA models. For this purpose, we consider access charges and payments in the classical solution. The access rates that ISP$_k$ charges ISP$_i$ and ISP$_j$ are

$$\check{a}_{ik} = c_i^k + c_j^k + \check{b}_j \qquad \check{a}_{jk} = c_i^k + c_j^k + \check{b}_i$$

where $\check{b}_i$ and $\check{b}_j$ ($\check{b}_i \geq b_i$, $\check{b}_j \geq b_j$) are access fees set by the customer providers correspondingly. The net payments of the customer providers are given by

$$\check{f}_{ik} = \check{a}_{ik}t_{ik} \qquad \check{f}_{jk} = \check{a}_{jk}t_{jk} \qquad (19)$$

Proposition 5. *The interconnection payments of the customer providers are less in DTIA than these in the classical model.*

Proof. Examining the payments of ISP$_i$ and ISP$_j$ given by (17) and (19) follows

$$\check{f}_{ik} - f_{ik} = t_{ik}^{nat}(\check{a}_{ik} - a_{ik}) + t_{ik}^{str}(\check{a}_{ik} - b_{ik}) > 0 \qquad (20)$$
$$\check{f}_{jk} - f_{jk} = t_{jk}^{nat}(\check{a}_{jk} - a_{jk}) + t_{jk}^{str}(\check{a}_{jk} - b_{jk}) > 0 \qquad (21)$$

□

4.3   Discussions

Tables 1-3 report the results of analytical studies, which examined how beneficial is the determination of a transmission initiator to the providers of different layers. The comparison results between unilateral settlement models are presented in Table 1. Tables 2 and 3 demonstrate the comparison of bilateral settlement arrangements with symmetric and asymmetric access charges. The analyses considered all available market states in terms of providers' market shares, where ISP$_i$ >ISP$_j$. The following parameter values were imposed to calculated the specific outcomes: $c_i^k = 0.4$, $c_j^k = 1.5$, $c_k = 0.9$, $b = 0.5$, $b_i = 0.3$, $b_j = 0.5$, $x = 35$, $N = 100$, and $M = 60$. In order to simplify analyses we assume that $\check{b} = b$, $\check{b}_i = b_i$, and $\check{b}_j = b_j$. The parameters are chosen to satisfy a condition that providers operate in different cost areas. However, the specification is clearly arbitrary. It is important to note, that our conclusions do not heavily depend on the chosen parameter values. The results obtained for a number of other parameter sets have not produced significant changes. Network i's total incremental

Table 1. Comparative Results of DTIA with Unilateral Settlements

Case	α_i	β_i	t_{ik}^{nat}	t_{jk}^{nat}	t_{ik}	t_{jk}	f_{ik} DTIA	f_{ik} TF	f_{jk} DTIA	f_{jk} TF	π_k DTIA	π_k TF
I	0.5	0.9	300	2700	94800	13200	55080	43200	150120	162000	205200	205200
$\alpha_i = \alpha_j$	0.5	0.8	600	2400	84600	23400	66960	43200	138240	162000	205200	205200
$\beta_i > \beta_j$	0.5	0.7	900	2100	74400	33600	78840	43200	126360	162000	205200	205200
II	0.9	0.5	2700	300	13200	94800	150120	43200	55080	162000	205200	205200
$\alpha_i > \alpha_j$	0.8	0.5	2400	600	23400	84600	138240	43200	66960	162000	205200	205200
$\beta_i = \beta_j$	0.7	0.5	2100	900	33600	74400	126360	43200	78840	162000	205200	205200
III	0.9	0.8	1080	480	17880	38280	65232	22464	41472	84240	106704	106704
$\alpha_i > \alpha_j$	0.8	0.7	1440	840	30840	51240	89856	32832	66096	123120	155952	155952
$\beta_i > \beta_j$	0.7	0.6	1680	1080	39480	59880	106272	39744	82512	149040	188784	188784
$\alpha_i > \beta_i$	0.6	0.55	1620	1320	47820	58020	106488	42336	94608	158760	201096	201096
IV	0.9	0.9	540	540	19440	19440	36936	15552	36936	58320	73872	73872
$\alpha_i > \alpha_j$	0.8	0.8	960	960	34560	34560	65664	27648	65664	103680	131328	131328
$\beta_i > \beta_j$	0.7	0.7	1260	1260	45360	45360	86184	36288	86184	136080	172368	172368
$\alpha_i = \beta_i$	0.6	0.6	1440	1440	51840	51840	98496	41472	98496	155520	196992	196992
V	0.9	0.2	4320	120	8520	151320	235008	63936	68688	239760	303696	303696
$\alpha_i > \alpha_j$	0.8	0.25	3600	300	14100	126300	198720	56160	68040	210600	266760	266760
$\beta_i < \beta_j$	0.7	0.35	2730	630	24780	96180	156492	48384	73332	181440	229824	229824

Table 2. Comparative Results of Bilateral Settlement Arrangements (Reciprocal Access Charges)

Case	f_{ik} DTIA	f_{ik} TF	f_{jk} DTIA	f_{jk} TF	f_{ki}	f_{kj}	r_k DTIA	r_k TF	r_i DTIA	r_i TF	r_j DTIA	r_j TF
I	133020	227520	21180	31680	6600	47400	100200	205200	126420	220920	-26220	-15720
	119040	203040	35160	56160	11700	42300	100200	205200	107340	191340	-7140	13860
	105060	178560	49140	80640	16800	37200	100200	205200	88260	161760	11940	43440
II	21180	31680	133020	227520	47400	6600	100200	205200	-26220	-15720	126420	220920
	35160	56160	119040	203040	42300	11700	100200	205200	-7140	13860	107340	191340
	49140	80640	105060	178560	37200	16800	100200	205200	11940	43440	88260	161760
III	26112	42912	54072	91872	19140	8940	52104	106704	6972	23772	45132	82932
	44616	74016	72576	122976	25620	15420	76152	155952	18996	48396	57156	107556
	56952	94752	84912	143712	29940	19740	92184	188784	27012	64812	65172	123972
	68568	114768	82548	139248	29010	23910	98196	201096	39558	85758	58638	115338
IV	27756	46656	27756	46656	9720	9720	36072	73872	18036	36936	18036	36936
	49344	82944	49344	82944	17280	17280	64128	131328	32064	65664	32064	65664
	64764	108864	64764	108864	22680	22680	84168	172368	42084	86184	42084	86184
	74016	124416	74016	124416	25920	25920	96192	196992	48096	98496	48096	98496
V	16248	20448	211968	363168	75660	4260	148296	303696	-59412	-55212	207708	358908
	23340	33840	177120	303120	63150	7050	130260	266760	-39810	-29310	170070	296070
	37422	59472	135282	230832	48090	12390	112224	229824	-10668	11382	122892	218442

Table 3. Comparative Results of Bilateral Settlement Models (Non-reciprocal Access Charges)

Case	f_{ik}		f_{jk}		f_{ki}	f_{kj}	r_k		r_i		r_j	
	DTIA	TF	DTIA	TF			DTIA	TF	DTIA	TF	DTIA	TF
I	133020	227520	18540	29040	3960	47400	100200	205200	129060	223560	-28860	-18360
	119040	203040	30480	51480	7020	42300	100200	205200	112020	196020	-11820	9180
	105060	178560	42420	73920	10080	37200	100200	205200	94980	168480	5220	36720
II	21180	31680	114060	208560	28440	6600	100200	205200	-7260	3240	107460	201960
	35160	56160	102120	186120	25380	11700	100200	205200	9780	30780	90420	174420
	49140	80640	90180	163680	22320	16800	100200	205200	26820	58320	73380	146880
III	26112	42912	46416	84216	11484	8940	52104	106704	14628	31428	37476	75276
	44616	74016	62328	112728	15372	15420	76152	155952	29244	58644	46908	97308
	56952	94752	72936	131736	17964	19740	92184	188784	38988	76788	53196	111996
	68568	114768	70944	127644	17406	23910	98196	201096	51162	97362	47034	103734
IV	27756	46656	23868	42768	5832	9720	36072	73872	21924	40824	14148	33048
	49344	82944	42432	76032	10368	17280	64128	131328	38976	72576	25152	58752
	64764	108864	55692	99792	13608	22680	84168	172368	51156	95256	33012	77112
	74016	124416	63648	114048	15552	25920	96192	196992	58464	108864	37728	88128
V	16248	20448	181704	332904	45396	4260	148296	303696	-29148	-24948	177444	328644
	23340	33840	151860	277860	37890	7050	130260	266760	-14550	-4050	144810	270810
	37422	59472	116046	211596	28854	12390	112224	229824	8568	30618	103656	199206

cost of connectivity is defined by $r_i = f_{ik} - f_{ki}$. Network k's profit obtained from interconnection is $r_k = (f_{ik} + f_{jk}) - (f_{ki} + f_{kj})$.

Comparative results obtained for arrangements with unilateral settlements demonstrated that in the presented model the payments are decreased for the smaller ISP and are increased for the larger ISP. This is achieved by different access rates charged for the distinguished traffic flows. Specifically, the payments of ISP_i are increased due to native traffic compensation, while the payments of ISP_j are decreased due to stranger traffic compensation. Further, the results showed that in DTIA the more outgoing traffic the lower costs of the provider. In particular, incoming and outgoing native traffic are directly proportional. Hence, the network that sends more native traffic incurs higher costs than the network that receives this traffic. This is explained by the higher access charges for native than stranger traffic. The costs of both customer ISPs are equal only in the case when their native and stranger traffic volumes are symmetric correspondingly. Finally, the results indicated that the revenues of the transit provider in the classical model based on the traffic flows compensation and DTIA are equal.

The key consequences provided below are based on the analytical studies, which explored bilateral settlement arrangements. In DTIA, the payments paid by the customer providers are decreased and these of transit provider remain the same. Specifically, providers ISP_i and ISP_j compensate based on the differentiated traffic flows, where the access charge for stranger traffic flow is lower than the access charge set in the classical model. As a consequence, the total incremental costs of the customer providers (r_i and r_j) are also decreased. On the other side, profits of ISP_k obtained from the interconnection (i.e., the differences between received and paid payments, r_k) are lower than these in the traffic flow based compensation model. However, as mentioned earlier, it was argued that traffic flows provide a poor basis for Internet interconnection cost sharing.

The provided studies examined a model consisting of one transit and two customer ISPs. One question that arises here is on the robustness of the obtained results for more realistic scenarios, which consider more transit and customer ISPs. From Propositions 2, 4 and 5 it can be noticed that the results depend only on the access charges of both DTIA and classical models. More specifically, in the unilateral settlement arrangements, the results rely on the inequality $(a_k - b_k) > 0$. Analogously, the results given by (15) and (16) depend on the inequalities $(\breve{a}_k - a_k) > 0$ and $(\breve{a}_k - b_k) > 0$, while results expressed by (20) and (21) are based on $(\breve{a}_{ik} - a_{ik}) > 0$ and $(\breve{a}_{ik} - b_{ik}) > 0$. Hence, the provided conclusions remain the same. Obviously, in the extended scenarios, access charges are obtained by solving a system of linear equations.

5   Summary and Conclusions

In this paper we proposed models and their analysis, which are based on the DTIA strategy for inter-provider cost compensation. The goal for this was to explore how the determination of a transmission initiator affects different providers, operated in different cost areas and arranged interconnection with unilateral and bilateral settlements (Tables 1-3). The results obtained from the analytical studies showed that DTIA was able to find better results (in terms of interconnection payments) than the classical solution for the both models. More specifically, the proposed scheme diminishes the existing inequity in allocation of the interconnection costs. From the comparison between unilateral settlement models follows that the costs of the smaller provider are decreased. This stimulates the retail prices fall in the market, where the provider operates and consequently, the development of the infrastructure in terms of subscribed customers. The growth of the smaller ISP leads to balance the volumes of a particular traffic type, and as a result, reduces the imbalance in cost allocation. Obviously, that revenue of the larger ISP obtained from retail market will be increased. From the perspective of a transit provider, its revenues obtained from the customer providers remain the same in the DTIA and classical models.

In the bilateral settlement arrangements, the net payments of both customer ISPs in the DTIA model are decreased. This leads to the decrease in the incremental revenue obtained by the transit provider. Finally, the comparison between the existing model with unilateral settlement and DTIA with bilateral settlement showed that our approach generally performed better for both smaller and larger ISPs in terms of reduced net payments. For the smaller provider, DTIA is dominated in all cases over the classical model, and for the larger provider only in Cases II and V. Obviously, that the profits of the transit provider in bilateral settlement model are decreased, since it along with other ISPs shares the interconnection costs. Resuming, the provision of a model, which compensates providers while exploiting their infrastructures, is advantageous for sustainable environment. From this point of view the proposed DTIA model is beneficial.

References

1. Laffont, J.J., Tirole, J.: Competition in telecommunications. MIT Press, Cambridge (2000)
2. DeGraba, P.: Bill and Keep at the central office as the efficient interconnection regime, FCC, OPP, Working Paper 33 (December 2000)
3. Federal Communications Commission: In the matter of developing a unified intercarrier compensation regime, Notice of Proposed Rulemaking (April 27, 2001)
4. Bezzina, J.: Interconnection challenges in a converging environment, the World Bank, Global Information and Communication Technologies Department (2005)
5. Dymond, A.: Telecommunications challenges in developing countries - asymmetric interconnection charges for rural areas, World Bank working paper (December 2004)
6. Lie, E.: International Internet interconnection next generation networks and development, GSR (2007)
7. Armstrong, M.: The theory of access pricing and interconnection. In: Handbook of Telecommunications Economics, vol. 1. North-Holland, Amsterdam (2002)
8. Noam, E.: Interconnection practices. In: Handbook of Telecommunications Economics, vol. 1. North-Holland, Amsterdam (2002)
9. Kende, M.: The digital handshake: connecting Internet backbones, FCC, OPP Working Paper No. 32 (September 2000)
10. Davoyan, R., Altmann, J.: Investigating the influence of market shares on interconnection settlements. In: Proceedings of IEEE Globecom (November 2008)
11. Davoyan, R., Altmann, J.: Investigating the role of a transmission initiator in private peering arrangements. In: Proceedings of IFIP/IEEE IM (June 2009)
12. Davoyan, R., Altmann, J., Effelsberg, W.: Intercarrier Compensation in Unilateral and Bilateral Arrangements. In: Proceedings of IEEE ICCCN (August 2009)
13. Laffont, J.J., Marcus, S., Rey, P., Tirole, J.: Internet interconnection and the off-net-cost pricing principle. RAND Journal of Economics (2003)
14. Weiss, M.B., Shin, S.: Internet Interconnection Economic Model and its Analysis: Peering and Settlement. Netnomics 6(1) (2004)
15. Besen, S., Milgrom, P., Mitchell, B., Srinagesh, P.: Advancing in routing technologies and Internet peering agreements. American Economic Review (2001)
16. Shrimali, G., Kumar, S.: Bill-and-Keep peering, in Telecom. Policy (2006)
17. Huston, G.: Interconnection, peering, and settlements, Part II. Internet Protocol Journal, Cisco Publications 2(2), 2–23 (1999)
18. Yoon, K.: Interconnection economics of all-IP networks. Review of Network Economics 5 (2006)
19. Huston, G.: Interconnection and peering (November 2000)
20. Economides, N.: The economics of networks. International Journal of Industrial Organization 14(6), 673–699 (1996)
21. Huston, G.: ISP survival guide: strategies for running a competitive ISP. Wiley, Chichester (1998)
22. Davoyan, R., Altmann, J., Effelsberg, W.: A new bilateral arrangement between interconnected providers. In: Reichl, P., Stiller, B., Tuffin, B. (eds.) ICQT 2009. LNCS, vol. 5539, pp. 85–96. Springer, Heidelberg (2009)

Application of Secondary Information for Misbehavior Detection in VANETs*

Ashish Vulimiri[1], Arobinda Gupta[1], Pramit Roy[1],
Skanda N. Muthaiah[2], and Arzad A. Kherani[2]

[1] Indian Institute of Technology, Kharagpur - 721302, India
agupta@cse.iitkgp.ernet.in
[2] GM India Science Lab, Bangalore, India

Abstract. Safety applications designed for Vehicular Ad Hoc Networks (VANETs) can be compromised by participating vehicles transmitting false or inaccurate information. Design of mechanisms that detect such misbehaving nodes is an important problem in VANETs. In this paper, we investigate the use of correlated information, called *"secondary alerts"*, generated in response to another alert, called as the *"primary alert"* to verify the truth or falsity of the *primary* alert received by a vehicle. We first propose a framework to model how such correlated *secondary* information observed from more than one source can be integrated to generate a "degree of belief" for the *primary* alert. We then show an instantiation of the model proposed for the specific case of Post-Crash Notification as the *primary* alert and Slow/Stopped Vehicle Advisory as the *secondary* alerts. Finally, we present the design and evaluation of a misbehavior detection scheme (MDS) for PCN application using such correlated information to illustrate that such information can be used efficiently for MDS design.

Keywords: VANET, Misbehavior detection, Primary alert, Secondary information.

1 Introduction

A vehicular ad hoc network (VANET) is an ad hoc wireless communication system setup between multiple vehicles in a neighborhood. The communication can either be between vehicle-to-vehicle (V2V) or may also involve some roadside infrastructures in which case it is termed as vehicle-to-infrastructure (V2I) communication. Several applications such as safety, traffic aid, infotainment, financial and navigational aid [1] etc., have been proposed for use in VANETs.

Typically, a V2X[1] based safety application triggers an alert in response to a specific event. For example, a crashed vehicle may trigger a Post Crash Notification (PCN) alert, a vehicle braking hard may trigger an Emergency Electronic

---

* Part of this work was supported by the GM Collaborative Research Lab at IIT Kharagpur, India.

[1] A generic term used to refer to V2V or V2I.

M. Crovella et al. (Eds.): NETWORKING 2010, LNCS 6091, pp. 385–396, 2010.
© IFIP International Federation for Information Processing

Brake Light (EEBL) alert, or a deceleration beyond a certain threshold may trigger a Slow or Stopped Vehicle Advisory (SVA) alert. In the presence of misbehaving vehicles, such alerts can also be raised even if the corresponding event has not happened, or the information sent in the alert can be wrong. For example a misbehaving vehicle may raise a PCN alert even in the absence of a crash, or report false information about the position of the crash. Note that authentication schemes are not sufficient to handle this as even authenticated users can turn malicious or misbehave due to faulty modules. Hence, upon reception of an alert from another vehicle, the receiving vehicle needs to verify if the event corresponding to a particular alert is true or false. This process will be referred to as *misbehavior detection*. Misbehavior detection is particularly important since it is expected that if such alerts are shown to a driver, the driver will respond to these alerts by taking some necessary action. A false action taken can have serious implications on the safety of drivers. On detecting a misbehavior using some Misbehavior Detection Scheme (MDS), a vehicle's On Board Unit (OBU) reports the misbehavior to a backend *Certificate Authority (CA)* so that the misbehaving vehicle's certificate may be revoked and the vehicle evicted from the network by the CA. In the rest of this paper, the alert whose truth or falsity is to be verified is termed as the *primary alert*.

The event corresponding to a primary alert may have effects on its neighboring vehicles and can cause other alerts to be sent. As an example, if a PCN alert is true, i.e., a vehicle has indeed crashed, it may cause other vehicles nearby to slow down for some time, possibly causing a series of SVA or EEBL alerts to be transmitted. Hence, the receipt of an appropriate number of SVA alerts following the reception of an PCN alert may be useful to strengthen a node's belief in the truth of the PCN alert and vice-versa. Such correlated information is termed as *secondary* information in this paper.

This work investigates the use of such information correlated to or generated due to a *primary* alert in designing misbehavior detection schemes. Specifically, we propose a probabilistic framework that models the use of secondary information for misbehavior detection. We then instantiate the model with a specific example of PCN and SVA alerts, and discuss how some of the necessary probabilities can be estimated empirically. We finally present the design and evaluation of a Misbehavior Detection Scheme (MDS) for PCN alerts to show that the probability estimations can be used effectively for misbehavior detection. Note that these secondary information will be generated irrespective of whether a vehicle is doing misbehavior detection or not, and are not caused by the misbehavior detection process of any vehicle.

2   Related Works

The need for security in VANET applications has been well-established, and several works [2][3][4][5] investigate the requirements and challenges involved in securing V2V communications. The IEEE 1609.2 standard [6] defines the functionalities of a security layer in V2V communication. However, though some of these works stress the need for misbehavior detection, no specific scheme for misbehavior detection is given for any application.

Among applications that can benefit from MDS, [7] and [8] discuss collision warning systems, both cooperative and autonomous. However, none of these papers propose any misbehavior detection scheme. Golle et al. [9] present a model to integrate information from different sensors and use it to identify malicious information and malicious nodes. However, the algorithm for actual detection is only sketched through examples. Also, the information used is local to within the vehicle and secondary information from other vehicles is not considered. A similar idea has been proposed by Schmidt et. al. [10]. They have given a trust calculation based scheme where a car calculates the trust in vehicles in its near vicinity from the various sensor values obtained. The paper lists only the scheme and its performance in various road scenarios have not been investigated. Ghosh et al. [11][12] have proposed misbehavior detection schemes for PCN application; however their schemes also do not consider any secondary information.

3 Model for Integrating Secondary Information

The event corresponding to a primary alert can cause other secondary events to occur, which can be detected by the same vehicle receiving the primary alert. Occurrence of such secondary events can act as supporting evidence in support of the primary alert. On the other hand, lack of such secondary events following a primary alert can indicate that the event corresponding to an alert may be false. The degree to which the available evidence supports the veracity of the primary alert (which we shall henceforth refer to as "degree of belief") is quantified by a numerical value β, $\{\beta \in [0,1]\}$. The degree of belief is 1 when we are certain that the event corresponding to the primary alert has occurred, and is 0 when we are certain that it has not occurred.

In the rest of this section, we first define events and their attributes more formally and propose a model to define the probability of the primary alert being true given an individual secondary information. We then propose a combining rule to obtain a final degree of belief β for the primary alert given all the secondary information present.

Event Model. An "event" is any observation that provides some information about the likelihood of the primary alert. The set of all events is partitioned into a finite number of "event classes", each class being some category of events. Each event class is characterized by a set of attributes that define the information content of each event in the class. If $\{a_1, a_2, \ldots, a_n\}$ is the set of attributes defining an event class, any single event in that class is merely an assignment of values to these attributes. The following two attributes are common to all event classes: t, the time of occurrence of an event and (x, y), the position of the source of the event.

We now formally define event classes. An event class $\mathcal{E}$ is an ordered tuple $< D, E, W >$ consisting of:

$D = D_1 \times D_2 \times \cdots \times D_n$: The set of all possible events. The D_i are the domains of each attribute in the event class. In the model considered in this work, the attribute space is composed of the following:
- $D1$: Time t: denotes the time at which the event happens.
- $D2, D3$: Locations x, y: denote the x, y coordinates at which the event occurs (we consider a 2D road topology for ease of understanding).
- u: $D_4, D_5, \ldots, D_n$: denotes the values of all other attributes

    We assume a model where the domains of t, x, y and u are discrete. The values of these attributes then define a 4-dimensional grid.

$R_E(t, x, y, u)$: A function that maps to each distinct tuple $\langle t, x, y, u \rangle$ a binary random variable that indicates if an event with attribute value u is generated by a car at time t at the position (x, y). The range of each random variable is $\{0, 1\}$. We denote by $GRID$ an observation of 0 or 1 values for each $R_E(t, x, y, u)$ in the event class. $GRID(t, x, y, u)$ is 1 if the event has been observed at the grid location $\langle t, x, y, u \rangle$, and 0 otherwise.

$W(t, x, y, u)$: A function that assigns a numeric weight indicating relative importance to each grid element. We impose the condition that

$$\left(\sum_t \sum_{(x,y)} \sum_u W(t, x, y, u) \right) = 1$$

Let $R_{prim}(t, x, y, v)$ denote a function that defines binary random variables indicating if the event corresponding to a primary alert with value v occurs at (x, y) at time t.

Probability of Events. Let t_a, x_a, y_a, v_a denote the time of occurrence, (x, y) coordinates, and the value of the primary alert. We wish to define a scheme that would allow the OBU to compute the posterior probability of the primary event being true, given the description of the secondary information (the set of alerts) received over the network. We compute this probability in several steps. First, we take each event class, and compute the probability of the event corresponding to the primary alert (we refer to this event as the *primary event* in the rest of this paper) being true given all observed events of this class. We then combine these probabilities computed for each class using a combining rule to get a final probability over all event classes. This final probability is the required "degree of belief" β.

    Let the posterior probability of the primary event being true given all events of a single event class $\mathcal{E}$ be denoted by $\beta_\mathcal{E}$. By definition,

$$\beta_\mathcal{E}(t_a, x_a, y_a, v_a) = P(R_{prim}(t_a, x_a, y_a, v_a) = 1 | GRID) \tag{1}$$

To compute this, we first compute the posterior probability of each individual event, and then combine them using a weight function W for the event class to get a single posterior probability as follows.

$$P(R_{prim}(t_a, x_a, y_a, v_a) = 1|GRID)$$
$$= \sum_{t} \sum_{(x,y)} \sum_{u} \{W(t, x, y, u) \times IND(t, x, y, u)\} \quad \text{where} \tag{2}$$
$$IND(t, x, y, u)$$
$$= P(R_{prim}(t_a, x_a, y_a, v_a) = 1|R_E(t, x, y, u) = GRID(t, x, y, u))$$

The probability $P(R_{prim}(t_a, x_a, y_a, v_a) = 1|R_E(t, x, y, u) = 1)$ is the probability of the primary event being true given the single random variable $R_E(t, x, y, u)$. Using Bayes' rule,

$$P(R_{prim}(t_a, x_a, y_a, v_a) = 1|R_E(t, x, y, u) = 1) =$$
$$\frac{P(R_{prim}(t_a, x_a, y_a, v_a) = 1) \times P(R_E(t, x, y, u) = 1|R_{prim}(t_a, x_a, y_a, v_a) = 1)}{P(R_E(t, x, y, u) = 1)}$$
$$\tag{3}$$

Thus, for each secondary event, a probability that the primary event has happened given the secondary event has happened can be calculated in terms of the probabilities on the right hand side. (We will later discuss and show an example of how all the probabilities on the right hand side can be estimated). The function W is an application specific weight assignment that reflects the relative importance assigned to each grid location. The highest weight would be assigned to the cells where the occurrence of an event provides the strongest evidence for or against the truth of the primary alert.

Computing β. In order to combine the $\beta_\mathcal{E}$ for all the event classes into a single estimate β of the value of the alert, a weighted combination is used.

$$\beta(t_a, x_a, y_a, v_a) = \sum_{\mathcal{E}} w_\mathcal{E} \beta_\mathcal{E}(t_a, x_a, y_a, v_a) \tag{4}$$

where $\sum_\mathcal{E} w_\mathcal{E} = 1$. The weight $w_\mathcal{E}$ assigned to a class $\mathcal{E}$ indicates the relative trust placed in that class with respect to all the other classes.

We thus have a method that allows us to compute the probability of truth or falsity of a primary alert given the secondary information observed. However, the probabilities defined above depend on several factors like the congestion model, the mobility models of the cars, the actual primary and secondary events considered and the nature of the correlation between them, and the chance of the safety condition occurring. Hence these probabilities may be hard to obtain analytically. In the next section, we show how this model may be developed for a specific application scenario and show how these probabilities may be estimated empirically through simulation.

4   An Example: PCN and SVA Alerts

We consider the PCN application and demonstrate how the required probabilities may be estimated in a system consisting of two classes of events: (i) the class containing the PCN alert, and (ii) the class containing the SVA alerts. The PCN alert is the primary event, and the secondary information is comprised of zero or more SVA alerts raised.

The PCN event class consists of a single event, the first PCN notification. This event has no distinct attributes apart from t, the time at which the alert was raised, and (x, y), the location of the crash site. Let $CRASH(t, x, y)$ denote the random variable corresponding to the primary event $(R_{prim}(t, x, y, true))$.

The SVA event class also consists of a single event, an SVA alert. An SVA alert may originate from a vehicle at any distance to the crash site. This event class again has no distinct event attributes other than t and (x, y). Let the random variables R_E be denoted by SVA. The range of x is $\{1, 2, \ldots, D\}$ for some $D > 0$, and that of y is $\{1, 2, \ldots, l\}$, l being the number of lanes on the road.

The component probabilities needed to compute the final estimate $\beta(true)$ of the PCN alert being true are estimated from a combination of historical data of driver behavior and experimental results obtained by simulating different crash scenarios.

$P(CRASH(t_a, x_a, y_a) = 1)$:

In general, this depends on the congestion on the road and on driving habits. This can be estimated for a given time duration and road segment based on historical data collected by the relevant authorities, such as state or national transportation authorities. Note that crashes are not very frequent in general, and the probability value can be taken to be low even in a congested scenario.

$P(SVA(t, x, y) = 1 | CRASH(t_a, x_a, y_a) = 1)$:

These values may be obtained by simulating a crash at time t_a at location (x_a, y_a) using a vehicular traffic simulator, and then recording the average number of SVA alerts generated at each grid cell (t, x, y) per simulation run.

$P(SVA(t, x, y) = 1)$:

Note that an SVA alert might be raised even when there is no crash due to road congestion. Applying Bayes' rule, the required probability is given by

$$P(SVA(t, x, y) = 1)$$
$$= P(SVA(t, x, y) = 1 | NOCRASH) \times P(NOCRASH)$$
$$+ \sum_{t,x,y} P(SVA(t, x, y) = 1 | CRASH(t, x, y) = 1) \quad (5)$$
$$\times P(CRASH(t, x, y) = 1)$$

Thus, the probability value can be obtained by simulating the no-crash as well as various crash scenarios and observing the average number of SVA alerts at each grid cell.

$W(t, x, y)$:

The weights may be obtained by simulating the crash and no-crash scenarios and computing the average number of extra alerts generated at each cell in the crash scenario. Multiple approaches might be taken to obtain the weight values from this information. One strategy might be to assign the weights in the proportion of the alerts raised, with the highest weight assigned to the cell generating the most alerts. An alternate approach might be to fix a certain threshold based on the shape of the alert distribution across the grid, assign a weight 0 to all the cells having fewer alerts than this threshold, and distribute the total weight equally among the remaining cells. We emphasize again that this is just one possible choice of W and many other variations are possible.

5   A Misbehavior Detection Scheme for PCN Alerts

In this section, we show how an MDS for PCN alerts can be built from the estimated probabilities. In order to estimate the probabilities, we simulated a number of traffic scenarios using the mobility simulator *VanetMobiSim* [13]. The simulator was modified to allow cars to be stopped at any time, thus simulating a crash, and to collect traces of the points of time when the conditions for raising an alert were satisfied in a certain car. In order to simulate a crash at time t_a, a car was forced to stop at time t_a, and the behavior of the other vehicles on the road was observed for time duration T_{af}.

The traffic scenario simulated to estimate the probabilities involved three groups of 50 cars each, with speeds between 30-50, 40-60, and 60-100 km/hour respectively. The minimum inter-vehicle separation is set at 2m, and the *politeness factor*, which controls the aggressiveness of the drivers in enforcing a lane-change in VanetMobiSim, is set to 1.0, 0.7, and 0.3 respectively (higher speed drivers have lower politeness factors, and hence are more aggressive in their lane-changing behavior). The speed for raising the SVA alert was set to 20 km/hour, meaning that a car raised an SVA alert if its speed fell below 20 km/hour. A random car was made to stop in the middle of the simulation and the alerts generated in its immediate neighborhood were noted for the next $T_{af} = 25$ time slots, each of duration 10 seconds. The road segment observed was 25 slots of length 10 meters each from the crash site. We ignored the lane number, so that the observed grid is 2-dimensional representing the time from crash and distance from crash site $(t \times d)$ of size 25×25.

The number of alerts at each grid cell is recorded and these are averaged out over a large number of simulation runs to compute the average probabilities. The computed probabilities (of an SVA alert in a grid cell given that a crash has actually occurred) are shown in Figure 1.

The distributions of the average number of alerts over distance from the crash site and over time from the crash time are shown in Figure 2. As expected, it is seen that most of the SVA alerts happen close to the crash site. However, the number of SVA alerts very close to the crash time is low, increasing in number after some time as cars slow down, and then decreasing again if cars get time to change lanes or if number of cars decreases. The distribution of the alerts can be used to define the following simple weight function:

Weight Function W: *Let x_d denote the number of alerts at distance d from the crash site (cumulative over time) and x_t denote the number of alerts at time t from the crash time (cumulative over distance). Let X^D be the maximum number*

0.31	0.16	0.18	0.11	0.06	0.07	0.04	0.03	0.03	0.05	0.03	0.03	0.04	0.04	0.03	0.06	0.02	0.08	0.09	0.03	0.05	0.03	0.05	0.03	0.04
0.48	0.27	0.26	0.13	0.13	0.11	0.13	0.11	0.08	0.07	0.09	0.1	0.09	0.07	0.06	0.07	0.06	0.07	0.06	0.04	0.05	0.06	0.05	0.05	0.04
0.58	0.45	0.35	0.31	0.22	0.21	0.16	0.11	0.11	0.16	0.14	0.12	0.11	0.06	0.12	0.1	0.14	0.11	0.13	0.17	0.14	0.14	0.13	0.15	0.18
0.65	0.48	0.5	0.46	0.4	0.4	0.36	0.32	0.35	0.27	0.26	0.24	0.28	0.16	0.23	0.16	0.2	0.2	0.27	0.24	0.28	0.27	0.27	0.28	0.27
0.8	0.64	0.62	0.59	0.47	0.49	0.43	0.41	0.39	0.45	0.36	0.4	0.38	0.35	0.36	0.34	0.3	0.34	0.31	0.31	0.36	0.39	0.3	0.35	0.35
0.77	0.7	0.68	0.64	0.64	0.51	0.51	0.49	0.47	0.46	0.42	0.44	0.45	0.4	0.38	0.48	0.36	0.45	0.43	0.42	0.46	0.45	0.45	0.42	0.38
0.87	0.78	0.82	0.8	0.71	0.69	0.66	0.63	0.62	0.61	0.6	0.53	0.54	0.55	0.53	0.51	0.55	0.49	0.47	0.52	0.49	0.48	0.54	0.51	0.48
0.93	0.89	0.89	0.84	0.84	0.81	0.79	0.72	0.69	0.62	0.64	0.62	0.57	0.61	0.55	0.61	0.56	0.56	0.64	0.56	0.64	0.57	0.6	0.64	0.63
0.97	0.93	0.9	0.9	0.9	0.88	0.87	0.83	0.76	0.76	0.74	0.73	0.75	0.77	0.75	0.74	0.71	0.74	0.73	0.75	0.74	0.67	0.71	0.76	0.74
0.99	0.94	0.96	0.94	0.94	0.92	0.92	0.9	0.91	0.9	0.88	0.92	0.89	0.86	0.84	0.85	0.87	0.89	0.89	0.87	0.87	0.88	0.85	0.84	0.91
0.99	0.96	0.98	0.98	0.98	0.98	0.96	0.96	0.94	0.96	0.95	0.93	0.95	0.94	0.93	0.89	0.91	0.9	0.89	0.91	0.91	0.92	0.89	0.94	0.9
1	1	1	1	1	0.97	0.98	0.97	0.98	0.97	0.96	0.96	0.97	0.92	0.97	0.96	0.96	0.95	0.94	0.93	0.92	0.94	0.93	0.95	0.93
1	1	1	1	1	1	1	1	1	1	1	1	0.99	1	1	1	0.94	1	1	0.98	0.99	0.95	0.99	0.87	0.95
1	1	1	1	1	1	1	1	1	1	1	1	1	0.99	1	1	1	0.94	1	0.98	0.82	0.82	0.78	0.78	0.56
1	1	1	1	1	1	1	1	1	1	1	1	1	1	1	0.99	0.99	1	0.97	0.9	0.88				
1	1	1	1	1	1	1	1	1	1	1	1	0.99	1	1	1	0.96	1	0.99	0.97	0.9	0.9	0.82	0.79	
1	1	1	1	1	1	1	1	1	1	1	1	1	0.98	0.94	0.91	0.88	0.82	0.82	0.78	0.78				
1	1	1	1	1	1	1	1	1	1	1	0.98	0.96	0.91	0.86	0.82	0.81	0.8	0.78	0.76	0.73	0.68			
1	1	1	1	1	1	1	1	0.98	0.95	0.91	0.9	0.82	0.82	0.82	0.77	0.7	0.68	0.62	0.61	0.56	0.53	0.52	0.47	0.45
1	1	1	1	1	0.99	0.96	0.91	0.85	0.82	0.82	0.82	0.74	0.71	0.67	0.66	0.59	0.56	0.55	0.5	0.48	0.44	0.35	0.28	0.25
1	1	1	0.95	0.91	0.89	0.82	0.82	0.81	0.74	0.7	0.7	0.63	0.59	0.56	0.55	0.53	0.48	0.47	0.35	0.3	0.23	0.21	0.18	0.14
1	0.94	0.91	0.82	0.82	0.79	0.74	0.74	0.7	0.69	0.59	0.57	0.55	0.54	0.51	0.45	0.35	0.3	0.3	0.19	0.16	0.12	0.09	0.09	0.08
0.91	0.82	0.82	0.77	0.74	0.73	0.69	0.57	0.57	0.55	0.54	0.52	0.45	0.42	0.35	0.28	0.23	0.16	0.12	0.08	0.04	0.08	0.04	0.01	0.01
0.82	0.74	0.74	0.7	0.65	0.57	0.55	0.55	0.51	0.48	0.42	0.33	0.28	0.25	0.19	0.12	0.09	0.08	0.04	0.04	0.03	0.01	0.01	0	0
0.74	0.68	0.65	0.57	0.55	0.53	0.49	0.45	0.36	0.28	0.25	0.19	0.12	0.11	0.1	0.08	0.04	0.04	0.01	0.01	0	0	0	0	0

Fig. 1. Probabilities of SVA alert in the presence of a crash

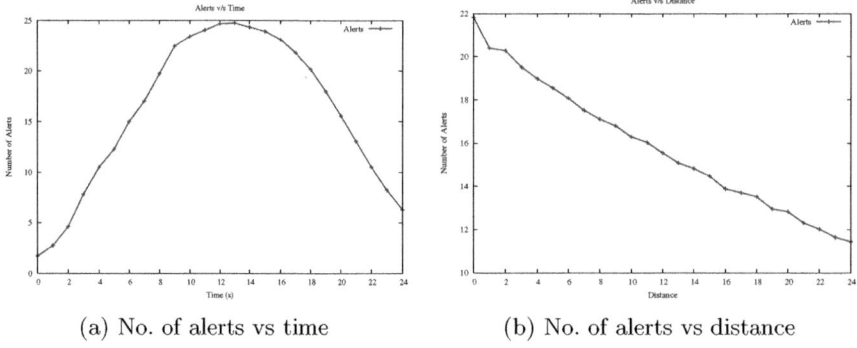

(a) No. of alerts vs time   (b) No. of alerts vs distance

Fig. 2. Alert distributions over time and distance slots

of alerts in any distance slot (cumulative over time), and X^T be the maximum number of alerts in any time slot (cumulative over distance). First assign weight $= 0$ to any grid cell (d_i, t_i) if $x_{d_i} < \frac{2}{3}X^d$ or if $x_{t_i} < \frac{2}{3}X^t$. Let N be the remaining number of grid cells (which are not yet assigned weight 0). Assign weight $\frac{1}{N}$ to each of these grid cells.

Thus, the weight function chosen just assigns 0 weights to cells if the cell is at a larger distance from the crash site or if it is at time at which the total number of alerts is small. The rest of the cells are assigned equal weights. The weight function values for the individual grid cells are shown in Figure 3. We emphasize that this is just one possible specification of the weight function chosen for illustration, many other variations are possible.

Similarly, the same road segment is observed for the same duration in the absence of a crash. The probability of an SVA alert in each grid cell is then computed by averaging out the alerts received over a large number of simulation runs. It was noted that all the grid cells have a probability of 0, i.e., no SVA

0	0	0	0	0	0	0	0	0	0	0	0	0	0	0	0	0	0	0	0	0	0	0	0
0	0	0	0	0	0	0	0	0	0	0	0	0	0	0	0	0	0	0	0	0	0	0	0
0	0	0	0	0	0	0	0	0	0	0	0	0	0	0	0	0	0	0	0	0	0	0	0
0	0	0	0	0	0	0	0	0	0	0	0	0	0	0	0	0	0	0	0	0	0	0	0
0	0	0	0	0	0	0	0	0	0	0	0	0	0	0	0	0	0	0	0	0	0	0	0
0	0	0	0	0	0	0	0	0	0	0	0	0	0	0	0	0	0	0	0	0	0	0	0
0	0	0	0	0	0	0	0	0	0	0	0	0	0	0	0	0	0	0	0	0	0	0	0
0.005	0.005	0.005	0.005	0.005	0.005	0.005	0.005	0.005	0.005	0.005	0.005	0.005	0.005	0.005	0	0	0	0	0	0	0	0	0
0.005	0.005	0.005	0.005	0.005	0.005	0.005	0.005	0.005	0.005	0.005	0.005	0.005	0.005	0.005	0	0	0	0	0	0	0	0	0
0.005	0.005	0.005	0.005	0.005	0.005	0.005	0.005	0.005	0.005	0.005	0.005	0.005	0.005	0.005	0	0	0	0	0	0	0	0	0
0.005	0.005	0.005	0.005	0.005	0.005	0.005	0.005	0.005	0.005	0.005	0.005	0.005	0.005	0.005	0	0	0	0	0	0	0	0	0
0.005	0.005	0.005	0.005	0.005	0.005	0.005	0.005	0.005	0.005	0.005	0.005	0.005	0.005	0.005	0	0	0	0	0	0	0	0	0
0.005	0.005	0.005	0.005	0.005	0.005	0.005	0.005	0.005	0.005	0.005	0.005	0.005	0.005	0.005	0	0	0	0	0	0	0	0	0
0.005	0.005	0.005	0.005	0.005	0.005	0.005	0.005	0.005	0.005	0.005	0.005	0.005	0.005	0.005	0	0	0	0	0	0	0	0	0
0.005	0.005	0.005	0.005	0.005	0.005	0.005	0.005	0.005	0.005	0.005	0.005	0.005	0.005	0.005	0	0	0	0	0	0	0	0	0
0.005	0.005	0.005	0.005	0.005	0.005	0.005	0.005	0.005	0.005	0.005	0.005	0.005	0.005	0.005	0	0	0	0	0	0	0	0	0
0.005	0.005	0.005	0.005	0.005	0.005	0.005	0.005	0.005	0.005	0.005	0.005	0.005	0.005	0.005	0	0	0	0	0	0	0	0	0
0.005	0.005	0.005	0.005	0.005	0.005	0.005	0.005	0.005	0.005	0.005	0.005	0.005	0.005	0.005	0	0	0	0	0	0	0	0	0
0	0	0	0	0	0	0	0	0	0	0	0	0	0	0	0	0	0	0	0	0	0	0	0
0	0	0	0	0	0	0	0	0	0	0	0	0	0	0	0	0	0	0	0	0	0	0	0
0	0	0	0	0	0	0	0	0	0	0	0	0	0	0	0	0	0	0	0	0	0	0	0
0	0	0	0	0	0	0	0	0	0	0	0	0	0	0	0	0	0	0	0	0	0	0	0
0	0	0	0	0	0	0	0	0	0	0	0	0	0	0	0	0	0	0	0	0	0	0	0

Fig. 3. Weights assigned to grid elements using weight function W

alerts were generated. Repeating the experiment for several other road segments gave similar results, some of which resulted in only a few SVA alerts.

The a-priori probability of a crash $P(CRASH(t_a, x_a, y_a) = 1)$ can only be estimated from observed data. We take it to be a small value, say, 0.001, indicating that a crash is reported in a particular place at particular time in around 1 in 1000 samples (for ex., a sample can be a day). Thus, we have now estimated $P(SVA(t, x, y) = 1|CRASH(t_a, x_a, y_a) = 1)$ and $P(SVA(t, x, y) = 1|NOCRASH)$. The a-priori probability of a SVA alert, $P(SVA(t, x, y) = 1)$, can now be easily computed noting that only one alert is considered and hence all but one of the terms in the summation in Equation 5 will be 0.

Hence, given any $GRID$ (i.e., an actual observation of 0-1 values for each of the random variables R_E corresponding to each grid cell), $P(R_{prim}(t_a, x_a, y_a, v_a) = 1|R_E(t, x, y, u) = 1)$ for each grid cell can be computed by Equation 3, and the weight function can be fixed to allow aggregating them into a single value. This final value gives β, the belief in the truth or falsity of the PCN alert, with a value closer to 1 indicating a higher belief in the occurrence of the crash.

To validate the correctness of the estimation process and show that such estimation can actually be used in an MDS, we use the estimated probabilities and the weight function for detecting both a true alert and a false alert. We first simulate a crash in VanetMobiSim, and then observe the alert pattern over the grid. We then use the individual probabilities of each of the grid cells and the weights to compute a final degree of belief in the alert being true. This is repeated for 10 runs (i.e., 10 alert patterns observed for 10 independent crashes) and the average degree of belief of these 10 runs is computed. This final degree of belief comes out to be 0.93. Note that the final degree of belief is high, indicating a high degree of belief that the crash actually occurred. This validates the correctness of the estimation process when a crash actually happens.

We repeated the same experiment with no crash in VanetMobiSim. The computed degree of belief comes out to be 0. This indicates that the MDS detects that the crash has not occurred (hence the alert is false), thereby validating the correctness of the estimation process when no crash actually happens.

Thus, it is seen that secondary information can be effectively used to design an MDS for PCN alerts. The false positive rate is seen to be 0 and the false negative rate is seen to be very low $(1 - 0.93 = 0.07)$ for the cases simulated.

5.1   Robustness of Estimation

In the above section, we have estimated the probabilities for a particular traffic model, and then applied it for misbehavior detection for scenarios following the same traffic model. However, in practice, the probabilities will be estimated once and stored in the OBU a-priori. The actual secondary information observed may come from a traffic scenario that is not exactly the same as the traffic model with which the probabilities were estimated. To evaluate the robustness of the proposed scheme over multiple traffic models, we perform the following experiments: We choose 4 traffic scenarios as shown in Figure 4.

Scenario 1	3 node groups each with 50 cars, with speeds 30-50 km/hr, 40-60 km/hr, 60-100 km/hr and politeness factors 1.0, 0.7, and 0.3 respectively
Scenario 2	1 node group with 100 cars, speed 30-50 km/hr, politeness factor 1.0
Scenario 3	1 node group with 100 cars, speed 40-60 km/hr, politeness factor 0.7
Scenario 4	1 node group with 100 cars, speed 60-100 km/hr, politeness factor 0.3

Fig. 4. Traffic scenarios simulated

Thus, Scenarios 2, 3, and 4 represent homogeneous drivers with progressively more aggressiveness, while Scenario 1 (which is the same as the scenario simulated earlier) is a mixture of the three. In the first set of experiments, we estimate the probabilities from simulating traffic from Scenario 1 only, and then apply it to detect misbehavior, if any, for traffic generated for all four scenarios. The final value reported is once again the average of 10 runs. The final value obtained for the 4 cases for both the cases of when there is a crash and when there is no crash is shown in Figure 5.

It is seen that the false positive rate is still 0 for all cases. However the false negative rate increases with increase in the aggressiveness of the driver. This is

	When there is crash	When there is no crash
Scenario 1	0.93	0.0
Scenario 2	0.83	0.0
Scenario 3	0.59	0.0
Scenario 4	0.43	0.0

Fig. 5. Belief values obtained with probabilities estimated from Scenario 1

expected as with higher politeness factor, drivers tend to change lane less often and hence, may have to slow down more and thus cause more SVA alerts. Thus, a majority of the SVA alerts in Scenario 1 are generated by cars with higher politeness factor, and hence the nature of alerts generated in Scenario 1 and Scenario 2 are similar. As the politeness factor decreases, the number of SVA alerts also decreases. This makes the scenario with which the probabilities are estimated very different than the scenario in which it is used for misbehavior detection, causing an increase in the false negatives.

The above arguments suggest that it may be better to estimate the probabilities from a scenario that represents a traffic model that is not very dissimilar with the traffic models for which misbehavior detection is needed. Hence, we next estimated the probabilities from Scenario 3, which intuitively represents a traffic model in between Scenario 2 and 4, and then applied it to detect misbehavior for all the four models. The final value obtained for the four cases for both the cases of when there is a crash and when there is no crash is shown in Figure 6.

	When there is crash	When there is no crash
Scenario 1	0.82	0.0
Scenario 2	0.94	0.0
Scenario 3	0.98	0.0
Scenario 4	0.98	0.0

Fig. 6. Belief values obtained with probabilities estimated from Scenario 3

It is seen that estimating the probabilities with Scenario 3 (a median scenario among the four) gives very good results for the MDS for detecting misbehaviors for all the scenarios. It is thus important to chose the right traffic model to estimate the probabilities.

6   Conclusion and Future Work

In this paper, an MDS based on a probabilistic framework of using correlated or secondary information to verify an alert in VANETs has been presented. An application of the proposed framework in identifying misbehavior for the specific case of PCN by using secondary information based on SVA alerts has been detailed and its performance evaluated.

This work can be extended in several directions. The PCN alert is a periodic alert and is sent periodically until the crash is cleared. Hence, the PCN alert that other vehicles receive may not be the first such alert sent. Hence the grid that vehicles observe may be shifted in time from the grid that is considered in this framework. Extending our model to incorporate these observations is an interesting problem. Also, characterizing and designing MDS for other applications using secondary information is another important activity.

References

1. Bai, F., Krishnan, H., Sadekar, V., Holland, G., ElBatt, T.: Towards characterizing and classifying communication-based automotive applications from a wireless networking perspective. In: 1st IEEE Workshop on Automotive Networking and Applications (2006)
2. Hubaux, J., Čapkun, S., Luo, J.: The security and privacy of smart vehicles. IEEE Security and Privacy 2(3) (2004)
3. Raya, M., Papadimitratos, P., Hubaux, J.: Securing Vehicular Communications. IEEE Wireless Communications Magazine, Special Issue on Inter-Vehicular Communications 13(5) (2006)
4. Torrent-Moreno, M., Killat, M., Hartenstein, H.: The challenges of robust inter-vehicle communications. In: IEEE 62nd Vehicular Technology Conference, vol. 1 (Fall 2005)
5. Gerlach, M., Festag, A., Leinmuller, T., Goldacker, G., Harsch, C.: Security architecture for vehicular communication. In: 5th International Workshop on Intelligent Transportation (2007)
6. IEEE Trial-Use Standard for Wireless Access in Vehicular Environments - Security Services for Applications and Management Messages, IEEE Std 1609.2-2006 (2006)
7. Tan, H.S., Huang, J.: DGPS-based vehicle-to-vehicle cooperative collision warning: Engineering feasibility viewpoints. IEEE Transactions on Intelligent Transportation Systems 7(4) (December 2006)
8. ElBatt, T., Goel, S.K., Holland, G., Krishnan, H., Parikh, J.: Cooperative collision warning using dedicated short range wireless communications. In: VANET '06: 3rd ACM International Workshop on Vehicular Ad Hoc Networks (2006)
9. Golle, P., Greene, D., Staddon, J.: Detecting and correcting malicious data in vanets. In: VANET '04: 1st ACM International Workshop on Vehicular Ad Hoc Networks (2004)
10. Schmidt, R.K., Leinmüller, T., Schoch, E., Held, A., Schäfer, G.: Vehicle behavior analysis to enhance security in vanets. In: 4th Workshop on Vehicle to Vehicle Communications, V2VCOM 2008 (2008)
11. Ghosh, M., Varghese, A., Kherani, A., Gupta, A.: Distributed misbehavior detection in VANET. In: IEEE Wireless Communication and Networking Conference (2006)
12. Ghosh, M., Varghese, A., Gupta, A., Kherani, A., Muthaiah, S.: Misbehavior detection scheme with integrated root cause detection in VANET. In: VANET'09, 6th ACM International Workshop on Vehicular Internetworking (2009)
13. Härri, J., Filali, F., Bonnet, C., Fiore, M.: VanetMobiSim: generating realistic mobility patterns for VANETs. In: VANET'06, 3rd ACM International Workshop on Vehicular Ad Hoc Networks (2006)

Path Attestation Scheme to Avert DDoS Flood Attacks

Raktim Bhattacharjee, S. Sanand, and S.V. Raghavan

Dept. of Computer Science & Engg.
Indian Institute of Technology Madras,Chennai, 600036
{raktim,sanand,svr}@cs.iitm.ernet.in

Abstract. DDoS mitigation schemes are increasingly becoming relevant in the Internet. The main hurdle faced by such schemes is the "nearly indistinguishable" line between malicious traffic and genuine traffic. It is best tackled with a paradigm shift in connection handling by attesting the path. We therefore propose the scheme called "Path Attestation Scheme" coupled with a metric called "Confidence Index" to tackle the problem of distinguishing malicious and genuine traffic in a progressive manner, with varying levels of certainty. We support our work through an experimental study to establish the stability of Internet topology by using 134 different global Internet paths over a period of 16 days. Our Path Attestation Scheme was able to successfully distinguish between malicious and genuine traffic, 85% of the time. The scheme presupposes support from a fraction of routers in the path.

Keywords: DDoS mitigation, Unspoofable Identity, Cascaded Filters.

1 Introduction

"Distributed Denial of Service" (DDoS) can be defined as an attempt made by malicious users to deny resources and services to legitimate users. DDoS attacks are relatively simple, yet powerful enough to bring down a Critical Infrastructure (CI). DDoS identification schemes can broadly be classified into (1) Behavioral based and (2) Identity based. Though behavioral based schemes can detect unknown attacks, it suffers from several drawbacks as listed in [5]. The main drawback of identity based techniques is the possibility of source IP spoofing. The first part of our work is the creation of an unspoofable identity. The core idea is that the path taken by a packet from a source is dictated by the destination IP and routing behaviour of the Internet, and hence is unspoofable. Spoofing the source IP of a packet will not change its path to destination. The novelty of the proposed technique is that it can mark packets as malicious, with a varying degree of confidence. We propose a metric called *Confidence Index* to quantify this. The scheme is effective against flood attacks, which make use of IP Spoofing.

The second part of this work is a router-level flood attack mitigation scheme. Each router has a set of priority queues and each packet is assigned to one of the

M. Crovella et al. (Eds.): NETWORKING 2010, LNCS 6091, pp. 397–408, 2010.
© IFIP International Federation for Information Processing

queues based on its *Confidence Index*. Unlike other similar efforts, in Path Attestation Scheme (PAS), small changes in route will not lead to packet drop but only de-prioritization of packets. When cascaded, such routers can exponentially reduce the attack volume reaching the CI. This scheme is particularly effective against bandwidth-depletion attacks.

The last part of our work is a study on different types of errors and performance of proposed technique. The system is prone to false positives occurring due to path changes. Based on an experimental study of 134 routes over a period of 16 days, we have modeled the probability of path changes in the Internet. The probability of false negatives through collision of path identities is modeled analytically. Finally, we study the cascaded effect of a sequence of filtering routers by simulations.

The remainder of the paper is organized as follows: Section 2 deals with the background and related works. In Section 3 we present the Path Attestation Scheme (PAS) framework. In Section 4 we analyze the effect of change in network topology in PAS. Section 5 deals with security and performance analysis of the proposed system and finally Section 6 concludes the paper.

2   Background and Related Works

Limitations of the Current Internet Architecture. Most of the research effort in the past was directed towards improving the performance and scalability of the Internet. No attention was paid to make it safe and secure. Internet today is susceptible to DDoS because of the lack of Authenticity, Accountability and Uniformity. Without authentication any user can claim any identity and there is no means to trace these malicious users. Infact there is very little that present day routers can do to improve the situation. Routing and forwarding protocols are designed to be destination oriented. Routers are designed to forward the packet without bothering about where it has come from. Lastly the resources in the internet are not uniformly distributed, a lot of potential is concentrated in the core of the network which can be evenly distributed to the edge network.

DDoS Prevention Techniques. All DDoS prevention schemes can be classified into three classes based on the place of deployment as (1) source based, (2) host based and (3) network based. Deploying DDoS solution at the source itself is the ideal case because it saves the network resources from unwanted traffic. Ingress Filtering [1] is one such type of solution where ingress routers block packets that arrive with source addresses having prefixes that do not match the customer's network prefixes. DWARD [4] is another solution that performs a proactive identification and filtering of suspicious flows originating from a customer network. The impact of the source based prevention cannot be felt directly by the deploying network, because of which there is little motivation for Internet Service Providers to deploy source based schemes in their network.

The host based DDoS mitigation schemes are preferred because the benefit of DDoS prevention is felt directly by the deploying system or network. Hop-Count Filtering [8] is a type of host based scheme. Other spoofing prevention method like history-based filtering [7] and Packet-score [3] have also adopted this approach. Though these solutions are good for preserving server resources, they do not prevent the abuse of network resources. Moreover, there is a possibility of launching an attack against the prevention system in which case the prevention system itself becomes a single point of failure.

The network based solutions require support from routers as well as wide scale deployment to be effective. The Pushback scheme [2] view flooding by DDoS as a congestion problem. This scheme requires router modification to detect and drop packet belonging to an attack flow. Further, network based solutions like Pi [10] and SIFF [9] use path based identification to filter out attack packets. These methods are prone to false positives due to frequent route changes and load balancing, in which case legitimate traffic may get filtered even in the absence of an attack.

Considering the different mitigation schemes we can say that Network based solution is a must if DDoS needs to be nipped in the bud. Thus we have proposed PAS that not only allows router level differential filtering but also does not suffer the drawbacks of other network based mitigation schemes.

3   Path Attestation Scheme (PAS)

The objective of the PAS is to mitigate flood attacks and give service to the legitimate users. The flooding problem is relatively difficult to handle because there is no way to differentiate between spoofed and genuine traffic. This differentiation is possible if we can attach an element to the packet that the attacker cannot modify. One such element that an attacker cannot modify is the path of the packet from its source to destination.

Any packet moving from a particular source to a destination follows a path. It is highly probable that the same path will be followed by subsequent packets between those two systems. Based on the amount of deviation from the older path each packet is given a *Confidence Index*. Packets with less deviation will have high *Confidence Index* compared to packets with more deviation. Packets with very low or zero *Confidence Index* are considered to be suspicious or malicious. Based on the *Confidence Index* packets are segregated in different priority queues. The packets in the higher priority queues will be processed before any packet in lower priority queues. When there is a DDoS flooding, lots of spoofed packets will be generated and these packets will have low *Confidence Index*. These packets will be assigned to the default/lowest priority queue. Thus, when there is a congestion (caused due to flooding) the lowest priority queue will get filled with spoofed packets. This will inturn result in packet drop, which reduces the number of attack packets leaving the router. The whole process in described in detail in the following subsection.

3.1   ID List Creation

Each router in the path puts its signature (ID) in the packet. The sequence of router IDs of a path forms an *ID List*. The ID, which router puts in the packet is the function of (1) source IP address, IP_{src}, and (2) Interface Identity Layer 2 Address, ID_{inf}, of the router outgoing interface.

$$ID = hash(IP_{src}|ID_{inf}) \tag{1}$$

The hash function proposed to use is MD5. MD5 generates 128 bit hash value but only the most significant 'n' bits will be taken as ID.

The purpose of selecting the router outgoing interface for the hash is to have invariability and uniqueness properties of path into the ID. Packets from different sources will have different path to a destination and thus different IDs. The reason for the ID to be a function of Source IP is that packets with different source IPs should have different IDs. In the event of an IP spoofing attack, a series of packets will be generated with random source IP address to a destination. Since ID is a function of Source IP also, each of these packets will have different ID value because of the change in source IP.

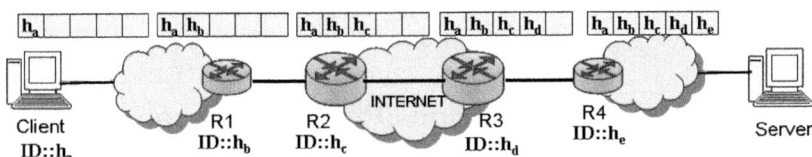

Fig. 1. ID List creation process

At the beginning of a TCP connection, i.e. when the first ACK is sent from client to the server, the *ID List* is generated. As the packet moves from source to destination each of the routers in the path calculates the 'n' bit ID and puts it into the packet as shown in Fig.1. The forty byte option field of IP packets is used to carry this *ID List* . When the server receives this, it keeps the *ID List* for subsequent use.

3.2   ID List Delivery

Once the *ID List* of a client is received at the server, it is kept in its memory. This *ID List* needs to be transferred to the client after verifying its authenticity. This is done at the end of a genuine TCP connection, piggybacked in the FIN packet. A connection is considered to be genuine when there is transfer of a certain amount of data between the client and the server. The rationale behind sending the *ID List* to client at the end of a genuine TCP connection is that; one cannot establish a TCP connection using a spoofed IP and thus cannot receive the *ID List* back. This *ID List* delivery process is shown in Fig.2(a).

3.3   ID List Verification

Once the client receives the *ID List* from the server it keeps it in its file system. From the next connection onwards client inserts this *ID List* to every packet destined to that server. All the routers in the path verifies this *ID List*. Each router

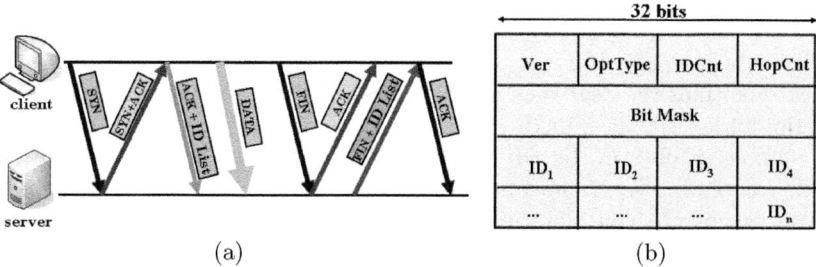

32 bits			
Ver	OptType	IDCnt	HopCnt
Bit Mask			
ID₁	ID₂	ID₃	ID₄
...	...	...	IDₙ

Fig. 2. (a) ID List delivery process. (b) Proposed IP Option Field.

in the path calculates the 'n' bit ID using the IP_{src} and ID_{inf}. The proposed IP option (Fig.2(b)) field carries a Bit Mask field. Each bit of the Bit Mask field is used to specify whether the corresponding ID is valid or not. By default all the bits in this Bit Mask is zero, indicating invalid. Now, if the calculated ID matches to the ID carried by packet then it marks the corresponding bit of the Bit Mask as 1 otherwise 0, as explained in Fig.3.

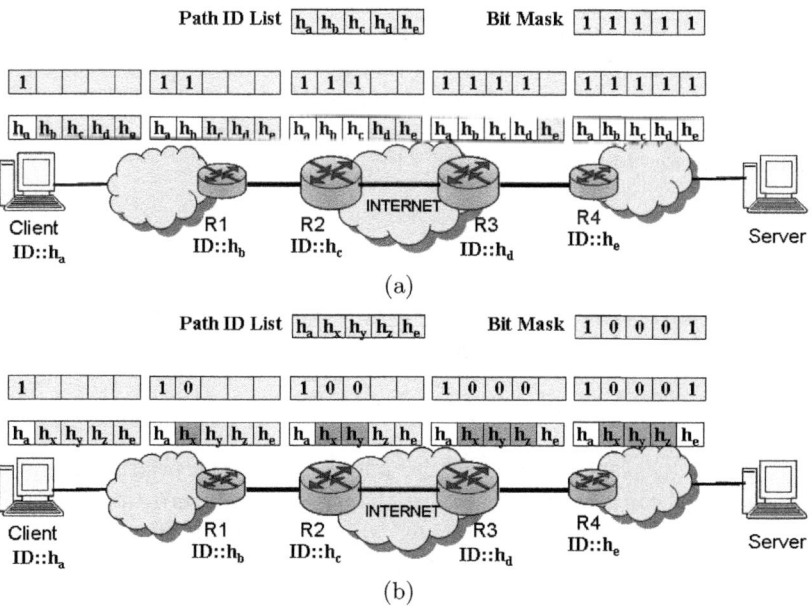

Fig. 3. *ID List verification process*. In Case (a) all the IDs in the List are correct. Router verifies and marks all the Bit Mask as 1. In Case (b) IDs h_x, h_y and h_z are invalid, router marks corresponding Bit Mask as 0.

3.4   Packet Classification

Packets are segregated by the router based on a metric called *Confidence Index* which quantifies the degree of maliciousness. *Confidence Index* of a packet is a function of its Bit Mask field. Higher the number of IDs matched, the higher is the number of ones in the Bit Mask and thus higher is the *Confidence Index*. While calculating the *Confidence Index*, the position of the routers in the path is also taken into consideration. The routers near the destination server are considered to be more trusted than the routers near the source. Thus we provide a weightage to each bit position. The weight is a linear function of the router position in the path. The *Confidence Index* of a packet in router 'r' is determined by this expression.

$$C_r = \frac{\sum_{i=0}^{r}(W_i \times V_i)}{\sum_{i=0}^{r} W_i} \tag{2}$$

Where,

C_r is *Confidence Index* of the packet in router 'r', W_i is Weight of ith Bit Mask position ($W_i = i$), V_i is Bit Mask in i^{th} position (1 implies Valid, 0 implies Invalid) and r is the Number of routers passed by the packet.

In the example, in Fig.3, based on the Bit Mask value the *Confidence Index* is calculated. In the first case where all the Bit Mask are 1 the *Confidence Index* is 1 but in the second case the *Confidence Index* is 0.4. When there is a genuine path change, at least a part of the *ID List* will match and the packet will have higher *Confidence Index* than a spoofed packet.

3.5   Router Queuing Process

Each router in the system maintains N_r priority queues. The queue to which a packet belongs is determined based on the value of its *Confidence Index*. The queue selection for a packet is done by this expression.

$$Q = \lceil C_r \times N_r \rceil \tag{3}$$

Based on this 'Q' value packet will be put in one of the priority queues. If a packet does not come with any *ID List* then it will have zero *Confidence Index* and will be put into the default queue, i.e. lowest priority queue. The packets having a valid *ID List* will be put into high priority queues. Spoofed packets will have very low *Confidence Index* and will be put into the lowest priority queue. When there is a congestion, due to DDoS attack, the lowest priority queue will get filled with spoofed packets. This will result in packet drop, which inturn reduces the number of attack packets leaving the router.

Fig.4 illustrates the temporal variation in incoming, outgoing and dropped traffic in a router in the event of an attack. During the normal phase, when there is no packet drop, the incoming and outgoing bandwidths are same. But

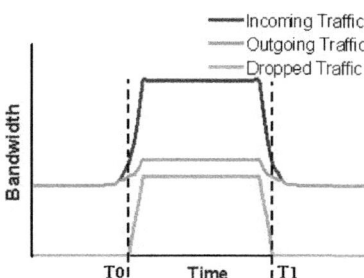

Fig. 4. Bandwidth profile of a router in the event of an attack

when the attack packets start flowing in the lowest priority queue starts getting filled. At one point, T_0, the lowest priority queue becomes completely filled and attack packets starts getting dropped. Beyond this point only a small fraction of attack packets will go out of the router.

4   Experimental Study

The effectiveness of PAS depends on the stability of paths in the Internet topology. According to a study [6] done during 1994-1995, about 2/3 of the Internet paths were having routing persistence of either days or week and most variation was either in one or two routers. Since this result is considerably old, we undertook an experiment to characterize the path change between client and server in the present day Internet.

We selected a set of 134 IPs from traceroute.org, which allows traceroute to their servers. These servers were geographically distributed over 40 countries which gives a more or less true representation of the Internet. A system in Network System Lab, IIT Madras, which has a public IP, was selected as the client. A Perl script was written which will do repeated traceroute operations to these servers at a regular interval. In all 257280 traceroutes were made and the data thus collected were stored in MySQL database to make analysis easier.

4.1   Temporal Variations of Path Stability

The aim of this analysis is to characterize the variation of path stability against different observation intervals. For each observation interval we plot the percentage of paths having hop variations ranging from 0 to maximum hop count. We derive the expression for path stability as follows. Let the set of all paths be 'P'. Let 'p' denote a path and 'h_p' be the number of hop variations of a path 'p'. Then the number of paths having 'h' hop variations is given by

$$n_h = |\{p|p \in P, h_p = h\}| \tag{4}$$

The total number of paths is given by

$$n = |P| \tag{5}$$

The percentage of path having 'h' hops variations is given by

$$\acute{n}_h = \frac{n_h}{n} \times 100 \qquad (6)$$

In the Fig.5(a) we have plotted Percentage of Path having 'h' hop variations '$\acute{n}_h$' for different observation intervals. The observation interval is varied from 1 to 8. The value plotted against an interval 'w' is the average of '$\acute{n}_h$' for all possible intervals of size 'w'. From the graph we observe that number of paths having 0 hop variation ($n_h = 0$) is asymptotically approaching a minimum value. We also observed that number of paths having hop variation less than 5 remains above 65%, even for large observation interval. Next, we define a term which is the number of paths having atmost 'h' path variation.

$$\hat{n}_h = \sum_{x=0}^{h} n_x \qquad (7)$$

We define path stability as follows,

$$P_t = \frac{\hat{n}_h}{n}, where, h \leqslant t \qquad (8)$$

Here,'t' is the maximum number of hop variations which can be sustained by the system. Now suppose that atmost 3 hop variation is acceptable, in that case, for w=1 we have $\acute{n}_h = 0.85$ for w=2 $\acute{n}_h = 0.84$ and for w=3 $\acute{n}_h = 0.76$. Thus it can be observed that if the observation interval is one or two days and at most three hops variation is allowed then we can achieve path stability above 0.8.

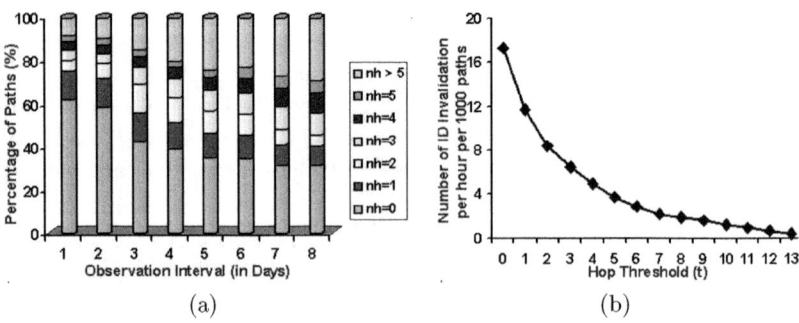

(a)   (b)

Fig. 5. Graph (a) between the observation Intervals in Days Vs. Percentage of Paths with different hop variations. Graph (b) between the number of ID Invalidations per hour per 1000 paths Vs. Hop Threshold.

4.2 Frequency of ID Invalidation

The validity of ID depends on the stability of the path. As long as the path is stable, the C_r value will be higher than user defined threshold T_h and *ID List* will be valid. But due to path changes C_r value will decrease and once it falls below T_h the *ID List* is renewed. Now the frequency of *ID List* change depends on the stability of the path P_t and specifically on the value of t. Fig.5(b) shows a variation of the number of ID invalidation per hour per thousand paths plotted against the acceptable hop variations. From the graph we can infer that if the invalidation algorithm does not accept any hop variation at all, it results in around 17 IDs invalidation per hour per thousand paths. But if we increase the threshold to 3 the number of ID invalidation drops to seven.

5 Security and Performance Analysis

5.1 Analysis of Brute Force Attack

There is a possibility that an attacker may try to spoof the *ID List* so as to give higher priority to his packets. By this analysis we are able to show that such a brute force attack is practically impossible and its success rate is of the order of $(2)^{-m}$ where 'm' is of the order of hundreds. Router prepares an ID which is an 'h' bit hash as explained in equation (1). The probability of an attacker selecting a correct match by brute force is $(\frac{1}{2})^h$. Now, suppose there are 'r' routers between the source and destination, then probability of getting a match in all the router is $\{(\frac{1}{2})^h\}^r$. The higher the size of the ID the higher is the probability of being safe, but at the same time IDs of the entire routers in the path should be accommodated in the 40 bytes IP option field. It is known that the maximum number of hops count in Internet is 30 and average is 14-19 [8]. Thus maximum of 30 router IDs should be accommodated in the option field. If we consider the size of ID as 8 bit, for 30 IDs, space required is 30 bytes which can be accommodated in the option field of 40 bytes. Now considering ID to be 8 bits the probability of getting the entire match in maximum case of 30 routers is $\{(\frac{1}{2})^8\}^{30} = (\frac{1}{2})^{240}$, and the probability of getting the entire match in average case of 17 routers $\{(\frac{1}{2})^8\}^{17} = (\frac{1}{2})^{136}$. These are very small values which implies that the probability of getting the entire IDs match is practically impossible.

5.2 Performance Analysis

Experimental Setup. In order evaluate the performance of our system we conducted some experiments considering the multilink topology of the network. The unique feature of PAS is that it can coexist with legacy routers that do not support it.The network setup (Fig.6) consists of two good sources (G) and ten attacking sources (A). There are a number of routers between the source and CI. Some of these routers support PAS (R_p) and some does not (R_L). The two genuine sources are connected to one router where as the attacking sources are distributed over the whole network. Each of the genuine sources and attacking source are generating traffic at the rate of 1 Mbps each. The routers are designed to handle traffic at the rate of 2 Mbps. The number of attacking sources and numbers of routers between source and destination are varied based on our experimental requirement.

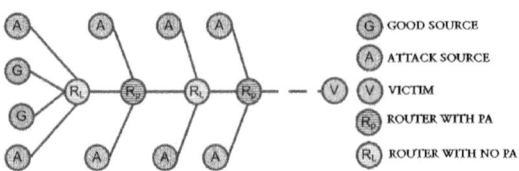

Fig. 6. Network Topology for the Experiment

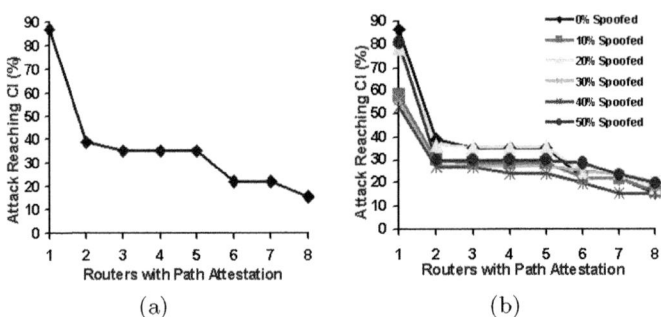

Fig. 7. Graph between the number of Routers with PAS Vs. percentage of Attack Packets reaching the CI. In case (b) the probability of ID spoofing varied from 0 to 0.5.

Performance vs. Router Support. The first experiment was conducted to quantify how effective our PAS is in mitigating DDoS flood attack. As this scheme is a multi-level filter, its effectiveness depends on the number of routers that supports it. For this purpose, in the path between source and critical Infrastructure, varied the number of routers with PAS from 1 to 8. The result obtained after this experiment was plotted in the graph in Fig.7(a) . It can be observed from the graph that as number of routers with PAS support increases, the percentage of attack packets reaching the CI decreases exponentially. With the support of 8 routers in the path we are able to mitigate DDoS flood attack reaching CI upto 85%. The same experiment was conducted assuming that attacker is trying to spoof the *ID List*. The *ID List* is generated with probability of match varying from 0 to 0.5. The same graph (Fig.7(b)) was plotted and was observed that increase in the probability of ID match does not change the nature of curve significantly.

Performance vs. Buffer Size. The second experiment was conducted to see how the number of buffers in the router queues affects the DDoS mitigation scheme. In this case the number of routers supporting PAS is kept constant, i.e. 8, but the numbers of buffers in the router queues were varied. The queue size was varied from 200 packets to 500 packets and the graph (Fig.8(a)) is plotted. It can be observed that with the increase in queue size the percentage of attacks reaching the CI increases exponentially. Thus smaller the queue size the better it is to overcome

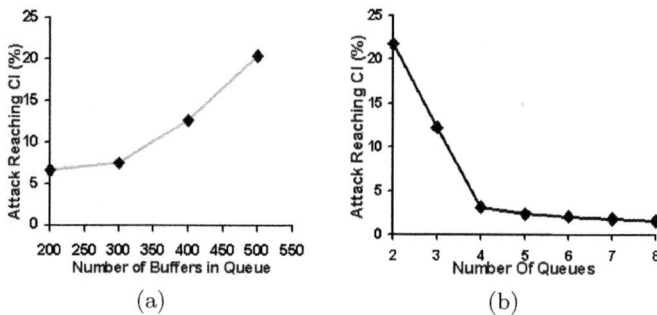

Fig. 8. Graph (a) between number of buffers in Queue Vs. percentage of Attack Packets reaching CI. Graph (b) between number of Queues in router Vs. percentage of Attack Packets reaching the CI.

flood attack. But at the same time if the queue size is reduced too much there is a chance of legitimate packets being dropped. Thus the challenge is to fix the queue in such a way that it drops only the spoofed packet in case of DDoS.

Performance vs. Number of Queues. This experiment was conducted to see how the number of queues in the router affects our DDoS mitigation scheme. This experiment was conducted by varying the number of queues from two to eight. The number of buffers in the router and the number of routers supporting PAS is kept constant. The graph is plotted between the number of queues and percentage of attack reaching the CI. From the graph in Fig.8(b) it can be observed that with the increase in numbers of queues, the percentage of attack reaching CI decreases. Further there is no significant drop in percentage of attack packet reaching CI when the number of queues is four and when it is increased to eight. Thus four priority queues are considered to be optimal in this scenario.

6   Conclusion and Future Work

The PAS addresses one of the key limitations of the existing DDoS mitigation schemes viz. the lack of router based differential filtering. In order to bring properties of Invariability, Uniqueness and Unspoofability we have used a sequence of router IDs as the basis of Identity. This sequence of router IDs helps in identifying a malicious packet with varying levels of certainty. The proposed metric called *Confidence Index* makes router-level differential filtering mechanisms possible. We have used multiple queues and a *Confidence Index* based scheduling algorithm to filter out attack packets from normal traffic at the routers. The performance analysis shows that we have achieved a success rate of 85% with the support of a very few routers in the path. Based on the study of Internet path stability we were able to derive the functional relationship between *Confidence Index* threshold and filtering accuracy. We were also able to predict the average number of ID renewals that can be expected in a system after it is deployed.

408 R. Bhattacharjee, S. Sanand, and S.V. Raghavan

The limitation of PAS is that it is effective only against a particular type of DDoS attack viz. flooding with spoofed source IPs. Being a novel concept PAS paves way for multiple research directions. Firstly, one can analyze the impact of various weightage functions in the definition of *Confidence Index* and its subsequent effect on filtering accuracy. Further, a study of the impact of various queuing techniques on the mitigation process is to be conducted. A lot more insights can be derived by further analysis on the data collected using traceroute. To conclude, a lot more engineering and technological studies need to be carried out before PAS can be deployed widely over the Internet.

References

1. Ferguson, P., Senie, D.: Network ingress filtering: Defeating denial-of-service attacks which employ IP source address spoofing, RFC 2827 (May 2000)
2. Ioannidis, J., Bellovin, S.M.: Implementing Pushback: Router-Based Defense Against DDoS Attacks. In: Proc. Network and Distributed System Security Symposium, San Diego, CA (February 2002)
3. Kim, Y., Lau, W., Chuah, M., Chao, J.: PacketScore: A statistical-based overload control against DDoS attacks. In: Proc. IEEE INFOCOM 2004, China (March 2004)
4. Mirkovic, J.: D-WARD: Source-End Defense against Distributed Denial-of-Service Attacks, PhD. Thesis, UCLA (August 2003)
5. Mirkovic, J., Reiher, P.: A taxonomy of DDoS attack and DDoS defense mechanisms. ACM SIGCOMM Computer Communication Review 34(2), 39–53 (2004)
6. Paxson, V.: End-to-end routing behavior in the Internet. In: Conference proceedings on Applications, technologies, architectures, and protocols for computer communications, Palo Alto, California, United States, August 28-30, pp. 25–38 (1996)
7. Peng, T., Leckie, C., Ramamohanarao, K.: Protection from Distributed Denial of Service Attack Using History-based IP Filtering. In: Proc. of IEEE ICC 2003, Anchorage, AK (May 2003)
8. Wang, H., Jin, C., Shin, K.G.: Defense against Spoofed IP Traffic Using Hop-Count Filtering. IEEE/ACM Trans. Networking 15(1), 40–53 (2007)
9. Yaar, A.P., Song, D.: SIFF: A Stateless Internet Flow Filter to Mitigate DDoS Flooding Attacks. In: IEEE Symposium on Security and Privacy (2004)
10. Yaar, A., Perrig, A., Song, D.: Pi: A Path Identification Mechanism to Defend against DDoS Attacks. In: Proc. of the 2003 IEEE Symposium on Security and Privacy, May 11-14, pp. 93–107 (2003)

Author Index

GPSR Compliance

The European Union's (EU) General Product Safety Regulation (GPSR) is a set of rules that requires consumer products to be safe and our obligations to ensure this.

If you have any concerns about our products, you can contact us on ProductSafety@springernature.com

In case Publisher is established outside the EU, the EU authorized representative is:

Springer Nature Customer Service Center GmbH
Europaplatz 3
69115 Heidelberg, Germany

Batch number: 09478804

Printed by Printforce, the Netherlands